Chlorine and Chlorine Compounds in the Paper Industry

Edited by

Victor Turoski

Ann Arbor Press
Chelsea, Michigan

Library of Congress Cataloging-in-Publication Data

Chlorine and chlorine compounds in the paper industry : developed from a symposium sponsored by the Division of Environmental Chemistry at the 210th National Meeting of the American Chemical Society. Chicago, Illinois, August 20–24, 1995 / edited by Victor Turoski.
p. cm.
Includes bibliographical references and index.
ISBN 1-57504-066-2
1. Papermaking--Chemistry--Environmental aspects. 2. Chlorine--Industrial applications. 3. Chlorine compounds--Industrial applications. 4. Wood-pulp--Bleaching. 5. Substitution (Technology) I. Turoski, Victor, 1940– . II. American Chemical Society. Meeting (210th : 1995 : Chicago, Ill.)
TS1120.C48 1998
676.042--DC21 97-22723
CIP

ISBN 1-57504-066-2

ANN ARBOR PRESS
121 South Main Street, Chelsea, Michigan 48118
Ann Arbor Press is an imprint of Sleeping Bear Press

PRINTED IN THE UNITED STATES OF AMERICA
10 9 8 7 6 5 4 3 2 1

Acknowledgment

I would like to acknowledge Jesus Christ, my friend and guide, for His help and encouragement.

About the Editor

Victor Turoski is a graduate of Wilkes University, Wilkes-Barre, Pennsylvania, and has pursued a career in analytical chemistry at both Carter Wallace and the American Can Company. Turoski presently manages the Corporate Analytical Laboratory at the James River Technical Center in Neenah, Wisconsin, and is currently involved in both fundamental and applied research in papers and various coatings. He has published several articles in the field of environmental chemistry, primarily in the area of gas chromatographic/mass spectrometric identification and quantification of resin acids and priority pollutants in complex matrices. He also edited the ACS book *Formaldehyde—Analytical Chemistry and Toxicology.* Turoski is actively involved with the Environmental Division of the American Chemical Society at the national level and currently serves the Division as Program Chairperson for Environmental Analytical Chemistry. He also performs article peer review for *Environmental Science and Technology.*

Preface

Chlorine, in some form, affects the lives of us all. It is used to disinfect drinking water, to bleach clothes and clean household surfaces, to bleach wood pulp for the manufacture of white paper, to serve as an essential component in some plastics and medications, and in many other applications. Chloroorganic compounds can be formed as a by-product of some industrial processes and naturally occurring phenomena. A very small percentage of these compounds are toxic, at least to certain organisms at some dosages. Because of the concern about these materials, research is being conducted throughout the world on the origin, fate, and potential hazards presented by chloroorganic compounds. Many industrial operations are being modified to minimize the potential formation of these compounds, and some environmental advocacy groups are seeking the elimination of chlorine from industrial processes. For the past several years the United States Environmental Protection Agency has been considering draft rules regarding pulp and paper industry discharges of chlorine compounds.

With all the concerns and questions being raised about chlorine, it seems appropriate and timely to try to increase scientific knowledge with respect to chlorine issues and to disseminate the latest research findings. This has been achieved, in part, by means of an international symposium, *Chlorine and Chlorine Compounds in the Paper Industry,* held in August, 1995 at the National Meeting of the American Chemical Society (ACS) in Chicago. The symposium was sponsored by the American Chemical Society's Division of Environmental Chemistry. The purpose of the symposium was to provide a platform for in-depth discussions of the scientific, technical, and regulatory issues surrounding chlorine, with the many viewpoints on chlorine being presented in a fair and open manner. Forty speakers presented papers over a four-day period. The symposium speakers included respected toxicologists as well as noted representatives from EPA, FDA, academia, and several paper companies. Speakers were invited from many environmental advocacy groups, including Greenpeace, The Sierra Club, The Natural Resources Defense Council, and the Environmental Defense Fund. Greenpeace and the Sierra Club accepted invitations. The Greenpeace speaker withdrew his paper due to another commitment; the Sierra Club paper was presented at the symposium but was withdrawn from the book because of time and resource constraints surrounding the rewrite of the manuscript.

The quality of the papers presented at the symposium was exceptional. Virtually all aspects of chlorine use in the paper industry were discussed. Current and proposed bleaching processes were explored. Regulatory and public perception issues were presented, and the toxicology aspects of chlorine and chloroorganics were discussed. As part of this symposium, a Risk Assessment Panel of toxicologists was convened to express opinions and to address questions from the audience. The symposium was the subject of international press coverage. Several of the speakers were asked by the media to give press conferences, and many requests were received for reprints of abstracts. Because of the continued and widespread interest in chlorine compounds, the ACS Division of Environmental Chemistry subsequently asked me to serve as editor of a book based on the symposium.

I wish to thank everyone who worked so hard and so long to make both the symposium and this book a reality, in particular the speakers at the symposium, the authors in the book, and the peer reviewers. I would especially like to thank Frank Vincent, Director of Product Safety and Compliance at James River, who helped me organize the symposium and preside at the sessions. It is my hope that the peer-reviewed information presented here will be helpful to individuals in government regulatory agencies, the paper industry, environmental groups, and the interested public.

Victor Turoski
Neenah, Wisconsin

Foreword

The forest products industry, as one of the world's top ten manufacturing industries, plays a key role in both the North American and world economies. The U.S. industry represents about one-third of global capacity, supplies one-third of the world's paper and paperboard, and employs a significant proportion of the nation's labor force. Furthermore, it is a capital-intensive industry, owing to the complexity of its manufacturing facilities and the enormity of the amounts of material processed. The industry spends approximately 15% of its sales dollars on capital expenditures — far more than the corresponding expenditures of all other manufacturing industries, which average less than 6% of sales. This capital-intensive nature is a disincentive to manufacturing process changes, a fact which has earned the industry a reputation, perhaps undeserved, of being overly conservative. Because the industry is so large, its potential for impacting the environment is correspondingly large. Consequently, pollution prevention and control are foremost among its technical concerns, and command enormous capital expenditures. Approximately one-fifth of total capital expenditures are environment-related.

Pulp Bleaching and the Environment: A Historical Perspective

Within the paper manufacturing process, it is possible to identify several sources of discharges having potential to affect the environment. In mills that manufacture white paper or bleached market pulp, an important source is the bleach plant. Unbleached kraft pulp is made by chemically degrading and dissolving the lignin component of the wood raw material, allowing the wood cells to separate into a fibrous pulp suspension. Since exhaustive lignin removal at this stage would result in damage to the fiber's strength-giving cellulose, a small portion is allowed to remain. This residual lignin, which has been severely discolored by the pulping process, imparts a brown color to the fibers and must be removed before the pulp can be used to make white paper products. Furthermore, the process used for its removal must be more selective than the pulping process, since the strength of the fiber must be retained to make high-quality paper.

The bleaching process has evolved from a single-stage hypochlorite (H) treatment, used at the turn of the century, to elaborate multistage sequences involving chlorine, chlorine dioxide, oxygen, hydrogen peroxide and ozone. Removal of residual lignin is accomplished by oxidizing it to soluble products that can be washed out with water or alkali. In 1920, bleaching was a simple batch process. Calcium hypochlorite, made by bubbling chlorine gas into a lime slurry, was added to a suspension of unbleached pulp at a consistency of 7% or less and the temperature was raised to 35°C. The stock was slowly agitated until the desired degree of bleaching had been achieved and the pulp was washed.

A two-stage (HH) version of this process followed. About two-thirds of the total hypochlorite charge was added and allowed to react until none remained. The pulp was then washed, the remainder of the hypochlorite was added, and the bleaching was completed in the second stage. This simple modification of the process had two beneficial effects. First, the removal of dissolved but still oxidizable impurities reduced chemical consumption. Second, a reduction in the maximum and time-averaged concentrations of oxidizing agent reduced the tendency for it to attack and degrade the carbohydrate portion of the pulp, which would result in low pulp strength. Both represented significant advances over the single-stage process.

The next major advance was the introduction of chlorine gas (C) as a bleaching agent. This occurred commercially in about 1930. Chlorine is a vigorous oxidizing agent for lignin, being much more reactive than hypochlorite. It is also more selective than hypochlorite, having less tendency to attack the cellulose and other carbohydrate components of the pulp. Chlorine does not brighten the pulp like hypochlorite does, but it extensively degrades the colored lignin, allowing much of it to be washed out with the spent chlorination liquor and most of the rest to be removed by subsequent alkaline extraction (E). As a result, the amount of hypochlorite required for final bleaching was very much smaller than in the single-stage process. In addition, the higher selectivity of chlorine and the reduction in the amount of hypochlorite used gave much higher pulp strength. Finally, it proved possible to reach higher brightness levels in the multistage (CEH) process. This development marked the advent of the modern multistage bleaching sequence, in which the early stages remove the bulk of the residual lignin, usually without significant brightness increase, and the later stages cause large gains in brightness by removing the last traces of residual lignin.

The commercialization of chlorine dioxide (D) use in the early 1940s took all of these advances one step further. Chlorine dioxide was shown to be a powerful brightening agent and immeasurably more selective than hypochlorite. By adding a chlorine dioxide stage after the hypochlorite stage to give a CEHD sequence, much higher brightness levels could be achieved without compromising pulp strength.

Hydrogen peroxide is similar to chlorine dioxide, in that it can be used to selectively attain high brightness levels, provided that it is applied to pulp from which most of the residual lignin has already been removed by other bleaching agents. During the period when the applications of chlorine dioxide were being developed, much development work was also done on the use of peroxide in the final brightening stages, as a rival to chlorine dioxide. Both were eventually used, but the generally greater cost-effectiveness and selectivity of chlorine dioxide resulted in its being more widely adopted. It progressively replaced hypochlorite, and by the 1960s the CEDED sequence had become the standard for the industry.

During the period of chlorine dioxide's ascent, research was being conducted on two other promising bleaching agents — oxygen and ozone. Oxygen was shown to be capable of removing a substantial part of the residual lignin in unbleached kraft pulp, but difficulties were encountered due to its low solubility and its tendency to be nonselective, attacking and degrading cellulose as well as lignin. The more reactive ozone was shown to be capable of rapid and complete removal of residual lignin, but it exhibited even poorer selectivity than oxygen.

Oxygen was the first of the two to become commercialized, principally as a result of the discovery of cellulose protectors by Robert and co-workers in France. They found that adding small amounts of magnesium salts to the pulp before oxygen bleaching allowed about half of the residual lignin to be removed without detrimental effect on pulp strength. Soon after this discovery, in 1970, the first commercial oxygen stage was started up at SAPPI's Enstra, South Africa mill. The significance of this development lay in its potential for decreasing the environmental impact of the bleach plant. Unlike chlorination effluents, oxygen stage effluents can be recycled to the kraft chemical recovery system and destroyed. Since the quantities of material discharged from the bleach plant to the environment are roughly proportional to the amount of lignin removed, 50% delignification in the oxygen stage results in a roughly 50% reduction in bleach plant discharges.

Oxygen bleaching was implemented rapidly in Sweden, becoming widespread there during the 1970s and early 1980s. All bleached kraft mills in Sweden now practice oxygen bleaching. Implementation in North America has been slower, and oxygen bleaching is still far from universal in the United States and Canada. The pace of implementation is, however, increasing, largely due to increased regulatory and market pressures to lessen the perceived effect of pulp bleaching on the environment.

Ozone has greater potential for lessening effects of the bleach plant on the environment, since, unlike oxygen, it is capable of shouldering the full task of residual lignin removal. In principle, therefore, ozone bleaching should allow all of the material removed in the bleach plant to be recycled to the recovery system and destroyed instead of being discharged to the environment. Nevertheless, the obstacles are greater than in the case of oxygen, and only recently has it become possible to achieve the required degree of selectivity in ozone bleaching. This was accomplished commercially for the first time by Union Camp Corporation at its Franklin, Virginia mill. In September 1992, the mill started up a 1,000 ton/day bleach plant incorporating both ozone and chlorine dioxide stages. Since then, approximately 15 commercial ozone stages have started up, mostly in Europe.

Chlorine dioxide, which is both more selective and more expensive than chlorine, was initially used only as a brightening chemical, applied in the latter part of the multistage bleaching sequence, after the bulk of the residual lignin had been removed by earlier chlorination and caustic extraction stages. It was recognized early, however, that chlorine dioxide could replace part of the chlorine in the chlorination stage, and that such replacement had several advantages. One was that the selectivity of the chlorination stage was improved at low substitution levels. At higher substitution levels, the overall oxidation efficiency was increased, so that fewer oxidizing equivalents were required to achieve the same degree of lignin removal. An additional advantage — a decrease in the formation of chlorinated organic by-products — assumed heightened importance in the mid-1980s when the discovery of dioxin formation during pulp chlorination focused attention on this aspect of pulp bleaching. As a result, it has become common to use chlorine dioxide to completely displace molecular chlorine from the bleaching sequence. This practice is known as elemental chlorine–free (ECF) bleaching.

Perhaps no other event has led to so much public attention being focused on the paper industry as the finding, in 1985, that fish collected downstream from paper mills contained detectable levels of dioxins. The finding was unexpected, since studies at some of the same locations only two years earlier had failed to show the presence of dioxins at levels which exceeded the then current detection limits in the parts per billion range. Advances in analytical methodology enabled the 1985 discovery at levels in the parts per quadrillion to parts per trillion range.

The extreme smallness of the concentrations detected went unnoticed by environmentalists and the general public. The alarming fact was that substances viewed as being among the most toxic materials on earth were being unnecessarily released to the environment. Such fears were reinforced by the knowledge that another class of chlorinated organic compounds, the polychlorinated biphenyls (PCBs), had adversely affected the reproductive success of some Great Lakes birds and animals. Unfortunately, the distinction between chlorinated dioxins and other chlorinated organic by-products was not appreciated by the general public, nor was it appreciated that PCBs are not among the by-products of pulp bleaching. Consequently, there grew a widespread perception that all chlorinated organic compounds in pulp bleaching effluents were toxic. The bulk of the scientific evidence, on the other hand, fails to support this view. With the exception of the highly chlorinated phenolics (which are absent from ECF effluents), few, if any, of the compounds present are demonstrably toxic. Repeated attempts have failed to establish a link between the total concentration of chlorinated organic compounds and toxicity.

Sum parameters reflective of the total amounts of chlorinated organic compounds in pulp bleaching effluents nevertheless gained new importance. One such parameter, total organically bound chlorine, or TOCl, relied on adsorption of the chlorinated organic compounds to a resin column to separate them from other effluent components, followed by their elution and determination. This method had the disadvantage of being experimentally complicated, and tended to give low values because of incomplete recovery from the resin column. An alternative method soon became available, the result being known as adsorbable organic halide (AOX). In this procedure, the effluent was passed through columns of activated carbon, which were then washed with potassium nitrate solution to rid them of inorganic chlorides. The carbon was then burned and the resulting hydrogen chloride determined by microcoulometric titration. The determination of AOX, being relatively straightforward, held much appeal for regulatory agencies, and it has been incorporated into most regulations governing the effluents from pulp bleaching.

In the United States, proposed guidelines for regulation of pulp mill emissions have been published by the U.S. Environmental Protection Agency. These are sometimes referred to as the "Cluster Rule," a reference to the inclusion of both air and water standards in a single set of rules. The part of the proposed rules that is most relevant to this discussion concerns effluents from mills producing paper grade bleached kraft pulp. Dioxins, specifically 2,3,7,8-tetrachlorodibenzodioxin (TCDD) and 2,3,7,8-tetrachlorodibenzofuran (TCDF) are included in the list, the requirements being that TCDD be nondetectable and TCDF be present at a daily average level no greater than 359 ng/ton, which is equivalent to about 7 parts per quadrillion (ppq) in the effluent from a typical bleach plant. The lowest level of either of these two compounds that can be reliably detected is about 10 ppq. The remaining compounds singled out for specific regulation are two halocarbons, two ketones, and nine chlorophenols.

In addition to these limitations on specific compounds in the effluent from the bleach plant, new limits have been proposed to apply to the whole mill effluent at the final discharge point. They govern the levels of AOX, biochemical oxygen demand (BOD), total suspended solids (TSS), chemical oxygen demand (COD) and color. As proposed, the EPA limits, especially those on AOX (0.156 kg/t) and COD (25 kg/t), are very demanding. Compliance would require not only the elimination of elemental chlorine but also either extended delignification or oxygen delignification (and perhaps both), together with an efficient secondary biological treatment system.

The stringency of these and similar regulations proposed in other jurisdictions, combined with market pressures arising from public reaction to the dioxin and PCB issues, has created commercial interest in a new set of pulp bleaching technologies, known as totally chlorine-free (TCF) bleaching. Markets in Western Europe have been particularly receptive to TCF pulps, and their Scandinavian suppliers have responded by starting up several large TCF bleach plants. In contrast, the North American industry has been unable to justify significant TCF production. Higher production costs and lower pulp quality are often cited as reasons not to pursue the TCF alternative.

Nature and Impact of Pulp Bleaching Effluents

Untreated effluents from chlorine-based bleaching typically contain about 5 kg of AOX per metric ton of pulp, about 1% of which is extractable by nonpolar solvents (EOX). The significance of the extractability of EOX is that it implies lipophilicity and the likelihood that it will concentrate in the fatty tissues of fish. A significant fraction of the AOX represents material of high molecular weight. The remaining low-molecular-weight portion comprises numerous chlorinated compounds, of which more than 300 have been identified. Complete replacement of chlorine by chlorine dioxide sharply reduces AOX, EOX, and the number of chlorinated compounds produced. These changes are attributable to chlorine dioxide's lower chlorine content, greater oxidizing power, and inability to undergo reactions that directly result in chlorine substitution.

Studies of the impact of effluents from mills employing high-substitution and ECF bleaching effluents on the aquatic life in the receiving waters have generally shown that their impact is low or nearly nonexistent, provided that the effluent has undergone the normal secondary biological treatment that is in place at virtually all U.S. bleached pulp mills. Such studies have shown that the effluents lack acute and chronic toxicity, embryotoxicity, carcinogenicity, and teratogenicity. Induction of detoxifying enzymes is, however, observed, and there is evidence of changes in population demographics of fish. These latter effects have been shown to be present in waters downstream of both bleached and unbleached mills, and cannot be attributed to the bleaching process. This is an area of continuing research.

It is notable that several studies have demonstrated a lack of correlation between AOX and toxicity, and that chlorinated organic compounds are found in nature.

Toxicity and Risk

Regulating agencies are charged with protecting the public from unnecessary risks that may otherwise be incurred as a result of industrial activity. In a given situation, they do this by estimating both the toxicity of a material and the public's exposure to it. Exposure is the product of concentration and intake, and risk is the product of exposure and potency (toxicity). Consequently, at least in principle, risk can be regulated by regulating the discharge levels that determine concentration.

This simple regulatory model assumes that toxicity can be unambiguously evaluated. In reality, this is seldom true. This is exemplified by dioxins. These compounds are singularly toxic toward laboratory animals, resulting in high levels of cancer incidence, but evidence exists for the supposition that toxicity toward humans is much less than the animal studies would indicate. These are matters of continuing debate and research.

Technology, Progress, and Cost

Environmental concerns and their ramifications have fueled a period of vigorous research, development, and implementation activity in the technology of bleached pulp manufacture. Implementations of extended delignification, oxygen delignification, ECF bleaching, and alkaline stages enhanced with oxygen and peroxide have resulted in sharp reductions in effluent loadings. Dioxins in bleached pulp mill effluents have been reduced by more than 92% since their discovery. Since 1985, there have also been sharp reductions in AOX, COD, color, highly chlorinated phenolics, chloroform, and effluent volume.

An unresolved question concerns the relative costs and benefits of further reductions. Divergent views are held by industry, environmentalists, and regulatory bodies. Even within these sectors, one finds a variety of positions. Within the industry, for example, there are varying degrees of commitment to the goal of the minimum effluent mill. Within the environmental community, organizations such as the Environmental Defense Fund have chosen dialogue with the industry and its customers as a route to improved performance, while other environmental groups have adopted more adversarial stances. A final resolution will require a realistic assessment of risks in relation to the incremental costs of reducing them.

The Future

It appears likely that the massive accumulation of data that has been prompted by traumatic occurrences such as the dioxin discovery and the Cluster Rule proposal will have results that are, in the end, beneficial to all parties to the debate on environmental effects of pulp bleaching. Increased dialogue between the industry and the environmental community has resulted in greater levels of understanding on both sides, and to more constructive agreements. There are strong signs that regulatory agencies are committed to developing rules that meet the criterion of "common sense." The Cluster Rule, for example, is about to be reproposed in a form that has been modified to take account of new data.

In terms of technology, it is a foregone conclusion that bleaching with molecular chlorine will be phased out entirely within a very few years. Chlorine dioxide use, in the form of ECF bleaching, will continue to increase during the chlorine phase-out period, and decline only gradually thereafter, as the penetration of extended delignification and oxygen delignification increases. Water use and effluent volumes will steadily decrease as the technology for dealing with nonprocess elements develops and the ideal of the "minimum-effluent mill" is approached.

Thomas J. McDonough
Institute of Paper Science and Technology

Keynote Address

Although this is my first opportunity to appear before your distinguished organization, I feel very much at home, having already seen a number of friends and colleagues here at your Chicago convention.

Let me make it clear at the outset: I am neither a chemist, scientist, nor an engineer. I do, however, have much more than a passing familiarity with the issues that affect you and your livelihood. That familiarity is derived from almost 15 years of service as president of the national trade associations for two major industries — plastics and paper — each with exceptionally strong foundations rooted in the field of chemistry.

While drawing on my experiences in helping lead two of our nation's great industries, I stand before you today in my role as the leader of the government affairs efforts involving the Alliance for Reasonable Regulation.

However, before getting into the heart of my remarks, I hope you will permit me a moment's commercial plug for my current organization. The American Plastics Council is the plastic industry's national trade association on resource conservation and benefits issues. APC, as our organization is referred to in shorthand, works on behalf of the overall plastics industry to enhance the integrity of plastics, focusing on resource-related environmental issues to ensure that plastics are a contributor to a safer and cleaner environment for the future. In pursuit of its objectives, APC works closely with all elements within the broader plastics industry.

And now to the message at hand.

Ladies and gentlemen, the integrity of your work is increasingly being called into question — a trend whose roots can be traced back to the late 1970s.

Advocacy groups and so-called "public interest" groups have become increasingly effective at pressuring government to take action to protect health, safety, and the environment. That pressure, oftentimes, however, results in government taking action before the science is in hand to make clear what actions are needed.

One of the more notable of these instances involves the issue of acid rain and its impact on our forests. In addition to the millions of research dollars earlier expended, the federal government in the late 1980s launched a Congressionally mandated $500 million research program — referred to as NAPAP — to determine the impacts of acid rain on our nation's forests and streams.

Even with all this money committed and a solid research program underway, the government preempted any findings and adopted a series of prescriptive measures of questionable value and cost-effectiveness.

I am sure many of you, from your own experience, could describe similar such instances.

It should be clear to you — it certainly is to me — that this type of cavalier disregard for scientific rigor is slowly but surely making a mockery of the interplay of sound public policy and good science. If this trend is allowed to continue unchecked, it is only a matter of time before good science — and, therefore, the integrity of your work — is marginalized in the development of public policy. Can we as individuals — can we as a nation — afford to have science disregarded in selecting our nation's funding priorities in a time of ever-increasing demand on limited resources? The simple answer is no!

There is good news on the way, however. We are not alone. Over the past three years, there has been a slow but, nonetheless, gradual increase in support for the employment of risk assessment/cost-benefit analysis in government's promulgation of regulations involving health, safety, and the environment.

I am here today to urge you to sign up as a soldier in this battle to ensure that the role of good science in the development of public policy be strengthened, not marginalized. You can do just that by supporting — and doing so in a public fashion whenever practical — the application of risk assessment/cost-benefit analysis in public policy.

I would like to use the remainder of my remarks to brief you on what is already under way regarding regulatory reform and risk assessment — why it is important, why legislation is needed, what the real issues are in the current debate, and where matters stand at present.

First, what is under way:

The lead national organization working to advance the principles of regulatory reform and risk assessment is the Alliance for Reasonable Regulation (ARR). ARR was formed last fall as a broad-based business and agriculture coalition supporting Congressional enactment of regulatory reform and risk assessment.

ARR, as the Alliance is called, consists of more than 2600 individual businesses, large and small, and industry groups representing almost two-thirds of our nation's gross domestic product and employment.

ARR has been conducting its work in concert with other groups, such as:

- Project Relief — a large coalition of business and grassroots organizations.
- Public interest groups, such as the Citizens for a Sound Economy.
- The Risk Policy Council — a bipartisan public/private partnership representing governors and mayors.
- Many other groups representing such entities as school boards, neighborhoods, and think tanks.

Why is regulatory reform and, specifically, legislation needed?

At issue is the future of the relationship between government, those it regulates, and how best to protect public health, safety, and the environment. It's all about how we, as a nation, get more for the time, resources, and money we spend on health, safety, and environment.

Let me share with you an example. The Harvard School of Public Health estimates that if our nation regulated smarter, we could save 60,000 *more* lives per year. For the $6,000 per year per family the United States spends on regulatory compliance, we can no longer afford to let matters continue along a "business as usual" path.

This debate is also all about how to re-engineer an internally inconsistent regulatory system that was created in an ad-hoc fashion over the past twenty-five years. Think for a moment about the shortcomings of the current system, which was designed when:

- Our science, technology, and analytical capabilities were less sophisticated;
- States had little experience or capacity to act;
- We didn't know which regulatory systems would or wouldn't work;
- The private sector hadn't learned how to apply its creativity to the challenge of protecting public health, safety, and the environment;
- Concepts like W. Edward Demming's quality manufacturing principles, including the need to reduce waste of all kinds, were not appreciated nor applied; and — most important to many of us —
- The voter and consumer had a different set of expectations for the manufacturers of the products we all buy.

This debate is all about what will happen to our international competitiveness, our quality of life, and the growing national dissatisfaction with government, if we don't successfully reinvent how we go about protecting our health, safety, and environment.

Intuitively, people relate to the notion of reform. Almost every businessperson I've encountered can give you an example of how they could reduce emissions beyond government requirements and do it for less money and hassle. Most homeowners, factory workers, municipal officials, and parents have some personal experience with a regulatory requirement that just doesn't make sense. Plain and simple, it's easy to understand why 80+% of the American public supports regulatory reform.

This reform won't happen easily, however. The opposition has mounted an all-out effort, principally by distorting the facts. What regulatory reform is *not* about is rollbacks of safeguards and protections, smoke billowing out of stacks, rivers on fire, and sludge flowing from our drinking water taps. That's what the opposition has claimed — trying to create a personalized sense of fear in peoples' minds.

Why is legislation needed? Can't the president just fix it himself?

The last five presidents have all issued Executive Orders on Regulatory Reform — Clinton, Bush, Reagan, Carter, and Ford. None of these Executive Orders had the force of law. That turns out to have been a repeated flaw.

To help you to better understand the problem, I'll use the current administration's performance as an example.

In September 1993, President Clinton issued Executive Order 12866, which has many similarities to S. 343, The Dole–Johnston Regulatory Reform Bill currently before the Senate.

Agencies aren't doing — and, in many cases, can't do by law — what Executive Orders say, regardless of political party in power or administration. Presidents from both parties have been unable to reinvent the current system. Reagan had as much trouble as Carter.

Under the Clinton Administration, for all its emphasis on voluntary government/business cooperation, the problem hasn't gone away either. A recent study of the Clinton Administration's adherence to its own regulatory Executive Order shows some interesting facts. Over a six-month period last year:

- 45 of 510 rules reviewed by OMB were considered "major" or "significant," and therefore subject to the tests and requirements set forth in the Executive Order.
- 26 of the 45 failed to justify the need for a rule.
- Only 28 of 45 assessed costs.
- Only 9 of 45 evaluated alternatives.
- Only 4 of 45 claimed the benefits justified the costs.
- None of rules examined by OMB were returned to the agencies for lack of compliance with the Executive Order, although the large majority clearly did not comply.

Although the administration claims they were using "negotiations," a check of the public files where negotiated changes are kept will reveal no changes. In short, nothing happened.

It's abundantly clear that the Executive Orders on regulatory reform are no more than paper tigers to be disregarded, without peril, by any agency. And we wonder why government regulations run amuck!

What are the basics of effective regulatory reform legislation?

The fundamental basis must be the use of good science, reasonable estimates of risk, and sound economics. By that, I mean that any legislative reform should include the following:

Risk Assessment

- We need to do a better, more consistent, more reasonable job of evaluating and characterizing risk. And, this includes not creating the "scare-of-the-day" by waving around poorly substantiated risk assessments.

Benefit-Cost Analysis

- We need to do a better job of understanding the benefits and costs of a legitimate range of alternative approaches to solving the problem.
- Benefits don't have to necessarily exceed or outweigh the costs. Analysis isn't that precise. But, clearly, there should be some reasonable relationship between the two; they should be in the same ball park.

Risk-Based Priority Ranking

- We need to do a better job of allocating federal agencies' resources toward those activities where we, as a nation, get the greatest benefits from the dollars spent. And, this doesn't have to mean we ignore sensitive or small subpopulations. These priorities should be incorporated into agency budgets, strategic plans, and research activities.

Apply to a "Major" Rule

- The analytical requirements don't need to be applied to every rulemaking.
- Obviously the scope of the analysis should be tailored to the scope of the problem. The Dole–Johnston bill now sets this level at $100 million.

Peer Review

- The quality of the agencies' risk assessments and benefit-cost analyses should be subject to independent, balanced outside review by expert peer review panels.

Applicability to Current Law

- The risk assessment and benefit-cost analyses must apply to new regulations issued under current laws to ensure they provide benefits commensurate with the costs.
- Under the Dole–Johnston bill, the risk assessment and benefit-cost analysis used in setting regulatory standards would supplement — but not supersede — the factors provided in the enabling statute.

Petitions

- There should be some mechanism — with the burden of proof on the petitioner — to require review of existing major rules with enough procedural protection to ensure agencies don't get tied in knots.
- Not every regulation should be on the books forever, especially if Congress adds new factors into the decisional criteria or, for example, new science shows we've learned something since the rule was originally written, or we've discovered a better way to meet the goal.
- Nuisance filings will be kept to a minimum. The cost of putting forward a well-documented case will be too high, the opportunities too limited, the compliance investments already made, and company reputations at risk for many faulty petitions.
- Some cite the petition process as the all-evil "rollback," but it is not.
- Petition outcomes would most likely result in a change in the flexibility for compliance more than anything else.

Judicial Review

- Without some form of an enforcement mechanism, agencies will ignore this law, as I described earlier in the case of the Executive Order.
- Real reform requires a law with teeth to make it effective. That's why this section of the bill is the one the opposition wants most to eliminate.
- The way judicial review is structured in the bill, there aren't a thousand and one opportunities to sue, as claimed by the opposition.

Emergency Exemption

- Exempt situations are permitted when the head of an agency considers there to be an emergency, or other activities are involved, such as military readiness.

States

- State actions are not directly covered.
- To the effect a state is implementing a federally delegated program, the guidelines may have some follow-on effect.

Generally

- Reform should create a common base for all environmental, health, and safety programs.
- Some argue that if there are problems with the mishmash of current laws, and changes are needed, then each law should be fixed, one at a time. I can only answer by saying it took 25 years

to put the current rules in place. It would take at least that long to fix them all, given the political and substantive battles that would ensue.

- As we take a look at each regulation, we need a common basis to start from — a common, organic statute that, for the first time, will give some unifying theme to all our health, safety, and environmental laws.
- This legislation currently under consideration in the U.S. Senate is a good start.

What happens from here on out?

A solid risk assessment/regulatory reform bill passed the House by a wide margin last spring.

Its Senate counterpart, S. 343 (Dole–Johnston), has been held hostage since just after Memorial Day. On three occasions, the Democratic leadership has successfully withheld sufficient votes to allow the Senate to vote on S. 343. All 54 Republicans and four Democrats — Senators Johnston and Breaux of Louisiana, Heflin of Alabama, and Nunn of Georgia — have also voted with the Republicans for cloture. Cloture is the necessary step to stop a filibuster when the opposition blocks a bill from being voted upon. It takes 60 votes to invoke cloture, so regulatory reform is trapped in the Senate for want of the support of two more Democrats.

If the two votes are obtained, then the House and Senate bills would be reconciled, and a joint bill returned to both Houses of Congress for a final vote. If passed, the bill goes to the President, who will either sign the bill or veto it. A two-thirds vote of the Members present in either chamber is needed to overturn a Presidential veto — a possibility, given the administration's strong opposition to date.

The outcome of the Senate bill will be decided in the weeks immediately following Congress' return after Labor Day.

Months of debate have resulted in a very balanced Senate bill that enjoys wide public support, including from some sympathetic to environmentalists' interests. For example, environmental groups' Gregg Easterbrook, *Newsweek* correspondent and author of the best-selling new book, *A Moment on the Earth: The Coming Age of Environmental Optimism,* has publicly declared the Senate bill to be a balanced one.

The hope of those of us working to gain enactment of regulatory reform/risk assessment legislation is that members of the U.S. Senate will hear of sufficient support for the legislation from their constituents to ensure passage can be effected in September. With only two more votes needed, the prospects are good, if the constituent interest level is sufficiently high. We remain hopeful.

In closing, please allow me to underscore again why this seemingly arcane subject and legislation is so important.

Without the force of law to ensure that good principles of risk assessment and cost-benefit analysis are employed in the development and enforcement of federal regulations, we — as a country — will be forced to continue down a path where the rule of science and reason is less and less a factor in public policy, and where the public confidence in government will continue eroding.

This is not an outcome any of us should find attractive. All of us are at risk in the face of a weak government that has lost the confidence of its citizens. A strong government, but a fair and reasonable one, is the best protector of our liberties. Let us all go forward in the hope that good science, reasonable estimates of risk, and sound economics become the law of the land for regulations aimed at protecting human health, safety, and the environment.

Red Cavaney
President & CEO
American Plastics Council

Contributing Authors

Lesa L. Aylward
Karch & Associates, Inc.
1701 K Street, NW, Suite 1000
Washington, DC 20006

Kathleen M. Bennett
James River Corp.
Corporate Headquarters
120 Tredegar Street
P.O. Box 2218
Richmond, VA 23218

Michael Bolger
U.S. Food & Drug Administration (HFS-308)
200 C Street, SW
Washington, DC 20204

S.J. Borghoff
Chemical Industry Institute of Toxicology
P.O. Box 12137
Research Triangle Park, NC 27709

T.M. Bousquet
National Council of the Paper Industry for Air and Stream Improvement, Inc. (NCASI)
720 S.W. 4th Street
P.O. Box 458
Corvallis, OR 97333

Byron E. Butterworth
Chemical Industry Institute of Toxicology
P.O. Box 12137
Research Triangle Park, NC 27709

Tashia V. Caughran
Oak Ridge Institute for Science and Education
c/o U.S. EPA
Athens, Georgia 30605

Red Cavaney
American Plastics Council
1801 K Street, NW
Suite 710L
Washington, DC 20006

Timothy W. Collette
U.S. Environmental Protection Agency
960 College Station Road
Athens, GA 30605

R.B. Conolly
Chemical Industry Institute of Toxicology
P.O. Box 12137
Research Triangle Park, NC 27709

D.L. Cook
National Council of the Paper Industry for Air and Stream Improvement, Inc. (NCASI)
720 S.W. 4th Street
P.O. Box 458
Corvallis, OR 97333

Tudor T. Davies
Director, Office of Science and Technology
Office of Water
U.S. Environmental Protection Agency
401 M Street SW
Washington, DC 20460

Thomas L. Deardorff
International Paper
6285 Tri-Ridge Blvd.
P.O. Box 7910
Loveland, OH 45140

Richard L. Erickson
Weyerhauser Company
Tacoma, WA 98477

John L. Festa
American Forest & Paper Association
1111 19th Street, NW
Washington, DC 20036

Robert P. Fisher
Vice President, Health Effects
National Council of the Paper Industry for Air and Stream Improvement, Inc. (NCASI)
P.O. Box 13318
Research Triangle Park, NC 27709-3318

William J. Gillespie
Vice President
Risk Assessment–Exposure
National Council of the Paper Industry for Air and Stream Improvement, Inc. (NCASI)
P.O. Box 13318
Research Triangle Park, NC 27709-3318

Robert J. Golden
Environmental Risk Sciences
1000 Thomas Jefferson Street, NW
Suite 506
Washington, DC 20007

Gordon W. Gribble
Department of Chemistry
6128 Burke Laboratory
Dartmouth College
Hanover, NH 03755

Sara Hale Henry
U.S. Food & Drug Administration (HFS-308)
200 C Street, SW
Washington, DC 20204

Marquita K. Hill
Department of Chemical Engineering
University of Maine
Orono, ME 04469-5737

Stewart E. Holm
Georgia-Pacific Corporation
Suite 775
1875 I Street, NW
Washington, DC 20006

Clifford T. Howlett, Jr.
1300 Wilson Blvd.
Arlington, VA 22209

Nathan J. Karch, Ph.D., D.A.B.T.
Karch & Associates, Inc.
1701 K Street, NW, Suite 1000
Washington, DC 20006

G.L. Kedderis
Chemical Industry Institute of Toxicology
P.O. Box 12137
Research Triangle Park, NC 27709

J.R. King
DynCorp, Inc.
Alexandria, VA 22314

Jaana Koistinen
Department of Chemistry
University of Jyväskylä
P.O. Box 35
FIN-40351
Jyväskylä
Finland

Larry LaFleur
National Council of the Paper Industry for Air and Stream Improvement, Inc. (NCASI)
720 S.W. 4th Street
Corvallis, OR 97333

Reino Lammi
Wisaforest Ltd.
P.O. Box 68601
Pietarsaari
Finland

Lindsay M. Lancaster
International Paper
Manufacturing Technology Center
Tri-Ridge Blvd.
Loveland, OH 45140-8318

Karl-Johan Lehtinen
Environmental Research Group–MFG
Tekniikantie 12
FIN-02150
Espoo
Finland

R.M. Litow
DynCorp, Inc.
Alexandria, VA 22314

Jeff Louch
National Council of the Paper Industry for Air and Stream Improvement, Inc. (NCASI)
720 S.W. 4th Street
Corvallis, OR 97333

Benjamin W. Lykins, Jr.
U.S. Environmental Protection Agency
Cincinnati, OH 45268

Thomas J. McDonough
Institute of Paper Science and Technology
500 10th Street, NW
Atlanta, GA 30318

A.B. McKague
Department of Chemical Engineering and Applied Chemistry
Pulp & Paper Centre
University of Toronto
200 College Street
Toronto, Canada M5S 3E5

Paula Maatela
Defence Forces Research Centre
P.O. Box 5
FIN-34111
Lakiala
Finland

Jaakko Paasivirta
Department of Chemistry
University of Jyväskylä
P.O. Box 35
FIN-40351
Jyväskylä
Finland

Helena Palm
Department of Chemistry
University of Jyväskylä
P.O. Box 35
FIN-40351
Jyväskylä
Finland

Kathleen S. Patterson
U.S. Environmental Protection Agency
Cincinnati, OH 45268

W. Harvey Persinger
Weyerhauser Company
Tacoma, WA 98477

Richard B. Phillips
International Paper
6283 Tri-Ridge Blvd.
Loveland, OH 45140-8318

A. Poland
McCardle Center for Cancer Research
University of Wisconsin–Madison
Madison, WI 53706

Tiina Rantio
Department of Chemistry
University of Jyväskylä
P.O. Box 35
FIN-40351
Jyväskylä
Finland

D.W. Reeve
Department of Chemical Engineering and
Applied Chemistry
Pulp & Paper Centre
University of Toronto
200 College Street
Toronto, Canada M5S 3E5

Jean J. Renard
International Paper
P.O. Box 160707
Mobile, AL 36616

Susan D. Richardson
U.S. Environmental Protection Agency
960 College Station Road
Athens, GA 30605

M.A. Rouf
Department of Biology and Microbiology
University of Wisconsin–Oshkosh
Oshkosh, WI 54901

D.R. Rushneck
Interface, Inc.
Ft. Collins, CO 80526

Robert J. Scheuplein
The Weinberg Group, Inc.
1220 Nineteenth Street, NW, Suite 300
Washington, DC 20036-2400

X. Shen
Department of Chemical Engineering and
Applied Chemistry
Pulp & Paper Centre
University of Toronto
200 College Street
Toronto, Canada M5S 3E5

Lucinda B. Sonnenberg
Director, Millar Wilson Laboratory
Jacksonville University
2800 University Blvd., North
Jacksonville, FL 32211

Thomas B. Starr
ENVIRON Corporation
7500 Rainwater Road
Raleigh, NC 27615

Jukka Tana
Environmental Research Group–MFG
Tekniikantie 12
FIN-02150 Espoo
Finland

William A. Telliard
U.S. EPA Office of Water
Washington, DC 20460

M.V. Templin
Chemical Industry Institute of Toxicology
P.O. Box 12137
Research Triangle Park, NC 27709

Peter T. Thomas, Ph.D.
Covance Laboratories, Inc.
P.O. Box 7545
Madison, WI 53707-7545

Alfred D. Thruston, Jr.
U.S. Environmental Protection Agency
960 College Station Road
Athens, GA 30605

Rudolph N. Thut
Weyerhauser Company
Tacoma, WA 98477

Terry Troxell
U.S. Food & Drug Administration (HFS-308)
200 C Street, SW
Washington, DC 20204

B. Frank Vincent
Environmental Sciences
James River Corporation
1915 Marathon Avenue
Neenah, WI 54956

D.C. Wolf
Chemical Industry Institute of Toxicology
P.O. Box 12137
Research Triangle Park, NC 27709

Contents

Part 1
Current and Proposed Bleaching Alternatives

Part 2
Analytical and Environmental Aspects

Part 3
Public Perceptions

Part 4
Regulatory Aspects

Part 5
Toxicology and Mechanisms

Chlorine and Chlorine Compounds in the Paper Industry

Edited by

Victor Turoski

Part 1

Current and Proposed Bleaching Alternatives

Chapter 1

The Revolution in Bleached Chemical Pulp Manufacturing Technology

T.J. McDonough

ABSTRACT

Environmental concerns and their ramifications have fueled a period of vigorous research, development, and implementation activity in the technology of bleached chemical pulp manufacture. Extended delignification is undergoing continued research and development and is gaining the status of established technology as operating experience accumulates. Oxygen delignification, though in some respects a mature technology, has been considerably advanced by recent work on two-stage systems and effects of additives, including hydrogen peroxide and activators. Biotechnology applications appear poised for a breakthrough in the form of practical mediated lignin-degrading enzyme systems, and specialized hemicellulases are appearing with increasing frequency. Elemental chlorine–free bleaching is being advanced through studies of chlorine dioxide delignification, kappa factor minimization, and sequence development. Totally chlorine-free bleaching, until recently a laboratory curiosity, has been extensively commercialized in Europe and is enjoying continued rapid development at the laboratory and pilot plant stages. This chapter discusses recent work and future directions in each of these areas.

INTRODUCTION

Chemical pulp manufacturing technology is in the midst of a period of rapid change driven by concern for the environment. The effects are seen both in the marketplace and in increasingly stringent government regulations. Although it may seem too dramatic to characterize this period of change as revolutionary, numerous examples may be cited in support of such a statement. The amount of chlorine used to bleach pulp in Sweden decreased from 150,000 tonnes per year in 1987 to zero in 1993.[1] Dioxins in North American pulp mill effluents have been virtually eradicated since their discovery in 1987.[2] Production of pulp bleached in totally chlorine-free (TCF) processes in Sweden and Finland grew from close to zero in 1990 to more than 3 million tonnes annually in 1994, and North American production of pulp bleached in chlorine dioxide–based (elemental chlorine–free, ECF) processes grew from less than two million tons annually in 1990 to nearly 17 million in 1995.[3] A final example is the sudden appearance of ozone as a commercial bleaching agent; the number of commercial ozone bleaching operations has increased from zero in 1992 to at least 11 in 1995.[4]

This chapter highlights newly introduced and rapidly evolving technologies for bleached chemical pulp (BCP) production with examples from the recent conference literature. It does not, however, attempt to present a comprehensive review of relevant developments; to do so would require more time and space than are currently available, given the rapid pace of research and development activity in this area.

EXTENDED DELIGNIFICATION

An attractive way to reduce bleach plant effluent discharges is to modify the pulping process to allow more lignin to be selectively removed in the digester, leaving less to be removed by bleaching. Many methods of doing this have been developed over the past two decades. Adding anthraquinone (AQ) to kraft digesters and simply extending the cook by manipulating pulping variables are two ways suggested in 1980.[5] Both ideas have recently resurfaced because they offer the possibility of signifi-

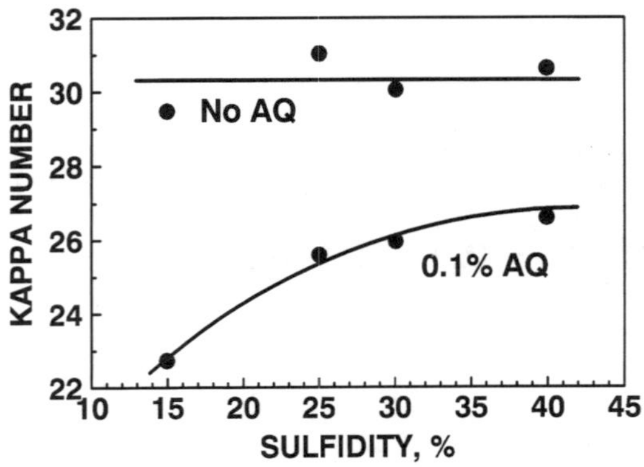

Figure 1.1. Unbleached softwood kraft kappa number vs. sulfidity with and without addition of 0.1% anthraquinone.[6]

cant reductions in unbleached kappa number without the substantial capital cost of digester replacement or retrofitting. Figure 1.1 shows the result of adding 0.1% AQ to softwood kraft cooks at different sulfidities, with alkali charges chosen to give constant kappa number in the absence of AQ.[6] At 25% sulfidity, AQ reduces the unbleached kappa number of softwood kraft pulps by 18% with no loss in yield and therefore no increase in loading to the recovery furnace. The reduction increases with decreasing sulfidity and is smaller for hardwoods. Extended cooking to kappa 20 in conventional digesters without AQ can be done without incurring a strength penalty, as has been recently demonstrated by Tikka.[7]

Another idea proposed in the early 1980s was the combined use of polysulfide (PS) and AQ to lower the unbleached kappa number.[8] This, too, has attracted renewed interest.[9,10,11] Unbleached kappa numbers can be reduced from 30 to approximately 18 without yield loss (but with a slight loss in tear strength) by oxidative conversion of part of the sulfur in 30% sulfidity white liquor to PS sulfur. Figure 1.2[11] illustrates this. Conversion corresponding to a sulfidity reduction to 12% gives a PS sulfur charge of slightly greater than 1% (o.d. wood basis). The sulfidity reduction has been shown not to affect kappa number or viscosity in hardwood cooks.[12] In extended modified continuous cooking (EMCC) of softwood, the combined use of PS and AQ can decrease the kappa number from 20 to 12 without yield loss but, again, with a slight loss in tear strength. The strength loss is the same as that incurred by cooking to kappa 12 with neither PS nor AQ.[10] The effectiveness of polysulfide is maximized by adding it early in the cook, as in the impregnation stage of the EMCC system.[12,13] Such a modification of EMCC pulping would be facilitated by a process, such as green liquor crystallization, for concentrating the total sulfide in a stream comprising only about half of the total white liquor volume.[12]

While PS and AQ addition may be considered low-capital approaches to decreasing unbleached kappa numbers, alternatives requiring more extensive equipment additions or modifications are available. Digester modifications for extended delignification are based on four ideals, each of which must be approached as closely as possible, given constraints imposed by practicality and the need to accomplish the desired extent of delignification within a reasonable time. They are (a) a hydroxide ion concentration that is as low as possible, especially at the beginning of the cook, and as nearly constant as possible throughout the cook; (b) a hydrosulfide ion concentration that is as high as possible, especially in the early stages of the cook; (c) a dissolved lignin concentration that is as low as possible, especially in the later stages of the cook; and (d) a temperature that is as low as possible throughout. While these apply equally to continuous and batch digesters, an additional requirement for batch systems is provision of some method for gently discharging the digester to preserve pulp strength. Such measures include "cold blow" — displacing the contents of the cooled digester by compressed air or steam — and pumped discharge. Their success is measured in terms of "strength delivery" — the ratio of the strength of the discharged pulp to that of a corresponding laboratory

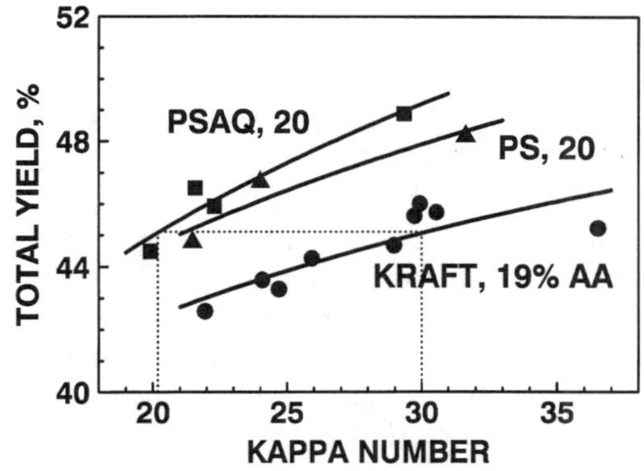

Figure 1.2. Yields of kraft, kraft-PS, and kraft-PSAQ pulps vs. kappa number.[11]

cook or basket cook.[14] The cold blow concept has been an integral part of the design of continuous digester systems since their commercial introduction many years ago.

Modifications of batch digester systems for extended delignification are marketed under several trade names, including Rapid Displacement Heating (RDH), SuperBatch, and Enerbatch. The first two are described in a review by Tikka[7] and the last in a recent conference presentation.[15] All incorporate multiple liquor accumulators and a series of liquor changes within each cook, whereby the liquor in the digester is displaced to one of the accumulators by a liquor of different composition and temperature. The RDH system employs a warm black liquor preimpregnation of the uncooked chips, displacement of this liquor by hot white and black liquors, and heating, cooking, and displacement of the black liquor in two stages by washer filtrate. The SuperBatch system differs principally in regard to the conditions of the black liquor preimpregnation; it uses more severe conditions and claims thereby to achieve improved selectivity. The Enerbatch system sacrifices the black liquor preimpregnation in favor of low-temperature white liquor preimpregnation for improved uniformity. Recent evolution of the RDH system has been toward the use of higher preimpregnation temperatures, with improved performance.[16] Results obtained during the first few months of operation of the first two mill-scale SuperBatch systems have recently been described.[17] Both achieved routine production of softwood pulp at kappa number 15 with 90% strength delivery (as compared to 75% from conventional systems). One of them, at Enocell Oy in Finland, has produced softwood pulp of kappa number 10 with good quality. Laboratory bleaching of such pulp has shown that it can be fully bleached in an ECF sequence with very low chemical consumption. The bleached pulp quality was equivalent to that of the corresponding fully bleached conventional pulp.[18] The same studies showed that the SuperBatch pulp could be bleached to 90 brightness in a totally chlorine-free (TCF) sequence with strength equivalent to that of the fully bleached conventional pulp.

The well-known modification of continuous digesters to “modified continuous cooking” (MCC) employs multiple white liquor addition points to decrease the maximum alkali concentration and smooth out the concentration profile. It also provides for countercurrent flow of liquor in the latter part of the cooking zone to decrease the dissolved lignin concentration at the end of the cook. Further development has led to “extended MCC” (EMCC), in which an additional countercurrent zone is provided, the temperature of the countercurrent zones is raised, and some of the white liquor is introduced at the bottom of the digester. More recently, “isothermal cooking” (ITC), in which the temperature of the cocurrent and first countercurrent cooking zones is decreased and that of the second countercurrent zone (formerly the wash zone) is increased to the same value, gives a uniform and relatively low temperature throughout the cook. Benefits include higher strength at low kappa numbers and lower rejects.[19] An even more recent modification is the inclusion of a black liquor preimpregnation step in the preimpregnation vessel, which gives better selectivity and increased tear strength at low kappa numbers.[19,20] All of these modifications have the effect of allowing the unbleached kappa number to be decreased without adverse effect on viscosity or strength. The retention of pulp yield, on the other hand, has not been demonstrated. A slight, but significant, yield loss may be incurred.

A proposed, but not yet commercial, alternative to extended kraft delignification systems is the alkaline sulfite-anthraquinone-methanol (ASAM) process. The pulping liquor contains about 15% sodium sulfite, 5% sodium hydroxide (both as Na_2O, o.d. wood basis), and 0.1% AQ (o.d. wood basis), all dissolved in a mixture of 80% water and 20% methanol (by volume). Compared to kraft MCC, ASAM requires higher temperature, comparable total cooking times, and lower alkali charges. The ASAM process requires a methanol recovery system, a sulfide-sulfite conversion process, digesters rated for higher pressure, and more washing capacity to compensate for the loss of the washing that occurs in the MCC digester. Such changes may be worth considering because the ASAM process, as applied to southern U.S. softwood over the kappa number range 15–25, has a number of advantages.[21] These include 3% higher yield over the entire kappa number range, 20% higher tear strength at a given tensile strength, a significant ultimate tensile strength advantage, and easy bleachability in both ECF and TCF sequences. Figure 1.3[21] shows that, relative to kraft pulping, ASAM would permit the unbleached kappa number of softwood pulp to be decreased from 26 to 16 with no yield loss. A similar reduction by simply extending the kraft cook would result in a yield reduction from 44.5 to 40.9%. It should be noted, however, that the latter yield loss would be much smaller if AQ were added to the kraft cook.

OXYGEN DELIGNIFICATION

Oxygen delignification is a well-established technology for removing a large fraction of the residual lignin in unbleached pulp and routing the dissolved lignin to the recovery furnace instead of to the bleach plant. Although it is a mature technology, research and development efforts continue to result in improvements and new applications. Subjects of recent development activity include two-stage oxygen delignification, use of hydrogen peroxide and other additives, improved pulp washing strategies, using magnesium oxide as base, and using an open (filtrate-discharging) oxygen stage to permit low AOX (adsorbable organic halide) bleaching.

Two-stage oxygen delignification has been developed in response to the need to remove as much lignin in the oxygen stage as possible, usually 40–60%, without adversely affecting pulp quality. This is often difficult, however, particularly in medium-consistency systems. Failure to disperse all of the oxygen as sufficiently small bubbles and coalescence of bubbles result in a slowing of the reaction as the pulp progresses upward through the reactor. Attempts to increase delignification by increasing the oxygen charge are often unsuccessful because the gas dispersing ability of the mixer decreases and channeling becomes more likely as the volumetric fraction of gas in the pulp suspension is increased.

Early two-stage medium-consistency systems[22] differed from their single-stage predecessors only by the addition of a mixer and a second reactor between the first reactor and the blow tank. More recently, the two-stage concept has been extended to include alkali and oxygen addition between stages, with optional interstage washing. Experiments in which a softwood kraft pulp was delignified without interstage washing gave a total extent of delignification of 70% in two successive oxygen stages, each with 3% NaOH, as compared to 53% in a single stage with 3% NaOH.[23] When interstage washing was used, the extent of delignification decreased but selectivity improved. Such a system could be part of a strategy for extended delignification, in which the kraft cook is terminated at a high kappa number and the pulp is extensively delignified with oxygen. Compared to the conventional route (50% delignification of kappa 30 pulp in a single oxygen stage), terminating the kraft cook at a kappa number of 40 and subsequently delignifying to kappa 13 in two oxygen stages would increase yield, decrease cost, decrease the solids load on the recovery furnace, decrease AOX generation in the bleach plant, and slightly decrease pulp strength. A similar system is now operating in a mill in Moss, Norway.[24] The medium-consistency sequence consists of a refiner used as a mixer, a 13 m^3 first-stage reactor, another mixer, and a second, larger (46 m^3) reactor, without interstage washing. The brownstock is 60 kappa polysulfide pulp; the first-and second-stage oxygen charges are 2.5% and 1.25%; and 5–6% alkali is used, all of it being charged to the first stage. The process incorporates

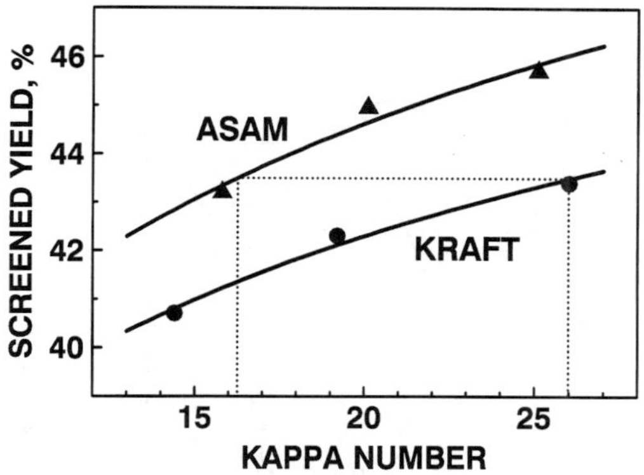

Figure 1.3. ASAM and kraft yields at low kappa number; the dotted lines indicate the kappa number reduction that may be expected without yield reduction.[21]

another interesting innovation — pretreatment of the brownstock with sulfite liquor to improve the selectivity of the oxygen stages. Delignification as high as 66% has been achieved.

Two-stage oxygen systems offer considerable flexibility in the choice of process conditions. For example, laboratory results indicate that it may be possible to significantly increase overall selectivity, and therefore extent of delignification, by using a lower temperature in the first stage than in the second.[25] Intermediate treatments with chelating or oxidizing agents are also possible, as are additions of additives to either stage. Chelation prior to the second stage has been shown to significantly improve selectivity in the laboratory.[26]

Most mills installing oxygen stages justify the cost partly on the basis of environmental benefits resulting from recycling the oxygen stage dissolved solids to the recovery system. This is not always the case, however, as illustrated by another recent mill installation.[27] The mill was required to produce an extremely low AOX grade that demanded very low unbleached kappa number. This could be achieved by extending the kraft cook, but the resulting recovery boiler and causticizing system overloads caused costly decreases in production. The solution to this problem consisted of a medium-consistency oxygen ("mini-oxygen") stage from which the dissolved solids were discharged instead of being recycled to recovery. This gave the required kappa number decrease, with no increase in the load on the recovery system.

Magnesium-based sulfite mills have also been engaged in implementing oxygen delignification systems, despite the complications introduced by having to use MgO, a weaker alkali than NaOH, in the oxygen stage. A German mill has successfully started up a single-stage bleaching process using a combination of oxygen and hydrogen peroxide with magnesium oxide from the recovery system as base.[28] Spruce pulp is delignified from kappa number 18 to kappa number 5, with a corresponding brightness increase to 82. Although the conditions required are quite severe (6 hours at 95°C when 1% MgO is used), they could easily be accommodated by the equipment that had been used for the previous three-stage sequence. A similar system in an Austrian mill has a two-hour retention time and can delignify magnesium-based sulfite pulp from kappa 24 to kappa 14, with no increase in brightness.[29] Mills having MgO-based oxygen systems may have difficulty reaching high final brightness targets because of the limited delignifying power of the oxygen stage. One possible solution is to use an acidic nonchlorine delignifying agent, such as peroxymonosulfuric acid, following the oxygen stage.[30]

Addition of hydrogen peroxide to oxygen delignification stages appears to have potential for kraft pulp applications.[31] Selectivity and extent of delignification are both significantly improved by adding hydrogen peroxide and cyanamide, the latter compound serving to activate the peroxide toward delignification and stabilize it toward decomposition. Repeated (PO) stages can be combined for even more extensive and selective delignification. Figure 1.4[31] illustrates the improved selectivity and low kappa numbers accessible by this technique.

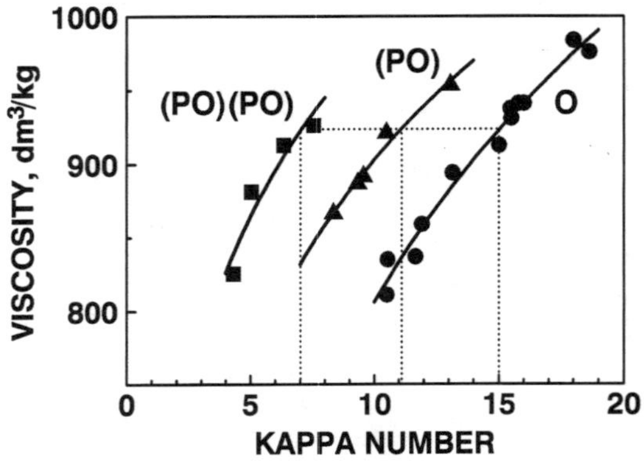

Figure 1.4. Effects of addition of hydrogen peroxide and cyanamide in single- and multistage oxygen delignification.[31]

BIOTECHNOLOGY

The beneficial effects of treating pulp with hemicellulose-degrading enzymes before bleaching[32] are well-known. When applied to kraft pulp under appropriate conditions of pH, temperature, and mixing, xylanases can reduce active chlorine consumption in both chlorine-based and ECF sequences. The savings typically amount to 10–15% for softwood pulps and 20–25% for hardwood pulps. The greater effect with hardwoods is presumably due to their higher xylan content, which may also be blamed for a small but significant yield loss. The mechanism has been presumed to involve removal of reprecipitated xylan from the fiber surface, enlargement of the intrafiber pore structure, or release of lignin fragments that are chemically bonded to xylan.

Interest in enzymes has not been lessened by the industry's movement toward ECF bleaching, in which mill trials have demonstrated their effectiveness. In one report,[33] the kappa factor (ratio of active chlorine charge to kappa number) in the first stage of a softwood D(EO)D(EP)D sequence was decreased from 0.165 to 0.125 with no loss in brightness and good pulp strength retention. A laboratory investigation[34] showed that the kappa factor could be decreased from 0.13 to 0.08 by xylanase pretreatment. The total chlorine dioxide savings was 0.5% (as ClO_2, o.d. pulp basis). Perhaps more significantly, the pretreatment allowed a kappa factor as low as 0.08 to be used without significant increases in the total requirement for chlorine dioxide, allowing significant reductions in AOX output.

TCF xylanase applications are exemplified by a study showing that addition of an enzyme stage to a hardwood OQPZP sequence gave a 20% reduction in ozone requirement and an 11% increase in viscosity at 89 brightness.[35] Similar results were observed at the 91 brightness level. Another investigation showed that, at a given level of ozone consumption, enzyme treatment before TCF bleaching of kraft pulps increased the final brightness by 6 points for eucalyptus and 4–5 points for softwood.[36] The strengths of the TCF pulps were lower than those of the ECF controls. Still another investigation of xylanase effects in TCF hardwood bleaching gave less encouraging results.[37] Although modest beneficial effects on oxygen, peroxide, and ozone bleaching of eucalyptus pulp were observed, yield losses of approximately 3% were attributed to the enzyme.

In spite of generally good laboratory results and numerous successful mill trials, commercial use of enzymes is far from universal. One negative factor is the decrease in enzyme effectiveness associated with conversion to MCC pulping processes. This has been recently documented by Allison and co-workers,[38] as illustrated in Figure 1.5. Other reasons for the lack of widespread commercial use of enzymes, as well as insight into the motivation of those mills that are using them, were recently sought in a comprehensive survey of Canadian bleached kraft mills.[39] Six of the 40 mills were found to be regular users, the leading reasons being to increase pulp throughput, to obtain a marketing advantage, to reduce cost, or to improve pulp quality. Problems experienced by users and reasons cited for nonuse were led by corrosion by the acid needed to bring the brownstock into the neutral pH range of enzyme effectiveness and problems associated with providing the needed residence time,

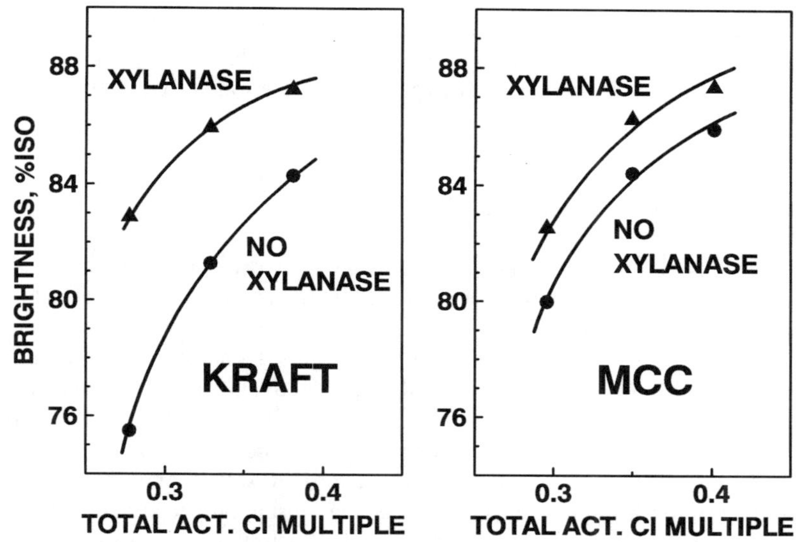

Figure 1.5. Effects of pretreatment with two commercial xylanases on the DED brightening of kraft and MCC pulps.[38]

generally in the high-density storage tower. Running the tower full curtails its ability to serve as a buffer between the pulp mill and the bleach plant. Fear of yield loss was cited by nonusers. Suppliers of enzymes are addressing the corrosion issue by continuing to develop their products in the direction of increased alkali tolerance, as exemplified by a recently reported screening study.[40] Three new xylanases, all effective at pH 9, were discovered.

In addition to the xylanases used to enhance delignification during bleaching, more specialized preparations and applications have begun to appear. These include one that is specifically targeted at elimination of dioxin formation during the chlorine bleaching of kraft pulp[41] and one designed to enhance shive removal.[42]

The xylanases have no direct delignifying effect, the kappa number after enzyme treatment generally being equal to or only slightly smaller than before. On the other hand, lignin-degrading fungal systems are known to secrete enzymes that function by direct fragmentation of the lignin structure. These systems can, in fact, be used for pulp delignification, as illustrated by the report of a recent investigation.[43] The kappa number of hardwood kraft pulp was decreased from 15 to less than 6 by a three-day fungal treatment followed by washing. In another recent study,[44] 1200 white rot fungi were screened for their ability to biologically bleach eucalyptus kraft pulp. The highest brightness achieved after a 10-day incubation was 71. Although the long treatment time is an obvious drawback of this approach, there appears to be much scope for further investigation and improvement.

In principle, it should be possible to isolate the delignifying enzymes from fungi and other organisms that produce them and use the isolated enzymes to delignify pulp. The progress of research directed toward this goal has been the subject of a recent review by Paice and co-workers.[45] Three relevant enzyme activities are recognized: laccase, manganese peroxidase, and lignin peroxidase. The first two, but not the third, have been shown to be produced by a fungus that delignifies pulp. Neither is commercially available. Manganese peroxidase applications have not been developed to the point of commercial viability. Isolated laccase is ineffective as a delignifying agent, but recent reports[46,47] indicate that its inactivity can be overcome by addition of another compound or "mediator." The mediator (believed to be 1-hydroxybenzotriazole) is oxidized by the enzyme and, in turn, oxidizes lignin. Softwood kraft pulp was delignified by as much as 70% in a single stage in the presence of enzyme (2 kg/t), mediator (5–20 kg/t), and oxygen (10 atm). The retention time was 4 hours, and the pH and temperature were 4.5 and 45°C. The indicated degrees of delignification were presumably measured after an alkaline extraction stage. Subsequent pilot trials gave similar results at an oxygen pressure of only 2 atm and retention time of 2 hr. The enzymatically treated and alkali-extracted pulps could be bleached in the laboratory to a brightness of 89 with 3.5% H_2O_2. The combined laccase and alkali extraction stages thus appear to offer a low-pressure, low-alkalinity alternative to an extended (e.g., two-stage) oxygen delignification system, assuming that its economic viability can be demonstrated.

ELEMENTAL CHLORINE–FREE BLEACHING

Environmental pressures have caused a strong movement away from bleaching with chlorine in its elemental form (molecular chlorine, Cl_2). The most readily implemented alternative is to replace all of the chlorine with chlorine dioxide; this alternative has come to be known as "elemental chlorine–free" or ECF bleaching. The extreme rapidity with which this technology is being implemented is readily apparent in the bleached chemical pulp (BCP) production statistics for several countries.[3] In the United States, ECF bleaching accounted for approximately 33% of 1994 BCP production, in Canada 50%, and in Sweden and Finland (combined), 70%; the corresponding 1990 figures were 2%, 6%, and 29%. In Canada, 75% of the bleach plants produced ECF grades in 1994.[48]

Much of the impetus for this movement comes from the substantially lower amounts of chlorinated organic by-products formed by chlorine dioxide. This is illustrated by the results of a study of the amounts and characteristics of the materials dissolved in the effluents from chlorine dioxide and chlorine bleaching of both unbleached and oxygen-delignified softwood kraft pulps.[49] The chlorine-to-carbon ratio (Cl/C) of each of several fractions of each effluent type was measured as a predictor of environmental effect. As shown in Table 1.1, substitution of chlorine dioxide for chlorine sharply reduced the AOX and Cl/C of the whole effluents. More significant were the observations that the reductions were larger in the ether-soluble fraction and its phenolic subfraction and that these fractions were also reduced in size. They indicate that replacement of chlorine with chlorine dioxide is more beneficial than the reduction in whole effluent AOX would suggest.

Much research has been aimed at finding optimum conditions for the chlorine dioxide delignification stage. One parameter of interest is the pH of the stage, which affects the extent of delignification, the amount of AOX formed, and the characteristics of the AOX. In general, studies have shown that as the pH is increased delignification becomes less effective, less substitution of chlorine onto organic material occurs, and less AOX is formed. Notwithstanding these general observations, it has recently been shown that if, instead of operating at pH 2, the pH is adjusted to 4 at the point of chlorine dioxide addition and allowed to fall, delignification is more complete, with no additional AOX formation, and the AOX formed is more benign in character (as indicated by ether-soluble and phenolic Cl/C).[50] A similar observation was reported as part of the results of a study of the combined effects of pH and chloride ion addition.[51] In that study, increasing the D stage pH from 2 to 4 decreased both the kappa number and the amount of AOX produced. Unexpectedly, it was also observed that addition of chloride ions to the pH 4 stage caused further decreases in both kappa number and AOX. A practical implication of this, and one that was experimentally demonstrated, is that chlorine-containing chlorine dioxide solutions, such as those produced by many existing chlorine dioxide generators, form less AOX than pure chlorine dioxide, provided that the chlorine dioxide is applied at pH 4 in the presence of added chloride ion. In other research aimed at minimizing D stage AOX formation, it was shown that AOX formation is increased by chloride addition at low pH and by increasing the consistency.[52]

Another important variable is reaction time. In an investigation of the effect of D stage reaction time on D(EO) bleaching of oxygen-delignified softwood kraft pulp, it was shown that a D stage of only 3 seconds duration resulted in 67% of the delignification of a 30-min stage, while producing only 26% of the AOX and 34% of the Cl/C.[53] A 1-min D stage achieved 84% of the delignification, while generating only 42% of the AOX and 51% of the Cl/C of the 30-min stage. The Cl/C of all effluent fractions except the neutral fraction was much lower at short reaction times and the size of the ether-soluble fraction reached a minimum at a reaction time of 1 minute.

Reduced AOX may be obtained by limiting the amount of chemical applied in the first chlorine dioxide stage. The reason is that the proportion of the chlorine atoms applied that ends up as AOX is much lower for the first chlorine dioxide stage than for succeeding ones. Bleach sequence development work has been done to exploit this phenomenon and to evaluate the resulting cost, in terms of

Table 1.1. Effect of Complete Substitution of ClO_2 for Cl_2 on Filtrate Characteristics.[a]

	AOX, kg/T			TOC, kg/T			Cl/C_{100}[b]		
Fraction	Cl_2	ClO_2	% Reduction	Cl_2	ClO_2	% Reduction	Cl_2	ClO_2	% Reduction
Whole	3.48	0.40	89	16.0	10.5	34	7.35	1.28	83
Ether-Sol.	0.88	0.062	93	2.10	0.73	65	14.2	2.88	80
Phenolic	0.106	0.0044	96	0.196	0.057	71	18.3	2.59	86

[a] Data represent sums of separately determined C or D and (EO) stage effluent characteristics.

[b] Cl/C_{100} = chlorine-to-carbon ratio—number of Cl atoms per 100 C atoms.

the increase in total chemical charge required for fully bleached final brightness. It has been shown[54] that the kappa factor that minimizes total ClO_2 consumption by softwood kraft pulp in a five-stage ECF sequence (following oxygen delignification) is surprisingly low — approximately 0.12. Another important result was the finding that the use of a xylanase pretreatment allowed the kappa factor to be decreased to 0.08, for the purpose of AOX reduction, with only a slight (3 FIM) increase in total bleaching chemical cost. A similar study of the corresponding three-stage sequence following oxygen delignification showed that the total bleaching chemical cost was a minimum at kappa factor 0.20.[55] On the other hand, the combined use of an enzyme pretreatment and peroxide addition to the extraction stage allowed the kappa factor to be decreased to 0.10, with a concomitant 10% increase in total bleaching chemical cost. Mills with no oxygen delignification stage may also consider low kappa factor ECF bleaching as a route to low AOX emissions. The sequence D(EOP)PDP, for example, is capable of fully bleaching softwood kraft pulp with the production of only 0.4 kg AOX per ton of pulp.[56] A review of laboratory studies of low kappa factor bleaching of softwood and hardwood kraft pulps with and without oxygen delignification[57] is in general agreement with the conclusions of the work described above and suggests the additional possibility of "zero kappa factor bleaching." This involves replacing the D stage preceding the EOP stage with a chelation or acid wash, and is claimed to result in AOX reductions of 57–85%, depending on the type of pulp.

A variety of ECF sequences have recently been investigated in the laboratory and implemented in the mill. Conversion of a (C+D)(EOP)D sequence to ECF has been shown to be technically feasible for full bleaching of both softwood and hardwood pulps, but with potentially large increases in bleaching chemical cost.[58] To minimize the cost increase, the following changes were recommended: increasing the first stage consistency from 3% to 12%, increasing the extraction stage temperature from 75°C to 95°C, operating the final D stage at high temperature, and adding a peroxide stage at the end of the sequence. Another short sequence, OD(EP)P, is capable of bleaching softwood kraft pulp to a brightness of 89, provided that chelants are added at several stages and the peroxide stage is operated under severe conditions (2% H_2O_2, 30% consistency, 4 hr at 90°C).[59] A similar forcing final peroxide stage may be used to allow full bleaching of softwood kraft pulp in the OQPDP sequence (Q = chelation stage).[60] This is one step further along the road to TCF bleaching than the OQPD(EP)D sequence, which is capable of the same performance. The sequences DE(EO)DED and OD(EP)(EOP)DED have been implemented in a western Canadian mill.[61] In another mill, the use of ozone in the OAZ(EOP)D(EP)D sequence was successfully demonstrated to yield fully bleached pulp with similar properties to those of the ECF reference sequence without the ozone stage and at similar cost.[62]

Any of these sequences might benefit from a relatively recent development in chlorine dioxide generation technology. The new process, called O1, employs chloric acid as feedstock, and catalytically reduces it to a mixture of chlorine dioxide, water vapor, and oxygen. Significant features of the process include its freedom from chlorine and salt cake by-products, the simplicity of the generation equipment, and the ease of controlling generation rate by controlling heat input to the generator. The last feature facilitates the coupling of the generation process with a gas-phase bleaching process. Recent investigations[63,64] of such a process reveal that it possesses unusually high delignification efficiency and a low AOX generation tendency.

TOTALLY CHLORINE-FREE BLEACHING

The technology for bleaching chemical pulps, especially kraft pulps, without the benefit of any chlorine-containing compounds has reached the commercial stage only recently. Milestones in its implementation were the first commercial use of the Lignox® peroxide process in Sweden in the early 1990s and the startup of the first commercial ozone kraft pulp bleaching stage at Union Camp's Franklin, Virginia, mill. Since then, there has been rapid development activity in both areas, as well as in applications of other TCF technologies, especially the use of peracids.

Peroxide Sequences

The brightening power of hydrogen peroxide has been well-known for many years, in both mechanical pulp and chemical pulp bleaching applications. Its usefulness for bleaching chemical pulps first became apparent in the latter stages of the bleaching sequence, where for a brief period in the history of bleaching it was considered a rival to chlorine dioxide. Although the superior selectivity and delignifying power of chlorine dioxide caused it to prevail in these early years, it was also recognized that peroxide's combination of brightening power and brightness stabilizing properties made it a useful complement to chlorine dioxide in such sequences as CEHDP.

The use of hydrogen peroxide as a primary delignifying agent was not seriously considered until much later[65] when it was proposed as an alternative to oxygen for partial delignification prior to conventional bleaching. After a brief period of commercial application,[66] the use of this technology was discontinued, presumably for economic reasons. A key aspect of using peroxide to delignify pulp is the role of peroxide decomposition and its catalysis by transition metal ions, particularly Fe, Mn, and Cu. It has been demonstrated that the rate at which peroxide oxidizes either lignin model compounds[67,68] or the residual lignin in kraft pulp[69] is extremely low when the peroxide is stabilized by exhaustive prior removal of metal ions and addition of stabilizing agents. On the other hand, if metal ion removal and stabilization are inadequate, delignification is poor because much of the peroxide decomposes and is lost as oxygen to the atmosphere. This situation is further complicated by the slowness of the brightening reactions of hydrogen peroxide, which dictates long reaction times and severe conditions, both of which also increase the likelihood of peroxide loss through decomposition to oxygen. Achieving an appropriate balance between stabilization and peroxide activation through controlled decomposition was thus an obstacle to the development of the bleaching process.

A breakthrough came in 1991, when Basta, Holtinger, and co-workers realized the importance of "metal profile" control, or removal of transition metal ions under conditions that allow the peroxide-stabilizing metals, Mg and Ca, to be retained by the pulp.[70,71] This gave rise to the Lignox® process, which comprises treatment with a chelating agent at pH 4–7 (Q stage), followed by interstage washing, and one or more long, high-temperature peroxide stages. The chelation treatment effectively removes transition metals, but not Mg and Ca, setting the stage for the use of relatively severe peroxide stages to promote delignification and brightening without excessive peroxide decomposition. Figure 1.6[72] shows the brightness obtained and its sensitivity to chelant stage pH when oxygen-delignified softwood kraft (kappa no. 10.3) is bleached in a QP sequence with 3.5% H_2O_2 at 90°C for 4 hr.

Other important aspects of TCF sequences of the QP type are their sensitivity to incoming kappa number and to the conditions in the P stage. Both have been the subject of much research, of which only a few examples are presented here. Figure 1.7[73] shows the effect of incoming kappa number on the brightness obtained by bleaching softwood kraft pulp with 2% H_2O_2 at 80°C and 30% consistency for 4 hr. It is apparent that fully bleached pulps can be obtained by decreasing the kappa number to 4 or less, irrespective of the prebleaching method used to obtain that kappa number. The effect of varying the incoming kappa number by oxygen delignification is similar.[74] Figure 1.8[75] illustrates the effects of temperature and consistency in the P stage when softwood kraft pulp is bleached after predelignification to kappa number 4.2. The most forcing conditions give the highest brightness after suitable metal profile adjustment.

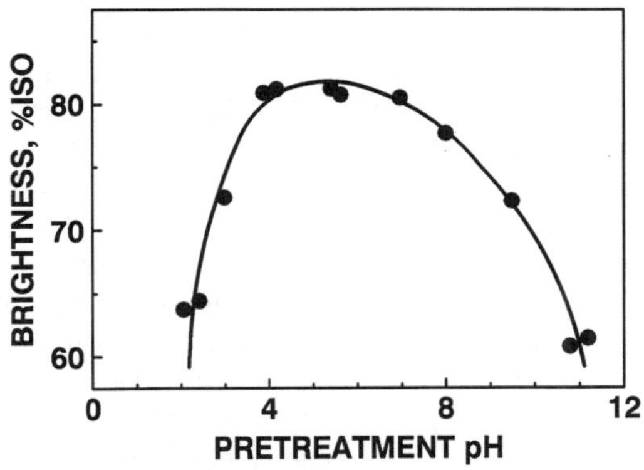

Figure 1.6. Brightness after P stage vs. EDTA pretreatment pH.[72]

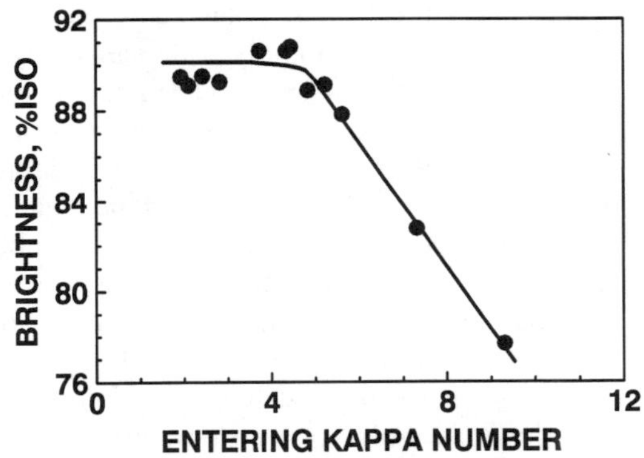

Figure 1.7. Effect of entering kappa number on brightness after P stage with 2% H_2O_2.[73]

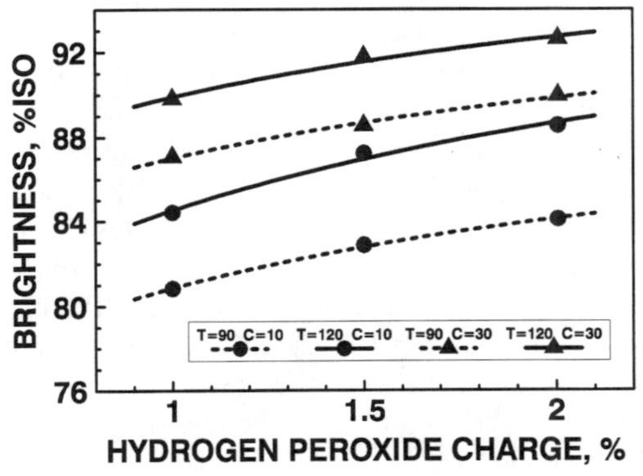

Figure 1.8. Effects of temperature and consistency on brightness after P stage.[75]

Activated Peroxide: Polyoxometalates and Nitrilamine

Peroxide can be made more reactive and selective toward lignin by adding certain compounds that convert the hydrogen peroxide to another active oxygen compound. Examples are polyoxometalates[76,77] and nitrilamine, H_2NCN, also called cyanamide.[78] Application of two nitrilamine-reinforced peroxide stages to an oxygen-delignified and (EPO)-treated pulp of kappa number 7.8 gave a bleached pulp of kappa number 3.5 and brightness 81.6; a control experiment without nitrilamine gave kappa number 6.9 and brightness 73.8. Viscosity retention was acceptable and the same in both cases, despite the greater degree of delignification in the former. Nitrilamine is believed to function by converting hydrogen peroxide to isourea peroxide, which is more strongly oxidizing and apparently more selective than hydrogen peroxide. During bleaching, isoureaperoxide is converted to urea, a harmless by-product.

Polyoxometalates are a class of discrete polymeric metal-oxide anions formed when simple oxides of vanadium, molybdenum or tungsten ions are combined in water under appropriate conditions. They differ from nitrilamine in that their reduced forms may be readily separated from the bleached pulp and regenerated by oxidation. This oxidation can be effected not only by hydrogen peroxide, but also by other oxidizers, including oxygen and air.[79] Compared to hydrogen peroxide, ozone, or oxygen, they possess a selectivity advantage for the residual lignin in softwood kraft pulps.

Ozone Bleaching

Like hydrogen peroxide, ozone is a chemical whose potential as a bleaching agent has been known for many years. Unlike peroxide, it is capable of rapidly and extensively removing residual lignin from kraft pulp. Indeed, this reactivity is the reason it remained uncommercialized until recently — it is so reactive that special precautions are needed to ensure that its tendency to attack cellulose is not allowed to result in a loss of pulp strength. Such precautions can be successful because ozone is more reactive toward lignin than toward cellulose, as demonstrated by the higher rate of depolymerization of bleached pulp by ozone than in the corresponding experiment with unbleached pulp.[80] Thus, one of the precautions to be observed in ozone bleaching is to ensure that the ozone charge remains below the level where an appreciable amount of residual ozone remains to attack the fiber when insufficient lignin remains to protect it. The second important precaution — to ensure good mixing and uniformity — is a corollary of this. Nonuniform exposure will result in locally high ozone concentrations and consequent fiber damage. Commercial ozone systems have arisen as a result of careful attention to these two principles, as well as to the more obvious economic need to minimize the requirement for ozone, a relatively expensive bleaching chemical, by making maximum use of oxygen delignification and other means of reducing the kappa number entering the ozone stage. Uniformity has been achieved in high-consistency systems by judicious reactor design, and in medium-consistency systems by the use of high-shear, fluidizing, medium-consistency mixers.

High-consistency systems are exemplified by the installation at the Franklin mill. The reactor design resulted from an intensive in-house research and development effort.[81] The major objectives, good uniformity and high turndown capability, were achieved in the final design, which consisted of a horizontal cylindrical reactor with paddles mounted on a rapidly rotating axial shaft. The paddles were designed to keep the fluffed pulp highly dispersed and in rapid movement, while at the same time causing the reactor flow pattern to approach the ideal of axially dispersed plug flow, thus establishing a narrow residence time distribution to achieve uniformity. The bleach plant that incorporates this reactor has been described by Clark.[82] Extensively oxygen-delignified pulp from the final postoxygen washer is cooled by dilution with cold pressate from the preozone press, acidified to pH 2, treated with chelating agent, pressed to 30% consistency, fluffed, contacted with ozone for 1–3 min, degassed, diluted, and washed.

An important consideration in the design of ozone bleaching systems is that of chemical supply. Ozone is generated onsite by passing pure, dry oxygen through a corona discharge in a commercial generator. The stream that emanates from the generator contains ozone at a concentration of 6–10%, the remainder being unconverted oxygen. Under these conditions, economics generally dictate recycling of the unused oxygen by collecting the gas stream from the bleaching reactor and then compressing, purifying, and drying it before feeding it to the ozone generator together with makeup oxygen. In the bleach plant just referred to, ozone is generated at 6% by weight and is transported to the reactor. Spent gas from the reactor is passed through a particulate removal device, after which residual ozone, carbon monoxide, and organics are destroyed by a combination of thermal and catalytic destruction units. The gas is then cooled, dried, and recycled to the ozone generator.[82] An alternative "once-through" system utilizes as much as possible of the spent reactor gas in other applications, such as oxygen delignification and (EO) stages, and vents any remaining gas to the atmosphere. This is more likely to be economically viable if the so-called "advanced technology" for ozone generation is used, since it allows generation of ozone at a concentration of 10%, lessening the amount of surplus oxygen to be disposed of.

Another reason for installing ozone generators that produce ozone at the highest possible concentration is to allow the use of an entirely different type of arrangement for contacting pulp with ozone — a medium-consistency system. According to this method, the entering pulp, instead of first being pressed to high consistency, passes directly to a medium-consistency mixer, to which the ozone is added. The amount of ozone that can be added in a single stage is limited by the capacity of the mixer; performance drops dramatically when the gas:liquid volume ratio exceeds 30%. This limitation makes it necessary to start with a high-concentration ozone stream and then to compress it to approximately 10 atm before adding it to the mixer. Multiple stages are necessary if the desired ozone consumption is greater than about 0.6%.

Selectivity and extent of delignification are similar for medium- and high-consistency systems. Sample data are shown in Figures 1.9 and 1.10.[83] Delignification of softwood kraft pulp from kappa number 11 to kappa number 6 may be achieved by consuming 0.6% ozone. The amount charged needs to be only slightly greater than this in well-operated systems because of the high reactivity of ozone. Viscosity decreases continuously as kappa number is decreased and tends to be lower at a given kappa number in high-consistency systems. Note, however, that the viscosity decrease observed in practice will depend on the uniformity of exposure and control of retention time. Operating experience at the Franklin mill has shown that adequate delignification and good strength are achieved when the high-consistency system is used as part of the OZED sequence.[84] Tables 1.2 and 1.3 summarize the operating conditions and performance of this installation.

A major incentive for installing ozone bleaching is to achieve environmental benefits deriving from the ability to recycle the filtrates from the bleach plant washers to the chemical recovery system. By evaporating and burning the bleach plant effluent in this way, the environmental problems associated with the discharge of conventional bleach plant effluent are avoided. The degree to which the ideal of complete recycle can be achieved may ultimately be limited by buildup of metals entering the mill with the wood. Since these materials have the potential to cause problems such as scaling and peroxide decomposition, it is necessary to limit their concentrations by means of an acidic purge stream, a metals removal process, or both. The ozone stage filtrate is acidic for bleaching selectivity reasons, and therefore represents an opportunity for purging metals from the system. The volume of any such purge stream must, of course, be limited in order to realize the environmental benefits of filtrate recycle.

Ozone stages have been incorporated into a variety of TCF sequences, both in the laboratory and commercially. Almost all of these begin with an oxygen stage, since this allows the application of ozone to be minimized, which is desirable for reasons of selectivity and economy. The OZP and OZEP sequences were among the earliest proposed, but were generally not capable of reaching the 88–90 brightness range without serious losses in pulp strength. These sequences have been developed into more efficient variants by (a) incorporating metals management stages such as chelation (Q) or acid wash (A) stages or stage modifications such as addition of magnesium salts; (b) modifying the peroxide stage by raising temperature, pressure, and/or consistency or by combining the peroxide stage with an oxygen stage to give a (PO) stage; or (c) incorporating another peroxygen compound stage such as a peracetic acid stage (Pa). Examples of such sequences are OAZQP,[85] OQP(ZQ)(PO),[86] and O(QZ)Q(PO).[87] The strength of softwood kraft pulps bleached to 90 brightness by these sequences is a controversial issue. Tear strength at a fixed tensile strength is often observed to be about 10% lower than that of the corresponding ECF pulps.[85]

Several commercial ozone stages are now in operation, and others are under construction. At least eight are currently operating,[88] and seven others are either operating or under construction.[89] These stages, most of which are in Europe, are located in kraft, sulfite, and recycled fiber mills; the furnishes being bleached include both hardwood and softwood pulps, and the bleach sequences of which they are a part include both TCF and ECF sequences.

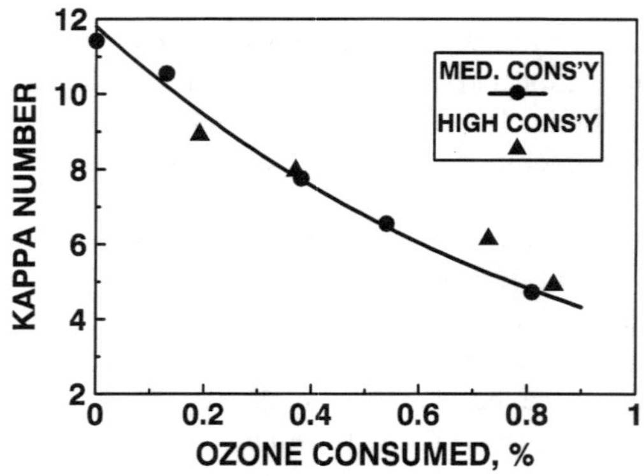

Figure 1.9. Ozone delignification of softwood kraft pulp.[83]

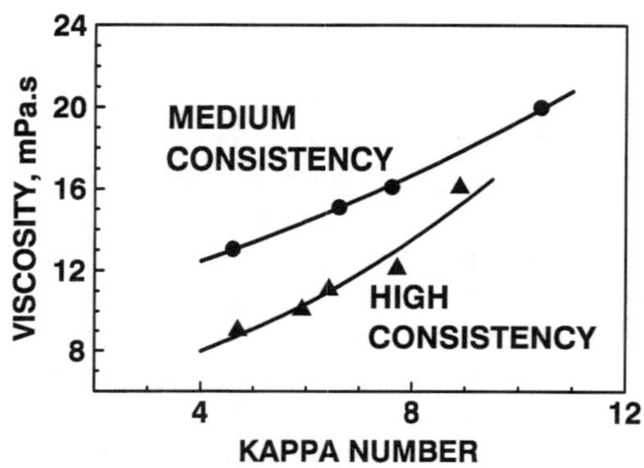

Figure 1.10. Selectivity in ozone delignification of softwood kraft pulp.[83]

Table 1.2. High-Consistency Ozone Stage Operating Conditions.[84]

Feed Consistency	35%–45%
pH	2.0–2.5
Temperature	40ºC–50ºC
Ozone Application	7–8 kg O_3/ton pulp

Table 1.3. High-Consistency Ozone Stage Performance.[84]

Ozone Consumption, % appl.	95–98
Kappa No. Reduction, $\Delta\kappa$	7.4
Ozone Efficiency, Pulp and Liquor[a]	1.0
Ozone Efficiency, Pulp Only[b]	1.3
Brightness Gain, points	15 (36→51)
Viscosity Decrease, dm^3/kg	106 (706→600)

[a] Normal wash with filtrate recycle; units: $\Delta\kappa$/kg O_3/admt.

[b] Clean pulp with no filtrate recycle; units: $\Delta\kappa$/kg O_3/admt.

Peracids

Peracids, like both hydrogen peroxide and ozone, were recognized as potentially viable bleaching agents a long time ago. The development of commercially viable technology for peracid bleaching of chemical pulps, however, has occurred only recently. Peracids, also known as peroxyacids, are acid analogs of hydrogen peroxide, bearing the same chemical relationship to their parent acids as hydrogen peroxide does to water. Thus, peracetic acid (or peroxyacetic acid) may be viewed as acetic acid with an extra oxygen atom inserted into its O-H bond, and peroxymonosulfuric acid (Caro's acid) is sulfuric acid with an extra oxygen atom inserted into one of its O-H bonds. Peracids can be made by an acid-catalyzed reaction between the parent acid and hydrogen peroxide, which produces water as a by-product. Since this is an equilibrium reaction, the resulting peracid concentration will be limited by the amount of water in the system. Peracids of high concentration can be made onsite by starting with concentrated or pure acid and concentrated hydrogen peroxide. Alternatively, peracetic acid can be obtained in relatively pure form by distilling it from the equilibrium reaction mixture at reduced pressure. Both procedures require safety precautions since both hydrogen peroxide and peracids are inherently unstable.

Onsite generation systems are commercially available for peracetic acid, Caro's acid, and mixtures of the two. All may be viewed as converters of hydrogen peroxide to more active form, so that one measure of their effectiveness is the conversion ratio, defined as the percentage of the peroxide's active oxygen that is converted to the peracid formed. It was discovered by Song and co-workers[90] that by adding acetic acid to equilibrium mixtures of Caro's acid, conversion ratio could be significantly increased. This effect has been quantified, and the effect of conversion ratio on bleaching has recently been studied.[91] Figure 1.11 illustrates the beneficial effect of acetic acid addition to Caro's acid mixtures. It is also apparent that a given conversion ratio can be obtained with different relative amounts of acetic acid and peracetic acid in the system. This flexibility is important because it provides some control over the bleaching effluent characteristics. If the effluent is to be discharged, it may be desirable to limit the use of acetic acid, which contributes to BOD. If the effluent is to be recycled to the recovery system, it may be desirable to limit the addition of sulfuric acid, which has implications for the mill's sodium-sulfur balance. Changing the relative amounts of the different types of peracids is without effect on the degree of delignification obtained, since it has been shown that the different peracids have the same bleaching power at any given active oxygen charge.[92]

The delignifying power of peracids is sufficient to allow a combination of a peracid stage with a following (EPO) stage to reduce the kappa number of kraft pulp by about 50%. This combination of stages therefore qualifies as a low-capital alternative to the installation of an oxygen delignification stage. Another promising application is to reduce the kappa number of pulp entering the kappa number after oxygen delignification in a sequence of the OQP type. Since the final brightness obtainable in such a sequence is very sensitive to the kappa number entering the peroxide stage, this has the effect of raising the brightness ceiling of the sequence or of reducing the necessary peroxide consumption to the point where the total bleaching chemical cost is reduced.[93] The peracids also have the ability to brighten pulp and may therefore be viewed as a hypochlorite replacement in ECF- or chlorine-based sequences, as well as candidates for inclusion in the later stages of TCF sequences.

Dimethyldioxirane

Dimethyldioxirane (DMD) is an organic peroxide that can be made by oxidizing acetone with the potassium salt of Caro's acid. It has a three-membered ring containing two oxygen atoms and one carbon atom; the strained ring structure and peroxidic character make DMD a powerful and reactive oxidizing agent. The potential of DMD for pulp bleaching applications was independently recognized by Lee and co-workers at the Pulp and Paper Research Institute of Canada[94] and Ragauskas at the Institute of Paper Science and Technology.[95] Lee et al. showed that the kappa number of a hardwood kraft pulp was reduced from 16.4 to 3.4 (79% delignification) by 2.5% isolated DMD, and that of a softwood kraft pulp was reduced from 31.5 to 5.4 (83% delignification) by 12.5% *in situ*-generated DMD followed by 3.3% NaOH. Ragauskas showed that the kappa number of a softwood kraft pulp

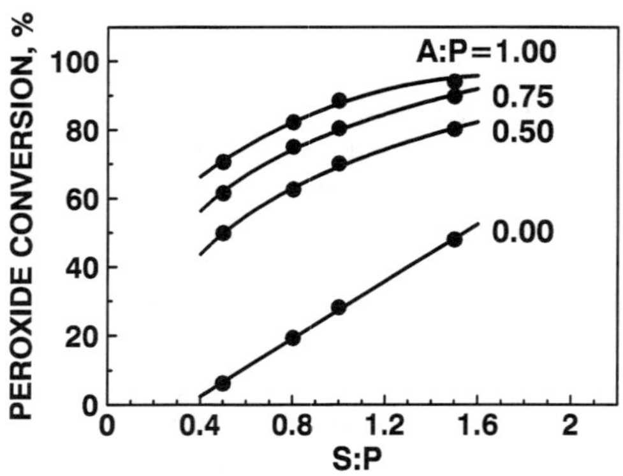

Figure 1.11. Conversion of hydrogen peroxide to peracids.[91]

was reduced from 39.5 to 12.5 (68% delignification) by 5% isolated DMD. *In situ* generation of DMD is accomplished by adding acetone, sodium bicarbonate, and potassium peroxymonosulfate to the pulp, the DMD charge being calculated from the amount of active oxygen added as peroxymonosulfate.

The studies cited above also showed that DMD is very selective as a delignifying agent. For example, in reducing the kappa number of a softwood pulp from 31.5 to 5.4, the viscosity was reduced only from 33.5 to 21.0. This reflects the relative inertness of DMD toward saturated carbohydrates. The literature on bleaching with dioxiranes and the relevant chemistry has been reviewed by Ragauskas.[96]

An evaluation of DMD as a stage in a TCF bleaching sequence showed that the sequence OIQP (I=DMD) appears to hold considerable promise.[97] A brightness in the mid-80s was reached with modest viscosity loss. The slight viscosity losses incurred are not manifested in decreased pulp strength. A comparison with the sequence OQPI showed that viscosity retention is far superior when the I stage is located early in the sequence. Similarly, the response of the final brightness to increased DMD charge was much better when the I stage appeared early in the sequence; a final I stage had either no effect or a slightly detrimental effect on brightness. Apparently, DMD is both effective and selective as a delignifying agent, but is neither as a brightening agent.

The economic viability of onsite DMD generation and use has yet to be demonstrated. A complicating factor is the necessity to handle, recover, and recycle a new chemical, acetone, which is volatile, combustible, and unfamiliar in a mill setting. Nevertheless, the exceptional delignifying power and selectivity of DMD may earn it a home in the rapidly changing arena of pulp bleaching technology.

SUMMARY AND CONCLUSION

Chemical pulp manufacturing technology is in the midst of a period of rapid change driven by concern for the environment. Pressure to reduce discharges of chlorinated organic compounds has focused attention on the bleach plant and has driven the development of technology for reducing the use of bleaching chemicals, especially chlorine. The resulting body of knowledge includes new applications in the categories of extended pulping, oxygen delignification, biotechnology, ECF bleaching, and TCF bleaching.

One way to reduce bleach plant effluent discharges is to modify the pulping process to allow more lignin to be selectively removed in the digester, leaving less to be removed by bleaching. Adding anthraquinone (AQ) to kraft digesters and simply extending the cook by manipulating pulping variables are two possibilities. Another is the combined use of polysulfide (PS) and AQ to lower the unbleached kappa number. Modifications of batch digester systems for extended delignification are marketed under several trade names, including Rapid Displacement Heating (RDH), SuperBatch, and Enerbatch. Modifications of continuous digesters include "modified continuous cooking" (MCC), "extended MCC" (EMCC), "isothermal cooking" (ITC), and inclusion of a black liquor preimpregnation step in the preimpregnation vessel. A proposed alternative to extended kraft delignification systems is the alkaline sulfite-anthraquinone-methanol (ASAM) process.

Research and development efforts continue to result in improvements to oxygen delignification technology. Two-stage oxygen delignification has been developed in response to the need to remove as much lignin in the oxygen stage as possible. The medium-consistency oxygen ("mini-oxygen") stage from which the dissolved solids are discharged instead of being recycled to recovery has been utilized to give a significant kappa number decrease with no increase in the load on the recovery system. Magnesium-based sulfite mills have implemented oxygen delignification systems using MgO in place of NaOH. Addition of hydrogen peroxide and cyanamide to oxygen delignification stages has been introduced to significantly improve both selectivity and extent of delignification.

Xylanases can reduce active chlorine consumption in both chlorine-based and ECF sequences. Enzyme treatment before TCF bleaching of kraft pulps increases the final brightness, but may decrease yield. Lignin-degrading fungal systems can be used for pulp delignification. Although the long treatment time is an obvious drawback of this approach, there appears to be much scope for further investigation and improvement. Other research has been aimed at isolating fungal enzymes to delignify pulp. One of these, laccase, is ineffective as a delignifying agent, but recent reports indicate that its inactivity can be overcome by addition of another compound or "mediator." The mediator is oxidized by the enzyme and, in turn, oxidizes lignin. The combined laccase and alkali extraction stages thus appear to offer a low-pressure, low-alkalinity alternative to an extended (e.g., two-stage) oxygen delignification system, assuming that their economic viability can be demonstrated.

A strong movement away from bleaching with chlorine is fueling rapid implementation of ECF bleaching, mainly as a result of the substantially lower amounts of chlorinated organic by-products formed by chlorine dioxide. Recent research has been directed toward finding optimum conditions for the chlorine dioxide delignification stage. Increasing the initial D stage pH from 2 to 4 decreases both the kappa number and the amount of AOX produced. Beneficial effects on effluent characteristics may also be obtained by limiting the reaction time or the amount of chemical applied in the first chlorine dioxide stage. ECF bleaching might benefit from new chlorine dioxide generation technology that catalytically reduces chloric acid to a mixture of chlorine dioxide, water vapor, and oxygen. The process is simple, easy to control, free from chlorine and salt cake by-products, and is easily coupled to a gas-phase bleaching process, which has high delignification efficiency and a low AOX generation tendency.

Significant developments in bleaching with peroxides, ozone, peracids, and dimethyldioxirane have contributed to progress in TCF bleaching. Recognition of the importance of "metal profile" control gave rise to the Lignox® process comprising oxygen delignification, controlled chelation, and one or more forcing peroxide stages. Important aspects of this process are its sensitivity to incoming kappa number and to the conditions in the P stage. Peroxide can be made more reactive and selective toward lignin by adding certain compounds that convert the hydrogen peroxide to another active oxygen compound. Examples are polyoxometalates and nitrilamine. Both are the subjects of continuing research. Ozone is capable of rapidly and extensively removing residual lignin from kraft pulp but is so reactive that special precautions are needed to ensure that its tendency to attack cellulose is not allowed to result in a loss of pulp strength. The chief precautions are to ensure that the ozone charge is kept low and to ensure good mixing and uniformity. Both high-consistency and medium-consistency systems have been designed to meet these requirements. Ozone stages have been incorporated into a variety of TCF sequences, both in the laboratory and commercially. Onsite generation systems are commercially available for peracetic acid, Caro's acid, and mixtures of the two. The delignifying power of peracids is sufficient to allow a combination of a peracid stage with a following (EPO) stage to reduce the kappa number of kraft pulp by about 50%. Dimethyldioxirane (DMD) is an organic peroxide that can be made by oxidizing acetone with the potassium salt of Caro's acid. DMD can selectively reduce the kappa number of kraft pulp by 70–85%, but is not effective as a brightening agent. DMD shows considerable promise, but its economic viability remains to be demonstrated.

The rate at which these new technologies have been introduced is unprecedented, and the level of research and development activity shows no sign of abating. Although a shift in emphasis from laboratory work to implementation may be expected, research will continue to progress at a level that will demand frequent revision of our concepts of the pulp mill of the future.

REFERENCES

1. Lagergren, S. "Swedish EPA and the Environment," Proceedings, 1994 Int. Nonchlorine Bleaching Conf., Miller Freeman Inc., San Francisco, Paper 1-2.
2. McDonough, T.J. 1993. "Dioxins and the Paper Industry," in Proceedings of the Fourth China Paper Technical Conference, Beijing (TAPPI PRESS, Atlanta).
3. Pryke, D.C. Course Notes, 1995 TAPPI Bleach Plant Operations Short Course, TAPPI PRESS, Atlanta, Chapter 2.1.
4. Albert, R.J. 1995. "Worldwide Status of Effluent-Free Technology for Bleached Kraft Pulp Production," Proceedings Int. Nonchlorine Bleaching Conf., Miller Freeman Inc., San Francisco, Paper 9-2.
5. McDonough, T.J. and V.J. Van Drunen. 1980. *Tappi J.* 63(11):83.
6. Sturgeoff, L.G. and Y. Pitl. 1993. Proceedings, Pulping Conf., TAPPI PRESS, Atlanta, p. 423.
7. Tikka, P.O. 1992. Proceedings, Pulping Conf., TAPPI PRESS, Atlanta, p. 699.
8. Pekkala, O. 1982. *Paperi ja Puu* 64(11):735.
9. Katz, G. 1993. "Anthraquinone and Anthraquinone Polysulfide Pulping for Extended Delignification," Proceedings, Pulping Conf., TAPPI PRESS, Atlanta, p. 323.
10. Jiang, J.E. 1993. "Extended Delignification of Southern Pine with Polysulfide and Anthraquinone," Proceedings, Pulping Conf., TAPPI PRESS, Atlanta, p. 313.
11. Jameel, H., J. Gratzl, D.Y. Prasad, and S. Chivukula. 1994. "Extending Delignification with AQ/Polysulfide," Proceedings, Pulping Conf., TAPPI PRESS, Atlanta, p. 781.
12. Jiang, J.E., K.R. Crofut, and D.B. Jones. 1994. "Polysulfide Pulping of Southern Hardwood Employing the Moxy and Green Liquor Crystallization Processes," Proceedings, Pulping Conf., TAPPI PRESS, Atlanta, p. 799.
13. Lloyd, J.A., R.W. Allison, and S.H. Wrathall. 1995. "MCC Pulping with Additives and Modified Sulfidity Profiles," Proceedings, APPITA General Conf., p. 1.
14. MacLeod, J.M., M. Cyr, D.F. Embley, and P. Savage. 1987. *J. Pulp Paper Sci.* 13(3):J87.
15. Kubelka, V., W. Wizani, G. Neubauer, and J. Kappel. 1994. "Enerbatch Extended Delignification for TCF Pulp," Proceedings, Int. Nonchlorine Bleaching Conf., Miller Freeman Inc., San Francisco, Paper 3-2.
16. Sezgi, U.S., G. Rangamannar, N.H. Shin, and M.J. Abuhasan. 1995. "RDH Kraft Pulping Process; Evolution and Current Status of Process Technology," Proceedings, APPITA General Conf., p. 9.
17. McDonald, S., P. Uusitalo, and S. Hiljanen. 1993. "Performance of Mill-Scale SuperBatch Displacement Batch Cooking Systems," Proceedings, Pulping Conf., TAPPI PRESS, Atlanta, p. 415.
18. Nordén, S. and R. Boman. 1993. "Bleaching of Pulps from the First Commercial SuperBatch Cooking Plant," Proceedings, Pulping Conf., TAPPI PRESS, Atlanta, p. 711.
19. Hjort, A. 1994. "Effects on Pulp and Paper Properties from Isothermal Cooking (ITC) and Black Liquor Impregnation in a Continuous Digester," Proceedings, Int. Nonchlorine Bleaching Conf., Miller Freeman Inc., San Francisco, Paper 3-1.
20. Hoglund, O., K. Pehu-Lehtonen, and A. Hjort. 1994. "Kraft Pulping with Black Liquor Pretreatment," Proceedings, Pulping Conf., TAPPI PRESS, Atlanta, p. 1225.
21. Borgards, A., R. Patt, O. Kordsachia, J. Odermatt, and W.D. Hunter. 1993. "A Comparison of ASAM and Kraft Pulping and ECF/TCF Bleaching of Southern Pine," Proceedings, Pulping Conf., TAPPI PRESS, Atlanta, p. 629.
22. Greenwood, B.F. 1990. "Oxygen Delignification Stage Design and Performance — A Kamyr Perspective," Bleach Plant Operations Short Course Notes, TAPPI PRESS, Atlanta, p. 63.
23. Prasad, B., J. Gratzl, A. Kirkman, H. Jameel, T. Rost, and V. Magnotta. 1993 "Extended Delignification by Kraft Pulping Followed by Oxygen/Alkali Treatment: Technical and Economic Evaluation," Proceedings, 1993 Pulping Conf., TAPPI PRESS, Atlanta, p. 297.
24. Kleppe, P.J. 1994. "Mill Scale Multistage Oxygen Delignification," Proceedings, 1994 Int. Nonchlorine Bleaching Conf., Miller Freeman Inc., San Francisco, Paper 8-1.
25. Svanberg, J. 1995. "Oxygen Delignification — Process Considerations for Production of ECF/TCF Pulp," Proceedings, APPITA General Conf., p. 49.
26. Kiviaho, I. 1995. "Optimizing Oxygen Delignification," Proceedings, 1995 Int. Nonchlorine Bleaching Conf., Miller Freeman, San Francisco, Paper 2-2.
27. McKenzie, D.J. 1994. "The Evolution of Pulp Bleaching Practices at Harmac Pacific," Papers, 1994 Int. Pulp Bleaching Conf., CPPA, Montreal, p. 163.
28. Nimmerfroh, N., H.U. Süss, H.-P. Böttcher, W. Lüttgen, and A. Geisenheiner. 1994. "The German Approach to the Closed Cycle Sulfite Mill," Papers, 1994 Int. Pulp Bleaching Conf., CPPA, Montreal, p. 247.
29. Merzeder, K., E. Bargfrieder, and H. Stockinger. 1994. "Leykam's Road to Becoming the World's Largest Integrated TCF Pulp and Paper Producer," Proceedings, 1994 Int. Nonchlorine Bleaching Conf., Miller Freeman Inc., San Francisco, Paper 8-3.

30. Devenyns, J., N. Troughton, and F. Desprez. 1994. "Strategies for High Brightness, Low Effluent TCF Bleaching of Sulphite Pulps," Proceedings, Pulping Conf., TAPPI PRESS, Atlanta, p. 481.
31. Odermatt, J., R. Patt, O. Kordsachia, and N. Troughton. 1994. "TCF Bleaching of Conventional and EMCC Softwood Kraft Pulp," Papers, 1994 Int. Pulp Bleaching Conf., CPPA, Montreal, p. 77.
32. Dunlop-Jones, N. and V. Gronberg. 1995. "Recent Developments in the Application of Xylanase Enzymes in Elemental Chlorine–Free (ECF) and Totally Chlorine-Free (TCF) Bleaching," Preprints, 80th Annual Meeting, Technical Section, CPPA, Montreal, p. A191.
33. Jean, P., J. Hamilton, and D. Senior. 1995. "Mill Trial Experiences with Xylanase: AOX and Chemical Reductions," Preprints, 80th Annual Meeting, Technical Section, CPPA, Montreal, p. A229.
34. Malinen, R., T. Rantanen, and R. Rasimus. 1993. "ECF Bleaching of Oxygen Delignified Softwood Pulp with the Minimum Charge of ClO_2," Proceedings, Pulping Conf., TAPPI PRESS, Atlanta, p. 925.
35. Ledoux, P., R. Detroz, E. deBuyl, N. Troughton, J. Shetty, and J.R. Presley. 1993. "Use of Bacterial Xylanase in Chlorine Free Bleaching Sequences," Proceedings, Pulping Conf., TAPPI PRESS, Atlanta, p. 1057.
36. Yang, J.L., D. Cates, V.M. Sacon, S.E. Law, and K.-E.L. Eriksson. 1993. "Bleaching of Kraft Pulps with Enzyme and Oxygen-Based Chemicals," Proceedings, Pulping Conf., TAPPI PRESS, Atlanta, p. 1033.
37. Nelson, P.J., C.W.J. Chin, S.G. Grover, and H. Ryynanen. 1995. "Elemental Chlorine–Free (ECF) and Totally Chlorine-Free (TCF) Bleaching of Eucalypt Kraft Pulps," Proceedings, APPITA General Conference, p. 179.
38. Allison, R.W., T.A. Clark, and A. Suurnakki. 1995. "Effect of Modified Kraft Pulping on Enzyme-Assisted Bleaching," Proceedings, APPITA General Conference, p. 157.
39. Tolan, J.S., D. Olson, and R.E. Dines. 1995. "Survey of Xylanase Enzyme Usage in Canada," Preprints, 81st Annual Meeting, Technical Section, CPPA, Montreal, p. B91.
40. Ball, D., R.W. Dunlop, C.J. Falk, P. Kumar, R.D. Latham, A.B. Roullo, and B. Wang. 1995. "Novel High Temperature Alkaline Bleach Boosting Xylanases," Proceedings, APPITA General Conference, p. 151.
41. Tolan, J.S. and B.E. Foody. 1994. "Novel Enzyme Treatment for Dioxin-Free Bleaching of Kraft Pulp with High Kappa Factor and Low ClO_2 Substitution," Papers, 1994 Int. Pulp Bleaching Conf., CPPA, Montreal, p. 263.
42. Tolan, J.S. and M. Guenette. 1995. "The Use of an Enzyme Treatment to Improve the Efficiency of Shive Removal by Bleaching," Preprints, 80th Annual Meeting, Technical Section, CPPA, Montreal, p. B41.
43. Kakezawa, M. 1994. "Biobleaching Process Using Hyperlignolytic Fungus Strain IZU-154 and Its Mutants," Papers, Int. Pulp Bleaching Conf., CPPA, Montreal, p. 183.
44. Pearce, M.H., R.W. Dunlop, L.A. Horgan, C.J. Falk, and A.B. Roullo. 1995. "Screening White Rot Fungi for Their Ability to Biologically Bleach Eucalypt Kraft Pulp," Proceedings, APPITA General Conference, p. 167.
45. Paice, M.G., R. Bourbonnais, I.D. Reid, F.S. Archibald, and L. Jurasek. 1994. "Oxidative Bleaching Enzymes: The Next Generation?" Papers, Int. Pulp Bleaching Conf., CPPA, Montreal, p. 211.
46. Call, H.P. and I. Mücke. "State of the Art of Enzyme Bleaching and Disclosure of a Breakthrough Process," Ref 3., Paper 0-3.
47. Call, H.P. and L. Mücke. 1995. "Further Improvements of the Laccase-Mediator System (LMS) for Enzymatic Delignification (Bleaching) and Results from Large Scale Trials," Proceedings, Int. Nonchlorine Bleaching Conf., Miller Freeman, San Francisco, Paper 4-3.
48. Pryke, D.C. 1995. "Substitution of Chlorine Dioxide for Chlorine in Canadian Bleached Chemical Pulp Mills," Preprints, 81st Annual Meeting, Technical Section, CPPA, Montreal, p. B255.
49. Schwantes, T.A. and T.J. McDonough. September 1994. "Characterization of Effluent Fractions from ClO_2 and Cl_2 bleaching of Unbleached and O_2 Bleached Softwood Kraft Pulps," *Tappi J.* 77(9):145.
50. Schwantes, T.A. and T.J. McDonough. 1995. "The Effect of D Stage pH on Effluent Quality: Characterization of Effluent Fractions from OD(EO) Bleaching of Softwood Kraft Pulp," Preprints, 80th Annual Meeting, Technical Section, CPPA, Montreal, p. A95.
51. Teder, A. and A. Törngren. 1994. "Chloride Ions and the Formation of AOX in DC Bleaching," Papers, Int. Pulp Bleaching Conf., CPPA, Montreal, p. 177.
52. Reeve, D.W., K.M. Weishar, and L. Li. 1994. "Process Modifications to Decrease Organochlorine Formation During Chlorine Dioxide Delignification," Papers, Int. Pulp Bleaching Conf., CPPA, Montreal, p. 281.
53. Schwantes, T.A. and T.J. McDonough. 1994. "The Effect of D Stage Reaction Time on the Characteristics of Whole Effluents and Effluent Fractions from D(EO) Bleaching of Oxygen Delignified Softwood Kraft Pulp," Papers, Int. Pulp Bleaching Conf., CPPA, Montreal, p. 217.
54. Malinen, R., T. Rantanen, and R. Rasimus. 1993. "ECF Bleaching of Oxygen Delignified Softwood Pulp with the Minimum Charge of ClO_2," Proceedings, Pulping Conf., TAPPI PRESS, Atlanta, p. 925.

55. Vegega, A.M., W.G. Strunk, D.D. Elm, P.P. Choma, and A. Sezgi. 1993. "Low Kappa Factor Bleaching of Oxygen Delignified Softwood Kraft Pulp," Proceedings, Pulping Conf., TAPPI PRESS, Atlanta, p. 1049.
56. van Lierop, B., N. Liebergott, and M. Faubert. 1994. "ECF Bleaching with Very Low ClO_2 Charges," Proceedings, Int. Non-Chlorine Bleaching Conf., Miller Freeman, San Francisco, Paper 2-1.
57. Klein, R., W.G. Strunk, and A.M. Vegega. 1994. "Low Kappa Factor Bleaching — 'A Low Capital Strategy to Achieve EPA Guidelines'," Proceedings, Int. Nonchlorine Bleaching Conf., Miller Freeman, San Francisco, Paper 2-3.
58. Parthasarathy, V.R., G.F. Rudie, and A.E. Detty. 1993. "Total Chlorine Dioxide Substitution in Short Sequence Bleaching and Its Effect on Pulp Brightness and Effluent Characteristics," Proceedings, Pulping Conf., TAPPI PRESS, Atlanta, p. 933.
59. Desprez, F., J. Devenyns, and N. Troughton. 1993. "P^*: One-Stage Hydrogen Peroxide Full Bleaching of Previously Delignified Softwood Kraft Pulp," Proceedings, Pulping Conf., TAPPI PRESS, Atlanta, p. 443.
60. Devenyns, J., F. Desprez, and N.A. Troughton. 1993. "Peroxygen Prebleaching and Bleaching Technologies for Stepwise Conversion from Conventional Chlorine Bleaching via ECF to TCF," Proceedings, Pulping Conf., TAPPI PRESS, Atlanta, p. 341.
61. McKenzie, D.J. 1994. "The Evolution of Pulp Bleaching Practices at Harmac Pacific," Papers, Int. Pulp Bleaching Conf., CPPA, Montreal, p. 163.
62. Helander, R., B. Nilsson, and G. Bohman. 1994. "Development and Progress in Ozone Bleaching at the Skoghall Mill," Papers, 1994 Int. Pulp Bleaching Conf., CPPA, Montreal, p. 289, and paper distributed at the conference.
63. Mendiratta, S.K., D.W. Cawlfield, L.R. Ward, and D. Breed. 1994. "Low AOX Process for Bleaching High Kappa Pulps; O1 Delignification," Proceedings, Pulping Conference, TAPPI PRESS, Atlanta, p. 821.
64. Parthasarathy, V.R., G.F. Rudie, S.K. Mendiratta, D.W. Cawlfield, and H.P. Bryan. 1994. "Gaseous Chlorine Dioxide Delignification of Softwood Pulps for High Brightness and Ultra Low AOX," Proceedings, Pulping Conference, TAPPI PRESS, Atlanta, p. 1031.
65. Lachenal, D., C. de Choudens, and P. Monzie. 1977. *Atip Rev.* 31(1):16.
66. Carles, J.E., H. LeMoyne, and W.R. Logan. 1980. "Peroxide Delignification of Unbleached Chemical Pulp," Proceedings, Pulping Conference, TAPPI PRESS, Atlanta, p. 325.
67. Agnemo, R. and G. Gellerstedt. 1979. *Acta Chem. Scand.* B33:337.
68. Smith, P.K. and T.J. McDonough. 1985. *Svensk Papperstidn.* 88(12):R106.
69. McDonough, T.J., R.C. Kirk, B. Backlund, and L. Winter. 1987. "Catalysis in Peroxide Delignification," Proceedings, Oxygen Delignification Symposium, TAPPI PRESS, Atlanta, p. 165.
70. Basta, J., L. Holtinger, and J. Höök. 1991. "Controlling the Profile of Metals in the Pulp Before Hydrogen Peroxide Treatment," Proceedings, 6th Int. Symp. on Wood and Pulping Chemistry, Melbourne, p. 237.
71. Basta, J., L. Holtinger, and P. Lundgren. 1991. "Reducing Levels of AOX — Part 3. Lowering of Kappa No. Prior to ClO_2 Bleaching," Preprints, Int. Pulp Bleaching Conf., SPCI, Stockholm, Vol. 3, p. 23.
72. Basta, J., L. Holtinger, W. Hermannson, and P. Lundgren. 1994. "Metal Management in TCF/ECF Bleaching. Part 1 — TCF Bleaching," Papers, Int. Pulp Bleaching Conf., CPPA, Montreal, p. 29.
73. Desprez, F., J. Devenyns, and N. Troughton. 1993. "P^*: One-Stage Hydrogen Peroxide Full Bleaching of Previously Delignified Softwood Kraft Pulp. Application to TCF Sequences," Proceedings, Pulping Conference, TAPPI PRESS, Atlanta, p. 443.
74. Troughton, N.A., F. Desprez, and J. Devenyns. 1994. "P 100 TCF Bleaching of Softwood Kraft Pulps to High Brightness," Proceedings, Pulping Conference, TAPPI PRESS, Atlanta, p. 807.
75. Devenyns, J., F. Desprez, and N. Troughton. 1994. "The Hydrogen Peroxide P^* Approach to TCF Final Brightening," Posters, Int. Pulp Bleaching Conf., CPPA, Montreal, p. 57.
76. Jakara, J., J. Patola, and S. Martikainen. 1995. "The Use of Activated Peroxide in ECF and TCF Bleaching of Kraft Pulp," Proceedings, International Nonchlorine Bleaching Conference, Miller Freeman, San Francisco, Paper 14-2.
77. Atalla, R., C.L. Hill, C.J. Houtman, R.S. Reiner, and I.A. Weinstock. 1995. 8th Int. Symp. on Wood and Pulping Chemistry, Helsinki, Proceedings, 1: Oral Presentations: pp. 369–376.
78. Sturm, W.G.J. and J.G. Kuchler. 1993. "The Nitrilamine Reinforced Hydrogen Peroxide Bleaching of Kraft Pulps," Nonchlorine Bleaching Conference, Miller Freeman, San Francisco, Paper 31.
79. Hill, C.L. and I.A. Weinstock. 1994. Delignification of Wood Pulp by Vanadium-Substituted Polyoxometalates. U.S. Patent 5,302,248.
80. Chirat, C. and D. Lachenal. 1995. "Fundamentals of Ozone Bleaching," Course Notes, Emerging Pulping and Bleaching Technologies Workshop, TAPPI PRESS, Atlanta.

81. White, D.E., T.P. Gandek, M.A. Pikulin, and W.H. Friend. 1993. "Importance of Reactor Design in High Consistency Ozone Bleaching," Preprints, 79th Annual Meeting, Technical Section, CPPA, Montreal, p. B201.
82. Clark, E.C.C., Jr. 1994. "Ozone Bleaching — Merits of High Consistency Operation," in "Advances in Forest Products Environmental and Process Engineering," AIChE Symposium Series No. 302, C.L. Verrill, Ed., American Institute of Chemical Engineers, New York, p. 205.
83. Funk, E., R. Szopinski, F. Munro, and A. Vilpponen. 1992. "Espanola Ozone Bleaching Pilot Plant: Status Report," Proceedings, Pulping Conference, TAPPI PRESS, Atlanta, p. 1091.
84. Steffes, F. and R. Reeves. 1994. "High Consistency Ozone Delignification — Mill Experience," Int. Non-Chlorine Bleaching Conf., Miller Freeman, San Francisco, Paper 4-3.
85. Ek, M., P. Axegard, E. Bergnor, and U. Ekholm. 1994. The Role of Metal Ions in TCF Bleaching of Softwood Kraft Pulps Proceedings, Pulping Conference, TAPPI PRESS, Atlanta, p. 1161.
86. Dahllöf, H. 1994. "Medium Consistency Ozone Bleaching. An Established Technology in Mill Scale," Int. Nonchlorine Bleaching Conf., Miller Freeman, San Francisco, Paper 4-2.
87. Germgård, U. and S. Nordén. 1994. "OZP-Bleaching of Kraft Pulps to Full Brightness," Papers, Int. Pulp Bleaching Conf., CPPA, Montreal, p. 53.
88. Liebergott, N. 1995. "Di-Per Process. Time for a Change. Peracid Delignification and Brightening," Preprints, 81st Annual Meeting, Technical Section, CPPA, Montreal, p. B157.
89. Liebergott, N. 1995. "Ozone Bleaching: New Developments," Course Notes, Emerging Pulping and Bleaching Technologies Workshop, TAPPI PRESS, Atlanta.
90. Song, J., D. Pan, H.-M. Chang, and H. Jameel. 1993. "Peroxymonosulfuric Acid and a Mixed Peracid Solution as Bleaching Agents," Proceedings, 7th Int. Symp. on Wood and Pulping Chemistry, Beijing, p. 34.
91. Wang, S., H.-M. Chang, and H. Jameel. 1994. "The Preparation of Peracids for Bleaching," Proceedings, Pulping Conference, TAPPI PRESS, Atlanta, p 531.
92. Amini, B. and J. Webster. 1994. "On-Site Peracids: Tools for Bleaching Strategies to Meet the Cluster Rule, and Considerations on How to Select Among Them," Proceedings, Pulping Conference, TAPPI PRESS, Atlanta, p. 371.
93. Seccombe, R., H. Martens, and A. Haakana. 1994. "The Use of Caro's Acid at an Existing Kraft Mill to Produce High Brightness TCF Pulps," Proceedings, APPITA General Conference, p. 139.
94. Lee, C.-L., K. Hunt, and R.W. Murray. 1993. "A Novel Non-Chlorine Agent for Chemical Pulp Bleaching," 1992 TAPPI Pulping Conference Proceedings, TAPPI PRESS, Atlanta, Book 3, p. 1237; "Dimethyldioxirane as a Nonchlorine Agent for Chemical Pulp Bleaching," *Tappi J.* 76(11):137.
95. Ragauskas, A.J. 1993. "Investigation of Dimethyldioxirane as a Bleaching Reagent for Kraft Pulp," *Tappi J.* 76(7):87.
96. Ragauskas, A.J. "Bleaching of Kraft Pulps via Dioxiranes," American Chemical Society, in press.
97. McDonough, T.J., A. Marquis, and A.J. Ragauskas. 1994. "Totally Chlorine Free Bleaching with Dimethyldioxirane," Papers, Int. Pulp Bleaching Conf., CPPA, Montreal, p. 47.

Chapter 2

Closing Up the Bleach Plant: Striving for a Minimum-Impact Mill

R.L. Erickson, R.N. Thut, and W.H. Persinger

SUMMARY

"Oxygen delignification" and "extended delignification" are becoming centerpiece technologies for the improved manufacture of bleached paper products on a global basis. These technologies facilitate the elimination of elemental chlorine as the primary bleaching chemical, allowing more emphasis on chlorine dioxide and oxygen-based chemicals. Although the debate around chlorinated organics and their impact on the environment continues in North America, the discussions elsewhere in the world are moving to another level — bleach plant effluent reduction. Acronyms like ECF (elemental chlorine–free) and TCF (totally chlorine-free) were common over the past five years. Now terms like BEF (bleach effluent–free) and "closed-mill" are being added to the list. This chapter discusses these topics from a Weyerhaeuser Company perspective.

INTRODUCTION

As the pulp and paper industry moves toward the future, it's useful to examine our progress and consider where our next steps will take us.

Weyerhaeuser operates 10 bleached pulp and paper facilities across North America (8 kraft and 2 sulfite). These facilities produce nearly 3.2 million metric tons annually of bleached products, including market pulp, fine paper, and bleached paperboard, and serve customers throughout the world. Relative to bleached kraft, this amounts to 7% of the capacity in North America and 4% of the global capacity (Figure 2.1).

Meeting the environmental preferences of customers throughout the world, and meeting expectations that increasingly exceed the process capabilities and current practices of the North American pulp and paper industry, is a Weyerhaeuser priority.

Our role in international markets gives us a clearer view of upcoming challenges. Currently, these challenges require us to reduce our environmental impact and maintain healthy, diverse forest ecosystems.

We have accepted that challenge.

We see a future in which we enhance the management of sustainable forestry and protect such forest resources as wildlife and water quality. Our vision incorporates efficient and, ultimately, closed-cycle manufacturing facilities that produce forest products without negatively impacting the environment. This vision is sustained by a financially successful company, satisfying worldwide customers while providing incomes for our employees, shareholders, and the communities in which we operate.

Achieving this vision requires a *holistic* perspective that considers not only environmental performance and ecological balance, but also social and economic balance. Key elements of this holistic approach include:

Meeting customer needs for product performance and purity.
Sustainable forestry through ecosystem management.
Clean manufacturing, including waste reduction.
Sustainable economics and solid shareholder returns.
Public involvement and open dialogue with regulatory bodies.

	World	Weyerhaeuser
U.S.	28,126	1,880 (7%)
Canada	11,465	1,060 (9%)
Sweden	4,771	
Finland	5,380	
Rest of World	26,005	
Total	75,747	2,940 (4%)

Figure 2.1. Weyerhaeuser overview: 1994 bleached kraft pulp and paper capacity.

The forest products industry has been asked to address three major environmental priorities: *forest management* (sustainability, biodiversity, and aesthetics), *recycling* (conservation and waste reduction) and *pollution reduction* (bleaching chemicals, toxics, and aesthetics). We've listened to our stakeholders, and we've heard consistent messages about what is expected.

These expectations are expanded in Figure 2.2 and can be expected to increase through time.

Specifically, in the area of pollution reduction, bleaching has received most of the attention. Much of the discussion started when trace levels of dioxin (TCDD and TCDF) were detected in pulp-bleaching effluent in the mid-1980s. This resulted from improved detection capabilities in the parts per trillion and parts per quadrillion (ppq) ranges. Finding dioxin was a surprise, and the source was not known at that time. The process of understanding and then working to eliminate dioxin prompted extensive analysis, experimentation, and process changes. Subsequently, since 1988, we have reduced the trace dioxin to nondetectable levels in final effluent at all 10 of our bleached pulp and paper facilities, based on a detection level of 5 ppq (Figure 2.3). These 10 facilities consist of 8 bleached kraft mills and 2 bleached sulfite mills.

Bleached products are vital to meeting the end use performance needs of pulp and paper industry customers throughout the world, including products with high purity and unique characteristics. Thus, bleaching has long been an integral part of pulp manufacturing. For the entire industry, the goal has always been to continuously improve the environmental performance of the bleaching process as well as to improve product quality and efficiency of the operations.

In recent years, most technical and regulatory attention has been focused on the type of bleaching chemicals used, specifically on chlorine and chlorine dioxide. Now, that focus is shifting to include bleach plant effluent reductions.

Technologies (chemicals and equipment) now exist for *new* and *existing* facilities to greatly reduce bleach plant effluent. For *existing* facilities, technologies must be incorporated carefully to achieve all benefits cost-efficiently.

Weyerhaeuser has just completed major modernizations totaling $1 billion at three of its bleached pulp and paper facilities: Plymouth, North Carolina; Longview, Washington; and Kamloops, British Columbia. Most efforts went toward upgrading the pulping and bleaching systems to improve environmental performance and product quality and to reduce costs.

Even as these modernizations conclude, plans are in the works to selectively target facilities for further improvements in bleaching. Reducing bleach plant effluent is an integral part of this effort.

MINIMUM-IMPACT MILL PHILOSOPHY

Environmental impacts of the pulp and paper industry can be viewed hierarchically (Figure 2.4). Acute toxic impacts already have been dealt with through secondary treatment and other practices. Chronic and more subtle effects are currently the focus. Through time, however, pulp and paper manufacturing will be expected to have a minimum impact on the environment, whatever the source of emissions. Bleach plants are but one of the areas that need to be addressed.

In describing a mill of the future, terms like minimum-impact mill (MIM), zero-impact mill (ZIM), no-effect mill, closed-loop mill, total effluent–free mill (TEF), bleached effluent reduction (BER) and bleached effluent–free (BEF) emerge. Some acronyms, like ECF (elemental chlorine–free) or

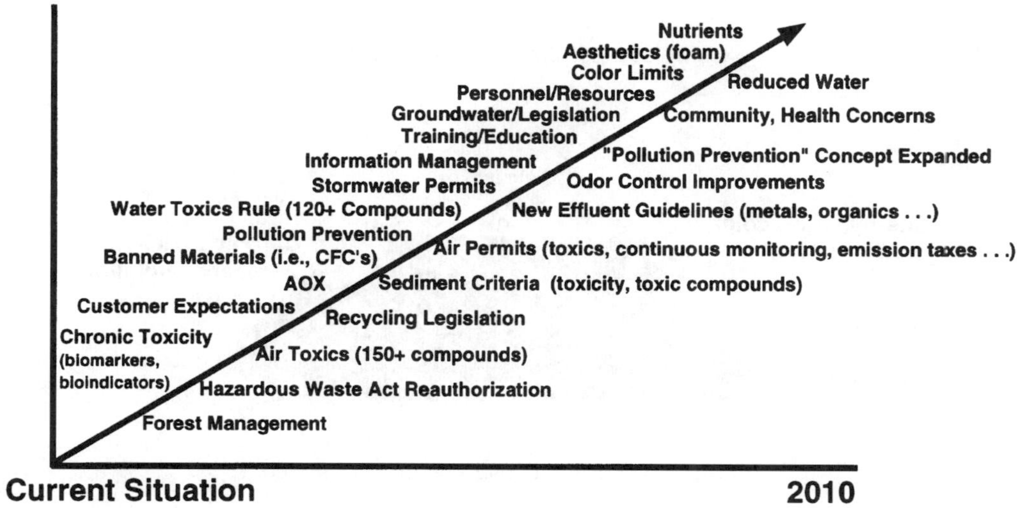

Figure 2.2. Stakeholder expectation.

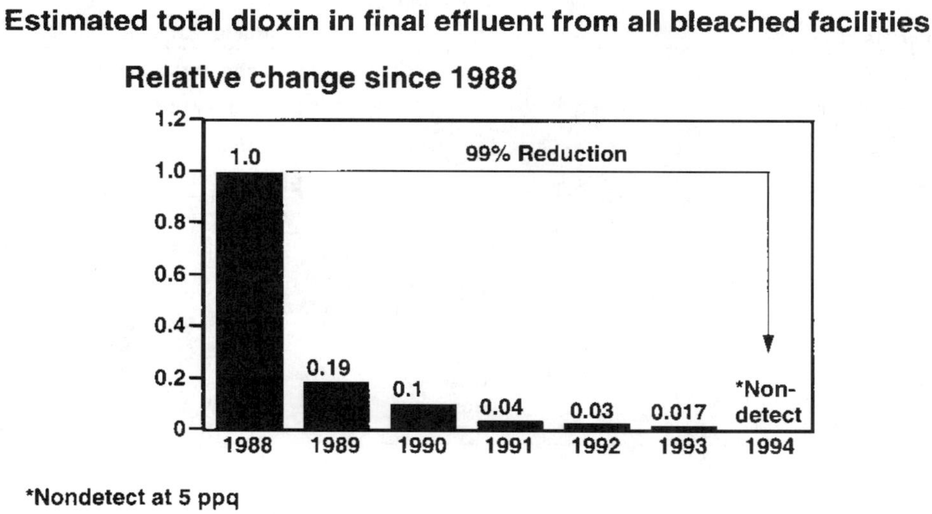

Figure 2.3. Weyerhaeuser dioxin reduction.

TCF (totally chlorine-free), may prove too narrow in context or won't adequately describe end-state environmental characteristics.

For now, we have adopted a "minimum-impact mill" philosophy as a way to describe that future state. For us, this template makes sense in that it is more "holistic" in nature. It recognizes that a "journey" will be required. And, as new information becomes available (scientific and public perception), it will help us understand what minimum-impact manufacturing truly requires. Attention must be paid to *all* facets of a mill's operations, not only single chemical or emissions streams.

We define a "minimum-impact mill" as one that has:

Accepted full accountability for the impact of its operations on the environment.
Committed to continuous reduction of air, water, and solid waste discharges.
Achieved discharge levels that are recognized as being "minimum-impact" using sound science.
Maintained product performance and purity.

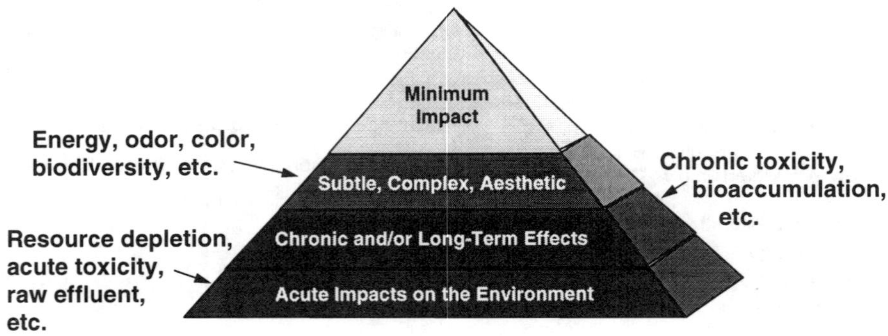

Figure 2.4. Environmental hierarchy of needs.

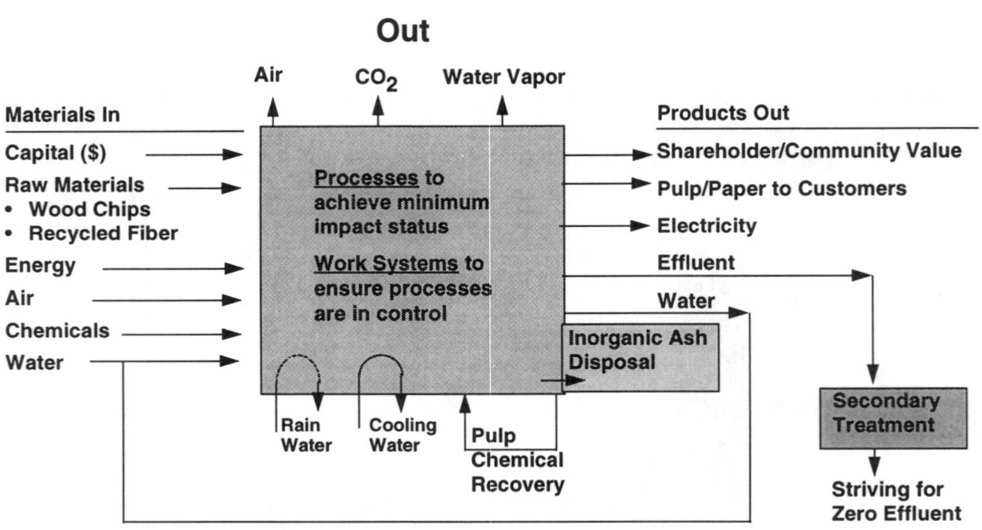

Figure 2.5. A minimum-impact mill model.

A model is illustrated in Figure 2.5.
Our model:

Uses materials and process to produce products, value-added by-products, and stakeholder value.
Recycles materials, chemicals, and water to the maximum degree possible.
Strives for zero effluents and emissions.

In order to monitor performance and look at benchmark information on a global basis, we use over 30 parameters covering water, air, solid waste, and other aspects as key indicators. These are identified in Table 2.1.

PULPING AND BLEACHING TRENDS

Pulping and bleaching technologies have evolved over the past several decades and will undergo further change (Figures 2.6, 2.7, and 2.8). Digesters for pulping have become more efficient. Re-

Table 2.1. Parameters Important to Demonstrating Continuous Improvement Toward a Minimum-Impact Mill.

Water	Air	Solid Waste	Other
Water usage	Particulate	Solid waste generated	Accidental releases
Bleach plant effluent volume	TRS	Solid waste disposition	Noncompliant events
	Methanol	• Landfill	SARA 313 releases
Final effluent volume	Chloroform	• Recycled	Energy use/energy export
BOD	Chlorine	• Energy	Transportation impacts
COD	Chlorine dioxide	Reduction plans/ achievements	Aesthetics
TSS	CO/CO_2		• Site appearance
Effluent AOX	NOx	Hazardous waste elimination	• Odor
Dioxin	SO_2		• Noise
Color	VOCs		Improvement projects
Chronic toxicity	Dioxin		Key environmental accomplishments
Nutrients (contamination)	Opacity		
	HAPS		

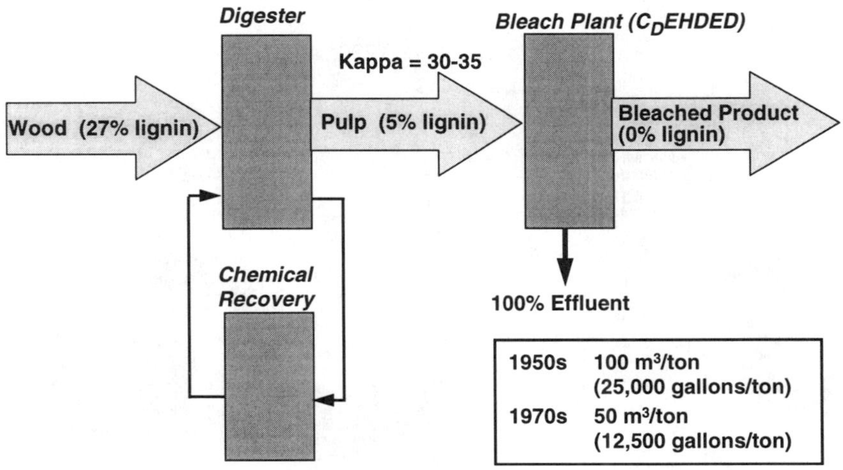

Figure 2.6. The "1970" mill.

sidual lignin levels after pulping, as measured by kappa number, can now be reduced without sacrificing product quality or operating efficiency. In addition, oxygen delignification has become a standard technology to further reduce the lignin content prior to bleaching. When properly employed together, these two technologies reduce bleaching chemical consumption by almost two-thirds. The effluent and residual organic material from these processes can flow back to the chemical recovery cycle. Although this sounds simple, for those facilities with a recovery boiler bottleneck, this can be a major concern and an economic penalty if not properly managed.

Currently, there is a strong shift away from elemental chlorine bleaching, with a move toward using more chlorine dioxide and peroxide. Ozone also is starting to gain favor, but is still undergoing development. Enzymes are in an even earlier stage of development.

These developments translate into bleaching chemical changes (Figure 2.9[1]). As can be seen, the pulp and paper industry is in transition. What is clear is that the chemicals used in 2010 will be significantly different than those used in 1980.

Weyerhaeuser's chemical trends are highlighted in Figure 2.10. Elemental chlorine has been virtually eliminated. This was done for several reasons: a series of facility modernizations, availability of

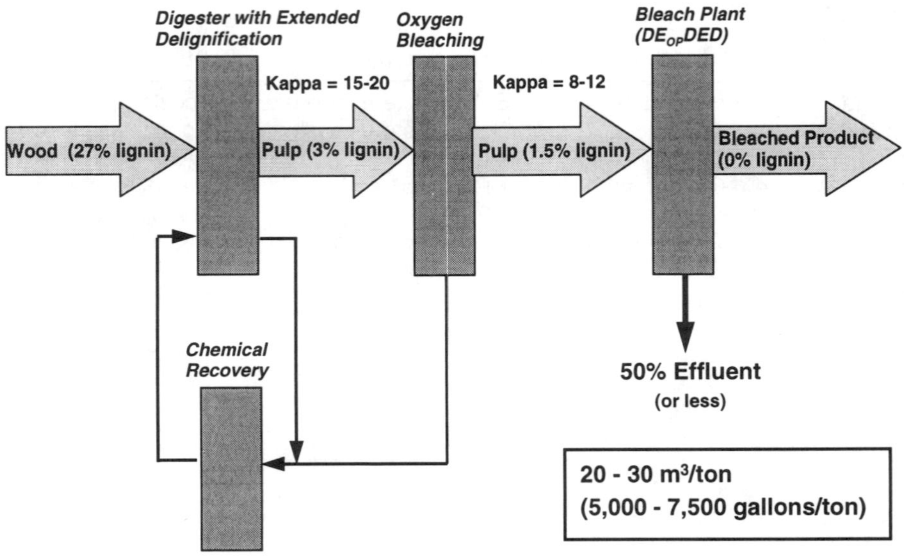

Figure 2.7. The "1990" mill.

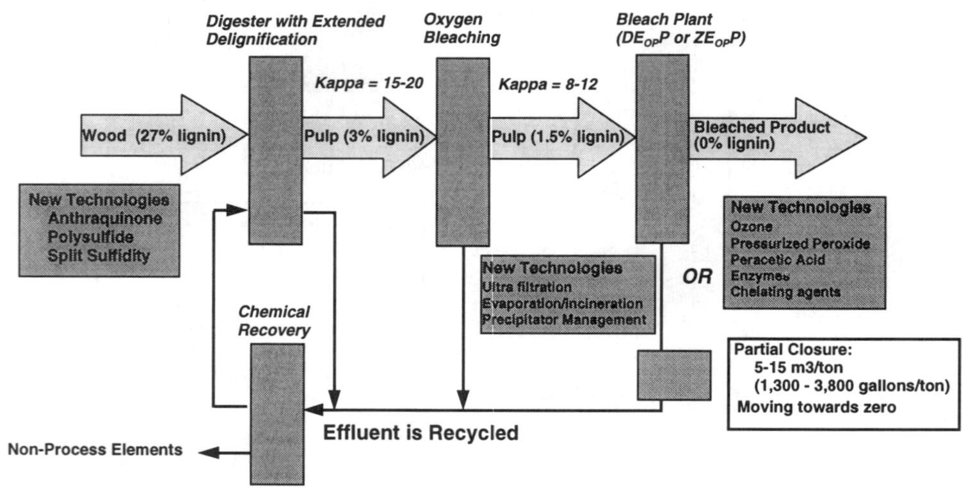

Figure 2.8. The "2010" mill.

proven technologies, customer requirements, a desire to reduce bleach plant effluent, and alignment with a "minimum-impact mill" vision.

Capital and operating costs of various technologies are summarized in Tables 2.2 and 2.3. This information suggests that as new fiberlines are built, it will be cost-effective to utilize new technologies. Care must be used in retrofitting older facilities to balance all parts of the equation.

It takes time to move new and innovative technologies into a capital-intensive industry such as pulp and paper. Figure 2.11[2] illustrates the evolution of oxygen delignification into the industry. Even though the technology was first developed with pilot plant capabilities in the 1960s, it took nearly 20 years to make it an accepted and widespread technology. With the success of installed oxygen systems, however, the direction is clear for the future — oxygen has become a technology of choice.

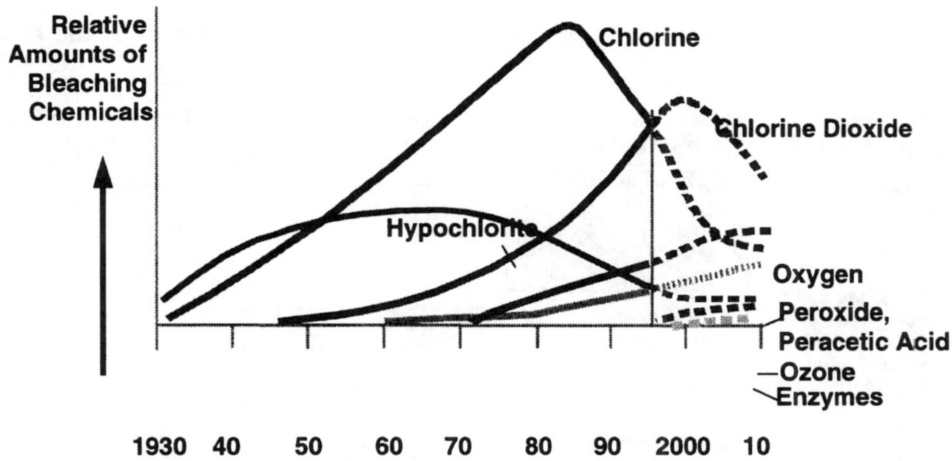

Figure 2.9. Bleaching chemical trends—a global perspective. (*From:* Croon Consult, Sweden, 1994.)

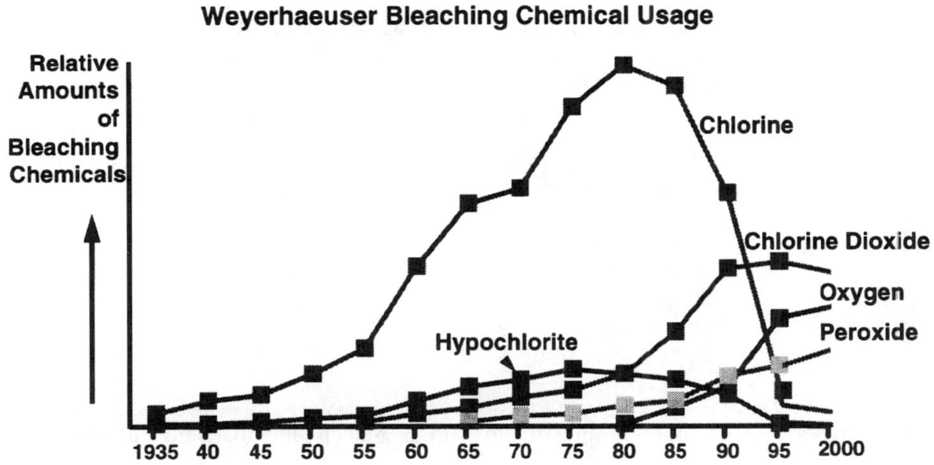

Figure 2.10. Bleaching chemical trends—Weyerhaeuser.

Indeed, oxygen delignification (OD) and/or extended delignification (ED) are becoming centerpiece technologies for the pulp and paper industry on a global basis to achieve improved strategies for pulping and bleaching. Both technologies likely will be standard for new bleached facilities and are being used to retrofit older facilities. They significantly reduce the unbleached pulp lignin content in an efficient manner prior to bleaching. Consequently, less chemicals are required for bleaching, and alternative chemicals — mostly oxygen-based — can be used.

These new technologies have proliferated around the world. Figure 2.12, developed from discussions with several sources,[3] illustrates the application of these technologies in the Nordic countries, North America, and at Weyerhaeuser. The trends are obvious.

BLEACH PLANT EFFLUENT-REDUCTION TRENDS

Effluent from the bleach plant can make its way back to the pulp mill chemical recovery cycle. However, chlorine compounds typically used in bleaching first must be eliminated or managed. This is now becoming possible.

Table 2.2. New Techniques — Capital and Operating Costs. Comparisons for a 1200 TPD Mill Producing 88–90 Brightness Pulp.

Mill Description: Technology Vintage:	Base Case Mill 1970's Technology	Modern Mill 1990's Technology	State-of-the-Art 1995+ Technology
Digester Type & Kappa:	Conventional to 30	Extended Delig. to 20	Extended Delig. to 18
O_2 Delig. & Kappa:	None	O_2 Delig. to 10	O_2 Delig. to 8
Bleach Sequence:	C_DEDED	DE_OD_ND	$QE_{OP}ZP_O$
Operating Cost:	Base Case		
For ECF Capability			
Sequence:	DE_ODED Retrofit	DE_OD_ND	
Product Capability:	ECF Product OK	Dedicated to ECF	
Incremental Capital:	+ $7 MM	None	
Operating Cost:	+ $30/ADT	Same as Base Case	
For TCF Capability			
Sequence:	$QE_{OP}QPP$ Retrofit	$QE_{OP}QP$ Retrofit	$QE_{OP}ZP_O$
Product Capability:	TCF not Marketable	Low End TCF Product	Dedicated to TCF
Incremental Capital:	+ $1 MM	+ $1 MM	None
Operating Cost:	+ $100/ADT	+ $60/ADT	Same as Base Case

Table 2.3. Capital Cost Comparison for the Various Components of a 1200 TPD Pulping and Bleaching System.

	Capital ($ MM)
Extended delignification	$60–65
O_2 delignification	30–35
Ozone	
contacting equipment	12±
ozone generation	12±
oxygen supply	4±
Pressurized peroxide	8±

There are several reasons for continuing to close up the bleach plant in a pulp and paper operation:

Water — Conservation and utilization (waste reduction) are becoming more important.
BOD — With ever tighter restrictions, reduced water flows to existing secondary-treatment systems will enhance efficiency.
Other chemicals — It will be prudent to recycle as much of the chemicals from the pulping and bleaching process as possible to reduce color and other emissions of by-product compounds.

Weyerhaeuser's total effluent and bleach plant effluent flows have been reduced continually since the early 1970s (Figures 2.13 and 2.14). Notable reductions in bleach plant effluent have occurred as a result of the recent modernizations. Further bleach plant effluent flow reduction projects currently are being scoped. Striving for zero bleach plant effluent is the long-term goal. This would be accomplished in a phased approach.

EFFLUENT QUALITY INDICATORS

Numerous quality indicators for effluent from bleached pulp and paper facilities exist; biological oxygen demand (BOD) and total suspended solids (TSS) have been traditional indicators. Weyerhaeuser trends for these two are illustrated in Figure 2.15.

Other nontraditional indicators for effluent quality, such as AOX, color, and COD, are being used more widely. These indicators are positively impacted by technologies such as oxygen delignification

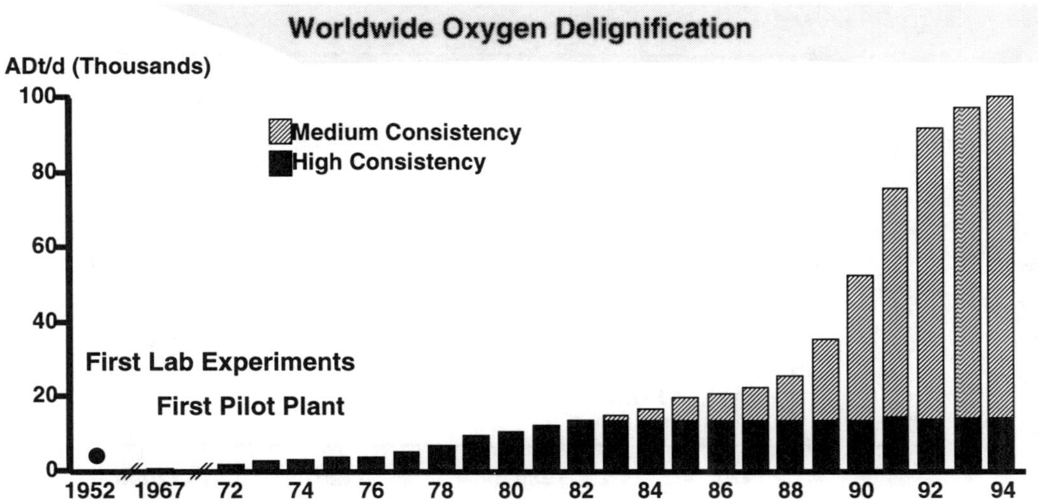

Figure 2.11. Worldwide production capacity for oxygen delignification. *Source:* Jaakko Poyry, 1994.

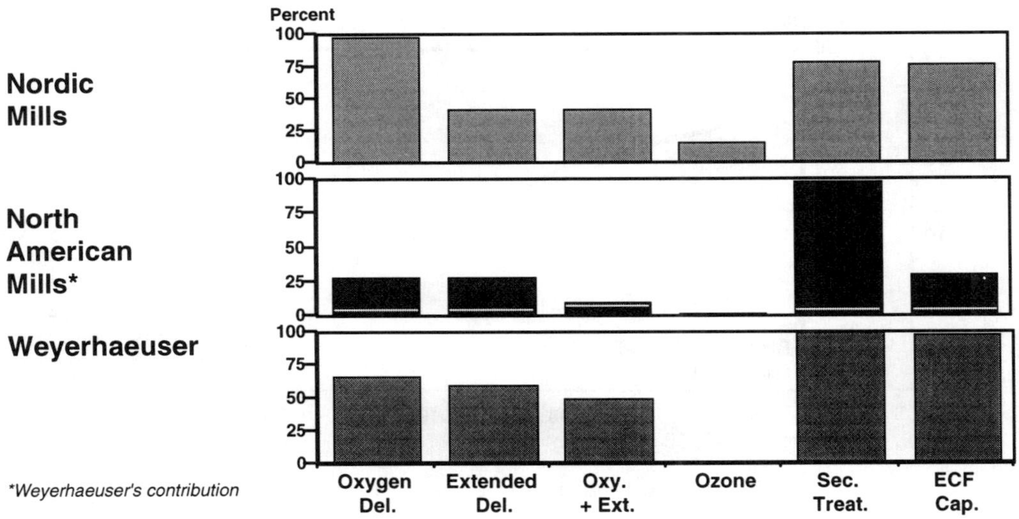

Figure 2.12. Bleached kraft technology goal perspective. (*From:* Jaakko Poyry, Weyerhaeuser.)

and extended delignification and the use of alternative bleaching chemicals. Weyerhaeuser's experiences in these areas are discussed below, and support our stated objective of "striving for a minimum-impact mill."

Besides averages, daily and monthly variability must be taken into account.

AOX (adsorbable organic halide) is a measurement associated with chlorine bleaching. It has been used as a way to describe the level of chlorinated organics in the effluent. Although there is still no cause-and-effect relationship between AOX levels and environmental performance, use of the measurement has become widespread. With the application of new technologies, AOX levels have dropped dramatically (Figure 2.16).

Independent of the AOX measurement, if bleach plant effluent is to be significantly reduced, or eliminated, by recycling it back to the chemical recovery process, all chlorine-containing

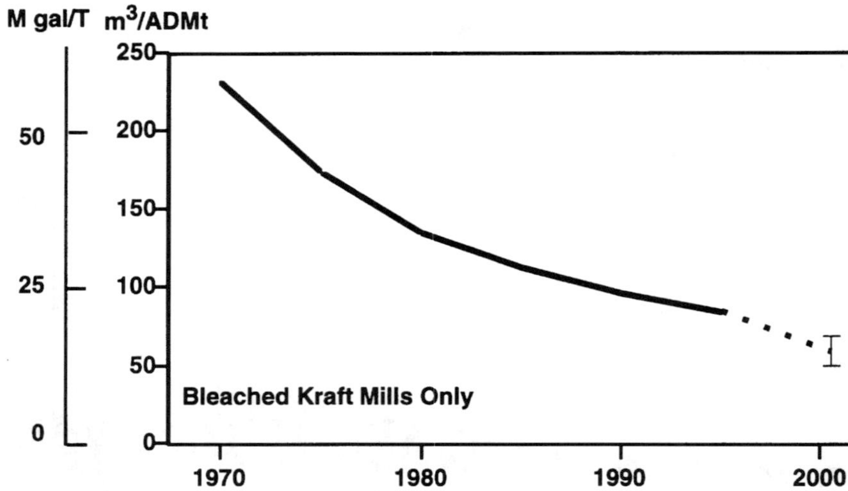

Figure 2.13. Total treated effluent discharged—Weyerhaeuser.

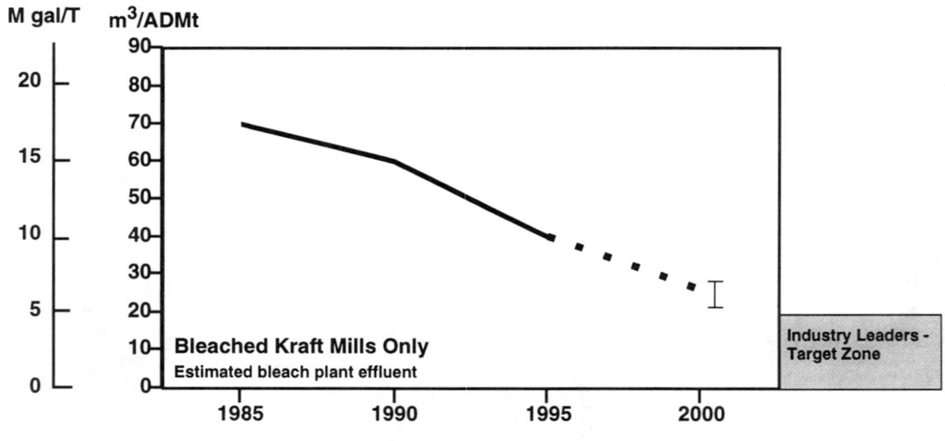

Figure 2.14. Bleach plant effluent—Weyerhaeuser.

compounds must be reduced. Chlorine compounds, whether as chlorinated organics or sodium chloride (table salt), can create major corrosion problems with conventional recovery boilers. There is also a risk that burning chlorine-based material may lead to undesirable air emissions.

It is known that compounds other than chlorinated organics contribute to the overall toxicity of pulp mill effluents. This was recently reinforced by McCubbin and Folke.[4] They indicate that there is no evidence supporting the correlation between AOX and effects on ecosystems. However, these other, nonchlorinated compounds likely will become of more interest in the future. Their studies suggest that mills that discharge very low AOX levels also have oxygen delignification. "In practice, this is accompanied by excellent brownstock washing and a closed screen room process, which results in a reduction of losses of black liquor organics to well under 25% of the losses typical of a mill without oxygen delignification." Consequently, oxygen delignification does more than merely reduce the level of chlorinated organics: it helps minimize other compounds of potential concern.

For this reason, other effluent indicators, such as color and COD, likely will be important indicators in the future:

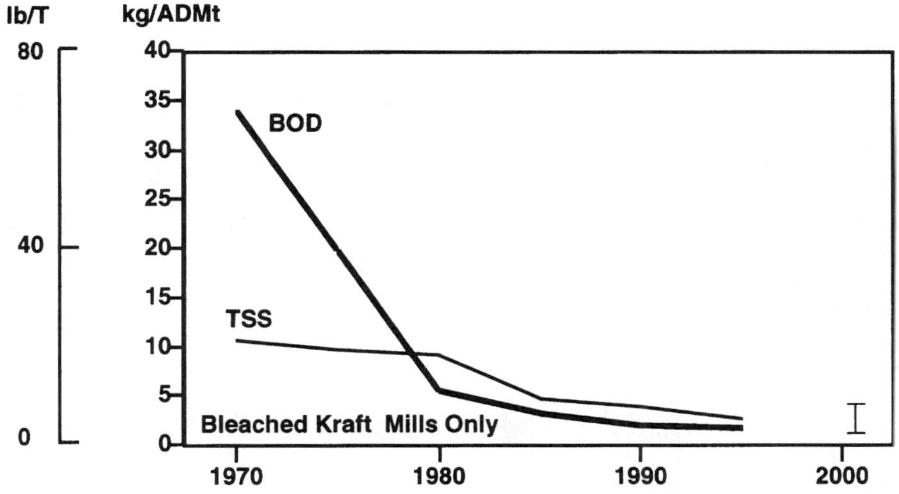

Figure 2.15. Final effluent BOD and TSS discharged—Weyerhaeuser.

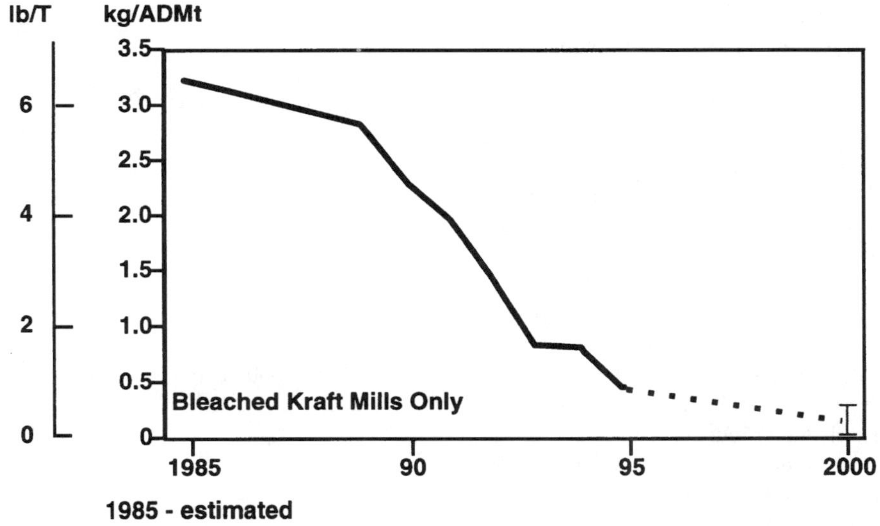

Figure 2.16. AOX—Weyerhaeuser.

Color is an inverse measure of the amount of light transmitted through a filtered sample of effluent. If the effluent color is much different than the receiving water into which the effluent flows, aesthetic issues are created which become a focal point for concern. Whether or not color has an environmental impact (from a minimum-impact mill perspective) continues to be investigated. Color can be significantly reduced with the use of these newer technologies (Table 2.4[5–8]). Trends are illustrated in Figure 2.17.

COD (chemical oxygen demand) measures the oxygen required to break down organic compounds and other oxidizable material in mill effluent. It is a measure more often used by the Nordic countries. COD is found to correlate with color. Since COD has not been used as a standard measure, only limited data are available from Weyerhaeuser mills. The technologies being discussed, however, do have a significant impact on COD, as illustrated in Table 2.4 and Figure 2.17.

Table 2.4. COD and Color for Different Technologies.

	COD		Color	
Mill	**lb/T**	**(kg/ADMt)**	**lb/T**	**(kg/ADMt)**
1950s type mill	170±	(85±)	500±	(250±)
1970s type mill	110±	(55±)	360±	(180±)
100% ClO_2 Substitution	80±	(40±)	220±	(110±)
100% ClO_2 Substitution plus oxygen	60±	(30±)	180±	(90±)
100% ClO_2 Substitution plus oxygen and extended delignification	30±	(15±)	90±	(45±)

Estimated ranges - bleached kraft only.
From: Weyerhaeuser; Malinen (1993); Bowen & Hsu (1990); Heimburger (1988).

STRIVING FOR A MINIMUM-IMPACT MILL

We are using a holistic approach to reduce emissions and effluent to ever lower levels while at the same time improving product quality and reducing manufacturing costs.

Weyerhaeuser's modernizations conform to trends and technology shifts that are occurring on a global basis. For our eight bleached kraft facilities, after the completion of the modernization effort in 1995, the positive effects of these technologies will prove to be striking.

On average, since 1985, these results have been achieved:

Kappa number to the bleach plant (residual lignin) has been reduced by 35%.
Chlorine (elemental) usage has been reduced by 97%.
Chlorine dioxide usage has increased by 110% (and peaked).
Hypochlorite usage has been reduced by 97%.
Oxygen usage has increased by a factor of 10.
Peroxide usage has increased by 350%.

The effect has resulted in reduced impact in the form of:

Dioxin (TCDD and TCDF) being reduced by at least 99% (to nondetectable).
Chloroform releases being reduced by 80%.
AOX levels being reduced by 85%.
Color being reduced by approximately 75%.
COD being reduced by approximately 75%.
Total treated effluent being reduced by 25%.
Bleach plant effluent being reduced by 45%.

In addition, improved quality and efficiency are achieved:

Product quality has been greatly enhanced.
Operating costs have been significantly reduced.

All in all, using minimum-impact manufacturing as a vision, Weyerhaeuser's set of operating facilities are well-positioned to compete globally into the next millennium.

THE FUTURE

As discussions around pulping and bleaching expand to include not only bleaching chemicals, but also bleach plant water use and effluent volume, the environment will benefit. Already, the volume of bleach plant effluent has been steadily dropping as technology shifts have occurred (Figures 2.6, 2.7,

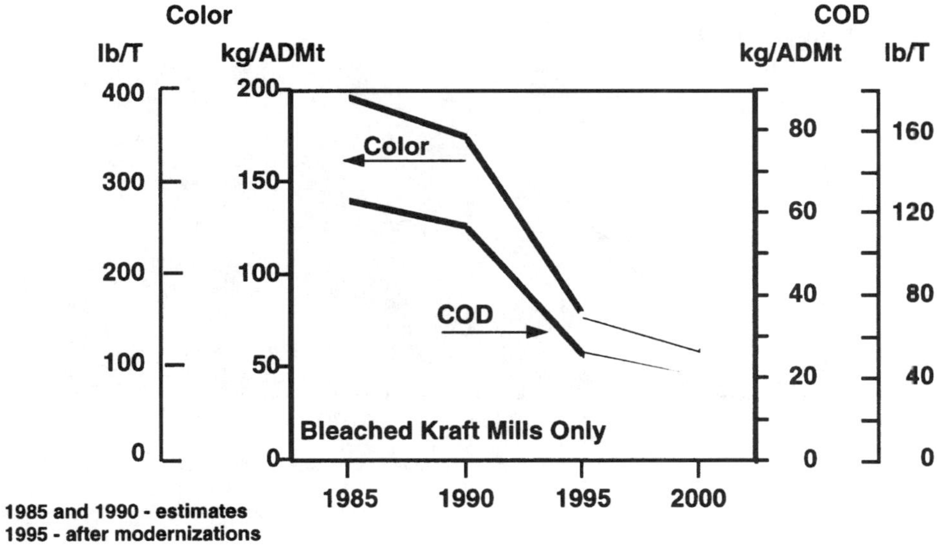

Figure 2.17. Color and COD—Weyerhaeuser.

and 2.8). At the low bleach plant effluent levels, the chemistry of the bleaching process will become increasingly important.

Some, like Ingemar Croon,[9] believe that the next profound technology shift will be in bleach plant effluent reduction. His projections for bleach effluent–free (BEF) facilities are illustrated in Figure 2.18.[10] If his projections are accurate, the transition will be dramatic over the next decade.

It is recognized that reduced bleach effluent flow concepts are growing. A number of facilities are beginning to talk about bleach plant effluent flows in the 5–15 cubic meters per ton range — a major reduction from traditional mills. A few facilities are experimenting with technologies which they hope will lead to totally closed bleach plants.

It is important to note that reduced bleach plant effluent flows likely will be achieved both with and without chlorine dioxide. Bleach effluent reduction can be achieved following either an ECF or a TCF pathway. It is too early to tell whether both paths can make it all the way to bleach effluent–free in a practical manner. However, oxygen delignification is anticipated to be a centerpiece to either approach.

Even though Weyerhaeuser has made significant improvements in its pulping and bleaching capabilities, especially with the completion of the recent modernizations, new projects are in the design phase to strive for even lower bleach plant effluent levels. The timing and sequencing of these future modernizations will be dependent on a variety of factors, including customer requirements, product quality and mill configuration. Even though it is not possible to have every facility at the same technology level at a given point in time, the direction and trends are becoming clear — we are moving toward minimum-impact mills.

Closing up the bleach plant is but one of the factors as we strive for a minimum-impact mill. By taking a holistic approach, other factors, such as air emissions, solid waste, etc., must be taken into account.

CONCLUSIONS

Pulping and bleaching chemicals and technologies have been evolving through time. This is evidenced by the rapid change in chemical usage in the past few years, specifically elemental chlorine. Continued technological changes are expected over the next decade. The evolution appears to be toward oxygen-based bleaching agents to allow the reduction of bleach plant effluent.

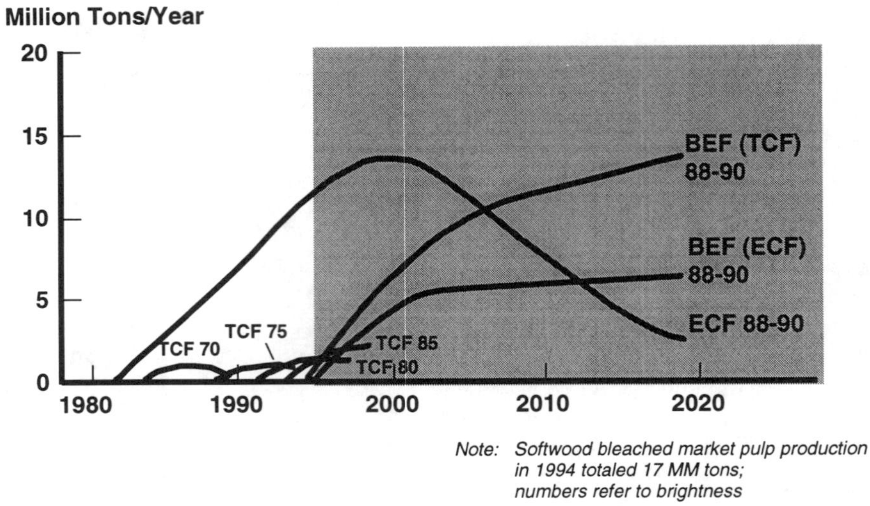

Figure 2.18. Softwood bleached kraft technology trends — a global prognosis. (*From:* Ingemar Croon, CroonConsult, Stockholm.)

It is clear that oxygen delignification and/or extended delignification will play important roles in the reduction or elimination of bleach plant effluent, even if chlorine dioxide is used in the future. Weyerhaeuser's view is that these lignin-reducing technologies applied prior to bleaching are "centerpieces" to the strategy. This view is supported by the evolution of technologies around the world.

With our vision, we can guide the application of technology on a site-by-site basis to achieve sustainable environmental and economic benefits.

REFERENCES

1. Croon, I. 1994. Future Challenges for the Pulp and Paper Industry, *Proceedings from Control Systems 94,* Swedish Association of Pulp and Paper Engineers, Stockholm, Sweden, pp. 11–12.
2. Silander, R. 1994. Jaakko Poyry *Know-How Wire.*
3. Personal communications, various Jaakko Poyry and Weyerhaeuser personnel.
4. McCubbin, N. and J. Folke. 1995. *Pulp and Paper Canada* 96(2):43–48.
5. Unpublished internal data.
6. Malinen, R. 1993. Future Development of Kraft Pulping, Scenarios on Pollution Loads, Proceedings of Eucepa International Environmental Symposium, Paris, France, pp. 97–113.
7. Bowen, I.J. and J.C. Hsu. 1990. Overview of Emerging Technologies in Pulping and Bleaching, *Tappi Journal,* 73(9):205–217.
8. Heimburger, S.A. 1988. Kraft Mill Bleach Plant Effluents: Recent Developments Aimed at Reducing Their Environmental Impact, *TAPPI Pulping Conference Proceedings* - Vol. 2, Atlanta, GA, pp. 285–305.
9. Personal communication, Croon, I., CroonConsult, Stockholm, Sweden.
10. Croon, I. 1994. *Strain Your Muscles,* Svensk Papperstidning/Norsk Cellulosa, 3:20–25.

Chapter 3

Overview of Ozone Bleaching

L.B. Sonnenberg

INTRODUCTION

The pulp and paper industry is exploring ways to reduce its impact on the environment. Reducing the formation of chlorinated organic compounds and the volume of discharged effluent are key areas for improvement. Bleaching with oxygen-based chemicals can serve the pulp and paper industry well in meeting both of these objectives. Ozone is one promising oxygen-based alternative.

Interest in ozone bleaching by the U.S. pulp and paper industry first emerged in the 1970s when environmental legislation in this country began in earnest. Considerable research was conducted in laboratory facilities and pilot plants at that time. Interest waned as it became clear that several problems could not be readily overcome with the existing technology, including poor selectivity, poor uniformity, and the high expense of generating ozone. However, interest in oxygen-based bleaching resurged in the 1990s due to the discovery of chlorinated dioxins and furans in pulp and paper mill effluents, increasing environmental regulation, and increasing public pressure. As a result, several ozone plants have recently been built in the United States, Canada, and Europe. Fourteen worldwide installations in 1995 have capacities ranging from 100 to 1600 tpd, and include three virgin fiber facilities in the United States.

The chemistry, applications, and environmental benefits of ozone bleaching are reviewed in this chapter.

OZONE CHEMISTRY

Ozone is an electrophilic, nonlinear, extremely potent oxidant with an oxidation potential of –2.07 eV ($O_3 + 2\,H^+ + 2\,e^- \rightarrow O_2 + H_2O$). By comparison, Cl_2 has an oxidation potential of –1.4 eV ($Cl_2 + 2\,e^- \rightarrow 2\,Cl^-$). The high degree of reactivity of ozone and the heterogeneity of pulp allow numerous types of chemical reactions to take place during bleaching. Ozone reacts rapidly at sites of high electron density, such as the side chain double bonds and aromatic rings found in lignin, and more slowly at less activated sites, such as the C-H bonds typical in carbohydrates. The generally fast rates of ozone reactions make mass transport to the desired sites of attack critical for efficient and selective bleaching.

Reactions with Carbohydrates

Carbon-hydrogen bonds in many saturated compounds are susceptible to cleavage by ozone, including the activated anomeric carbon-hydrogen bonds in carbohydrates. Highly reactive hydrotrioxide intermediates are produced from the ozonation of acetals,[1–3] including glucosidic acetals.[4–6] Ozonation of ethers,[7–9] aromatic aldehydes,[7,10,11] and saturated hydrocarbons[7,12] also produces hydrotrioxides. When there is an oxygen atom adjacent to the insertion site, the hydrotrioxide may be stabilized by an intramolecular hydrogen bond, forming a six-membered ring. Several mechanisms have been proposed for the formation of the hydrotrioxides, including a concerted 1,3 dipolar insertion (Figure 3.1a), hydride transfer (Figure 3.1b), hydrogen abstraction (Figure 3.1c), and an initial attack at acetalic or etheric oxygens (Figure 3.1d).

Most kinetic data, solvent effects, substituent effects, stoichiometry, and thermochemical calculations support either the concerted insertion,[1,7,10,13] or the hydride transfer.[7,12] However, attack by ozone on compounds with highly acidic α hydrogens of –CHOR groups, forming tetroxide intermediates, has also been reported for a ß-keto ether.[9] For compounds that do not have acetal oxygens,

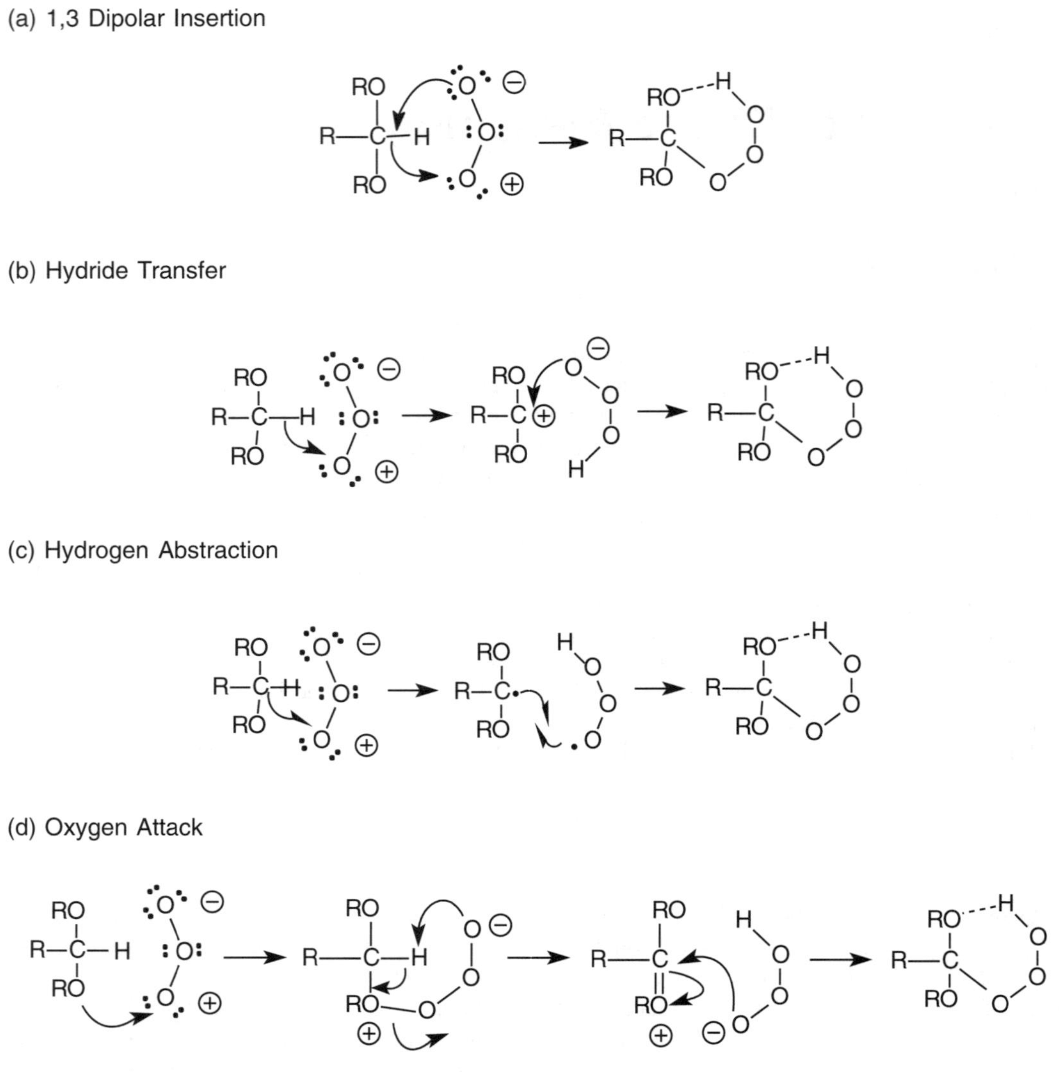

Figure 3.1. Potential mechanisms of hydrotrioxide formation.

substituent effects suggest a moderately positively charged carbon transition state consistent with an insertion mechanism. When acetals are ozonated, the acetal oxygen stabilizes the positively charged transition state, thereby masking any substituent effects.[1] Small solvent effects and net retention of configuration also support the insertion mechanisms; however, these phenomena could occur with the hydride transfer mechanism if there is a solvent-caged radical pair intermediate. While entropies of activation have been reported to be consistent with both mechanisms,[1] Giamalva et al.[7] and Nangia and Benson[12] suggest that the experimental values are more consistent with a hydride transfer. The hydride mechanism is also more consistent with the energies of activation for ozonation of C-H bonds than with the insertion mechanism, which requires a highly ordered and energetic pentavalent carbon transition state.[7,12] Although there is evidence that radicals are present during ozonation, they are not likely to arise from hydrogen abstraction by ozone to form the hydrotrioxides, but result rather from the decomposition of the hydrotrioxides or other reaction pathways.[7,12]

Once the hydrotrioxide intermediates are formed, they may decompose either heterolytically (Figure 3.2) or homolytically (Figure 3.3). The ionic decomposition pathways depend on the other substituents on the hydrotrioxide-bearing carbon, solvents, and other reactants. The products can be

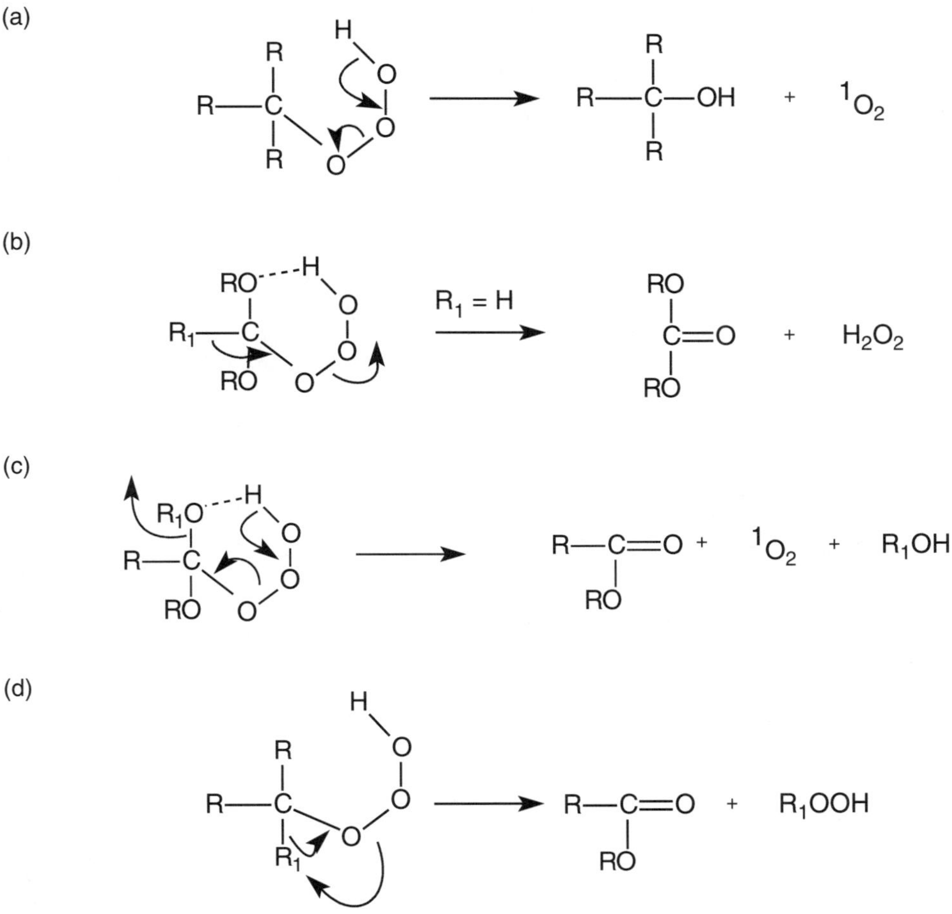

Figure 3.2. Ionic mechanisms of hydrotrioxide decomposition.

either carbonyls and hydrogen peroxide (Figure 3.2a), carbonyls and singlet oxygen (Figure 3.2b), alcohols and singlet oxygen (Figure 3.2c), or carbonyls and alkyperoxyl compounds (Figure 3.2d).[3,14] Homolytic cleavage of oxygen-oxygen bonds of the hydrotrioxide can occur at two sites to form either alkyperoxyl and hydroxyl radicals (Figure 3.3a), or alkoxyl and hydroperoxyl radicals (Figure 3.3b). These radical reactants may then further decompose, abstract a hydrogen, or couple. The radical decomposition reactions may not occur as readily as the ionic mechanisms in some instances;[3,14] however, they initiate the formation of radicals that may play an important role in subsequent chain reactions.

Deslongchamps et al. ozonated glycosides,[4] as well as other acetals,[2] primarily in nonaqueous solvents. They found that for conformationally rigid glycopyranosides that exist predominantly in the C1 conformation, the ß-anomers react with ozone while the α-anomers do not. The α-anomers can react with ozone only if they can attain a 1C conformation. The researchers proposed that reactivity with ozone is enhanced in structures with conformations in which a nonbonding orbital of each acetal oxygen is antiperiplanar to the insertion-point hydrogen. Stabilization of a positively charged transition state by correctly oriented nonbonding orbitals is required for the reaction to occur by the proposed 1,3 dipolar insertion mechanism. Either hydrotrioxide or hemiorthoester intermediates may be reaction intermediates. The sole products from ozonated methyl glycopyranosides were glyconate methyl esters. In these studies, the specificity of the decomposition of the intermediate by way of ring cleavage rather than glycoside cleavage was attributed to stereoelectronic requirements similar to those for hydrotrioxide formation. There must be two nonbonded electron pairs antiperiplanar to the

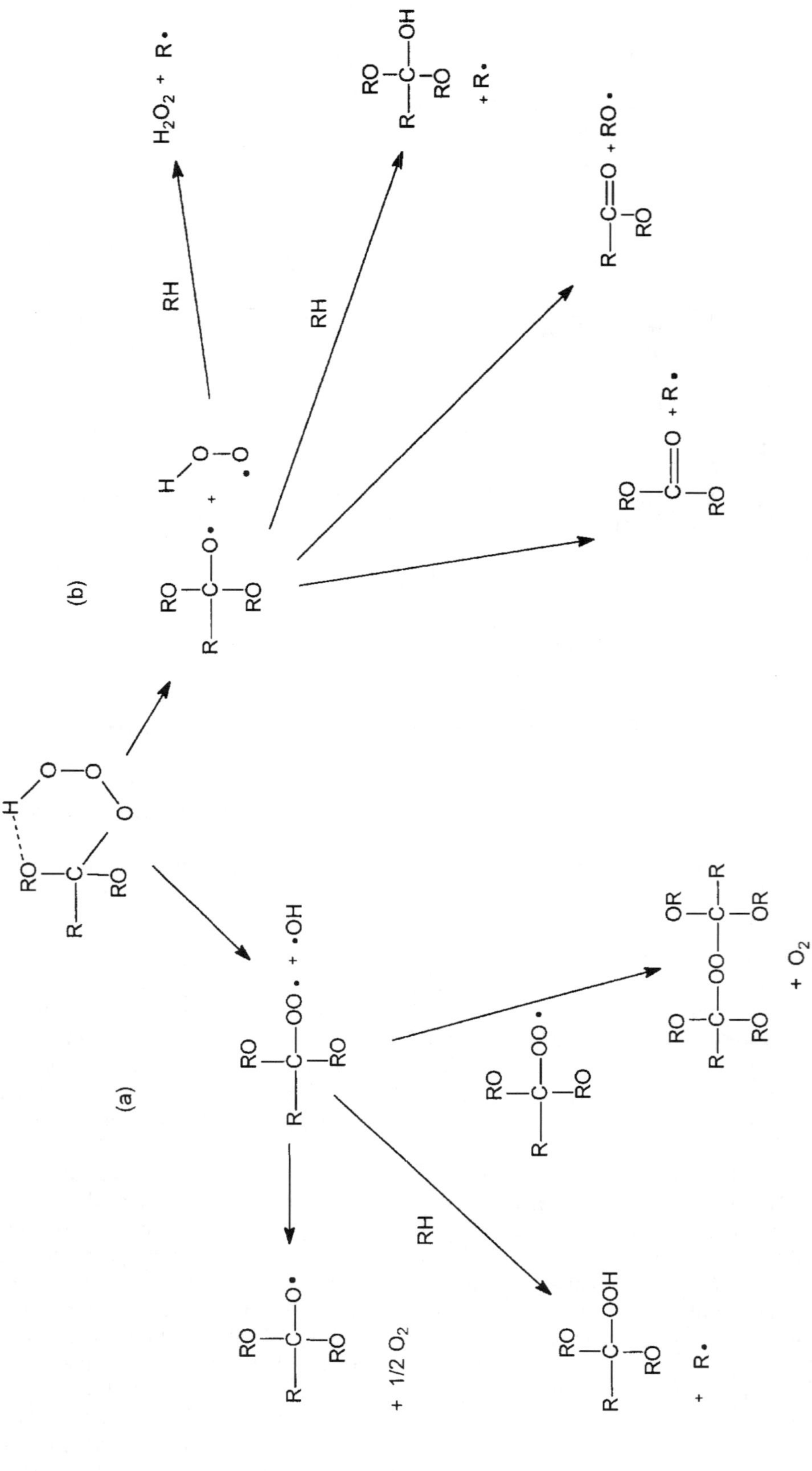

Figure 3.3. Radical mechanisms of hydrotrioxide degradation.

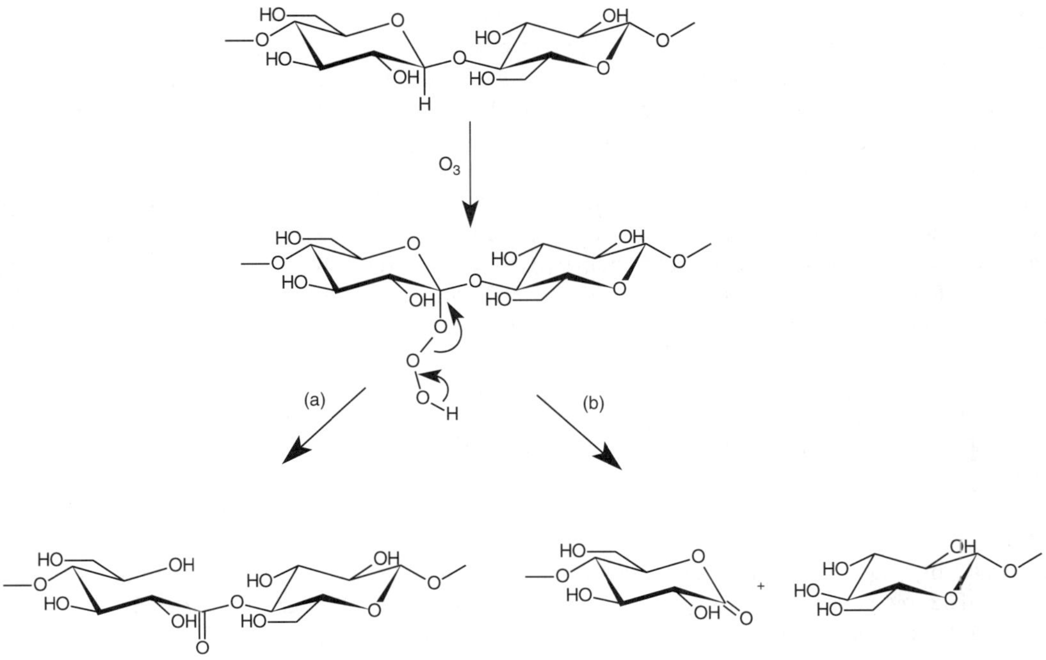

Figure 3.4. Ozone cleavage of glycosides.

departing alkoxy group in the tetrahedral intermediate, a condition which is fulfilled for the cleavage of the ring oxygen, but is not fulfilled for the glycosidic oxygen. Although these reactions by themselves would not result in chain cleavage, the ester linkage would cause the cellulose chains to be susceptible to cleavage by alkaline and acid hydrolysis.

Ozonations of glycosides in aqueous systems have produced different compounds, which suggests there are other significant mechanisms of carbohydrate degradation when water is present. The major products from ozonation of aqueous solutions of β-methyl glycoside are gluconolactone and gluconic acid.[5] A decomposition pathway for the hydrotrioxide was proposed in which the glycosidic bond is cleaved, rather than the ring oxygen bond, to form gluconolactone, which subsequently hydrolyzes to form gluconic acid. These two β-glycoside reaction mechanisms are illustrated in Figure 3.4.

Oxidation of C6 to form glucuronic acid during ozonation of glucose and cellobiose has also been reported.[16] Other products from cellulose ozonation include formaldehyde[16,17] and glucose oligomers;[16] arabinose is a reaction product as well. Its formation has been attributed to a radical Ruff-type degradation in which the C2 hydrogen of gluconic acid is abstracted, a *gem* diol is formed at C2, and subsequent decarboxylation of C1 occurs. When anomeric mixtures of mannose and xylose were ozonated, similar products were found.[17]

Researchers have reported that ozone reacts with α-anomers in aqueous solutions, albeit considerably slower than with β-anomers. Katai and Schuerch[15] proposed a mechanism similar to acid catalyzed hydrolysis of glycosidic bonds to explain the glucose that resulted as a major product from the ozonation of an α-methyl glucoside. They proposed that ozone directly attacks the glycosidic oxygen to form a tetroxide intermediate, which then decomposes to glucose anomers through a carbocation intermediate and unspecified degradation products of the tetroxide. The authors felt that the anomeric methyl group was an unlikely site of attack, although other investigators[7] have shown that methyl ethers react slowly with ozone. Pan et al.[5] ozonated β-methyl glucoside and also found glucose, but only as a minor product, and proposed the same mechanism.

It is interesting that the methyl ester of gluconic acid was not identified in any of the aqueous reactions of ozone with methyl glycosides, in contrast to the nonaqueous reaction results of Deslongchamps et al.[4] Possible reasons include a lack of identification of this compound in the reaction mixtures, acid hydrolysis of the ester with a subsequent lactonization of the gluconic acid, or a solvent-mediated shift of the optimal reaction pathway.

Measurements of functional group concentrations in ozonized cellulose from pulp have shown that ozone only moderately increases carboxylic acid groups[18–21] while carbonyl concentrations tend to fluctuate, but generally increase.[18–20] Most of the carbonyl compounds are ketones. While these general trends are similar in most studies, it should be noted that the relative abundances of carbonyl and carboxyl groups vary with different studies. For example, Chirat and Lachenal[18] report carboxyl concentrations ranging from 2.9–3.8 meq/100 g and carbonyl concentrations ranging from 3.2 to 3.8 meq/100 g after ozonating fully bleached pulp (0–0.85% O_3 charge). By contrast, Chandra and Gratzl[19] found higher values, 4.4–4.6 meq carboxyl and 9–23 meq carbonyl/100 g of high-cellulose pulp. Godsay and Pearce[20] reported values of 0.9 meq COOH/100 g and 2.0 to 4.9 meq carbonyl/100 g of high-cellulose pulp (0–10% O_3 charge). These differences illustrate the difficulties in ascertaining the dominant reaction pathways of carbohydrate degradation. The equilibrium between chain esters, lactones, and acids, coupled with the analytical difficulty in differentiating the carbonyl functional groups in complex mixtures, make elucidation of the direct mechanism of ozone degradation of cellulose challenging.

The introduction of carbonyl groups seriously affects the stability of carbohydrates in alkaline conditions, as well as color reversion. Carbonyl groups at C2 and C3 have been implicated in the susceptibility of ozonated pulp to degradation by β-elimination reactions during alkaline extraction stages. The low apparent viscosity of ozonated pulps, relative to strength properties, has also been attributed to carbonyl groups produced by ozone; depolymerization reactions occur during the alkaline viscosity assay. By contrast, oxidation of C1 and C6 to acid groups imparts a resistance to hydrolysis for carbohydrates and inhibits further degradation. Reduction of carbonyl groups to alcoholic groups also retards carbohydrate degradation by ozone.

Reactions with Lignin

The strong electrophilic properties of ozone allow it to react efficiently and rapidly with sites of high electron densities in lignin, such as double bonds in the side chain and the aromatic rings. In addition, ozone may also attack the C-H bonds in aryl ether–substituted side chains. Reactions of ozone with lignin are complex because of the diversity of potential sites of attack and because products of the initial attack are quite susceptible to further attack. Furthermore, other reaction by-products, such as hydrogen peroxide, are also reactive.

Reactions of ozone with alkenes proceed by way of the Criegee mechanism (Figure 3.5). The reaction begins with a 1,3 dipolar concerted addition reaction across the double bond to form a highly unstable 1,2,3-trioxolane, or primary ozonide (Figure 3.5a). Although the primary ozonide may directly decompose into a carbonyl and hydroperoxide by attack of participating solvents (Figure 3.5b), it is more likely to cleave at the C-C bond to produce a carbonyl and a carbonyl oxide zwitterion or radical pair (Figure 3.5c). The directionality of the cleavage depends on the carbocation or radical stabilization capabilities of the substituents on each of the alkenyl carbons; thus, a higher degree of alkyl substitution on a carbon allows it to become the carbonyl oxide. Once the two fragments are formed, several pathways are possible.

For aldehydic carbonyl compounds in inert solvents, the two fragments may recombine to form 1,2,4-trioxolane, or a secondary ozonide (Figure 3.5d), which can be readily isolated in many cases. When the secondary ozonide is exposed to water, it can hydrolyze to two carbonyl fragments and hydrogen peroxide, probably through a hydrotrioxide intermediate (Figure 3.5e). Alternatively, it may decompose to a carboxylic acid and a carbonyl compound (Figure 3.5f). In addition to recombining with the carbonyl compound, the carbonyl oxide can also undergo dimerization reactions to form bisperoxides.

Decomposition of the primary ozonide to the carbonyl and carbonyl oxide fragments is the reaction pathway that is most likely to predominate in the aqueous conditions normally used in ozone bleaching. The carbonyl oxide is trapped by water to form an α-hydroxy hydroperoxide (Figure 3.5g). The fate of the hydroperoxide depends on other substituents and solvents. The hydroxy group,

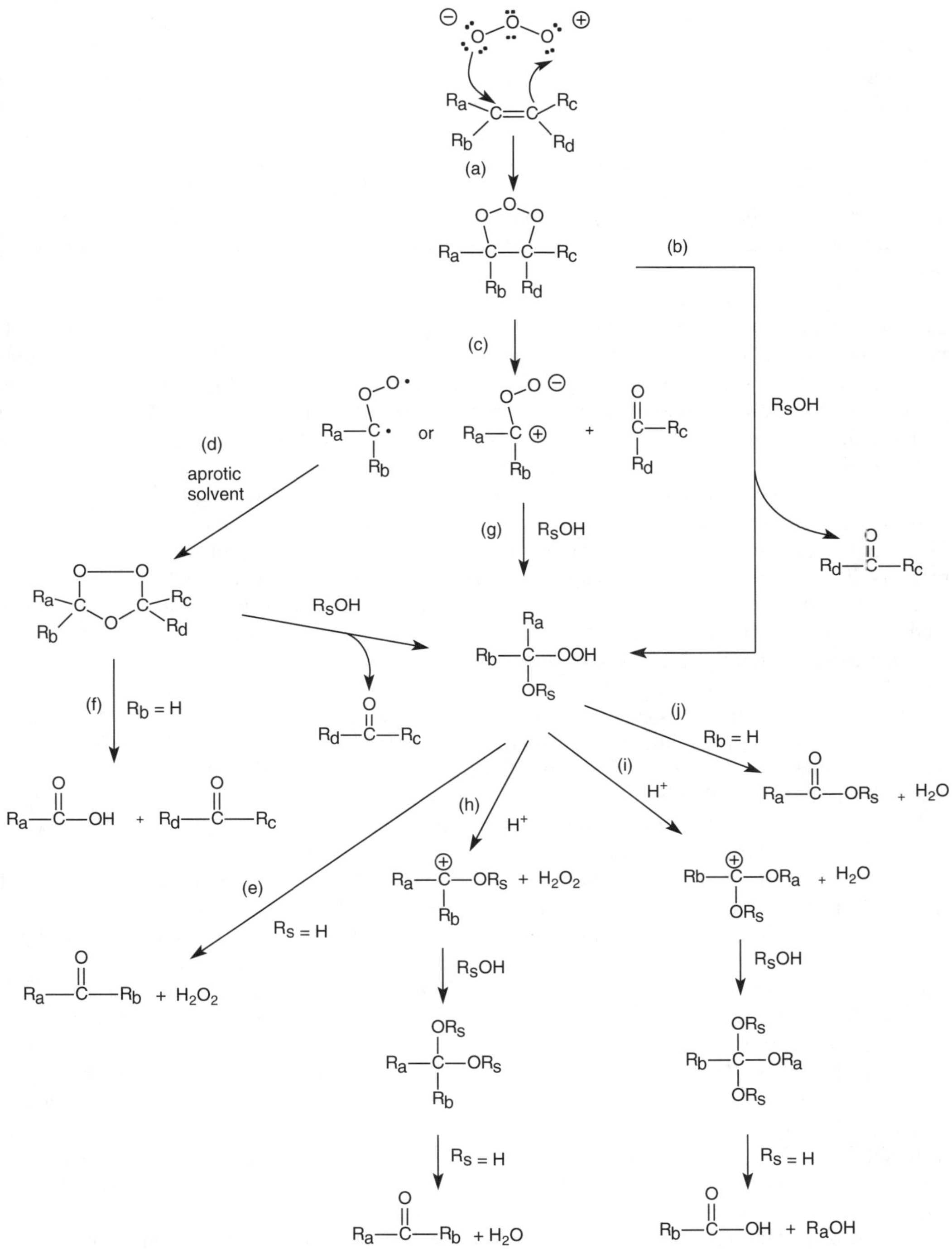

Figure 3.5. Common reactions of ozone with double bonds (R_s refers to solvent).

originally from the water, can become a carbonyl group, releasing hydrogen peroxide (Figure 3.5e). Under acidic conditions, the peroxy oxygen can become protonated, forming hydrogen peroxide and a carbocation. The carbocation then reacts with water to form a *gem* diol that is likely to become further oxidized to a carbonyl compound (Figure 3.5h). Also in acid, the terminal oxygen in the peroxyl group can become protonated. If there is a group on the carbon that migrates easily, a rearrangement can occur to form a carbonyl and an alcohol (Figure 3.5i). Finally, if one of the groups on the hydroxy hydroperoxide carbon is hydrogen, the peroxide can decompose to water and a carboxylic acid (Figure 3.5j).

Aromatic compounds also react by the Criegee mechanism, although at a slower rate than alkenes. There are three sites of attack on each aromatic ring. Preferential sites of attack and higher rates of reaction will occur at those carbons with the highest electron density. Therefore, aromatic compounds with the most electron-releasing groups react most rapidly with ozone, and the ozonide forms at the carbons bearing the electron-releasing groups. Once the ring cleavage product is formed — a muconic acid — the two aliphatic double bonds also become susceptible to attack.

At least three other reaction pathways have been proposed for aromatic compounds, particularly phenols. Phenols become hydroxylated and quinones have been reported as ozonation products. Standard electrophilic substitution mechanisms have been invoked to explain these products.[22,23] The electron-deficient ozone oxygen attacks the aromatic pi electrons; an arenium ion forms which, for the ozone electrophile, is a trioxide carbocation. The trioxide anion loses oxygen and the ring either rearomatizes so that the net result is ring hydroxylation, or quinones form if good leaving groups are present. Epoxidation can also occur via electrophilic substitution at double bonds.[24,25] Hydrogen abstraction from phenolic hydrogens has also been proposed as an initiating reaction for the formation of quinones, and for polymerization reactions in nonaqueous solvents.[26]

Nearly all of these reactions have been reported for the ozonolysis of lignin model compounds and lignin. Common pathways of ozonolysis proposed for lignin type compounds are summarized in Figure 3.6, which shows the complete oxidation of all unsaturated sites in an aromatic model. Ozonides, isolated from veratryl lignin model compounds ozonated in inert solvents, decompose to small aldehydes, acids, and methyl esters upon warming.[27] Lignin model dimers give similar small, oxygenated products as well as monomeric rings. Quinones and other "active oxygen" compounds have been reported as ozonation products of lignin and lignin model compounds.[27,28] The mechanisms discussed previously were used to explain the reaction products.

Most of the differences in reactivity between functional groups in lignin can be explained in terms of an initial reaction that is strongly electrophilic; the most electron-donating substituents enhance reactivity, while electron-withdrawing substituents inhibit it. These combined reactivity phenomena lead to interesting competitive pathways of ozone consumption. Eriksson and Gierer[24] report that the order of ozone reactivity of lignin models is stilbenes > styrenes > phenols > muconic acids > nonphenols > aldehydes. Therefore, ozone will cleave the double bonds in stilbenes and styrenes before attacking any aromatic rings. Such reactions are likely to be important for delignification and for decolorization[29] of kraft residual lignin, which is enriched in vinyl ethers compared to wood lignin. Once the side chain double bonds are cleaved, the aromatic rings are attacked. Phenolic lignin models react with ozone before nonphenolic lignin models; this is probably related to the electronic effects of the hydroxyl and methoxyl groups (e.g., Hammett constants σ_p for $-OH$ and $-OCH_3$ are -0.38 and -0.28, respectively). Other substituents, of course, will also have an impact. Kojima et al.[30] attributed the reactivity of dimeric model compounds toward ozone to the total number of substituents on the two rings. Thus, a guaiacyl biphenyl (2,2′dihydroxy-3,3′-dimethoxy-5,5′-dialkyl diphenyl) with a total of 8 activating ring substituents reacted faster than phenylcoumarin with a total of 7 groups (3 alkyl, 2 methoxyl, 1 alkoxyl, 1 hydroxyl), which in turn reacted faster than a β-aryl ether with a total of 5 alkoxyl or alkyl ring substituents. Furthermore, when dimeric compounds are ozonated, the ring with the most activation will react first; many times it will be the only ring that reacts — but not always.[31]

As with any electrophilic reactions, substituents will help direct lignin reaction pathways. For example, model compound work has shown that aromatic ring cleavage will occur primarily (but not exclusively[31]) between oxygen-bearing C3 and C4 for softwood lignin models that have an electron-donating side chain. However, if the side chain is electron-withdrawing, rates of ozonation decrease and cleavage increases at C2-C3 or C4-C5, producing more nonesterified small acids.[27]

The products of the initial ring attack, muconic acid derivatives, have opposing reactivity features: they are olefins, which promotes their reactivity over the parent aromatic, but they also have electron-withdrawing carbonyl groups as a result of the oxidation. These characteristics make the reactivity of muconic acid structures to ozone generally intermediate to phenolic and methoxylic aromatic compounds. Therefore, once ring cleavage occurs in dimeric compounds, if the remaining ring is phe-

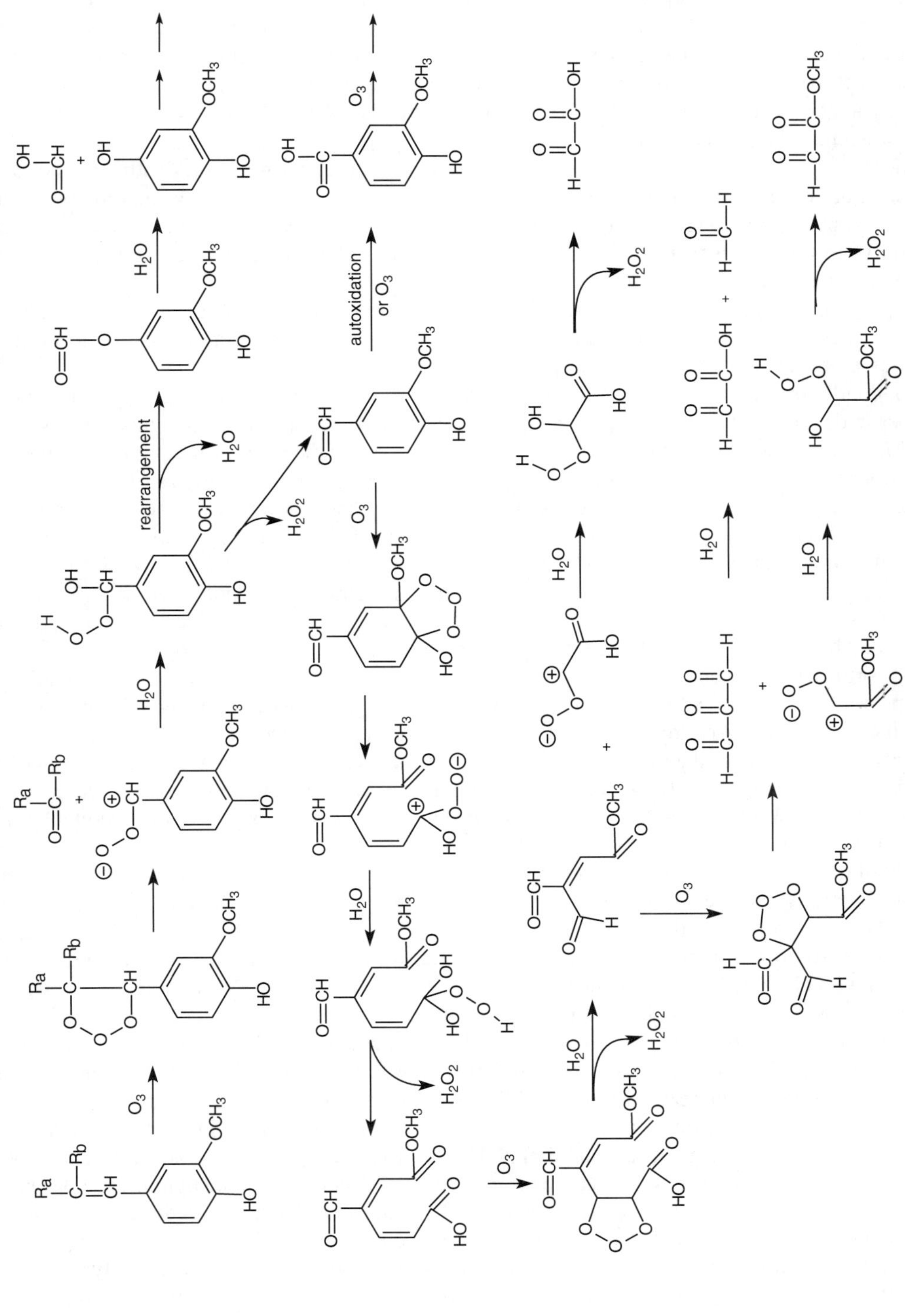

Figure 3.6. Reactions of ozone with aromatic lignin models.

nolic, the second ring will be attacked. Conversely, if the second ring is methoxylic, the side chain will continue to be attacked by ozone and the second ring will remain intact.[24] Ozonolysis of the two double bonds in muconic acid structures produce the two-carbon oxygenated fragments typically found in lignin and lignin model ozonations, including glyoxal, glyoxilic acid, glyoxilic acid methyl ester, oxalic acid, monomethyl oxalate, and dimethyl oxalate. The primary pathways to these products from lignin phenolic monomers is given in Figure 3.6. For the phenylpropane monomer shown, four moles of hydrogen peroxide can be generated, along with two moles of glyoxilic acid, one mole of glyoxylic acid methyl ester, one mole of formaldehyde, and one side chain carbonyl functional group, when the unsaturated sites are fully oxidized.

Under limiting ozone conditions, muconic acid derivatives may form lactones, rather than cleaving to smaller products. If there is an α-hydroxyl group on the side chain, the oxygen will bond with the C4-carbonyl carbon to form a δ-lactone for both phenolic and nonphenolic compounds.[30,31] If an α-hydroxyl group is not present on the side chain, a γ-lactone will form between C1 and the carboxylate oxygen of an oxidized phenolic at C4.[24]

The predominant pathways of lignin model ozonolysis are those that have been discussed. However, the potency of ozone and the complexity of both lignin and its reaction by-products allow numerous other reactions to occur. The prevalence of one pathway over another depends on the reaction conditions and the specific features of the lignin. For example, while it appears that most hydroperoxide intermediates decompose by losing hydrogen peroxide, rearrangements may also occur.[27,28] In another pathway, hydroperoxides decompose to produce the carboxylic acids that are prevalent in ozone mixtures. However, it is difficult to differentiate the latter pathway from a simple oxidation of aldehydes to acids. Substitution reactions may produce epoxides from muconic acid derivatives,[24] hydroxylate phenols,[31] and produce quinones.[28] The formation of high-molecular-weight material from ozonation of lignin model compounds indicates that ozone can initiate polymerization; Diels-Alder type reactions, radical coupling, and esterification have all been proposed as potential routes.[31,32]

Balousek et al.[31] ozonated a β-aryl ether model compound and found products that supported a reaction in which the ozone inserts into the β-carbon-hydrogen bond, forming a hydrotrioxide intermediate, similar to the reactions with glycosides discussed previously. They proposed a homolytic decomposition pathway to a hydroperoxy radical and a ketone, analogous to a mechanism reported by Plesnicar et al.[14] for α-methyl benzyl ethers and alcohols. It is noteworthy that the insertion reaction was competitive with the rapid reactions with electron-rich alkenes and aromatics; it might be predicted to have a considerably slower reaction rate. However, if the lignin insertion reaction occurs under bleaching conditions, it is likely to effectively depolymerize the portions of the residual lignin macromolecule that still retain β-aryl ethers after cooking.

Many of the reactions that have been proposed as a result of model compound studies have been borne out by changes in the concentrations of functional groups when lignin and pulp are exposed to ozone. There is a pronounced increase in carboxylic acids in lignin in residual lignin of ozonated kraft pulp,[18,21] dissolved lignin fragments from ozonated kraft pulp,[28] as well as ozonation products of isolated wood lignins,[28,33] and isolated kraft lignin.[27,34] Carbonyl groups also increase in number with ozonations,[18,28,29,34] although the extent is variable and limited.

As expected, ozone depletes phenolic groups in kraft residual lignin,[18,21,34] similar to oxygen and chlorine dioxide.[35] The aromatic content of ozonated wood lignins is less than that of unozonated lignins.[28,33,36] However, the reactivity of wood lignin appears to be different from kraft residual lignin. Chirat and Lachenal[18] found the aromatic carbon content of residual lignin from ozonated kraft pulp to be similar to the original residual lignin. The latter result suggests that most ring cleavages that occur during ozone bleaching kraft pulp ultimately cause dissolution of the lignin. This may be a function of the competitive nature of the muconic acid reactions versus the aromatic reactions. Such a conclusion is supported by the fact that molecular weight distributions of residual lignin of ozonated pulp are not substantially changed by moderate charges of ozone.[34,35] Furthermore, analyses of the dissolved fragments from ozonated pulp and groundwood show little intact aromatic compounds[28,37] and there appears to be more high-molecular-weight material in filtrates from ozonated pulp than from pulps treated with other bleaching reagents.[38,39] Together, the data suggest that the biggest impact of ozone on kraft lignin is the introduction of hydrophilic groups rather than substantial depolymerization.[35]

Methoxyl and methyl ester contents are somewhat variable for ozonated pulp, and the values depend on the substrate, conditions of ozonation, and the analytical methods used. Demethoxylation may arise from ester hydrolysis of cleaved rings, or from a radical demethoxylation pathway, both of which are likely to be highly dependent on the pH, metals content, and other variables of the reaction medium.

Freshly ozonated lignin appears to contain reactive intermediates that degrade over time. It has been suggested that this "active oxygen" consists of hydroperoxides, peroxides, ozonides, or quinones.[28,40] There is also evidence of free radicals in ozonated lignin mixtures,[40] and they may cause polymerization or other radical-initiated reactions.

Selectivity

Selectivity refers to the preferential attack of a reagent on lignin versus carbohydrates. The rates of carbohydrate reactions determine selectivity under various ozone bleaching conditions. Carbohydrates may react with ozone directly, or react with lignin or carbohydrate by-products, particularly radicals. Radicals, in turn, may be initiated from the decomposition of ozone, by-product hydrogen peroxide, hydroperoxidic lignin by-products, or intermediate hydrotrioxides. The contribution of each of these pathways of carbohydrate deterioration is currently unclear due to contradictory conclusions from numerous studies (e.g., References 41–44). When these processes are better understood, it may be possible to develop strategies to effectively prevent or minimize the most important pathways of carbohydrate deterioration. Furthermore, a clearer understanding of the maximum possible selectivity of ozone bleaching will be gained.

Direct Reactions

Selectivity depends, in part, on the reactivity of ozone directly with various functional groups. Second-order rate constants for direct reactions of ozone with several compounds that have functional groups relevant to pulp are given in Table 3.1. These rate constants and other kinetic parameters suggest that when lignin is present, ozone should be rapidly consumed by the lignin, and cellulose should be fairly unreacted.

The data in Table 3.1 indicate the double bond of styrene and the activated benzene rings are very reactive, followed by the aliphatic unsaturated acids. Only the very small acids react more slowly than the carbohydrates. The data suggest that carbohydrates are protected from direct reactions with ozone by the presence of lignin and lignin degradation products. There is considerable data supporting this contention. Even benzaldehyde, a deactivated lignin structure, completely degrades before β-methyl glucopyranoside begins decomposing in ozonated mixtures.[24] Guaiacol and syringol are nearly completely consumed before the β-glucoside is degraded when mixtures of the compounds are ozonated.[5]

The model reactivities are reflected in pulp ozonation. When initial kappa numbers of pulp are high, viscosity is retained better than when ozonation occurs in later stages of a bleaching sequence and the initial kappa numbers are low.[21,45,46] Once a critical level of lignin has been removed from the pulp, ozone will begin to react to a greater extent with the C-H bonds of carbohydrates. The protecting effect of lignin appears to exist only to a point; at lower kappa numbers, lignin appears to promote carbohydrate degradation.[46,47] It has been suggested that ozone may abstract phenolic hydrogen atoms, which initiates a radical pathway that forms the hydroxyl radical.[48] This may explain the superior selectivity of oxygen-bleached pulps over kraft pulps, since oxygen preferentially depletes phenolic structures in residual lignin.[44]

Radical Reactions

There are numerous species of radicals that may be associated with the ozonation of organic compounds in aqueous conditions. These radicals are involved in complex initiation, propagation, and termination reactions that depend upon several reaction parameters. The pH, time, and inorganic

Table 3.1. Second-Order Rate Constants for Direct Reactions with Ozone.[a]

Compound	Structure	pH	k $M^{-1} \cdot s^{-1}$	Scavengers
Styrene	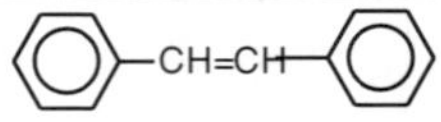	3 in CCl_4 [b]	3×10^5 1×10^5	t-Butanol
Resorcinol	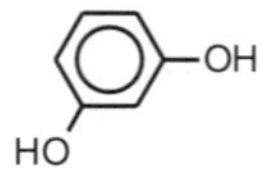	2	$> 3 \times 10^5$	Propanol
Phenol	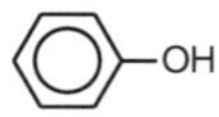	1.7–2	1.3×10^3	Propanol
Maleic Acid	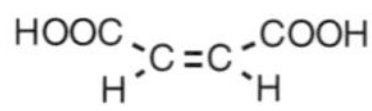	2 6	1×10^3 5×10^3	t-Butanol
Anisole	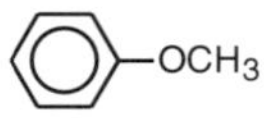		2.9×10^2	Propanol
Benzaldehyde	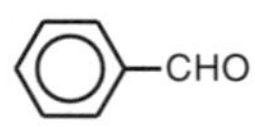	1.7	2.5	Propanol
Acetaldehyde	$\mathrm{CH_3\overset{O}{\overset{\|}{C}}H}$	2	1.5	—
Glucose	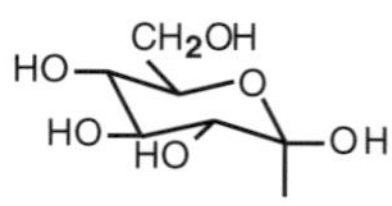	2	0.5	—
Methyl β-Glucoside[c]	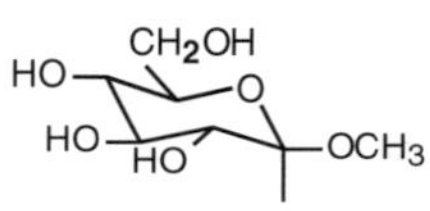	3	0.5	—
Glyoxilic Acid	$\mathrm{HOOC\overset{O}{\overset{\|}{C}}H}$	1.5–5	0.2	t-Butanol
Formaldehyde	$\mathrm{H\overset{O}{\overset{\|}{C}}H}$	0.2	0.1	—

[a] Values are taken from References 91 and 92 unless otherwise indicated.

[b] Reference 93.

[c] Reference 5.

and organic substrates, as well as other reaction conditions, can have a strong impact on the formation of radicals and the mechanisms by which they are destroyed.

The hydroxyl radical is clearly an important radical species that is formed directly from decomposition of ozone or reaction intermediates, or from chain reactions initiated by other radicals. The selectivity factor of the hydroxyl radical for lignin over carbohydrates is substantially lower than for the direct ozone reactions.[49] For this reason, the hydroxyl radical is often invoked as the species that is responsible for carbohydrate deterioration. Other species that may be formed include hydroperoxyl (HOO•), organic peroxyl (ROO•), organic oxyl (RO•), and ozonide (HOOO•) radicals. While these latter compounds are less reactive than the hydroxyl radical,[50] they enter into complex chain reactions that may ultimately produce hydroxyl radicals.

During ozone bleaching, it is possible that radicals may form directly from ozone, or they may arise from the homolytic decomposition of products from direct ozone reactions. The main pathways include metal-catalyzed decomposition of ozone, homolytic decomposition of lignin peroxides, metal-catalyzed decomposition of intermediate hydrogen peroxide, and hydrogen abstraction by ozone followed by decomposition of the resulting hydrotrioxide. Another potential source of radical initiation is the decomposition of the hydrotrioxide intermediates of carbohydrates and lignin. These pathways are briefly reviewed.

Radicals from Ozone Reactions

The decomposition of ozone to radicals is initiated by hydroxide ions and metals. Other pathways that produce radicals directly from ozone are reactions with other radicals and abstraction reactions. Under the acidic conditions used in pulp bleaching, the ozone reactions that are most likely to directly initiate radical formation are the metal-catalyzed decompositions (Equation 3.1) and hydrogen abstraction reactions (Equations 3.2–3.4). Radical chain reactions (Equations 3.5–3.7) will consume more ozone and propagate more radicals. The radicals will exist predominantly in their protonated form under typical bleaching conditions (pKa of HOO• = 4.9); however, low levels of the superoxide radical anion are likely to be present even at low pHs and may be sufficient to propagate radical pathways.[48,51] Many of the reactions proceed through ozonide intermediates.

Protonated phenols probably react primarily by nonradical mechanisms because of the rapidity of those reactions; however, as shown in Equation 3.4, abstraction of an electron by ozone from phenolates may be a source of radicals.[48,51] This pathway is probably only significant when the pH of the bleaching conditions is above 3 and the proportion of ionized phenols is sufficient for the reaction to take place. Below pH 3, phenols with an α-carbonyl group may react by the abstraction pathway because of their reduced pKa.[51]

It has been suggested that a key radical intermediate is the superoxide radical anion that propagates hydroxyl radical formation by its reactions with ozone (Equation 3.7), as well as reactions with hydrogen peroxide.[51] The superoxide arises from the decomposition of organic peroxyl radicals. It has been proposed that by controlling the level of the superoxide radical anion, the radical chain reactions may be terminated, and selectivity improved.[51]

$$O_3 + Mn^+ \; H^+ \rightarrow M^{(n+1)+} + HO\bullet + O_2 \tag{3.1}$$

$$O_3 + RH \rightarrow R\bullet + HO\bullet + O_2 \tag{3.2}$$

$$O_3 + ROOH \rightarrow R\bullet + HO\bullet + O_2 \tag{3.3}$$

$$O_3 + PhO^- + H^+ \rightarrow PhO\bullet + HO\bullet + O_2 \tag{3.4}$$

$$O_3 + HO\bullet \rightarrow HOO\bullet + O_2 \tag{3.5}$$

$$O_3 + HOO\bullet \rightarrow 2\ O_2 + HO\bullet \tag{3.6}$$

$$O_3 + O_2\bullet^- + H+ \rightarrow 2\ O_2 + HO\bullet \tag{3.7}$$

Another source of radicals is the decomposition of peroxide and hydrotrioxide intermediates produced from direct ozone reactions. Hydrogen peroxide arises primarily from the decomposition of hydroxy hydroperoxide intermediates of lignin (Figures 3.5e and 3.6). A comparatively minor source is the homolytic decomposition of hydrotrioxides (Figure 3.3b). The main radical initiation mechanism from hydrogen peroxide is the classic, metal-catalyzed Fenton reaction (Equation 3.8). High concentrations of H_2O_2 during high-consistency bleaching may cause this pathway to be more important than at low-consistency conditions.[52] Hydrogen peroxide may also propagate radicals by reacting with superoxide radical anions to form the highly reactive hydroxyl radical (Equation 3.9).[51] Although ozone will react with H_2O_2 at a high pH, the reaction is suppressed at low pH,[53] and probably is not a significant pathway for radical production during ozone bleaching.

$$H_2O_2 + M^{(n+)} + H^+ \rightarrow HO\bullet + H_2O + M^{(n+1)} \tag{3.8}$$

$$H_2O_2 + O_2\bullet^- \rightarrow HO\bullet + HO^- + O_2 \tag{3.9}$$

Organic peroxides are most likely to arise from reactions of ozone with lignin (Figure 3.5g). They may decompose to radicals by direct reactions with ozone (Equation 3.3) or by metal-catalyzed reactions (Equations 3.10–3.11). Propagation reactions in which the peroxides are involved include the abstraction of hydrogen from the peroxides by alkyl radicals (Equations 3.12–3.13). Peroxyl and alkoxyl radicals participate in additional chain carrier reactions (Equations 3.14–3.16). Equation 3.14 may be particularly important because it represents another pathway to the superoxide radical anion, a proposed key radical chain carrier.[48,51] Note that under oxygenated conditions an additional pathway to peroxyl radicals may be coupling of oxygen to organic radicals (Equation 3.17).

$$ROOH + M^{(n+)} \rightarrow ROO\bullet + H^+ + M^{(n+1)} \tag{3.10}$$

$$ROOH + M^{(n+)} + H^+ \rightarrow RO\bullet + H_2O + M^{(n+1)} \tag{3.11}$$

$$ROOH + R'\bullet \rightarrow ROR' + HO\bullet \tag{3.12}$$

$$ROOH + R'\bullet \rightarrow R'OH + RO\bullet \tag{3.13}$$

$$ROO\bullet \rightarrow R_{ox} + O_2\bullet^- \tag{3.14}$$

$$ROO\bullet + R'H \rightarrow ROOH + R'\bullet \tag{3.15}$$

$$RO\bullet + R'H \rightarrow ROH + R'\bullet \tag{3.16}$$

$$R'\bullet + O_2 \rightarrow ROO\bullet \tag{3.17}$$

Hydrotrioxides may arise from ozone insertion in C-H bonds in carbohydrates (Figure 3.4) and in the side chains of lignin. Although these reactions are expected to be slow compared to other ozone reactions, they may play a part in the radical initiation under some conditions.[54] Numerous homolytic decompositions of hydrotrioxides are outlined in Figure 3.3. The types of radicals formed are comparable to those already discussed (R•, RO•, ROO•, HO•) and enter into the same propagation reactions.

Hydroxyl Radical Reactions

Reactions of lignin with the hydroxyl radical are only marginally favored over carbohydrate reactions ($k_{lignin}/k_{carbohydrates} = 5–6$).[49] Hydroxyl radicals react readily with lignin in pulps with high lignin content, although, at least one study indicates there is little delignification due to these reactions.[43]

Model compound studies with hydroxyl radicals and lignin show that phenols and etherified phenols react differently. Radical coupling predominates when phenols are exposed to hydroxyl radicals, while hydroxylation and demethoxylation are the major reactions with etherified phenols.[41]

Hydroxyl radicals may react with carbohydrates by abstracting hydrogen atoms from either C1 or C2.[52,55] The resulting radicals may couple with oxygen to form intermediate organic peroxyl radicals. Depolymerization might occur as a direct result of fragmentation of the intermediate radicals. Alternatively, if carbonyl groups are introduced at C2 as a result of hydrogen abstraction and oxygen coupling at C2, then depolymerization may take place through β-elimination when the carbohydrates are exposed to alkaline conditions.[55]

Mass Transfer Limitations

Because of its high reactivity, ozone mass transfer to lignin is critical during ozone bleaching. Diffusion of ozone across the immobile water layer adjacent to the cell wall can be a significant rate-determining step for low-consistency and poorly mixed medium-consistency bleaching.[56] Diffusion within the cell wall is rate-limiting in high-consistency and well-mixed low-consistency ozone delignification. Griffin et al.[57] have proposed a model of ozone delignification whereby an advancing front of lignin-ozone reactions takes place across the cell wall to the lumen. They propose that the highly reactive hydroxyl radicals produced by ozone-lignin reactions immediately degrade the adjacent carbohydrates, and therefore all cellulose decomposition takes place within the ozone-lignin front.

APPLICATIONS

When ozone is applied to pulp, several process conditions must be optimized to maximize the desired reactions with lignin and minimize the harmful reactions with carbohydrates. Optimizing efficiency and selectivity generally focuses on: 1) minimizing ozone decomposition, 2) promoting uniform exposure of lignin to ozone, 3) minimizing exposure of carbohydrates to radicals and high ozone concentrations, and 4) minimizing reactions of ozone with dissolved by-products. Important controlling parameters are consistency, charge, pH, concentration of ozone in the gas, time, temperature, carryover, metals removal, mixing, additives, and position in sequences.

In general, ozone decomposition is reduced at conditions of low pH, medium or high consistency, effective metals removal, low charges, and low temperatures. Uniform bleaching is promoted by effective mixing and fluffing (at high consistency). Carbohydrates are protected from high ozone concentrations by using charges less than 0.5% by weight and by good mixing. Viscosity is sometimes better preserved at low and well-mixed medium consistency and by shorter reaction times. Exposure of cellulose to radicals can be reduced by additives that act as radical scavengers as well as by reducing ozone decomposition.

Bleaching efficiency can be reduced by carryover; for example, a COD load of 10 kg COD/ADt was found to double the ozone demand in medium-consistency pilot trials.[58] Lower kappa number pulps are generally considered suitable for ozone bleaching because of the relatively high cost of ozone. Oxygen delignification is often used prior to a Z stage because of the low expense of oxygen and because the residual lignin remaining from oxygen appears to react readily with ozone.

Excellent and extensive reviews of industrial applications of ozone bleaching have recently been published.[59–61] Important process variables in ozone bleaching and their impact on selectivity and efficiency are briefly reviewed below.

Consistency

Early work on consistency showed high consistency produced the strongest pulps. However, as mixing technology has improved, results at low and medium consistency have improved. All three conditions have been used and each has its disadvantages. Low consistency disadvantages include a large water demand, large effluent volume, and mixing limitations. In medium consistency, the gas-to-pulp ratio must be small enough to be adequately mixed, yet the ozone concentration must be high enough to delignify. Either the O_2/O_3 stream must be pressurized or a high concentration of ozone in the O_2/O_3 flow must be used; each option has its drawbacks.

Lindholm[62,63] found that low consistency is more selective than high consistency; this was attributed to better homogeneity at low consistency. Carbohydrates may degrade more during high-consistency bleaching because of the heterogeneity of the system, which causes localized areas of high ozone concentration and low lignin concentration. During low-consistency bleaching, the reaction medium is more homogeneous, the ozone can be more readily dispersed, and reaction products may contribute to carbohydrate-protecting effects.[64] In medium consistency pilot studies, selectivity was between that of high- and low-consistency bleaching.[58] However, lab results show best selectivity at medium consistency with high-intensity mixing.[65,66]

Delignification efficiency (the amount of ozone required per unit kappa number drop) has been found to be worst at low consistency in lab studies,[63,66–68] although one study reported delignification efficiency to be unaffected by consistency.[65] More ozone is consumed at lower consistencies.[63,66] Since the consumed ozone does not result in more delignification, more decomposition or secondary reactions must occur at the lower consistencies. The combination of improved delignification at high consistency and improved viscosity retention at low consistency has driven the interest in medium-consistency ozonation.

Charge and pH

Increasing the charge of ozone from 1–2% by weight decreases the kappa number and increases the brightness at the expense of viscosity. Without a cellulose protector, a 1% charge of ozone has been considered to be the maximum for optimal cellulose quality.[67,69] In practice, ozone charges generally do not exceed 0.7%, even for softwoods, which require higher charges.

A pH between 2 and 3 is optimal for pulp ozonation.[58,63,67] Delignification is improved at low pH and the strong acidity does not contribute to substantial losses of viscosity. Viscosity loss is minimal up to pH 4. Beyond pH 4, viscosity loss occurs and brightness is reduced.[58] Low pH may be optimal because of three factors. Decomposition reactions that produce radicals are inhibited; metals that promote ozone decomposition may be removed in acid washes; and primary ozone-lignin reactions may be favored.[63] The dependency of delignification on pH has been found to be more complicated when pulps have a higher residual lignin content.[46] Furthermore, recent work suggests that high pH can be tolerated during ozone bleaching under appropriate conditions.[43]

Concentration

The concentration of ozone produced in O_2/O_3 streams is important because it affects pulp quality and because it affects the optimization of various other process variables, particularly reaction time. In general, concentrations below 10% have been used due to safety considerations and limitations on ozone generation. More recently, higher ozone concentrations up to 15% have been effectively employed (e.g., Reference 70).

The dependency of pulp quality on concentrations of ozone varies depending on whether very low concentrations are used (<3%) or higher concentrations are used (10%). When increasing the concentration up to 2–3% by weight, lower concentrations improved viscosity using high-consistency conditions.[68] However, a 10% concentration was found to be more selective than a 2.2% concentration, also using high consistency; the higher ozone concentrations resulted in greater delignification.[71] Medium-consistency ozonations are also more selective at higher ozone concentrations.[58] Other researchers have found little effect of ozone concentration on kappa number or viscosity,[59,60,72] but higher concentrations reduced reaction times for delignification.[59,60]

Time

Reaction time is particularly important at low and medium consistencies. When constant ozone charges were applied to softwood pulps at different reaction times, longer reaction times resulted in higher ozone consumption and improved delignification for low- and medium-consistency ozonations. With long enough reaction times, delignification at low and medium consistency nearly matched that achieved under high-consistency conditions.[65] At high consistency, longer reaction times suppress carbohydrate degradation but do not affect the drop in kappa number.[68]

Hardwood pulps appear to have a different time dependency than softwoods. Laboratory experiments using medium consistency showed that hardwood pulp ozone consumption increases with reaction times, but brightness and delignification actually decrease.[65] Pilot studies using medium-consistency ozone bleaching of hardwood dissolving pulp also found that longer reaction times consume more ozone, with no increase in delignification.[58]

Selectivity is best in the beginning of high-consistency ozone bleaching and drops off later.[21,71] Increases in the concentrations of radicals or radical precursors, coupled with a decrease in the amount of available lignin to consume the radicals, are possible reasons for this observation.

It is important to note that the dependence of bleaching behavior on reaction time may be a significant variable in comparing some laboratory studies to mill practices. While laboratory studies generally employ reasonably long reaction times, technical bleaching may be conducted in seconds (medium consistency) or minutes (high consistency).

Temperature

Lower temperatures promote selectivity probably by reducing carbohydrate reactions.[73] High temperatures promote ozone decomposition, which creates the cellulose-degrading radicals, although below 25°C, ozone decomposition is minimal.[74] Delignification is fairly constant up to about 55°C for medium-consistency hardwood ozonation; kappa numbers rise slightly at higher temperatures.[58] Liebergott et al.[59,60] reported slight increases in kappa and slight reductions in viscosity when temperatures were raised from about 20 to 80°C. These effects have been attributed to increased ozone decomposition with temperature. For low-consistency operations, temperature will also affect ozone solubility, which may in turn affect bleaching efficiency.[75]

Carryover/Filtrates

The effects on ozone efficiency of carryover and of dissolved organic material in filtrates are important because of the implications for brownstock washing and filtrate recycle, respectively. Although ozone reacts with organic material in ozonation filtrates, Lindholm and Vilpponnen[76] found that it will react with pulp preferentially. Consequently, ozonating pulp using recycled ozonation filtrates at high or low consistency does not affect delignification efficiency. In fact, viscosity protection at low consistency has been attributed partly to the presence of degradation products.[64] By contrast, a COD carryover load of 10 kg C/ADt was found to double the ozone demand in medium-consistency pilot trials.[58]

Mixing

Doubling mixing times reduced kappa numbers and increased efficiency in the medium-consistency ozonation of hardwood pulps.[58] Increasing mixing intensity improved viscosity protection and delignification (when adequate reaction times were used) in medium-consistency bleaching.[65] There is higher ozone consumption at higher mixer speeds when ozonating at medium consistency. As the ZE stage (after ozone and alkaline extraction) brightness increases, viscosity and lignin levels decrease with increasing mixer intensity for softwoods and hardwoods.[65]

Additives/Solvents

Large numbers and types of compounds have been evaluated for their potential as carbohydrate protectors and have been compiled by Liebergott et al.[59,60,67] Nothing that is commercially feasible has been found to be highly effective and Liebergott notes that the problem of finding a cheap, effective commercial inhibitor of carbohydrate depolymerization during ozone delignification remains to be solved.[67]

Methanol has been found to protect cellulose; DMSO protected carbohydrates at high consistency without inhibiting delignification in one set of studies,[68] but another study found no effect of DMSO at high consistency or low consistency.[64] It is interesting that some chemicals that protect cellulose in oxygen bleaching don't work with ozone. This suggests that some carbohydrate damage from ozonation does not arise from radicals.

The acids used to reduce pH may affect ozone efficiency. Although acetic acid inhibits metal-catalyzed decomposition of ozone, some studies have shown that there is no difference between sulfuric acid and acetic acid (at low consistency) on viscosity, strength[68] or delignification efficiency.[64] More recent studies suggest that ozonating in 90% acetic acid under low-consistency conditions improves selectivity, particularly when ozonating down to low kappa numbers.[44,77] Oxalic acid also improves selectivity at low and high consistencies at $pH < 2$.[64]

Protective effects of lignin and lignin fragments have been noted[64,77–79] and attributed to radical scavenging by these materials.[46] On the other hand, cellulose reacts with ozone more readily in the presence of small amounts of lignin than under lignin-free conditions, which may result from cellulose attack by radical intermediates generated by lignin oxidation.[46,47]

Sequences

Placement of an ozone stage in several different points in a bleach sequence has been investigated over the years. Most bleaching sequences now use an oxygen stage prior to ozonation and place the ozone stage early in the sequence. Examples of sequences recently implemented in pilot plants include OZP (Graz, Austria), OZEP (Baienfurt, Germany), OZEP, OZED, O(pZE)P (Paprican), and (O)Z with any sequence (Espanola, Ontario). Two recent commercial installations have employed (EOP)ZP (Lenzing, Austria) and OZ(EO)D or OZEP (Franklin, Virginia). Sequences such as ZPZP, ZEP, chelation followed by OO(ZE)P(ZE)P, OO(ZE)PP, ZEDED, and others have also been studied.

Comparison of the different positions of ozone in a bleaching sequence is influenced by the position of alkaline extraction stages. The carbonyls formed from cellulose ozonation are susceptible to further reaction under alkaline conditions, resulting in depolymerization and viscosity reduction. Some studies show that ozone selectivity is lost in an alkaline post-treatment stage,[46,71] resulting in a recommendation that ozone be introduced last or that a strong reducing agent be added to the alkali stage. Other studies show inconsistent or negligible effects on viscosity at charges of 1% ozone or below and 1 to 1.5% NaOH charge in the extraction stage.[59,60,65] Lachenal et al.[78,79] have noted that although ZE produces pulp with lower viscosity than Z, the latter is still better than an O stage at low kappas. Alkaline extraction reduced brightness when comparing an OZD to an OZED sequence.[59,60]

Washing between Z and E stages resulted in slightly higher kappas than when there was no washing between Z and E.[67] Washing between a D stage and a Z stage was also found to be unnecessary.[78,79]

Generally, oxygen bleaching prior to ozone application is better than the reverse sequence.[59,60,71,78,79] Because ozone does not react well with bark and shives, better quality pulp results when oxygen precedes ozone. Also, it is economically more feasible to use oxygen to remove the bulk of lignin and to follow up with ozone. A peroxide stage between an O stage and a Z stage has been shown to improve pulp bleachability.[68]

Ozone has been used to replace chlorine, chlorine dioxide, or hypochlorite stages in traditional sequences. Using a chlorine dioxide stage immediately adjacent to an ozone stage to achieve a constant kappa number improves selectivity over ozone alone. The high selectivity of chlorine dioxide is utilized in this sequence. Furthermore, a DZ sequence results in higher-viscosity pulp than a ZD sequence.[78,79] Two possible reasons have been put forth. In a DZ sequence, the chlorine dioxide may modify the residual lignin so that lignin becomes more reactive to ozone. Alternatively, a DZ sequence may result in fewer cellulose-degrading radicals arising from lignin oxidation, thereby reducing carbohydrate degradation from that source. An O(DZ)ED sequence produced good viscosity pulp with a very low AOX in effluent.[78,79] Hypochlorite or ozone followed by peroxide resulted in similar viscosity losses in medium-consistency bleaching.[58] However, when one D stage is replaced by ozone in a DEDED sequence, the brightness ceiling is lowered.[46]

ENVIRONMENTAL BENEFITS

The environmental effects of wastewaters from oxygen-based bleaching, relative to those from chlorine-based bleaching, vary due to the influence of a number of factors. Some of these include differences in wood species, quality of pulp produced, the extent and specifics of water recycling, target species for toxicity assays, other bleaching stages used, the receiving water quality, and even the specific experimental techniques used (including sample preservation and storage). It is also important to recognize that the bleach plant is only one input into whole mill effluents; dilution and chemical transformations will take place. Furthermore, the ultimate environmental loading from a mill will depend on the effectiveness of biological treatment, which varies from site to site.

Despite the variability, it is clear that replacing a C or D stage with ozone drastically reduces the formation of AOX and specific chlorinated compounds. While effluent AOX ranges from 0.1 to 1.0 kg ptp for mills using chlorine dioxide without oxygen,[80] some values reported for AOX from ozone bleaching filtrates include nondetect for a (QZ)(Q)(PO) mill sequence,[81] 0.0002 kg/tonne for a Z(PO) laboratory sequence,[82] and 0.09 kg/ADMT (D filtrate) from a mill OZED sequence.[83] The actual benefit to the environment from low AOX discharges is unresolved. In general, at discharges less than 2 kg/ton, AOX does not correlate with toxicity or other environmental effects, while discharges above 2 kg AOX/ton do correlate with environmental impacts.[80] Chlorodioxins, chlorofurans, and chlorophenols, which can have a significant environmental impact, are not detected; however, traces of chloroform remain in some TCF cases.[83] A potentially significant environmental advantage to TCF bleaching is the expected decline in the releases of chlorine and chlorine dioxide, two important Toxics Release Inventory chemicals. It is unclear how emissions of other volatile compounds from the bleach plants using TCF and ECF sequences differ.

The oxygen demand parameters in effluents will vary considerably depending on how much lignin and carbohydrates are removed from the pulp in the bleach plant, and on how much water is recycled back to recovery. An OZED sequence with nearly complete recycle of the OZE stages produces mill discharges of COD and BOD that are at least 90% less than discharges from a CEDED sequence and 70–90% less than an O(DC)ED sequence.[84] No water recycle with OZED sequences results in comparable or slightly higher levels of organic carbon,[85] possibly due to larger quantities of carbohydrates in TCF filtrates than in ECF filtrates. Laboratory studies in which TOC, COD, and BOD have been measured in untreated and treated filtrates suggest that the organic material produced in ozone-containing sequences is more easily degraded by microbes than ECF filtrates.[82] These findings are consistent with the higher BOD/COD ratios of wastewaters containing ozone filtrates than for wastewaters with C or D stage filtrates.[82,86]

The toxicity of ozone filtrates has been found to be higher, lower, and the same as conventional filtrates, depending on the specific study. For example, exposure of several marine organisms to untreated filtrates from D(EO)DD and Z(PO) sequences (using 50 and 22 m^3 water/tonne pulp, respectively), showed that ozone filtrates were generally more toxic than the untreated D filtrates.[82] Target organisms included scallops, algae, urchins, fish, and bacteria. Peroxides in the untreated ozone filtrates may have been a primary source of toxicity in the TCF filtrates. Similarly, laboratory Z filtrates in a TCF sequence had a higher chronic toxicity to fathead minnows and *Ceriodaphnia* than ECF filtrates, but no acute lethal toxicity. Fathead minnows tend to be more sensitive to chlorine-containing effluents, while *Ceriodaphnia* are more sensitive to OZ filtrates.[87] By contrast, other studies of hardwood mill effluents indicate filtrates from a QZEopAEP sequence were slightly less toxic than those from a DEopDED sequence (using 16 and 25 m^3 water/ton, respectively).[88] Cates et al.[89] recently showed Microtox toxicity decreased in the order: ECF softwood > ECF hardwood > TCF softwood > TCF hardwood. A clearer understanding of how the toxicological properties of ozone filtrates compare to those of conventional filtrates depends on the identification and quantification of the primary toxicants in all types of filtrates.

Despite the complexity of the issue of toxicity of effluents, it is clear that fewer potential toxicants and other compounds will reach the aquatic environment when most of the bleach plant waste is recycled back to the recovery, and zero or near-zero effluent is discharged. Ozone bleaching is one way to achieve those goals. It remains to be determined whether emerging alternative approaches to mill closure, such as Champion's bleach plant filtrate recycle (BFR) of ECF filtrates,[90] will offset some of the current environmental advantage of ozone bleaching.

ACKNOWLEDGMENTS

The author is greatly indebted to Professor Donald R. Dimmel, Institute of Paper Science and Technology, and Professor Göran Gellerstedt, Royal Institute of Technology, for their thoughtful, insightful, and painstaking reviews of the manuscript. Their efforts have very much improved this chapter. The member companies of the Institute of Paper Science and Technology are gratefully acknowledged for their support all throughout the author's sojourn into ozone chemistry.

REFERENCES

1. Taillefer, R.J., S.E. Thomas, Y. Nadeau, S. Flisczar, and H. Henry. 1980. The Mechanism of Oxidation of Acetals by Ozone. I. Stoichiometry, Order of the Reaction, Solvent Effects, and Substituent Effects. *Can. J. Chem.* 58, 1138.
2. Deslongchamps, P., C. Moreau, D. Frehel, and P. Atlani. 1972. The Importance of Conformation in the Ozonolysis of Acetals. *Can. J. Chem. Comm.* 50, 3402.
3. Kovac, F. and B. Plesnicar. 1979. The Substituent Effect on the Thermal Decomposition of Acetal Hydrotrioxides. Polar and Radical Decomposition Paths. *J. Am. Chem. Soc.* 101(10):2677.
4. Deslongchamps, P., P. Atlani, D. Frehel, A. Malaval, and C. Moreau. 1974. The Oxidation of Acetals by Ozone. *Can. J. Chem.* 52, 3651.
5. Pan, G., C.-L. Chen, H.-M. Chang, and J.S. Gratzl. 1981. Model Experiments on the Splitting of Glycosidic Bonds by Ozone. *Proceedings International Symposium on Wood and Pulping Chemistry*, Stockholm, Sweden, Vol. 2. p. 132.
6. Parthasarathy, V.R. and R.C. Peterson. 1990. Reaction Kinetics of Ozonolysis of D(+)-Mannose and D(+)-Xylose in Acidic and Neutral pH Aqueous Solutions. *TAPPI Symposium Notes. Oxygen Delignification Symposium*, Toronto, Ontario, p. 61.
7. Giamalva, D.H., D.F. Church, and W.A. Pryor. 1986. Kinetics of Ozonation. 5. Reactions of Ozone with Carbon-Hydrogen Bonds. *J. Am. Chem. Soc.* 108, 7678.
8. Price, C.C. and A.L. Tumolo. 1964. The Course of Ozonation of Ethers. *J. Am. Chem. Soc.* 86, 4692.
9. Bailey, P.S. and D.A. Lerdal. 1978. Ozonation of Nucleophiles. 10. Ethers. *J. Am. Chem. Soc.* 100(18):5820.
10. Erickson, R.E., D. Bakalik, C. Richards, M. Scanlon, and G. Huddleston. 1966. Mechanism of Ozonation Reactions. II. Aldehydes. *J. Org. Chem.* 32, 461.
11. White, H.M. and P.S. Bailey. 1965. Ozonation of Aromatic Aldehydes. *J. Org. Chem.* 30, 3037.
12. Nangia, P.S. and S.W. Benson. 1980. Thermochemistry and Kinetics of Ozonation Reactions. *J. Am. Chem. Soc.* 102(9):3105.
13. Batterbee, J.E. and P.S. Bailey. 1967. Ozonation of Carbon-Hydrogen Bonds. Anthrone. *J. Org. Chem.* 32, 3899.
14. Plesnicar, B., F. Kovac, and M. Schara. 1988. Chemistry of Hydrotrioxides. Preparation, Characterization, and Thermal Decomposition of Hydrotrioxides of Alkyl α-Methylbenzyl Ethers and α-Methylbenzyl Alcohol. Attempted Spin Trapping of Trioxyl Radicals. *J. Am. Chem. Soc.* 110, 214.
15. Katai, A.A. and C.J. Schuerch. 1966. Mechanism of Ozone Attack on α-Methyl Glucoside and Cellulosic Materials. *J. Polymer Science Part A-1.* 4, 2683.
16. Laurent, J., J. Mathieu, L. Van Nifterik, and J. Xu. 1993. Analysis of Cellulose and Kraft Ozonolysis Products by Anion-Exchange Chromatography with Pulsed Amperometric Detection. *Journal of Chromatography* 640, 335.
17. Nooney, S.M. 1993. Formaldehyde as a Byproduct from Ozone Bleaching. M.S. Report, Institute of Paper Science and Technology.
18. Chirat, C. and D. Lachenal. 1994. Effect of Ozone on Pulp Components: Application to Bleaching of Kraft Pulps. *Holzforschung* 48, 133 Suppl.

19. Chandra, S. and J.S. Gratzl. 1985. Kinetics of Carbohydrate and Lignin Degradation and Formation of Carbonyl and Carboxyl Groups in Low Consistency Ozonation of Softwood Pulps. *Proceedings International Pulp Bleaching Conference*, Quebec City, Quebec, p. 27.
20. Godsay, M.P. and E.M. Pearce. 1984. Physico-chemical Properties of Ozone Oxidized Kraft Pulps. *TAPPI Symposium Notes. Oxygen Delignification Symposium*, San Francisco, CA, p. 55.
21. Sonnenberg, L.B. and K. Poll. 1994. Studies of High Consistency Ozone Delignification of Oxygen Bleached Pulps. *Proceedings International Pulp Bleaching Conference*, Vancouver, BC, p. 101.
22. Decoret, C., J. Royer, B. Legube, and M. Dore. 1984. Experimental and Theoretical Studies of the Mechanisms of the Initial Attack of Ozone on Some Aromatics in Aqueous Medium. *Environmental Technology Letters* 5, 207.
23. Legube, B., S. Guyon, H. Sugimitsu, and M. Dore. 1986. "Ozonation du Naphtalene en Milieu Aqueux-I," *Water Research* 20(2):197.
24. Eriksson, T. and J. Gierer. 1985. Studies on the Ozonation of Structural Elements in Residual Kraft Lignins. *Journal Wood Chemistry and Technology* 5(1):53.
25. Bailey, P.S. 1978. *Ozonation in Organic Chemistry*; Vol. I, Academic Press, NY, pp. 198–219.
26. Razumovskii, S.D. and S.E. Zaikov. 1984. *Ozone and its Reactions with Organic Compounds*, Elsevier, Amsterdam, pp. 274–285.
27. Kratzl, K., P. Claus, and G. Reikchel. 1976. Reactions of Lignin and Lignin Model Compounds with Ozone. *Tappi Journal* 59(11):86.
28. Soteland, N. 1971. Some Attempts to Characterize the Oxidized Lignin after Ozone Treatment of Western Hemlock Groundwood. Part II. *Norsk Skogindustri* 5, 135.
29. Plonka, A.M., J. Rutkowski, and R. Szopinski. 1987. Changes of the Chromophore System in Sulphate Pine Pulp Treated with Ozone. *Cellulose Chemistry and Technology* 21, 535.
30. Kojima, U., K. Miura, and T. Kayama. 1978. Oxidative Degradation of Softwood Lignin Model Compounds with Ozone. *Res. Bull. Coll. Expt. Forests* Hokkaido Univ. 35(1):165.
31. Balousek, P.J., T.J. McDonough, R.D. McKelvey, and D.C. Johnson. 1981. The Effects of Ozone Upon a Lignin Model Containing the β-Aryl Ether Linkage. *Svensk Papperstidning* 84(9):R49.
32. Katuscak, S., A. Hrivik, and M. Mahdalik. 1971. Ozonation of Lignin. Part I. Activation of Lignin with Ozone. *Paperi Puu* 9, 519.
33. Haluk, J.P. and M. Metche. 1986. "Caracterization Chimique et Spectrographique de la Lignine de Peuplier Par Acidolyse et Ozonolyse," *Cellulose Chemistry and Technology* 20, 31.
34. Mansson, P. and R. Oster. 1988. Ozonation of Kraft Lignin. *Nordic Pulp and Paper Research Journal* 2, 75.
35. Lachenal, D. 1994. Behavior of Residual Lignin in Kraft Pulp During Bleaching. *Proceedings International Pulp Bleaching Conference*, Vancouver, BC, p. 41.
36. Katuscak, S., I. Rybarik, E. Paulinyova, and M. Mahdalik. 1971. Ozonization of Lignin. Part II. Investigation of Changes in the Structure of Methanol Lignin During Ozonization. *Paperi Puu* 11, 665.
37. Sonnenberg, L.B., K. Poll, G. Murphy, and R. LeLacheur. 1992. Characterization of Pulp Ozonolysis Products. *Proceedings TAPPI Environmental Conference*, Richmond, VA, p. 353.
38. Sonnenberg, L.B. 1994. Unpublished data.
39. Schwantes, T.A. and T.J. McDonough. 1994. Gel Permeation Chromatography of Effluents Produced Under Low AOX Process Conditions. *Proceedings TAPPI Pulping Conference*, San Diego, CA, p. 1039.
40. Katuscak, S., A. Hrivik, G. Katuscakova, and O. Schiessl. 1972. Ozonization of Lignin. IV. The Course of Ozonization of Insoluble Lignins. *Paperi Puu* 12, 861.
41. Gierer, J., E. Yang, and T. Reitberger. 1991. Model Studies of the Reaction Between Hydroxyl Radicals and Aromatic Lignin Structures. *Proceedings of the Sixth International Symposium on Wood and Pulping Chemistry*; Melbourne, Australia, Vol. 2, p. 157.
42. Ni, Y., G. Kang, and A.R.P. van Heiningen. 1996. Are Hydroxyl Radicals Responsible for Degradation of Carbohydrates During Ozone Bleaching of Chemical Pulp? *Journal of Pulp and Paper Science* 22(2):J53.
43. Chirat, C. and D. Lachenal. 1994. Role of Hydroxyl Radicals During Ozone Bleaching. *Extended Abstracts. Third European Workshop on Lignocellulosics and Pulp*, Stockholm, Sweden, p. 260.
44. Brolin, A., J. Gierer, and Y. Zhang. 1993. On the Selectivity of Ozone Delignification of Softwood Kraft Pulp. *Wood Science and Technology* 27, 115.
45. Chirat, C., D. Lachenal, C. Coste, and J.P. Zumbrunn. 1993. Effect of Ozone on Cellulose and Pulp. *Proceedings Seventh International Symposium of Wood and Pulping Chemistry*, Beijing, China, p. 368.
46. Jacobson, B., P.-O. Lindblad, and N.-O. Nilvebrant. 1991. Lignin Reactions Affect the Attack of Ozone on Carbohydrates. *Proceedings International Bleaching Conference*, Stockholm, Sweden, Vol 2. p. 45.
47. Kang, G., Y. Zhang, Y. Ni, A.R.P. van Heiningen, and J. McKenzie. 1995. Influence of Lignins on the Degradation of Cellulose During Ozone Bleaching. *J. Wood Chem. Technol.* 15(4):413.

48. Gierer, J. and Y. Zhang. 1993. The Role of Hydroxyl Radicals in Ozone Bleaching Processes. *Proceedings Seventh International Symposium of Wood and Pulping Chemistry*, Beijing, China, p. 951.
49. Ek, M., J. Gierer, and K. Jansbo. 1989. Study on the Selectivity of Bleaching with Oxygen Containing Species. *Holzforschung* 43(6):391.
50. Hoigné, J. 1982. Mechanisms, Rates, and Selectivities of Oxidations of Organic Compounds Initiated by Ozonation of Water. In *Handbook of Ozone Technology Applications.* Rice, R., Ed. Ann Arbor Science, Ann Arbor, MI, pp. 341–379.
51. Reitberger, T., T. Eriksson, M. Ek, and G. Ihren. 1995. On the Importance of Radical Species in TCF Bleaching. *Proceedings Eighth International Symposium of Wood and Pulping Chemistry*, Helsinki, Finland, p. 302.
52. Chirat, C. and D. Lachenal. 195. Effect of Hydroxyl Radicals on Cellulose and Pulp and Their Occurrence During Ozone Bleaching. *Proceedings Eighth International Symposium of Wood and Pulping Chemistry*, Helsinki, Finland, p. 293.
53. Staehelin, J. and J. Hoigné. 1982. Decomposition of Ozone in Water: Rate of Initiation by Hydroxide Ion and Hydrogen Peroxide. *Environ. Sci. Technol.* 16(10):676.
54. Reitberger, T., G. Ihren, and M. Ek. 1994. Advances in TCF Bleaching Chemistry. *Extended Abstracts. Third European Workshop on Lignocellulosics and Pulp*, Stockholm, Sweden, p. 106.
55. Gierer, J. 1985. Basic Principles of Bleaching — Part 1. *Holzforschung* 19, 2755.
56. Osawa, Z. and C. Schuerch. 1963. The Action of Gaseous Reagents on Cellulosic Materials. *Tappi J.* 46(2):79.
57. Griffin, R., Y. Ni, and A.R.P. van Heiningen. 1994. The Development of Delignification and Lignin-Cellulose Selectivity During Ozone Bleaching. *Proceedings 81st Annual Meeting — Technical Section*, Montreal, Canada, CPPA, p. A117.
58. Sixta, H., G. Gotzinger, A. Schrittwieser, and P. Hendel. 1991. Ozone Bleaching: Medium Consistency Technology and Mill Experience. *Workshop on Emerging Pulping and Chlorine-Free Bleaching Technologies*, North Carolina State University, Raleigh, NC.
59. Liebergott, N., B. van Lierop, and A.A. Skothos. 1992. Skothos Survey of the Use of Ozone in Bleaching Pulps, Part 1. *Tappi J.* 75(1):145.
60. Liebergott, N., B. van Lierop, and A.A. Skothos. 1992. Survey of the Use of Ozone in Bleaching Pulps, Part 2. *Tappi J.* 75(2):117.
61. Byrd, M.V., J.S. Gratzl, and R.P. Singh. 1992. Delignification and Bleaching of Chemical Pulps with Ozone: A Literature Review. *Tappi J.* 75(3):207.
62. Lindholm, C.-A. 1987. Effect of Pulp Consistency and pH in Ozone Bleaching, Part 1. General Aspects. *Paperi Puu* 69(3):21.
63. Lindholm, C.-A. 1987. Effect of Pulp Consistency and pH in Ozone Bleaching: Part 2. Lignin Removal and Carbohydrate Degradation. *Proceedings TAPPI International Oxygen Delignification Conference*, San Diego, CA, p. 155.
64. Lindholm, C.-A. 1990. Effect of Dissolved Reaction Products on Pulp Viscosity in Low Consistency Ozone Bleaching. *Paperi Puu* 72(3):254.
65. Hurst, M.M. 1992. The Effect of Pulp Consistency and Mixing Intensity on Ozone Bleaching. *Proceedings TAPPI Pulping Conference*, Boston, MA, p. 1061.
66. Funk, E., R. Szopinski, F. Munro, and A. Vilpponen. 1992. Espanola Ozone Bleaching Pilot Plant: Status Report. *Proceedings TAPPI Pulping Conference*, Boston, MA, p. 1091.
67. Liebergott, N., B. van Lierop, and A. Skothos. 1991. The Use of Ozone in Bleaching Pulps. *Proceedings TAPPI Pulping Conference*, Orlando, FL, p. 1.
68. Kassebi, A., J.S. Gratzl, C.-L. Chen, and R.P. Singh. 1982. Non-Conventional Kraft Pulp Bleaching — The Role of Ozone. *Proceedings TAPPI Pulping Conference*, Toronto, Ontario, p. 327.
69. Allison, R.W. 1991. Potential of Ozone in Kraft Pulp Bleaching. *Proceedings APPITA 45th Annual General Conference*, Australia, p. 201.
70. Berry, R.M., H. Barclay, J. Prins, G. Sacciadis, A. Skothos, N. Ayala, V. Magnotta, D. Breed, J. Rounsaville, and L. Shackford. 1994. Medium Consistency Bleaching with High Concentration Ozone Gas in Paprican Pilot Plant and a Comparison with Laboratory Bleaching. *Proceedings International Pulp Bleaching Conference*, Vancouver, BC, p. 87.
71. Hartler, N., V. Granlund, J. Sundin, and A. Tubek-Lindbolm. 1991. Ozone Bleaching of Chemical Pulps. *Proceedings International Pulp Bleaching Conference*, SPCI, Stockholm, Sweden, p. 75.
72. Singh, R.P. 1981. Advances in Ozone Bleaching — Part I. The Ozone Bleaching Process — Laboratory to Pilot Plant, *Proceedings TAPPI Annual Meeting*, Chicago, IL, p. 247.
73. Soteland, N. 1974. Bleaching of Chemical Pulps with Oxygen and Ozone. *Pulp and Paper Canada* 75(4):T153.

74. Parthasarathy, V.R. and R.C. Peterson. 1990. Ozone Bleaching Part 1. The Decomposition of Ozone in Aqueous Solution — The Influence of pH, Temperatures and Transition Metals on the Rate of Kinetics of Ozone Decomposition. *TAPPI Symposium Notes. Oxygen Delignification Symposium*, Toronto, Ontario, p. 24.
75. Schwarzl, K. 1991. Pulp Bleaching Using Ozone. *Workshop on Emerging Pulping and Chlorine-Free Bleaching Technologies*, North Carolina State University, Raleigh, NC.
76. Lindholm, C.-A. and A. Vilpponnen. 1991. Molar Mass Distribution of Ozone Bleaching Filtrates. *Paperi Puu* 73(5):441.
77. Lindholm, C.-A. 1988. Effect of Recirculation of Bleaching Filtrates in Low Consistency Ozone Bleaching. *Paperi Puu* 70(1):62.
78. Lachenal, D., M.T. Taverdet, and M. Muguet. 1991. Improvement in the Ozone Bleaching of Kraft Pulps. *Proceedings International Pulp Bleaching Conference*, SPCI, Stockholm, Sweden, p. 33.
79. Lachenal, D. and M. Muguet. 1991. Degradation of Residual Lignin in Kraft Pulp with Ozone. Application to Bleaching. *Proceedings Sixth International Symposium on Wood and Pulping Chemistry*, Melbourne, Australia, p. 107.
80. Axegard, P., O. Dahlman, I. Haglind, B. Jacobson, R. Morck, and O. Stromberg. 1993. Pulp Bleaching and the Environment — The Situation 1993. *Nordic Pulp and Paper Research J.* 4, 365.
81. Steffes, F. and U. Germgard. 1995. Cost and Environmental Impact Comparisons for ECF and TCF Mills Applying a Decade of Technological Development, *Proceedings International Nonchlorine Bleaching Conference*, Amelia Island, FL, p. 13.
82. Nelson, P.J., J.L. Stauber, L. Gunthorpe, J.G. Deavin, B.L. Mundy, R. Krassoi, and J. Simon. 1995. Study Shows ECF, TCF Effluents Have Long-Term Toxic Impact on Sea Life. *Pulp and Paper* 9, 103.
83. Bicknell, B., D.B. Spengel, and T.J. Holdsworth. 1995. Comparison of Pollutant Loadings from ECF, TCF, and Ozone/Chlorine Dioxide Bleaching, *Proceedings International Nonchlorine Bleaching Conference*, Amelia Island, FL, p. 23.
84. Nutt, W.E., S.W. Eachus, B.F. Griggs, and M.A. Pikulin. 1992. Development of an Ozone Bleaching Process. *Proceedings TAPPI Pulping Conference*, Boston, MA, p. 1109.
85. Lewis, R. 1995. Ozone-Mill Applications. *Workshop on Emerging Pulping and Chlorine-Free Bleaching Technologies*, North Carolina State University, Raleigh, NC.
86. Osborne, M. 1989. The Effects of Modifying Bleaching Processes on Environmental Factors. M.S. Report, Institute of Paper Science and Technology, Atlanta, GA.
87. O'Connor, B.I., T.G. Kovacs, P.H. Martel, and B.A. van Lierop. 1994. A Laboratory Assessment of Environmental Quality of Alternative Pulp Bleaching Effluents. *Pulp and Paper Canada* 95(3):47.
88. Kovacs, T.G., J. Tana, K.-J. Lehtinen, and O. Sangfors. 1995. Comparison of the Environmental Quality of Elemental Chlorine-Free (ECF) and Totally Chlorine-Free (TCF) Hardwood Bleach Plant Effluents. *Proceedings International Nonchlorine Bleaching Conference*, Amelia Island, FL, p. 25.
89. Cates, D.H., C. Effert, J.L. Yang, and K.-E. Eriksson. 1995. Comparison of Effluents from TCF and ECF Bleaching of Kraft Pulps. *Tappi J.* 78(12):93.
90. Maples, G., R. Ambody, J.R. Caron, S. Stratton, and R. Vega Canova. 1994. BFR: A New Process Toward Bleach Plant Closure. *Tappi J.* 77(11):71.
91. Hoigné, J. and H. Bader. 1983. Rate Constants of Ozone with Organic and Inorganic Compounds in Water. I. Non-Dissociating Organic Compounds. *Water Research* 17, 173.
92. Hoigné, J. and H. Bader. 1983. Rate Constants of Ozone with Organic and Inorganic Compounds in Water. I. Dissociating Organic Compounds. *Water Research* 17, 185.
93. Bailey, P.S. 1978. *Ozonation in Organic Chemistry*; Vol. I, Academic Press, NY, p. 21.

Part 2

Analytical and Environmental Aspects

Chapter 4

An Overview of NPDES Compliance Monitoring Methods for the Pulp and Paper Industry

W.A. Telliard and J.R. King

The U.S. Environmental Protection Agency (EPA) proposed effluent limitations guidelines and standards in the December 17, 1993 *Federal Register* (the Cluster Rule). Section 304(h) of the Clean Water Act directs the EPA to promulgate guidelines establishing test procedures (analytical methods) for the analysis of pollutants. These methods are used to determine the presence and concentrations of pollutants in wastewater, and are used for filing applications and for compliance monitoring under the National Pollutant Discharge Elimination System (NPDES) [40 CFR 122.41(j)(4) and 122.21(g)(7)], and for the pretreatment program [40 CFR 403.7(d)].

The EPA has promulgated analytical methods for monitoring discharges to surface waters at 40 CFR Part 136, and has promulgated methods for analytes specific to a given industrial category and for other purposes at Parts 400–480 of the CFR. In the Cluster Rule, methods not promulgated at 40 CFR Part 136 were proposed at 40 CFR Part 430 to support regulation of discharges from the pulp, paper, and paperboard industrial category. The list of analytes in Table 4.1 includes the methods that pulp, paper, and paperboard manufacturers will be required to use for the purpose of compliance monitoring. Table 4.1 lists methods already promulgated by the EPA in 40 CFR Part 136, updated to new versions where appropriate, as well as analytical methods not contained in 40 CFR Part 136. At a later date, the EPA may promulgate the methods contained in Table 4.1 as allowable methods under 40 CFR Part 136.

CHLORINATED DIOXINS AND FURANS

EPA Method 1613 is to be used to analyze pulp and paper wastewaters for the 17 tetra- through octa-substituted dibenzo-*p*-dioxin and dibenzofuran isomers and congeners that are chlorinated at the 2, 3, 7, and 8 positions. Method 1613 is also isomer-specific for the 2,3,7,8-TCDD and 2,3,7,8-TCDF isomers to be regulated under the Cluster Rule, and effluent limits for 2,3,7,8-TCDD and 2,3,7,8-TCDF in the Cluster Rule were developed using Method 1613.

Method 1613 uses isotope dilution and high-resolution gas chromatography combined with high-resolution mass spectrometry (HRGC/HRMS) for separation and detection of the individual 2,3,7,8-substituted tetra- through octa-isomers and congeners. Separate procedures are available for measurement of these analytes in water and solid matrices, although the method is proposed for use under the pulp, paper, and paperboard industrial category in treated and untreated wastewaters only.

In the procedure for water, a 1-L sample is passed through a 0.45-μ glass fiber filter. The filter is extracted with toluene in a Soxhlet/Dean-Stark (SDS) extractor. The aqueous filtrate is extracted with methylene chloride in a separatory funnel. Extracts from the SDS and separatory funnel extractions are combined and concentrated. To remove interferences, the combined, concentrated extract is cleaned up using various combinations of acid and base washes, acidic and basic silica gel, gel permeation chromatography (GPC), high-performance liquid chromatography (HPLC), and activated carbon. The cleaned-up extract is concentrated to 20 μL and a 1–2 μL aliquot is injected into the HRGC/HRMS. The labeled compounds serve to correct the variability of the analytical technique.

The detection limit of the method is usually dependent upon sample matrix interferences rather than instrument limitations. The MDL determined for 2,3,7,8-TCDD is 4.2 parts-per-quadrillion (ppq). Minimum levels determined for Method 1613 are 10 ppq for 2,3,7,8-TCDD and 2,3,7,8-TCDF, 100 ppq for the octa congeners, and 50 ppq for the penta- through hepta-substituted CDDs and CDFs.

Table 4.1. List of Analytes of Interest.

Analyte	CAS Registry No.	Method	Technique	Regulatory Status
2,3,7,8-TCDD	1746-01-6	1613	HRGC/HRMS	Proposed, 1990 56 FR 5090
2,3,7,8-TCDF	51207-31-9			
Volatile Organics		1624	LRGC/LRMS	Promulgated 40 CFR Part 136
Chloroform	67-66-3			
Acetone	67-64-1			
Methyl Ethyl Ketone	78-93-3			
Methylene Chloride	75-09-2			
Organic Halides (AOX)	—	1650	Titrimetric	Cluster Rule
Chlorinated Phenolics		1653	HRGC/LRMS	Cluster Rule
2,3,4,6-Tetrachlorophenol	58-90-2			
2,4,5-Trichlorophenol	95-95-4			
2,4,6-Trichlorophenol	88-06-2			
3,4,5-Trichlorocatechol	56961-20-7			
3,4,5-Trichloroguaiacol	57057-83-7			
3,4,6-Trichlorocatechol	32139-72-3			
3,4,6-Trichloroguaiacol	60712-44-9			
4,5,6-Trichloroguaiacol	2668-24-8			
Tetrachlorocatechol	1198-55-6			
Tetrachloroguaiacol	2539-17-5			
Trichlorosyringol	2539-26-6			
Pentachlorophenol	87-86-5			
Chemical Oxygen Demand (COD)	—	410.1 410.2	Titrimetric	Promulgated 40 CFR Part 136
Effluent Water Color	—	NCASI 253	Spectrometric	Cluster Rule

CHLORINATED PHENOLIC COMPOUNDS

Chlorinated phenolic compounds (chlorophenolics) are a group of toxic pollutants characterized by a benzene ring with one hydroxyl (OH) group (phenols) and at least one chlorine atom (chlorophenols). Ring substituents may be a second hydroxyl group at the ortho position (catechols), a methoxy (CH_3O) group at the ortho position (guaiacols), a methoxy group at the ortho position and an aldehyde (CHO) group at the para position (vanillins), and methoxy groups at both ortho positions and a hydroxyl or aldehyde group at the para position (syringols and syringaldehydes).

The chlorophenolics to be regulated under the Cluster Rule are measured using EPA Method 1653. Method 1653 measures compounds in water and wastewater amenable to *in situ* acetylation, extraction, and determination by HRGC combined with low-resolution mass spectrometry (LRMS). In this method, chlorophenolics are derivatized *in situ* to form acetic acid phenolates that are extracted with hexane, concentrated, and injected into the HRGC/LRMS where separation and detection occur. Detected chlorophenolics are quantified by isotope dilution if a labeled analog is available. Where labeled analogs are not available, detected compounds are quantified by the internal standard technique. In early data gathered for the Cluster Rule, National Council of the Paper Industry for Air and Stream Improvement (NCASI) methods were used. Effluent limits for the Cluster Rule, however, were developed using data gathered with Method 1653.

The detection limit of Method 1653 is usually dependent on interferences rather than instrument limitations. With no interferences present, MDLs of 0.09–1.39 μg/L (ppb) have been determined for chlorophenolics in water. Based on these MDLs and on calibration of the GCMS instrument, minimum levels have been determined to be 1.25, 2.50, or 5.00 μg/L, depending on the specific compound. Method 1653 is based on NCASI Method CP-86.01.

VOLATILE ORGANIC COMPOUNDS

Volatile organic compounds (VOCs) are determined in wastewater using EPA Method 1624. Method 1624 is the only method promulgated at 40 CFR Part 136 in which all four of the pollutants to be regulated under the proposed pulp and paper rule are listed. The method uses isotope dilution and GC/MS to separate and quantify the regulated analytes. VOCs are purged from a water sample onto a trap by a stream of inert gas. Subsequent heating of the trap introduces the concentrated VOCs into the GC/MS.

The detection limit of the method is usually dependent on interferences rather than instrumental limitations. With no interferences present, minimum levels of 10–50 μg/L can be achieved, depending on the specific compound.

ADSORBABLE ORGANIC HALIDES

Adsorbable organic halides (AOX) is a measure of halogenated organic compounds that adsorb onto granular activated carbon. AOX in water and wastewater is measured by EPA Method 1650, Revision A. This method is a microcoulometric technique that determines organically bound halides (chlorine, bromine, and iodine) dissolved or suspended in a water sample. The results are reported as organic chlorine even though other halides may be present.

The concentration of organic halides is determined by adsorption onto granular activated carbon, removal of inorganic halides by washing, and combustion of the organic halides (along with the carbon) to form hydrogen halides. The amount of hydrogen halides produced is determined by titration with silver ions in the microcoulometer.

The detection limit of Method 1650 is usually dependent on interferences rather than instrumental limitations. With no interferences present, an MDL of 6.6 μg/L (ppb) can be achieved. Based on this MDL and on calibration of the microcoulometer, the minimum level in Method 1650 has been determined to be 20 μg/L.

CHEMICAL OXYGEN DEMAND

Chemical oxygen demand (COD) is a measure of the oxygen equivalent of the organic matter in a sample that is susceptible to oxidation by a strong chemical oxidant. For samples from a specific source, COD can be related empirically to biochemical oxygen demand (BOD). Several methods for COD are incorporated by reference into 40 CFR Part 136, notably EPA Methods 410.1 through 410.4. Methods 410.1 and 410.2 are the only EPA COD methods that will be allowed for monitoring under the pulp and paper rule. In these methods, the organic and oxidizable inorganic substances in an aqueous sample are oxidized by a solution of potassium dichromate in sulfuric acid. The excess dichromate is titrated with standard ferrous ammonium sulfate using orthophenanthroline ferrous complex (ferroin) as an indicator. Method 410.2 covers COD concentrations in the range of 5–50 mg/L, whereas Method 410.1 covers COD concentrations from 50–2000 mg/L. Other methods incorporated by reference into 40 CFR Part 136 that use oxidation with potassium dichromate in sulfuric acid and subsequent titration of the excess dichromate (e.g., Standard Method 5520B) are also allowed.

Early studies of wastewaters in the pulp and paper industry were conducted with Method 410.4. In this method, COD is determined colorimetrically after digestion of the organic matter in a sample using hot chromic acid solution. However, the highly colored nature of some pulp and paper samples interferes with the colorimetric determination of COD. Therefore, Method 410.4 and other colorimetric methods incorporated by reference into 40 CFR Part 136 will not be allowed for monitoring under the proposed pulp and paper rule. Similarly, Method 410.3 is intended for high levels of COD in saline waters and will not be allowed.

COLOR

Color in water and wastewater may result from the presence of natural metallic ions (e.g., iron and manganese), humus and peat materials, and industrial wastes. Color is measured by the procedure described in NCASI Technical Bulletin 253, which was specifically designed for measurement of color in pulp bleaching effluents. Samples are adjusted to pH 7.6 and the turbidity of the solution is reduced by filtration. Color is then determined at 465 nm by spectrophotometric comparison of a wastewater sample and a standard solution of chloroplatinate ions. Color results are reported in either color units or mg/L relative to the platinum standard. The MDL for this method is approximately 1 mg/L.

The EPA has added quality control (QC) tests and specifications as an addendum to NCASI 253 in order to bring the QC in this method to the level of QC required in other EPA wastewater methods. Specifications for calibration linearity, calibration verification, and initial and ongoing precision and recovery are some of the QC tests required.

Chapter 5

Experiences with Isotopic Dilution Measurements of Chlorinated Phenolics Using EPA 1653

J.R. Louch and L.E. LaFleur

EPA Method 1653, "Chlorinated Phenolics in Wastewater by *In Situ* Acetylation and GC/MS,"[1] is a good "test" for isotopic dilution methodology because the natives are determined via measurement of a derivative which is generated *in situ* and then extracted into hexane. Traditionally,[2,3] analyses utilizing this *in situ* methodology have employed quantifications vs. calibration curves generated using acetylated standards, i.e., all calibration standards are acetylated (in reagent water) using the same procedure as used for samples. Thus, calibration curves account for compound-specific acetylation and extraction bias, and even external calibration quantifications affect some measure of correction for these effects. As an additional correction, all applications of this technique also employ internal standard quantification. Thus, NCASI's method CP-86[3] specifies that 3,4,5-trichlorophenol (3,4,5-TCP) be added to all samples and standards prior to acetylation, and that all quantifications be performed vs. the resulting 3,4,5-TCP derivative. EPA Method 1653 attempts to improve on this approach further by including eight ^{13}C-labeled surrogates, and specifying isotopic dilution quantifications for the corresponding eight native compounds. In this chapter we analyze the utility of these isotopic dilution analyses with respect to both the quality of the resulting data and overall cost-effectiveness.

RESULTS AND DISCUSSION

The EPA has proposed regulating 12 chlorinated phenolic compounds in pulp and paper industry combined bleach plant filtrates, and is proposing the 1653 "minimum levels" as the compliance point for most categories. In order to assess the ability of laboratories to make accurate measurements at these concentrations, NCASI performed and reported[4] an interlaboratory Compliance Monitoring Detection Limit/Quantitation Limit (CMDL/CMQL) study. For this study a number of pH preserved (pH 2) sample matrices were fortified with chlorophenolics at multiple concentration levels, and the samples distributed to laboratories for analysis by 1653.

When the NCASI lab performed these analyses it was noted that the recoveries of the spiked chlorocatechols from individual matrices varied greatly. Experiments were then initiated to examine this phenomenon. Of most interest, experiments were performed to characterize the effect of the prescribed pH adjustment (pH 2 preserved to pH 7 prior to the acetylation) on recoveries. In these experiments, unspiked pH 2 preserved aliquots of samples were analyzed as pairs. One aliquot was fortified with chlorophenols *prior* to bringing the pH to 7 (by addition of hydroxide). The other aliquot was fortified with chlorophenols *after* bringing the pH to 7. For each pair the resulting relative percent difference (RPD) in the recoveries of the analytes was calculated such that a positive RPD reflects greater recovery from the aliquot fortified *after* adjusting the pH. Table 5.1 summarizes the results of these experiments (all quantifications were performed vs. the instrumental internal standard 2,2'-difluorobiphenyl). These data clearly show that the pH adjustment step using (6N) sodium hydroxide results in varying degrees of chlorocatechol degradation, the degree of degradation being matrix-dependent. Note that the recoveries of the labeled chlorocatechols are unaffected because they were added after the pH adjustment in all cases (as per the method-specified spiking scheme).

Although the mechanism for the observed loss of chlorocatechols is not fully understood, the data clearly show that Method 1653 is not suitable for the determination of these compounds. Furthermore, these data provide an example of how surrogate recoveries can provide misinformation. In this spe-

Table 5.1. Mean Relative Percent Differences (RPD) Between Recoveries from Samples Fortified with the Target Chlorophenolics Before and After pH Adjustment.

	Relative Percent Difference		
	Mean	Standard Deviation	Range
4-Chlorophenol[a]	–0.6	5.6	–3.8–7.2
2,4-Dichlorophenol[a]	5.0	5.6	–2.6–11
2,6-Dichlorophenol[a]	2.9	5.4	–4.8–6.9
2,4,5-Trichlorophenol[b]	3.8	5.2	–4.2–9.9
2,4,6-Trichlorophenol[b]	1.9	4.7	–6.0–2.5
2,3,4-6-Tetrachlorophenol[b]	2.4	5.3	–8.2–5.5
Pentachlorophenol[b]	2.7	5.4	–7.0–9.6
4-Chloroguaiacol[a]	8.0	3.2	5.7–12
3,4-Dichloroguaiacol[a]	3.9	4.1	–1.0–9.0
4,5-Dichloroguaiacol[a]	4.9	3.4	0.4–8.8
4,6-Dichloroguaiacol[a]	6.7	2.1	3.7–8.3
3,4,6-Trichloroguaiacol[b]	3.0	2.6	0.0–6.6
3,4,5-Trichloroguaiacol[b]	3.8	4.0	–2.6–9.9
4,5,6-Trichloroguaiacol[b]	5.3	1.0	3.5–6.9
Tetrachloroguaiacol[b]	5.3	3.4	–1.3–8.0
4-Chlorocatechol[a]	36	23	16–68
3,4-Dichlorocatechol[a]	42	29	21–84
3,6-Dichlorocatechol[a]	68	52	19–140
4,5-Dichlorocatechol[a]	23	17	10–48
3,4,5-Trichlorocatechol[b]	41	36	14–110
3,4,6-Trichlorocatechol[b]	76	66	19–190
Tetrachlorocatechol[b]	63	62	12–180
5-Chlorovanillin[a]	3.0	4.4	–3.2–6.7
6-Chlorovanillin[a]	4.6	3.7	–0.5–8.0
5,6-Dichlorovanillin[a]	6.8	5.4	0.7–14
2-Chlorosyringaldehyde[a]	1.9	1.8	–0.6–3.4
2,6-Dichlorosyringaldehyde[a]	1.3	4.3	–4.7–5.2
Trichlorosyringol[b]	8.4	8.3	–1.2–24
Surrogate Compounds			
3,4,5-Trichlorophenol[b]	1.6	1.9	–1.3–4.2
2,4-Dichlorophenol-d3[b]	3.2	4.6	–2.1–10
Pentachlorophenol-13C6	4.4	2.9	–1.4–6.6
4-Chloroguaiacol-13C6[b]	3.9	2.3	0.7–7.2
4,5,6-Trichloroguaiacol-13C6[b]	4.0	1.1	2.5–5.7
Tetrachloroguaiacol-13C6[b]	3.4	3.8	–3.1–7.2
4,5-Dichlorocatechol-13C6[b]	–0.5	6.4	–11–9.0
Tetrachlorocatechol-13C6[b]	3.9	2.1	1.3–6.0
5-Chlorovanillin-13C6[b]	3.2	6.2	–7.5–11

[a] Mean RPD from 3 samples.
[b] Mean RPD from 5 samples.

cific case, the surrogate spiking scheme does not account for analyte (chlorocatechol) losses associated with the analytical procedure, and performing the analysis as specified can give "acceptable" surrogate recoveries even when significant losses of native analytes are occurring.

NCASI experience with the *in situ* acetylation suggests that native analytes other than the chlorocatechols may all behave similarly, implying that a single surrogate would be effective for recovery correcting quantifications for these compounds. In order to examine this hypothesis, 1653-labeled surrogate recoveries generated as part of a joint NCASI-EPA effluent variability study (EVS)[5]

Table 5.2. Average Relative Percent Recoveries of the 1653 Surrogates from 559 Analyses.

Surrogate	Percent
3,4,5-Trichlorophenol	100
2,4-Dichlorophenol-d3	98
Pentachlorophenol-13C6	95
4-Chloroguaiacol-13C6	96
4,5,6-Trichloroguaiacol-13C6	97
Tetrachloroguaiacol-13C6	96
4,5-Dichlorocatechol-13C6	83
Tetrachlorocatechol-13C6	72
5-Chlorovanillin-13C6	82

were normalized to the recovery of 3,4,5-TCP on a sample-specific basis. These normalized recoveries were then pooled and the means calculated. Table 5.2 lists these data.

From Table 5.2, the improvement in "accuracy" (as measured by labeled compound recovery) resulting from the use of isotopic dilution ranges from nominally 2% for 2,4-dichlorophenol to 28% for tetrachlorocatechol. Discounting the chlorocatechols and the chlorovanillins, the relative recoveries for the chlorophenols and chloroguaiacols range between 95% and 98%. These data show that the isotopic dilution quantifications for the chlorophenols and chloroguaiacols effect, at best, a 5% improvement in accuracy over quantification vs. 3,4,5-TCP. Note that these data represent 559 real sample analyses and, thus, reflect matrix effects.

Sojo[6] have demonstrated that quantitative accuracy for chlorophenolics improves when compounds are quantified vs. a chemically similar surrogate. This is a common sense position also supported by the data presented in Table 5.2. Thus, it seems logical to quantify *all* chlorocatechols vs. one of the labeled chlorocatechols, and quantify *all* chlorovanillins vs. the labeled 5-chlorovanillin. Disregarding the flaws in the 1653 spiking scheme, this quantification scheme makes scientific sense, and considering that a single ampule of the labeled surrogate mix costs $3500 (U.S.), it would also be more cost-effective. In general, the quantification scheme specified by 1653 is not an effective use of the isotope dilution technique, and is certainly not an effective use of the specified set of surrogates.

A final consideration when using labeled standards is their effect on an analyst's ability to confirm the presence of an analyte on review of spectral data. As an example, Figure 5.1 shows the background corrected mass spectrum at the GC peak apex of a (nominal) 5 mg/L pentachlorophenol spike. The mass spectrum of the native compound is effectively obscured by the mass spectrum of the labeled surrogate, and manual inspection would most likely fail to detect the presence of the native analyte. In this specific case, the qualitative identification of native pentachlorophenol under 1653 spectral criteria failed due to a (third) co-elutor throwing the m/z 167 (relative percent abundance) ratio out of specifications. Under these circumstances, the method allows for "expert" interpretation, which would be extremely tenuous considering Figure 5.1.

CONCLUSIONS

The data clearly show that Method 1653 is unsuitable for determining chlorocatechols in many matrices regardless of the isotopic dilution quantifications. Furthermore, the labeled catechol surrogates are not used effectively, with "acceptable" surrogate recoveries dropping to 0% and most catechols being quantified vs. 3,4,5-TCP. For the remaining analytes, the use of isotopic dilution for quantification of any analytes other than the chlorovanillins provides only minimal improvement in data quality vs. quantifications performed using 3,4,5-TCP as an internal standard. Thus, the use of the labeled surrogates is generally ineffectual and unwarranted with respect to accuracy. Furthermore, the levels at which the labeled surrogates are added to samples can hinder spectral identification of some analytes.

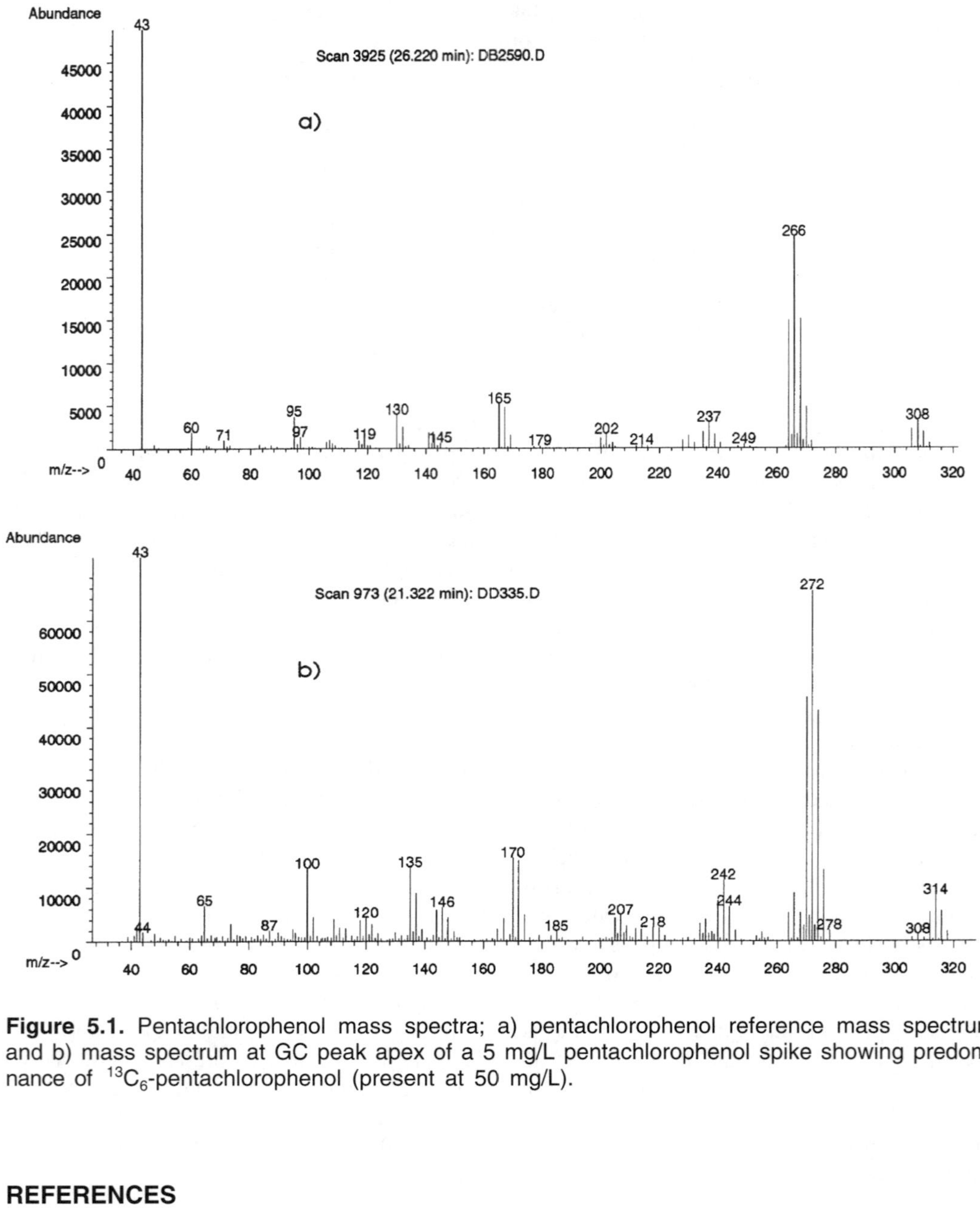

Figure 5.1. Pentachlorophenol mass spectra; a) pentachlorophenol reference mass spectrum, and b) mass spectrum at GC peak apex of a 5 mg/L pentachlorophenol spike showing predominance of $^{13}C_6$-pentachlorophenol (present at 50 mg/L).

REFERENCES

1. EPA. December, 1991. *EPA Revised Method 1653*, "Chlorinated Phenolics in Wastewater by In-situ Acetylation and GC/MS," United States Environmental Protection Agency, Washington, DC.
2. Voss, R.H., T. Wearing, and A. Wong. 1981. In Keith, L., Editor, *Advances in the Identification of Organic Pollutants in Water: Vol II*, Ann Arbor Science, Ann Arbor, MI, Chapter 53.
3. NCASI. 1986. NCASI Technical Bulletin No. 498, NCASI Methods for the Determination of Chlorinated Phenolics in Pulp Industry Wastewaters.
4. NCASI. April 1994. NCASI Special Report, Compliance Monitoring Detection and Quantitation Levels for Method 1653.
5. Unpublished data.
6. Sojo, L.E., D. Hoover, M.C. Hamilton, and B.R. Fowler. 1993. *Proceedings of the TAPPI 1993 Environmental Conference, Boston,* pp. 101–124, Analysis of Chlorophenols, Chlorocatechols, Chloroguaiacols and Related Compounds by GC/MS.

Chapter 6

Effect of Breakthrough on AOX Measurements*

W.A. Telliard, R.M. Litow, and D.R. Rushneck

ABSTRACT

Adsorbable organic halides (AOX) are a class of organic compounds that contain one or more halogen atoms (fluorine, chlorine, bromine, and/or iodine) that are partially or totally adsorbed from water onto granular activated carbon during analytical measurement using a column or a batch adsorption technique. In the column technique, two small columns (tubes) packed with granular activated carbon (GAC) are connected in series, and a sample containing AOX is forced through the column pair. When using columns in series, a phenomenon known as "AOX breakthrough" can occur whereby AOX is not captured on the first column. The amount of breakthrough is the percentage of organic halide captured by the second column relative to the total in both columns.

Based on recommendations by equipment manufacturers, the current EPA method (Method 1650) sets a 10% upper limit on breakthrough. This limit can cause a laboratory to have to reanalyze samples, at varying dilution levels, until the breakthrough criterion is met. This study was designed to examine breakthrough levels of less than and greater than 10% and to evaluate their effects on detected sample concentrations. Nine samples from pulp and paper mills were analyzed by each of four laboratories, using four types of columns and a batch procedure. Of the nine samples, four were influents to the treatment processes used by the facilities, and five were effluents from the treatment processes. The study also examined factors that might contribute to variable levels of AOX breakthrough, including adsorption procedure, column type, source type, dilution level, and laboratory.

Results of this study show that, in general, more dilute samples produced lower AOX breakthrough than did less dilute samples, measured AOX concentrations are not affected by breakthrough levels up to 40%, and the precision of measurements is not adversely affected by breakthrough. It is concluded that breakthrough does not adversely affect AOX measurements, and therefore, inclusion of the breakthrough specification at the current level in Method 1650 is not warranted and should be removed or adjusted upward.

INTRODUCTION

Method 1650

EPA Method 1650, *Adsorbable Organic Halides by Adsorption and Coulometric Titration*, incorporates the techniques specified in several existing methods for organic halide measurements (References 1–7). The method includes both batch and column adsorption procedures. The column method requires the use of 2 mm i.d. glass columns that are 4–5 cm long and specifies that the sample must be diluted and reanalyzed when the AOX value from the second column exceeds 10% of the total from the first and second columns. There is no breakthrough specification in Method 1650 for the batch adsorption procedure.

---

* This chapter is based on the report "The AOX Breakthrough Study" prepared by the authors for the USEPA Office of Water.

Purpose of the EPA's AOX Breakthrough Study

In the development of the AOX column technique, the manufacturer of one of the instruments used for the analysis (Dohrmann Instruments) specified the level of allowable breakthrough at 10%. The Dohrmann manual states that if the breakthrough exceeds 10%, results are suspect and the analysis needs to be repeated. Conversations with Dohrmann personnel have revealed that the 10% specification has no scientific basis but was arbitrarily and conservatively chosen by Dohrmann using the logic that if 90% is adsorbed on the first column (10% is not adsorbed) and 9% (90% of the remainder) is adsorbed on the second column, the total adsorbed on both columns will be 99%. It was reasoned that a loss of 1% of the AOX would not adversely affect any AOX measurement.

Prior to this study, data gathered by the EPA and the paper industry from analysis of wastewater samples suggested that an evaluation of the 10% breakthrough criteria was warranted. Further, the data and discussions with laboratory staff and instrument suppliers suggested that several factors might be related to the amount of breakthrough observed. These factors included the type of column used (e.g., column diameter, column material, and column packing material), the type of sample being analyzed, the sample dilution level, and the laboratory performing the analyses. The objectives of this study were to

1. Determine if concentration differences obtained where breakthrough was above and below 10% are statistically significant.
2. Examine the effect of column type, source type, dilution level, and laboratory on breakthrough results and concentration.
3. Determine if a breakthrough specification should be added to the batch adsorption procedure in EPA Method 1650.

AOX

Adsorbable organic halides (AOX) are a class of organic compounds that contain one or more halogen atoms (fluorine, chlorine, bromine, and/or iodine) that are partially or totally adsorbed from water onto granular activated carbon during analytical measurement using a column or a batch adsorption technique. AOX is a method-defined parameter that is also commonly referred to as total organic halide, or TOX. Analysis of this parameter is not specific to individual compounds or substances. The amount of an individual compound can be determined only if that compound is known to be the only organohalogen present in the sample. Since the AOX test is considerably less expensive than determining the presence and concentrations of individual compounds, it is commonly used to determine the bulk concentration of organic halides in industrial discharges that employ chlorine or another halogen in the manufacturing process.

AOX Breakthrough

The term "AOX breakthrough" is used to describe a phenomenon that occurs when AOX concentrations are determined in environmental samples. Traditionally, concern about AOX breakthrough has been limited to measurements made using the column technique. In this technique, two small columns (tubes) that are 2–3 mm inside diameter (i.d.), approximately 4–5 cm in length, and packed with granular activated carbon (GAC) are connected in series. A sample containing AOX is forced through the column pair. After washing to remove inorganic salts, the AOX in each column is measured by combustion of the organic compounds along with the GAC, and the halide content is determined using a microcoulometer. Ideally, all organic compounds are adsorbed on the first column; in practice, however, not all compounds are completely adsorbed from some samples. The amount of breakthrough is the percentage of organic halide captured by the second column relative to the total in both columns. If all of the organic halide is adsorbed on the first column and none on the second column, the percent breakthrough will be zero.

In the batch procedure, GAC is added to a sample in a flask and the flask is shaken or stirred to bring the AOX in contact with the GAC. After adsorption, the sample and GAC are filtered through a polycarbonate filter to trap the GAC. The GAC and filter are washed to remove inorganic salts, and are then combusted, and the organic halide is determined by microcoulometer. Since current methods require only a single adsorption, breakthrough is undefined when using the batch adsorption technique.

TECHNICAL APPROACH

Five factors that may affect adsorption of organic halides were tested. These were:

- Adsorption procedure
- Column type
- Source type
- Dilution level
- Laboratory

Evaluation of the effect of adsorption procedure focused on the use of the column versus the batch techniques. Evaluation of column type focused on the inside diameter of the columns used and the granular activated carbon used to prepare the column. Evaluation of source type focused on the differences or similarities between final effluent and in-process samples from the pulp and paper industry. Evaluation of dilution level focused on whether greater dilution resulted in lower breakthrough. Finally, the effects of laboratory variability were evaluated across the four other potential sources of variability.

Adsorption Procedure and Column Type

Adsorption was performed using either the column or batch adsorption procedure. In order to measure breakthrough with the batch procedure, Method 1650 was modified to require a second adsorption. The amount of breakthrough was then measured as the fraction of organic halide found in the second adsorption relative to the total in both adsorptions.

Prior to the study, Method 1650 specified the use of glass columns that were 4–5 cm long with a 2 mm i.d. Other methods for the determination of AOX, however, allowed the use of 2–3 mm i.d. columns as well as plastic columns. In order to evaluate the effect of these column diameters and construction materials on AOX results, Method 1650 was modified for this study to allow use and evaluation of several column sizes, types, and packing methods. These were:

- 2 mm i.d. prepacked glass columns (2MMPP)
- 2 mm i.d. hand-packed glass columns (2MMHP)
- 3 mm i.d. prepacked plastic columns (3MMPP)
- 3 mm i.d. hand-packed glass columns (3MMHP)

In order to minimize GAC as a potential source of variability, the GAC used in all tests was supplied from the same lot by CPI of Palo Alto, California. Prepacked columns were packed by CPI; hand-packed columns were packed in each laboratory using the GAC supplied by CPI.

Samples

A total of nine individual samples from pulp and paper mills were analyzed. Of the nine samples, four were influents to the treatment processes used by the facilities, and five were effluents from the treatment processes. Samples were preserved to pH < 2 (as required by Method 1650), packed in ice, and shipped via overnight courier directly to the participating laboratories. Sufficient sample volume was supplied for all tests. The laboratories were required to refrigerate the samples until analysis.

Not all adsorption types were used by all laboratories. The matrix of laboratories and adsorption types is as follows:

Laboratory	Batch	2MMHP	2MMPP	3MMHP	3MMPP
A				X	X
B	X			X	X
C		X	X	X	X
D	X	X	X	X	X

Ideally, each adsorption type should be tested in a minimum of three laboratories to retain the potential effects of interlaboratory variability in the results. As can be seen from the matrix above, the 3 mm hand-packed and prepacked types were tested in four laboratories, and each of the other types was tested in two laboratories. The small number of laboratories would likely result in an underestimation of the total variability of AOX measurements, but the effects of breakthrough should be discernible in this small number of laboratories if breakthrough is a predominant process.

Dilution Level and Breakthrough

The study design required that each sample be analyzed in duplicate at two breakthrough levels, one greater than and one less than 10%. Breakthrough was controlled by diluting varying amounts of a given sample using a constant volume of reagent water (water known to contain no organic halide). If breakthrough was greater than 10%, a smaller aliquot of sample was diluted; if breakthrough was less than 10%, a larger aliquot of sample was diluted. The objective was to bracket the 10% breakthrough level so that the uniqueness of that level could be studied.

When determining the amount of organic halide in a given sample, the microcoulometer reading for an accompanying blank is subtracted from the reading for the sample. This subtraction is necessary because it is virtually impossible to obtain a zero or nondetect reading from the instrument due to traces of organic halide that are typically present in the GAC used for adsorption and/or traces of organic vapors that are typically present in the laboratory atmosphere. The subtraction becomes unreliable, however, as the concentration of organic halide in a given sample, or at a given level of dilution of a sample, approaches the level in the blank. In order to ensure the reliability of the results in this study, the amount of halide detected in the sample was required to be at least three times that of the accompanying blank. If dilution of the sample to a concentration level of less than three times the blank value was necessary to achieve less than 10% breakthrough, the laboratory was to report the result at the lowest sample volume that produced a result above the factor of three level, regardless of the amount of the breakthrough.

Data Reporting and Validation

Laboratories reported data on diskette in a format supplied by the EPA Sample Control Center. The diskette format was used to elicit consistent reporting of results. Raw and summary data accompanied the diskette. Data were checked for completeness by comparing the number and types of analyses performed with the number that were required under each laboratory's contract. Laboratories were notified of incomplete or missing data and were requested to resubmit or provide clarification as necessary to address deficiencies. Data were validated by recalculating RPDs between the duplicates and by using calibration data to recalculate breakthrough. Recalculating results assured that all calculations were performed uniformly.

Statistical Analyses

Evaluation of the data consisted of (1) confirming that breakthrough levels reported by laboratories above and below the 10% criterion were significantly different, (2) examining the relationship between sample dilution and the amount of breakthrough, (3) examining the relationship between amount of breakthrough and AOX concentration detected in the samples, (4) examining the relationship between breakthrough and measurement error (as indicated by relative percent difference between duplicate samples), (5) examining the effects of adsorption type, laboratory variability, and sample variability on the concentration of AOX detected in sample, (6) examining the effects of the number of adsorptions on detected AOX concentration, and (7) examining the relative contribution of each column/batch run to the total detected concentration, and the precision of each column/batch run.

As stated above, each sample was analyzed in duplicate at two breakthrough levels, one greater than and one less than 10%. Each sample was divided into pairs consisting of one replicate analyzed at a breakthrough level below 10%, and one replicate analyzed at a breakthrough level above 10%. This allows extraneous factors such as the matrix and the mix of analytes present to be controlled. Paired t tests were performed to determine if any differences existed in detected concentration, amount of breakthrough, or measurement error. An Analysis of Variance (ANOVA) was performed to determine if any differences existed among laboratories, adsorption types, or sources of samples (i.e., influent or effluent). The results of specific statistical tests used to evaluate the data are described below.

RESULTS

Data Validation

An operational problem created by the blank subtraction requirement in Method 1650 precluded measurement of less than 10% breakthrough in some samples. To maximize the chance of determining if breakthrough affected measured concentrations, results for these samples were removed from further statistical analysis.

The batch adsorption data by Laboratory B showed much higher concentrations of AOX for many samples than the batch adsorption data by Laboratory D. These concentrations were also much higher than all of the column data (including the Laboratory B column data). The batch adsorption data by Laboratory B also showed breakthroughs as high as 90%. Because the high concentrations detected in the batch adsorption at Laboratory B are inconsistent with the batch adsorption at Laboratory D and inconsistent with the column adsorptions at all of the laboratories (including Laboratory B), it was concluded that the Laboratory B batch data may be the result of contamination. Further, because the percent breakthrough in the Laboratory B batch adsorption data ranged as high as 90%, it was concluded that some of these results may be attributable to reversal of results from the first and second adsorptions. Therefore, the Laboratory B batch data were eliminated from further analysis.

Confirmation of Breakthrough

The results of the paired t tests to confirm that the breakthrough levels achieved in this study were significantly different across all samples at the two dilution levels reported by the laboratories are shown in Table 6.1. The null hypothesis for these tests was that the difference between the breakthrough at the two dilution levels is zero. The probability of incorrectly rejecting the null hypothesis for each source and adsorption type is shown in Table 6.1 under the column labeled "Prob >|T|." Breakthrough is considered to be statistically different if the number in this column is less than or equal to 0.05 — that is, there is 95% probability that a difference exists.

The results in Table 6.1 show the differences without and with log-transformation of the data. Due to the heterogeneous nature of the variances, a log transformation was employed. This transformation tends to accentuate the differences by compressing the data and reducing the effects of possible outliers. It should be noted that this may also reduce the effects of variant, but legitimate, data. The data in Table 6.1 show that the overall differences in breakthrough achieved during this study are statistically significant. The only exceptions are the levels achieved in influent samples analyzed by the batch adsorption procedure. The low number of samples included in that data set (N = 2), however, precludes drawing conclusions with any degree of confidence.

Table 6.1. Confirmation of Different Breakthrough Levels.

a. Without Log Transformation

			Low Breakthrough Replicates			High Breakthrough Replicates			Difference				Prob
Source	Ads_Type	N	Mean BT[a]	Std Dev	RSD[b]	Mean BT	Std Dev	RSD	Mean	Std Dev	RSD	t-Value	> ITI
Effluent	2MMHP	7	7.36	1.40	18.99	19.73	7.30	37.02	12.37	7.01	56.64	4.67	0.00
Effluent	2MMPP	9	5.42	2.94	54.16	15.87	3.61	22.77	10.44	3.43	32.81	9.14	0.00
Effluent	3MMHP	12	6.08	2.69	44.20	19.60	5.79	29.55	13.53	6.55	48.46	7.15	0.00
Effluent	3MMPP	15	4.87	3.19	65.49	19.18	8.05	41.97	14.31	8.04	56.18	6.89	0.00
Effluent	Batch	4	6.95	1.22	17.60	13.23	1.93	14.56	6.28	3.01	47.97	4.17	0.03
Influent	2MMHP	7	5.69	2.55	44.85	17.81	4.99	28.01	12.13	4.93	40.66	6.51	0.00
Influent	2MMPP	7	4.93	3.27	66.44	13.46	2.58	19.15	8.53	2.24	26.25	10.08	0.00
Influent	3MMHP	5	4.96	3.92	78.97	20.04	4.84	24.14	15.08	7.48	49.60	4.51	0.01
Influent	3MMPP	10	6.31	2.24	35.50	16.34	4.36	26.66	10.03	4.42	44.04	7.18	0.00
Influent	Batch	2	7.85	1.06	13.51	13.20	1.98	15.00	5.35	0.92	17.18	8.23	0.08

b. After Natural Log Transformation

			Low Breakthrough Replicates			High Breakthrough Replicates			Difference				Prob
Source	Ads_Type	N	Mean BT	Std Dev	RSD	Mean BT	Std Dev	RSD	Mean	Std Dev	RSD	t-Value	> ITI
Effluent	2MMHP	7	2.02	0.14	6.76	2.86	0.37	13.00	0.84	0.33	39.01	6.78	0.00
Effluent	2MMPP	9	1.42	0.89	62.84	2.74	0.19	7.00	1.32	0.83	62.64	4.79	0.00
Effluent	3MMHP	12	1.29	1.20	92.98	2.92	0.32	10.87	1.63	1.24	76.49	4.53	0.00
Effluent	3MMPP	15	1.16	1.29	111.33	2.87	0.35	12.13	1.71	1.39	81.20	4.77	0.00
Effluent	Batch	4	1.94	0.16	8.36	2.58	0.16	6.11	0.63	0.29	46.11	4.34	0.02
Influent	2MMHP	7	1.71	0.45	26.47	2.90	0.26	8.79	1.19	0.41	34.64	7.64	0.00
Influent	2MMPP	7	0.96	1.84	192.29	2.60	0.17	6.38	1.65	1.75	106.21	2.49	0.05
Influent	3MMHP	5	1.36	1.08	79.23	3.01	0.22	7.34	1.65	1.19	72.20	3.10	0.04
Influent	3MMPP	10	1.69	0.45	26.58	2.76	0.25	9.24	1.07	0.38	35.63	8.88	0.00
Influent	Batch	2	2.00	0.26	13.14	2.62	0.06	2.22	0.62	0.20	33.06	4.28	0.15

[a] Percent breakthrough.
[b] Relative standard deviation (RSD) = 100×Std. dev/mean.

Relationship Between Sample Dilution and Breakthrough

One of the hypotheses concerning breakthrough is that more dilute samples will result in lower amounts of breakthrough than less dilute samples. In order to test this hypothesis, paired t tests were performed to compare the amount of breakthrough for the less dilute aliquot of each sample and adsorption type with that for the more dilute aliquot. The results of these tests are shown in Table 6.2. The null hypothesis for these tests was that the difference between the breakthrough at the two dilution levels is zero. The probability of incorrectly rejecting the null hypothesis for each source and adsorption type is shown in Table 6.2 under the column labeled "Prob >|T|." Breakthrough is considered to be statistically different at the two dilution levels if the number in this column is less than or equal to 0.05.

The data in Table 6.2 show that the overall differences in breakthrough between dilution levels are significant for 8 of 10 source/adsorption type combinations, and therefore that more dilute samples do result in lower breakthrough than less dilute samples. The only exceptions are the levels achieved in (1) influent samples by the batch adsorption procedure and (2) influent samples by the 2MMPP column procedure. Again, in the case of the influent samples by batch adsorption procedure, the number of samples analyzed (N = 2) was too small to produce a reliable t test result.

Relationship Between Breakthrough and Detected Concentration

The relationship between breakthrough and the detected concentration of AOX was evaluated in two ways. First, the difference in the concentration detected with low and high breakthrough was calculated for each low/high breakthrough pair. These calculated concentration differences were plotted as a function of the high percent breakthrough for each of the pairs. This plot is shown in Figure 6.1. If greater breakthrough levels produced a lower concentration because AOX was being lost, the graph would show a downward trend. However, such a trend does not exist. Therefore, the concentration determined in a sample appears to be independent of the amount of breakthrough, at breakthrough levels up to the highest seen in this study (40%).

A paired t test was performed as a second means of evaluating the relationship between breakthrough and detected concentrations of AOX. Specifically, the paired t test was performed for each type of sample and for each type of adsorption in order to determine if breakthrough above and below the 10% level produced statistically significant differences in detected AOX concentrations. As some samples had to be diluted to produce breakthrough less than 10%, the laboratory-reported AOX concentrations were normalized by dividing by the sample volumes used to bracket the 10% breakthrough level. Table 6.3 provides the results of the paired t tests with and without log-transformation.

The null hypothesis for this paired t test is that the difference between the normalized values is zero. The probability of incorrectly rejecting the null hypothesis for each source and adsorption type is shown in Table 6.3 under the column labeled "Prob >|T|." The difference between the detected concentrations at the low and high breakthrough levels is statistically different if the associated probability is less than or equal to 0.05. As can be seen from Table 6.3, the only sample and adsorption type that showed significant differences in detected concentration at different breakthrough levels was the 3MMHP column in final effluent samples. For all other column and stream types, breakthrough levels of greater than 10% do not result in significantly different AOX concentrations than breakthrough levels of less than 10%.

Relationship Between Breakthrough and Relative Percent Difference

A premise of the study was that measurement error would increase as the percent breakthrough increased. To evaluate the validity of this premise, the RPD between each duplicate pair analyzed at the same dilution level was evaluated as a function of percent breakthrough. This relationship is plotted in Figure 6.2, which suggests that RPD is independent of percent breakthrough. Independence of the relationship between RPD and breakthrough was verified through a correlation analysis ($r = -0.03$).

Table 6.2. Differences in AOX Breakthrough as a Function of Dilution Level.

Difference = Breakthrough (More Dilute) – Breakthrough (Less Dilute)

a. Without Log Transformation

			More Dilute Replicates			Less Dilute Replicates			Difference				Prob
Source	Ads_Type	N	Mean BT	Std Dev	RSD	Mean BT	Std Dev	RSD	Mean	Std Dev	RSD	t-Value	> ITI
Effluent	2MMHP	7	7.4	1.4	19	20	7.3	37	–12.0	7.0	–57	–4.7	0.00
Effluent	2MMPP	9	5.4	2.9	54	16	3.6	23	–10.0	3.4	–33	–9.1	0.00
Effluent	3MMHP	12	6.1	2.7	44	20	5.8	30	–14.0	6.6	–48	–7.2	0.00
Effluent	3MMPP	15	4.9	3.2	65	19	8.0	42	–14.0	8.0	–56	–6.9	0.00
Effluent	Batch	4	7.0	1.2	18	13	1.9	15	–6.3	3.0	–48	–4.2	0.03
Influent	2MMHP	7	5.7	2.6	45	18	5.0	28	–12.0	5.0	–41	–6.5	0.00
Influent	2MMPP	6	6.4	4.3	67	11	5.5	48	–4.9	8.0	–160	–1.5	0.20
Influent	3MMHP	5	5.0	3.9	79	20	4.8	24	–15.0	7.4	–50	–4.5	0.01
Influent	3MMPP	10	6.3	2.2	36	16	4.4	27	–10.0	4.4	–44	–7.2	0.00
Influent	Batch	2	7.9	1.1	14	13	2.0	15	–5.4	0.9	–17	–8.2	0.08

b. After Natural Log Transformation

			More Dilute Replicates			Less Dilute Replicates			Difference				Prob
Source	Ads_Type	N	Mean BT	Std Dev	RSD	Mean BT	Std Dev	RSD	Mean	Std Dev	RSD	t-Value	> ITI
Effluent	2MMHP	7	2.0	0.14	6.8	2.9	0.37	13.0	–0.84	0.33	–39	–6.7	0.00
Effluent	2MMPP	9	1.4	0.89	63.0	2.7	0.19	7.0	–1.30	0.83	–63	–4.7	0.00
Effluent	3MMHP	12	1.3	1.20	93.0	2.9	0.32	11.0	–1.60	1.20	–76	–4.5	0.00
Effluent	3MMPP	15	1.2	1.30	110.0	2.9	0.35	12.0	–1.70	1.40	–81	–4.7	0.00
Effluent	Batch	4	1.9	0.16	8.4	2.6	0.16	6.1	–0.63	0.30	–46	–4.3	0.02
Influent	2MMHP	7	1.7	0.45	26.0	2.9	0.26	8.8	–1.20	0.40	–35	–7.6	0.00
Influent	2MMPP	6	1.1	2.00	185.0	2.4	0.45	19.0	–1.30	2.10	–160	–1.5	0.18
Influent	3MMHP	5	1.4	1.10	79.0	3.0	0.22	7.3	–1.70	1.20	–72	–3.1	0.04
Influent	3MMPP	10	1.7	0.45	27.0	2.8	0.25	9.2	–1.10	0.38	–36	–8.9	0.00
Influent	Batch	2	2.0	0.26	13.0	2.6	0.06	2.2	–0.62	0.20	–33	–4.3	0.15

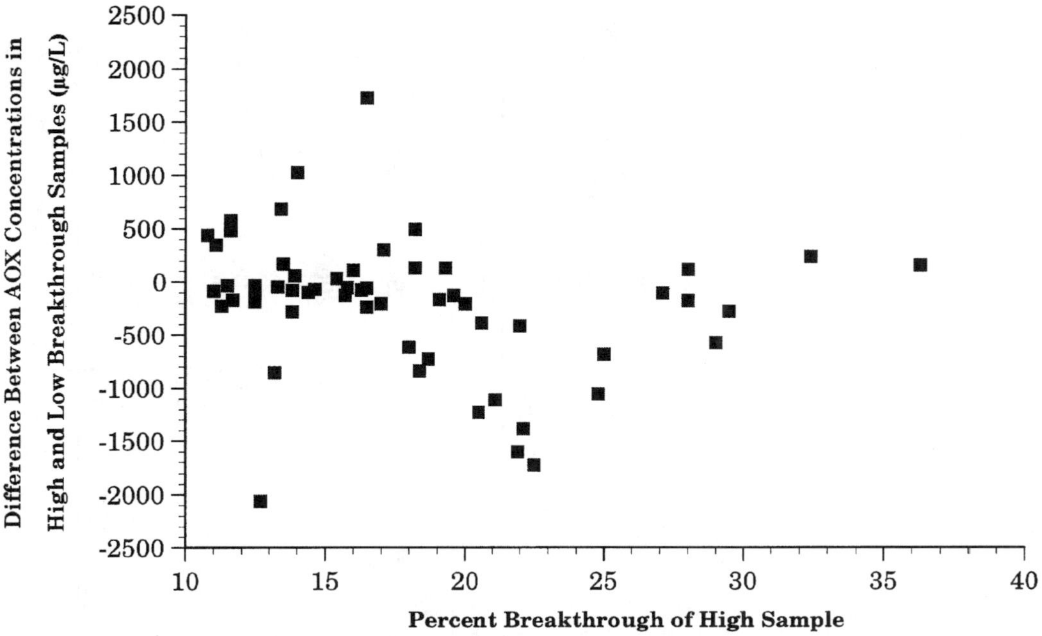

Figure 6.1. Differences in AOX concentration as a function of high breakthrough level.

Lastly, a paired t test was constructed with the null hypothesis that the RPDs at low and high breakthrough are identical. The results of the paired t test for non-log-transformed data and log-transformed data are shown in Table 6.4. The RPDs at low and high breakthrough are statistically different for a given source and adsorption type if the associated number in the "Prob >|T|" column is less than or equal to 0.05. The results of the paired t test show that, with the exception of effluent samples analyzed by the 2 mm prepacked glass columns (2MMPP), the differences in RPD at low and high breakthrough levels are not significant.

Differences in Concentration by Adsorption Technique, Laboratory, and Source Type

The data were examined for differences in the detected concentration of AOX among the five adsorption techniques, among the four laboratories, and between the two wastewater sources (influent and effluent). To test the effect of each of these differences and interactions among these differences, a three-way (adsorption technique × source × laboratory) analysis of variance (ANOVA) was performed to determine the following:

1. Differences among adsorption techniques (Batch, 2MMHP, 2MMPP, 3MMHP, 3MMPP)
2. Differences between source types (effluent vs. influent)
3. Interactions between source type and adsorption technique (e.g., is the relationship among adsorption techniques consistent across source types?)
4. Differences among laboratories
5. Interactions between laboratory and adsorption technique (e.g., is the relationship among adsorption techniques consistent across laboratories?)
6. Interactions between laboratory and source type (e.g., is the relationship among laboratories consistent across source types?)
7. Interactions between laboratory, adsorption technique, and source type combined (e.g., is the relationship among adsorption techniques consistent across laboratories and source types?)

Table 6.3. Paired T-test on Differences in AOX Concentration as a Function of Breakthrough.

Difference = Concentration (low breakthrough) – Concentration (high breakthrough)

a. Without Log Transformation

			Low Breakthrough Replicates			High Breakthrough Replicates			Difference				Prob
Source	Ads_Type	N	Mean Conc[a]	Std Dev	RSD	Mean Conc	Std Dev	RSD	Mean	Std Dev	RSD	t-Value	> ITI
Effluent	2MMHP	7	3900	1500	37	4000	1600	39	–61.0	190	310	–0.86	0.43
Effluent	2MMPP	9	3400	1600	48	3400	1700	51	–6.9	390	5700	–0.05	0.96
Effluent	3MMHP	12	4600	1800	39	4000	1600	40	500.0	550	110	3.20	0.01
Effluent	3MMPP	15	3900	1500	38	3800	1700	44	110.0	370	350	1.10	0.29
Effluent	Batch	4	4300	1560	37	4100	1600	39	140.0	96	68	3.00	0.06
Influent	2MMHP	7	7600	4000	53	7300	3700	51	250.0	630	250	1.10	0.33
Influent	2MMPP	6	7900	3700	46	7900	3700	46	4.4	670	15000	0.02	0.99
Influent	3MMHP	5	7600	4300	56	7200	4200	58	430.0	740	170	1.30	0.27
Influent	3MMPP	10	8400	4100	49	8100	4000	50	300.0	1100	370	0.86	0.41
Influent	Batch	2	9000	5900	65	8700	5600	64	220.0	240	110	1.30	0.42

b. After Natural Log Transformation

			Low Breakthrough Replicates			High Breakthrough Replicates			Difference				Prob
Source	Ads_Type	N	Mean Conc	Std Dev	RSD	Mean Conc	Std Dev	RSD	Mean	Std Dev	RSD	t-Value	> ITI
Effluent	2MMHP	7	8.2	0.45	5.5	8.2	0.47	5.7	–0.01	0.04	440	–0.60	0.57
Effluent	2MMPP	9	8.0	0.50	6.2	8.0	0.51	6.4	0.00	0.10	2000	0.15	0.89
Effluent	3MMHP	12	8.3	0.47	5.7	8.2	0.53	6.4	0.13	0.15	110	3.10	0.01
Effluent	3MMPP	15	8.2	0.44	5.4	8.1	0.53	6.5	0.06	0.14	240	1.60	0.12
Effluent	Batch	4	8.3	0.43	5.2	8.3	0.48	5.8	0.05	0.05	120	1.70	0.18
Influent	2MMHP	7	8.8	0.50	5.7	8.8	0.48	5.5	0.02	0.07	290	0.90	0.39
Influent	2MMPP	6	8.9	0.43	4.8	8.9	0.43	4.8	0.00	0.12	17000	0.02	0.99
Influent	3MMHP	5	8.8	0.55	6.3	8.8	0.58	6.6	0.07	0.17	250	0.90	0.42
Influent	3MMPP	10	8.9	0.50	5.6	8.9	0.50	5.7	0.04	0.14	340	0.90	0.37
Influent	Batch	2	9.0	0.71	7.9	9.0	0.69	7.8	0.02	0.01	67	2.10	0.28

[a] Mean AOX concentration (μg/L)

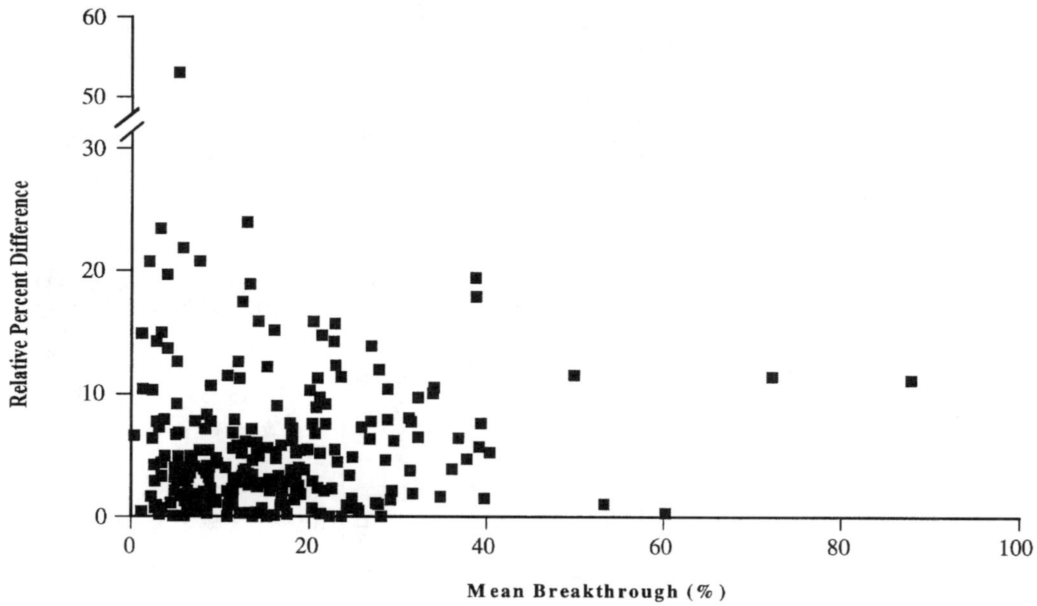

Figure 6.2. RPD as a function of breakthrough.

The results of this ANOVA are shown in Table 6.5. A value of less than 0.05 in the column labeled "Prob > F" denotes a significant difference for the associated effect or interaction. The results indicate that only source type (influent vs. effluent) yielded any statistically significant difference between the detected levels of AOX. Specifically, levels of AOX detected in influent waste streams were approximately twice the levels detected in effluent waste streams. No differences were detected among extraction techniques or among laboratories or among interactions of extraction technique, laboratory, and/or source type. The lack of interactions indicates that one or more of the variables is not masking a change in the other variables.

Number of Adsorptions

Table 6.6 shows the results of paired t tests on the effects of the number of adsorptions on detected AOX concentration. A difference between adsorptions is considered statistically significant if the value in the column "Prob > |T|" is less than or equal to 0.05.

The effects of the first, second, or both adsorptions on the precision and recovery of AOX determinations are shown in Table 6.7. The first adsorption recovers an average of 85% of the AOX found with two columns, whereas the second recovers an average of 15%. The percentage of recovery by the second adsorption is equivalent to the breakthrough.

DISCUSSION AND CONCLUSIONS

Sources of variability examined in this study included laboratory, sample source (influent or effluent), dilution level, adsorption procedure (column or batch), and column type (2 or 3 mm, hand-packed or prepacked). In retrospect, in order to evaluate the effect of the adsorption procedure and column type (2MMHP, 2MMPP, 3MMHP, 3MMPP, and batch), it may have been better to locate a sample susceptible to breakthrough, and then test that sample extensively in the laboratory capable of producing the most precise and accurate results. However, including all sources of variability has the practical value that the data are indicative of the use of multiple laboratories in the real world, and these data indicate that breakthrough is not a controlling source of variability in the AOX analysis.

Table 6.4. Differences in RPD as a Function of Breakthrough.

Difference = RPD (low breakthrough) – RPD (high breakthrough)

a. Without Log Transformation

			Low Breakthrough Replicates			High Breakthrough Replicates			Difference				Prob
Source	**Ads_Type**	**N**	**Mean RPD**	**Std Dev**	**RSD**	**Mean RPD**	**Std Dev**	**RSD**	**Mean**	**Std Dev**	**RSD**	**t-Value**	**> ITI**
Effluent	2MMHP	7	2.2	2.10	96	4.40	3.20	72.0	–2.20	3.0	140	–1.90	0.10
Effluent	2MMPP	9	2.0	2.10	110	3.30	1.30	40.0	–1.30	1.6	120	–2.50	0.04
Effluent	3MMHP	12	7.6	5.50	73	6.00	3.80	64.0	1.60	5.2	330	1.00	0.32
Effluent	3MMPP	15	5.1	5.80	110	5.90	6.00	100.0	–0.79	4.7	590	–0.66	0.52
Effluent	Batch	4	1.7	0.38	23	2.30	0.76	33.0	–0.62	1.1	180	–1.10	0.34
Influent	2MMHP	7	4.5	2.90	64	2.20	2.40	110.0	2.30	4.5	190	1.40	0.22
Influent	2MMPP	5	2.2	2.00	89	3.50	2.00	58.0	–1.20	1.8	130	–1.70	0.16
Influent	3MMHP	5	9.6	6.80	71	3.80	3.10	81.0	5.80	5.1	88	2.50	0.07
Influent	3MMPP	10	7.1	7.50	100	5.50	5.10	92.0	1.60	9.4	580	0.50	0.60
Influent	Batch	2	2.2	1.30	61	0.30	0.03	9.4	1.90	1.3	69	2.10	0.29

b. After Natural Log Transformation

			Low Breakthrough Replicates			High Breakthrough Replicates			Difference				Prob
Source	**Ads_Type**	**N**	**Mean RPD**	**Std Dev**	**RSD**	**Mean RPD**	**Std Dev**	**RSD**	**Mean**	**Std Dev**	**RSD**	**t-Value**	**> ITI**
Effluent	2MMHP	7	–0.03	1.80	6800	1.20	0.84	69.0	–1.20	1.70	140	–1.90	0.10
Effluent	2MMPP	9	0.05	1.40	2900	1.10	0.46	41.0	–1.10	1.20	120	–2.60	0.03
Effluent	3MMHP	12	1.80	0.60	35	1.60	0.72	46.0	0.25	0.70	280	1.20	0.24
Effluent	3MMPP	15	0.96	1.30	130	0.62	3.00	480.0	0.34	3.10	930	0.40	0.68
Effluent	Batch	4	0.48	0.29	54	0.77	0.38	50.0	–0.29	0.60	210	–0.97	0.40
Influent	2MMHP	7	1.30	0.67	51	–0.77	3.80	490.0	2.10	4.30	200	1.30	0.24
Influent	2MMPP	5	0.50	0.87	180	0.89	1.20	140.0	–0.39	0.84	210	–1.10	0.35
Influent	3MMHP	5	2.00	0.76	37	1.10	0.70	62.0	0.92	0.82	89	2.50	0.07
Influent	3MMPP	10	1.30	1.40	110	0.43	3.50	820.0	0.87	4.10	470	0.68	0.52
Influent	Batch	2	0.68	0.65	95	–1.20	0.09	7.7	1.90	0.56	29	4.80	0.13

Table 6.5. Analysis of Variance on AOX Breakthrough Data.

Dependent Variable: Concentration

Effect	D F	Sum of Squares	Mean Square	F Value	Prob > F
Ads_Type	4	5485062	1371266	0.20	0.94
Source	1	751578769	751578769	107.45	0.00
Ads_Type×Source	4	9820297	2455074	0.35	0.84
Lab	3	17897955	5965985	0.85	0.47
Ads_Type×Lab	5	3202554	640511	0.09	0.99
Source×Lab	3	8290300	2763433	0.40	0.76
Ads_Type×Source×Lab	5	1777227	355445	0.05	0.99
Error	199	1391915951	6994553		
Total	224	2189968116			

Source	Mean	N
Influent	7300	98
Effluent	3600	127

Key:
Ads_Type = adsorption type (batch vs. 2MMHP vs. 2MMPP vs. 3MMHP vs. 3MMPP).
Source = source type (influent vs. effluent).
Lab = laboratory (Laboratory A vs. Laboratory B vs. Laboratory C vs. Laboratory D).

Table 6.6. Paired t Tests on the Effect of the Number of Adsorptions on Detected AOX Concentration.

Batch Adsorptions	Column Adsorptions	Mean RPD[a]	RSD (%)	t-Value	Prob > ITI
1	1	0.76	1091	0.33	< 0.747
1	2	−15.30	50	−7.24	< 0.001
2	1	15.00	50	7.24	< 0.001
2	2	−1.00	493	−0.73	< 0.479

[a] RPD = relative percent difference in AOX concentration between batch and column adsorptions. A negative RPD indicates a higher concentration was detected with column adsorption.

Table 6.7. Effect of Adsorption on Recovery and Precision.

	Relative Recovery (% of both)	Precision (RPD Between Duplicates)
Both Adsorptions	100%	5.0
Column/Batch 1 alone	85 ± 9%	5.4
Column/Batch 2 alone	15 ± 9%	20.1

In summary, the results of this study were as follows:

1. Breakthrough levels of less than and greater than 10% were achieved for all adsorption types, including the batch procedure. This finding confirms that the goal of the study to bracket the 10% level was met.
2. Except for influent samples analyzed with 2 mm prepacked columns, more dilute samples produced lower AOX breakthrough than did less dilute samples. This finding confirms another premise of the study. The lack of significant difference for 2 mm prepacked columns may be due to the relative proximity of the mean breakthrough of the less dilute samples (11%) to the 10% cutoff.

3. In general, the concentration of AOX analytically determined in the sample is independent of AOX breakthrough. This finding was observed for all source types and adsorption techniques except for treated effluent samples analyzed with 3 mm hand-packed columns. It should be noted that the highest level of breakthrough observed in this study was 40%. Therefore, any conclusion about the independence of breakthrough and concentration may only be stated with certainty up to breakthrough levels of 40%. In addition, this conclusion only applies to AOX analyses using columns, as Method 1650 currently requires only one batch adsorption, and breakthrough cannot, therefore, be determined.
4. With the exception of the treated effluent samples analyzed with 2 mm prepacked columns, measurement error (quantitated as RPD between each duplicate pair analyzed at the same dilution level) was not related to the level of AOX breakthrough. This finding contradicted one of the hypotheses of the study.
5. AOX results generally agreed among laboratories.
6. Levels of AOX detected in the influent waste stream samples were approximately twice the levels detected in the treated effluent samples, regardless of laboratory or adsorption technique.

Breakthrough

Method 1650 requires that if column breakthrough exceeds 10%, results are to be considered suspect and the analysis needs to be repeated after sample dilution. The difficulty in analyzing some samples in this study at breakthrough levels of less than 10% (because the concentration of AOX detected in the sample at these breakthrough levels approaches the concentration detected in the blank) is consistent with similar difficulties encountered by both the EPA and the paper industry. The assumption has been that nearly all of the adsorbable organic halides in the sample are readily adsorbed by the carbon, and that organic halides are preferentially bound to the activated carbon. In reality, not all of the organic halides are adsorbed at the same rate, nor are they bound to the carbon to the same degree. Some species, such as chloroacetic acid, are only 5–15% adsorbed by the carbon. Such partially bound species of organic halides are also partially adsorbed by the second column, resulting in measurable breakthrough. As long as the proportion of organic halides absorbed by the column remains constant relative to the total amount of organic halides in the sample, the measured AOX concentration should be reliable. The constant proportionality assumption is supported, in this study, by the low relative percent difference in detected concentration (mean = 8.7%) between different dilution levels of each sample.

Column Diameter

Method 1650 also requires the use of 2 mm i.d. glass columns for the column adsorption technique, while similar methods allow the use of 2–3 mm i.d. glass or plastic columns. The results of this study do not indicate that column diameter or construction material have a significant impact on either the amount of breakthrough observed or on the concentration of AOX detected in the sample. Indeed, the data presented in Table 6.5 indicate that all of the column diameters and construction materials tested in this study produce similar results in terms of detected AOX concentrations. However, the uncertainties associated with the effects of column diameter and material on breakthrough* suggest that additional study is needed before it can be conclusively stated that the column requirements specified in Method 1650 should be revised.

* E.g., more dilute influent samples analyzed with 2 mm prepacked columns did not produce lower breakthrough than less dilute influent samples; breakthrough was not independent of concentration when effluent samples were analyzed with 3 mm hand-packed columns; and RPD was only related to breakthrough when influent samples were analyzed using 2 mm hand-packed columns.

Number of Adsorptions

As discussed earlier in this chapter, Method 1650 does not require the use of two batch adsorptions, and therefore does not include a breakthrough specification for the batch procedure. Other published methods are similar to Method 1650 in this respect. The finding of measurable breakthrough in samples analyzed by the batch procedure suggests that consideration of a dual adsorption and breakthrough specification for the batch technique may be appropriate. As shown in Table 6.6, the concentration of AOX detected is influenced by the number of adsorptions. Two adsorptions, whether by column or batch procedure, consistently yield greater concentrations than does one adsorption. Furthermore, AOX concentrations determined with columns are similar to those determined with batch procedures when the number of adsorptions is the same.

Precision, measured as the average relative percent difference in concentration between sample duplicates, is nearly identical for the first and both adsorptions, but is more variable for the second adsorption. This is consistent with the variability of the breakthrough, which is reflected in the second adsorption. Thus, according to these data, the only purpose served by the second adsorption is to increase the AOX concentration by approximately 15%. That the entire issue of breakthrough can be avoided makes elimination of the second adsorption worth considering. However, since the current Agency rulemaking is based on data obtained with two adsorptions for columns, a change in the number of adsorptions is not recommended without further investigation, under a study specifically designed to test this issue.

The findings presented in this report suggest that the breakthrough specification should either be removed from Method 1650, or should be set close to the highest breakthrough observed in this study (approximately 40%). The most conservative approach would be to set the breakthrough specification at 40% until further study of higher breakthrough levels indicates that the 40% breakthrough specification should be either adjusted upwards or removed entirely. Based on the results of this study, no other modifications to Method 1650 are warranted at this time.

Chapter 7

The Natural Production of Organochlorine Compounds

G.W. Gribble

ABSTRACT

Natural chlorine-, bromine-, iodine-, and fluorine-containing organic chemical compounds are abundant on our planet. Occurring on land, in the oceans, and in the atmosphere, they pervade every form of life. They are produced by marine and terrestrial plants, bacteria, fungi, insects, marine animals, and even mammals. Natural combustion processes, such as volcanoes and other geothermal events, and forest and brush fires contribute large quantities of halogenated compounds to the environment. In 40 years, the number of known natural organohalogens has multiplied 200 times, from a dozen in 1954 to nearly 2,800 today, a figure that includes more than 1,600 organochlorine compounds.

INTRODUCTION

In 1968, Fowden wrote: "Present information suggests that organic compounds containing covalently bound halogens are found only infrequently in living organisms."[1] While this observation was undeniable then, it is no longer true today. Indeed, the field of organohalogen natural products has flourished, and the first "International Conference on Naturally Produced Organohalogens" was held in The Netherlands in 1994. Not only are naturally occurring organohalogen compounds ubiquitous in our environment, but concentrations of some of these chemicals exceed their anthropogenic levels. Furthermore, new examples are continually being isolated and characterized from a variety of marine and terrestrial plant and animal sources. Although many examples of chlorine-, bromine-, iodine-, and even several fluorine-containing natural products exist,[2] the present chapter focuses principally on organochlorine compounds. Chlorine is in the political and environmental spotlight,[3] and an aggressive campaign by some environmental groups to ban summarily all production and uses of chlorine (and bromine) is underway. The number of known natural organohalogens currently stands at nearly 2,800, and is increasing monthly.[4] The present survey is an overview of the entire field of natural organochlorine compounds, with some coverage of the late 1995 literature. For a complete review of naturally occurring organohalogen compounds consult Gribble.[4]

A summary of the natural organohalogen compounds is as follows:

Compounds containing chlorine	1,650
Compounds containing bromine	1,450
Compounds containing iodine	85
Compounds containing fluorine	20

Many of these compounds contain both chlorine and bromine, and several contain iodine as well.

OCCURRENCE

Simple Alkanes and Related Compounds

The simplest natural organohalogen compounds — halogenated alkanes — are abundant on our planet (Table 7.1). The chloride ion is normally present in plants, wood, soil, and minerals,[5,6] and the combustion of this material inevitably leads to the formation of organochlorine compounds. Forest

Table 7.1. Biogenic Sources of Simple Haloalkanes.

Compound	Source(s)	Reference(s)
CH_3Cl	marine algae, giant kelp, wood-rotting fungus, ice plant, cultivated mushrooms, cedar, evergreen cypress, fungi, phytoplankton, forest fires, brush-burning, volcanoes, tobacco smoke, potato	5–10
CH_3Br	oceans, marine algae, giant kelp, volcanoes	8a,10
CH_3I	oceans, marine algae, volcanoes	6,8a,10
$CHCl_3$	marine algae, white cedar, lemon, orange, moss, barley, volcanoes, drill wells	6
$CHBr_3$	marine algae, oceans	12
CCl_4	marine algae, oceans, volcanoes, drill wells	6
CH_2Cl_2	volcanoes, barley	6
CBr_4	marine algae	12
CH_2ClBr	marine algae	2
CH_2ClI	marine algae	2
$CHBrCl_2$	marine algae, oceans	2
$CHBr_2Cl$	marine algae, oceans	12
CHI_3	marine algae	12
CH_2Br_2	marine algae, oceans	12
$CHBrICl$	marine algae	12

fires,[7] brush and vegetation burning,[8] and volcanoes[6,9,10] all produce significant and, in some cases, massive quantities of chloromethane. This global emission rate of chloromethane from the marine and terrestrial biomass is estimated to be 5 million tons per year,[11] whereas "man-made" chloromethane emissions are only 26,000 tons per year.[5] Several other simple haloalkanes also have natural sources (Table 7.1). Although chloroform, carbon tetrachloride, and methylene chloride all have natural sources, it is believed that these amounts are significantly less that the quantity of natural chloromethane that is produced.

Numerous halogenated alcohols, ketones, carboxylic acids and amides, aldehydes, epoxides, and alkenes have also been isolated and characterized from marine algae.[12] Indeed, nearly 100 different chlorinated, brominated, and iodinated compounds have been found in *Asparagopsis taxiformis*, the favorite edible seaweed of most Hawaiians.[12] The red alga *Marginisporum aberrans* produces the simple dichloroacetamide, the first report of this compound as a natural product.[13] The simple marine haloalkanes, such as chloroform and bromoform, probably arise in part from *in vivo* haloform reactions, a mechanism which may enable algae, for example, to secrete continuously these "anti-predator" chemicals.[12,14]

Volcanoes produce large quantities of both hydrogen chloride (up to 3 million tons/year) and hydrogen fluoride (up to 11 million tons/year),[6] which can react with organic compounds to produce organohalogens. Although the quantities may be small compared with anthropogenic levels, several chlorofluorocarbons (CFCs), such as $CHFCl_2$, CHF_2Cl, $CFCl_3$, CF_2Cl_2, $FClC{=}CF_2$, CCl_2FCClF_2, and others, have been detected in volcanic gases from the Santiaguito volcano in Guatemala[9] and the Kamchatka volcanoes in Siberia.[6] The concentrations of these chemicals as measured in and around the solfataric vents of the Kamchatka volcanoes range from 0.4–160 ppb of CFCs, which is some 400 times that of background levels. Estimates are that 75% of the world's 2000 active volcanoes have the proper mineral composition and geological configuration to produce these organochlorine compounds.[6]

Interestingly, both trichloroethylene and tetrachloroethylene are produced naturally by several subtropical species of marine macroalgae.[15] These dry-cleaning solvents were long thought to only have anthropogenic sources.

Terpenes and Related Compounds

Terpenes are ubiquitous in terrestrial organisms and are essential for life, but halogenated terpenes were only first isolated in 1963.[16] Examples of chlorinated terrestrial terpenes were first reported in 1969.[17] Since then, a legion of new chlorinated and brominated terpenes have been isolated and identified,[18,19] a few examples of which are shown here (Figures 7.1 and 7.2).

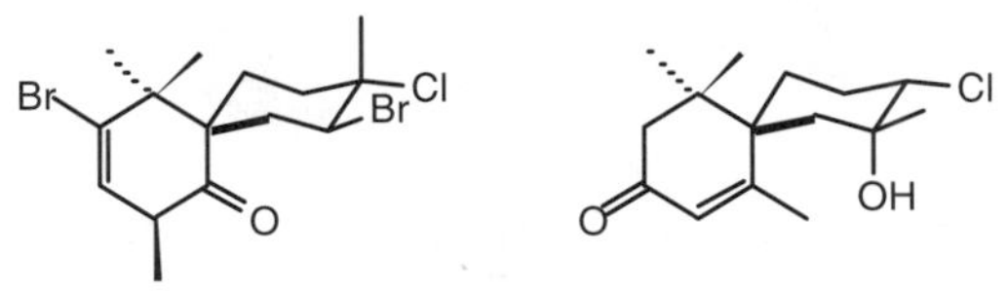

Figure 7.1. Halogenated sesquiterpenes from *Laurencia* red algae.

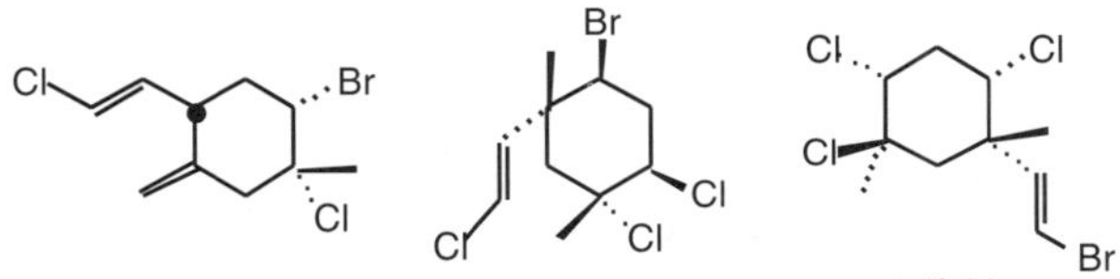

Figure 7.2. Halogenated monoterpenes from *Plocamium* red algae.

Numerous species of sea hare, a marine mollusk, also contain a varied collection of chlorinated terpenes, and, in many cases, these compounds originate in the animals' algal diet. For example, *Aplysia dactylomela* contains the novel compounds dactylomelol (**1**)[20] and (**2**).[21]

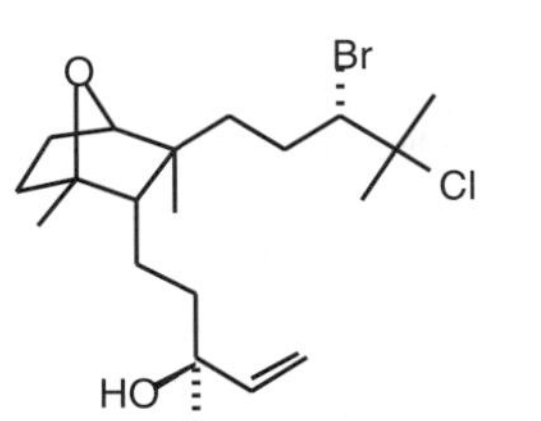

1 (dactylomelol)

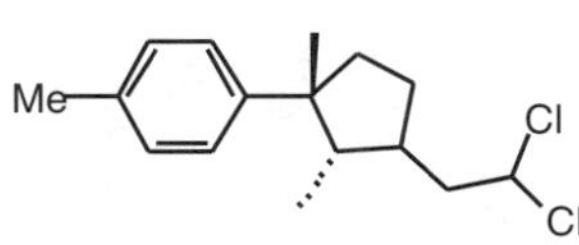

2

Nearly twenty chlorinated diterpenes and one tetraterpene have been isolated from several soft corals, and an example of the structural complexity of these compounds is erythrolide C (**3**).[22] A series of about ten chlorinated sesquiterpene indanones occur in species of ferns and related plants. For example, pterosin H (**4**) is found in *Microlepia*, *Dennstaedtia*, *Myriopteris*, and *Pteridium* spp.[23] The chlorinated 1,4-naphthoquinone **5** is produced by an Australian *Streptomyces* sp.,[24] and nearly 20 chlorinated iridoids, e.g., asystasioside E (**6**) (*Asystasia bella*),[25] have been isolated from several terrestrial plant species.

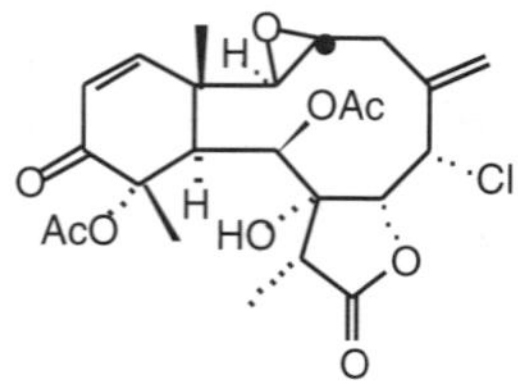

3 (erythrolide C)

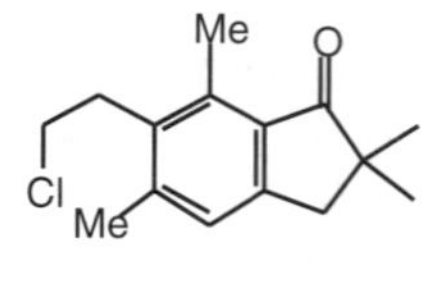

4 (pterosin H)

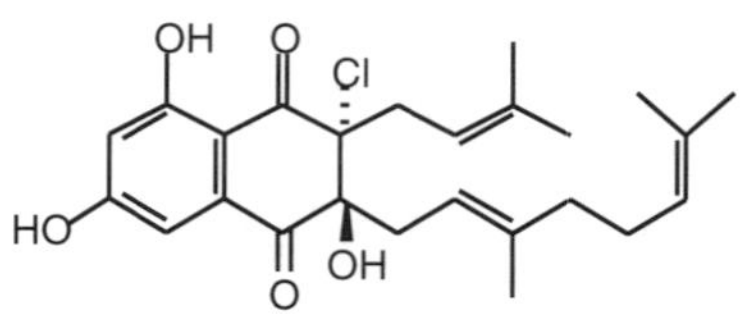

5

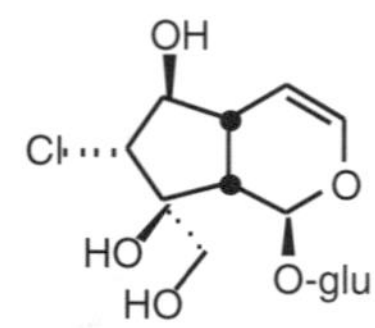

6 (asystasioside E)

Another very large group of chlorinated organic compounds is the "nonterpenes," compounds which do not arise from an obvious terpenoid pathway. The Guam "bubble shell" (*Haminoea cymbalum*) contains kumepaloxane (**7**),[26] a feeding-deterrent chemical which is discharged when this mollusk is disturbed by unwelcome carnivorous fishes. Marine red algae, especially *Laurencia*, are a source of fantastically complex nonterpenoid organohalogen chemicals. Thus, *L. implicata* has yielded six new cyclic ethers, some of which are allenes (e.g., **8**).[27]

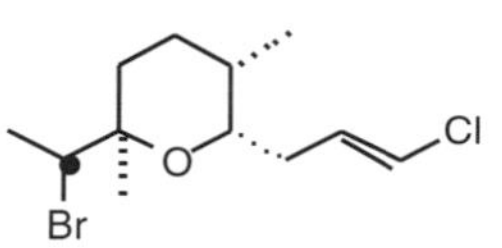

7
(kumepaloxane)

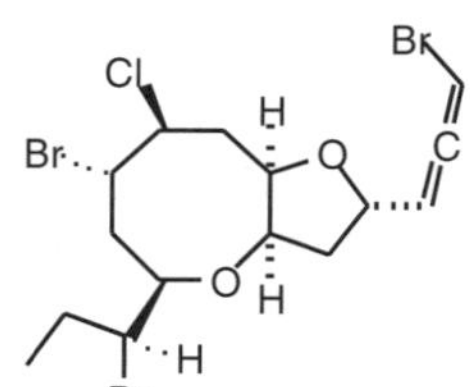

8

Nostoc blue-green algae are recognized as a threat to human health when they infest drinking water sources.[28] Cultivation and extraction of *Nostoc linckia* has yielded four unusual chlorinated paracyclophanes [e.g., nostocyclophane D (**9**)].[29] The structures of cryptophycins A (**10**) and C (**11**), which are produced by the blue-green alga *Nostoc* sp., have been established as shown, reflecting the presence of the D-configuration of the chloro-*O*-methyltyrosine unit.[30] It is of interest to note that cryptophycin A has extraordinary anticancer activity against solid tumors; clinical evaluation and future medical use are a strong possibility.

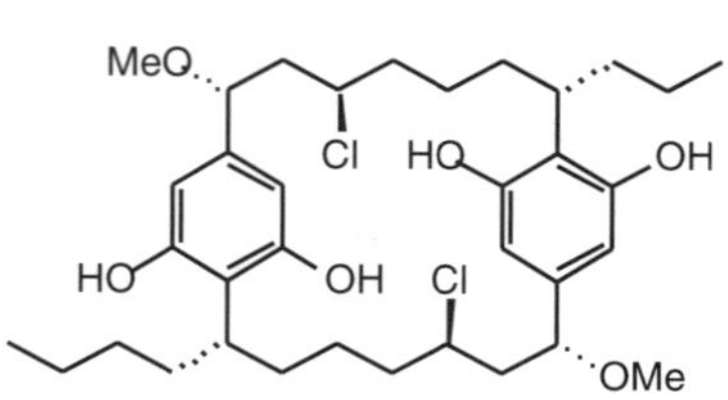

9 (nostocyclophane D)

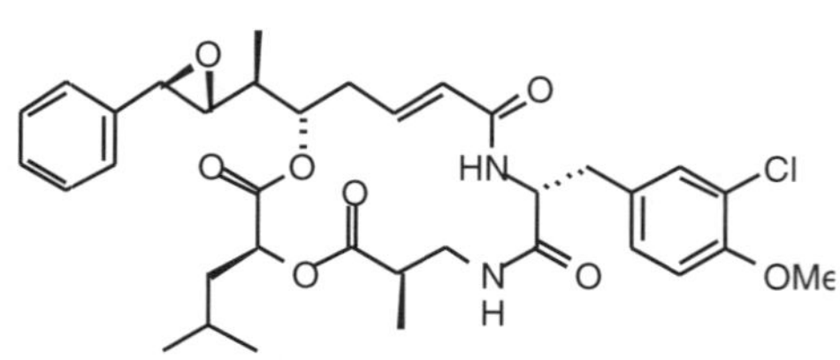

10 (cryptophycin A)

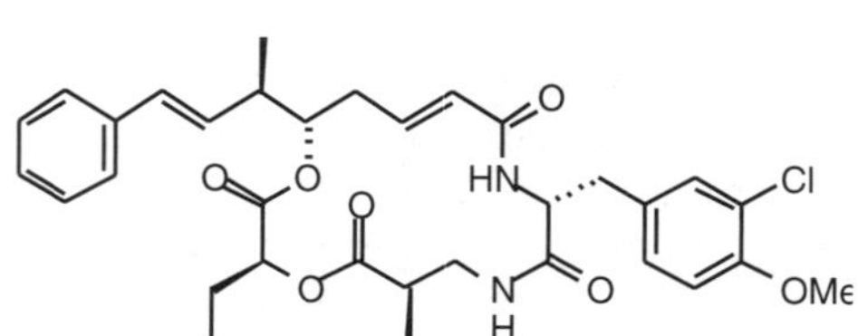

11 (cryptophycin C)

A new iridoid, phoyoside II (**12**), has been isolated from the plant *Phlomis younghusbandii*, bringing the total number of chlorinated iridoids to about 15.[31] The sesquiterpene lactone chloroscoparin (**13**) is a newly discovered metabolite of the plant *Centaurea scoparia.*[32] Similarly, *C. glatifolia* has yielded the novel guaianolide epicebellin J (**14**),[33] as well as the previously known chlorine-containing cebellin J, episolstiolide, 19-desoxy-15-chlorojanerin, centaurepensin, and epicentaurepensin (**15**).[34] In some cases, these chlorohydrins are probably artifacts and are formed from the corresponding epoxides by reaction with adventitious hydrogen chloride.

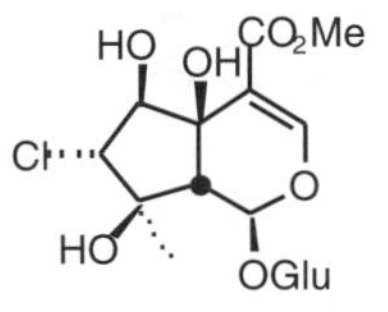

12 (phoyoside II)

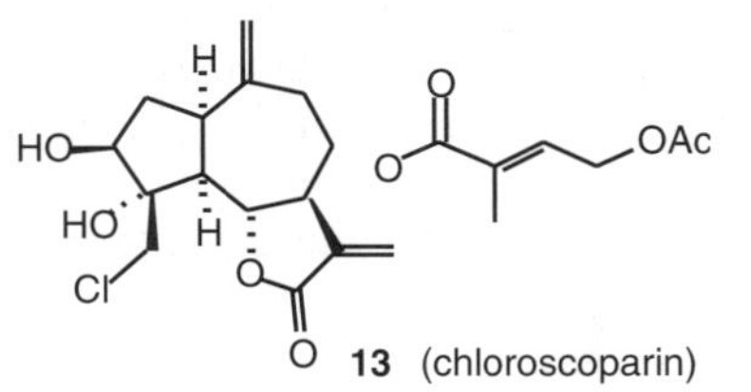

13 (chloroscoparin)

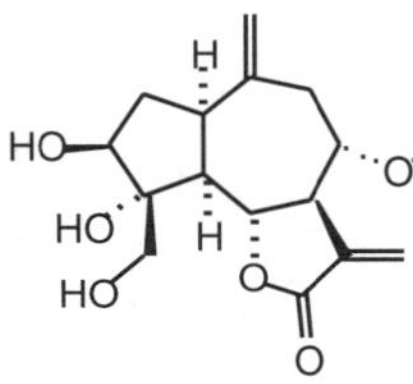

14 (epicebellin)

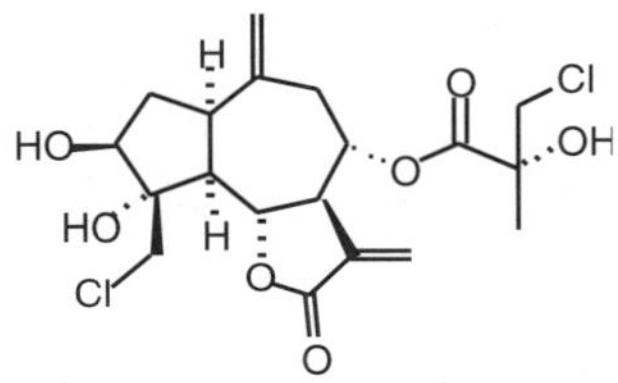

15 (epicentaurepsin)

The novel sesquiterpenoid 12-chloroillifunone C (**16**) has been found in the plant *Illicium tashiroi,*[35] and a later study of this same plant has revealed the two similar compounds 12-chloroillicinone E (**17**) and **18**.[36] The latter metabolite increases choline acetyltransferase activity and, thus, may find use in the treatment of Alzheimer's disease. The parasitic fungus *Cylindrocarpon olidum* produces 8-chlorocannabiorcichromenic acid (**19**), which has antifungal activity.[37] The known monoterpene **20** has been found for the first time in a blue-green alga, *Lyngbya majuscula.*[38] The North American "Cup Plant" (*Silphium perfoliatum*), which was used in folk medicine by Native Americans, contains the labdane diterpene chlorosilphanol A (**21**).[39] The plant *Triphasia trifolia* contains the novel coumarin derivative **22**.[40]

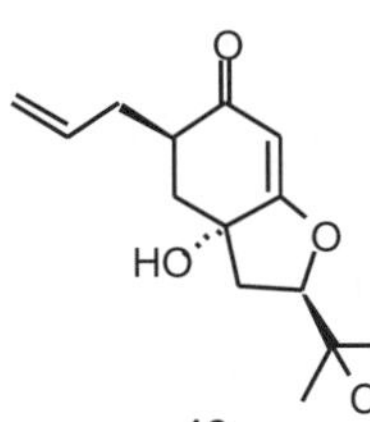

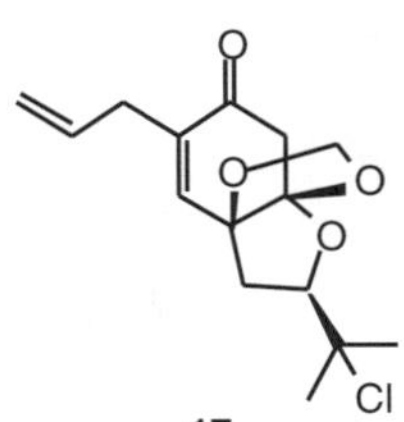

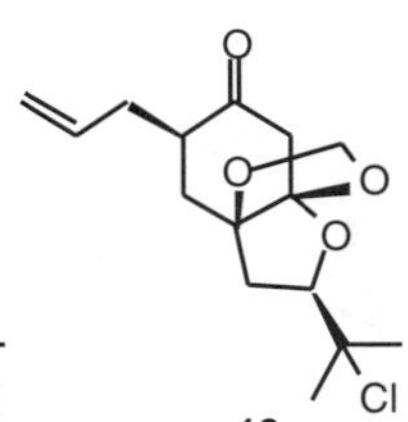

16 (12-chloroillifunone C) **17** (12-chloroillicinone E) **18**

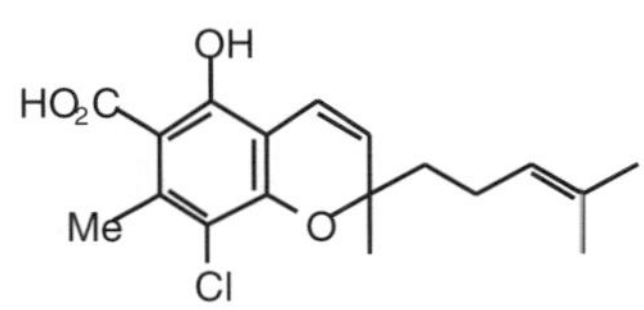

19 (8-chlorocannabiorcichromenic acid)

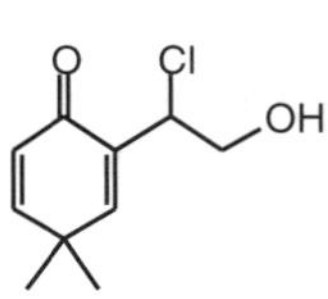

20

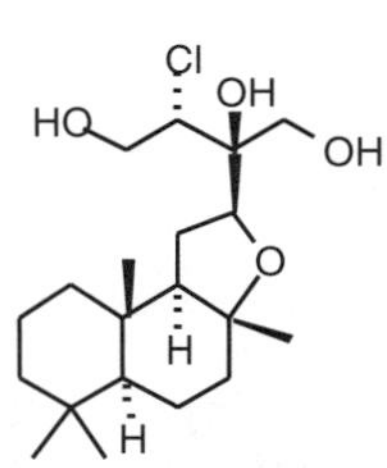

21 (chlorosilphanol A)

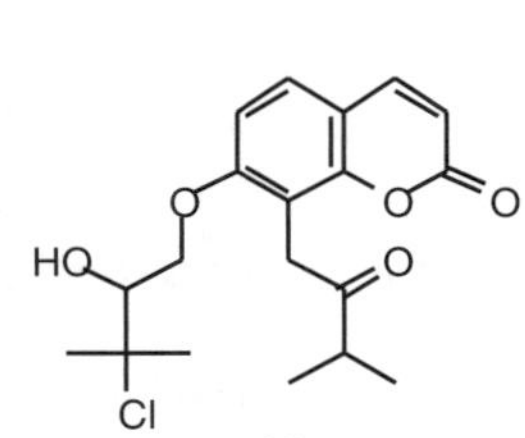

22

Amino Acids and Peptides

With the exception of the naturally occurring iodinated tyrosines, such as thyroxine, there are relatively few examples of halogenated amino acids and simple peptides. The marine sponge *Dysidea herbacea* contains four novel trichloromethyl metabolites, e.g., dysidin (**23**) and dysidenin (**24**).[41] Several *Streptomyces* and *Pseudomonas* spp. contain chlorinated amino acids and/or peptides, many of which have potent antibacterial activity. For example, *Streptomyces griseosporeus* produces γ-chloronorvaline[42] and *Pseudomonas syringae*, a bacterial pathogen of plants, produces several cyclic peptides, some of which contain the novel amino acid 4-chlorothreonine.[43] The terrestrial blue-green alga *Anabaena* sp. synthesizes the most unusual puwainaphycin C, a cardioactive cyclic peptide containing the unique amino acid 3-amino-14-chloro-2-hydroxy-4-methylpalmitic acid.[44] One of several antitumor cyclic peptides from the plant *Aster tataricus* has been isolated and shown to have the structure **25** (astin I).[45] This metabolite possesses the unusual amino acid β-hydroxy-γ-chloroproline.

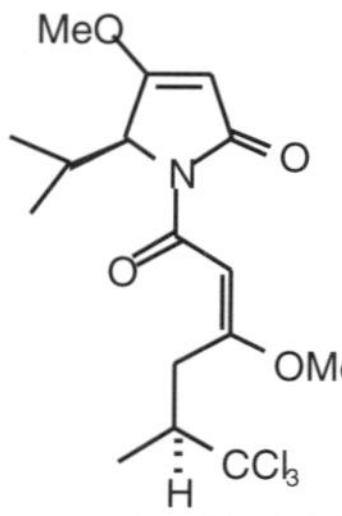

23 (dysidin)

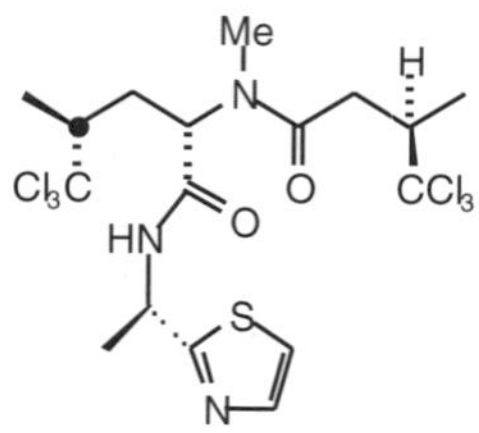

24 (dysidenin)

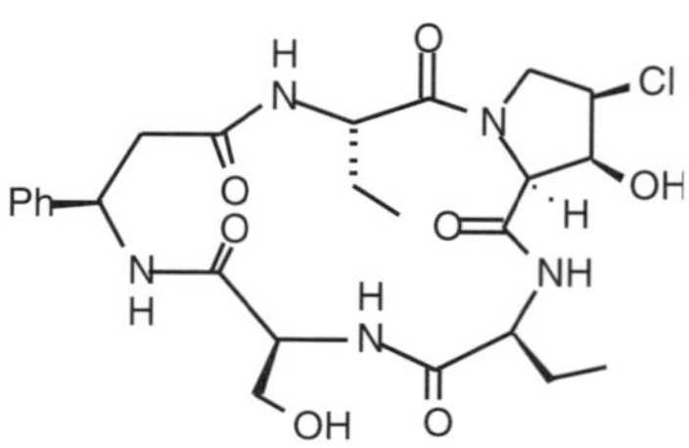

25 (astinI)

Steroids

A few examples of chlorinated steroids are found in the plants *Withania somnifera*, *Acnistus breviflorus*, *Physalis peruviana*, and *Jaborosa magellanica*, all of which contain withanolides [e.g., physalolactone C (**26**)] and related compounds [e.g., jaborochlorodiol (**27**)].[46] Control experiments indicate that these chlorohydrins are not isolation artifacts. The first three examples of chlorinated marine steroids, kiheisterones C, D, and E, were recently characterized from the Maui sponge *Strongylacidon* sp.,[47] and, as mentioned later, the German cockroach (*Blattella germanica*) utilizes two chlorinated steroid glycosides, blattellastanosides A and B, as aggregation pheromones.

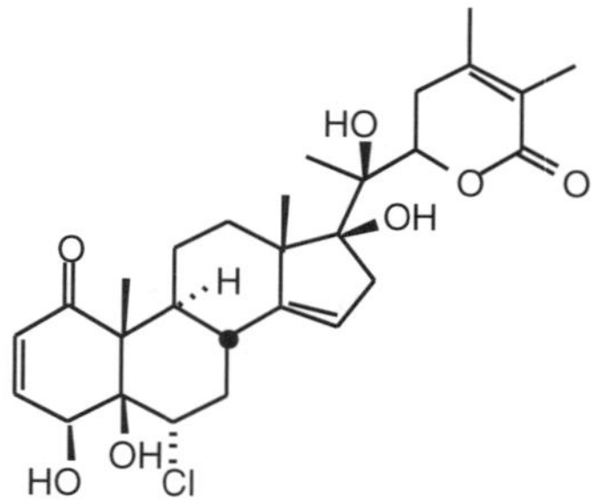

26 (physalolactone C)

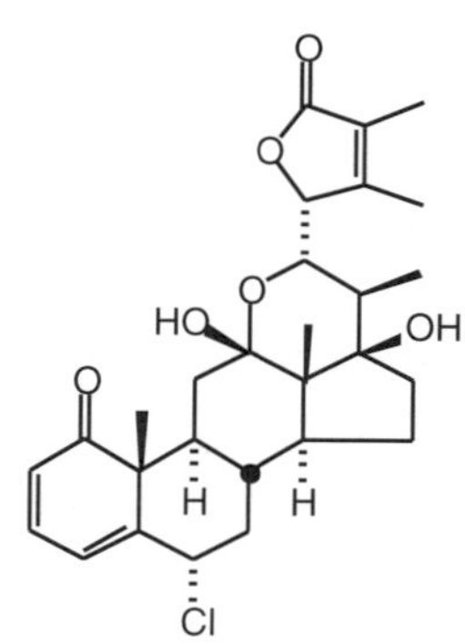

27 (jaborochlorodiol)

Fatty Acids, Prostaglandins, and Lipids

Although fluoroacetic acid and fluoro-fatty acids in certain terrestrial plants have been known for many years,[48] chloro-fatty acids were unknown until recently. The edible jellyfish and the white sea jellyfish (*Auritia aurita*) contain six fatty acid chlorohydrins.[49] Mammalian prostaglandins are of paramount importance in the regulation of physiological processes. Therefore, it was of enormous interest when chlorinated prostaglandins were discovered in the octocoral *Telesto riisei*[50] and in the stolonifer *Clavularia viridis,*[51] a marine animal. A total of about 30 novel chlorinated prostaglandins have now been identified [e.g., punaglandin 1 (**28**)], several of which have potent antitumor activity, and two of which contain bromine and iodine. Of some 22 freshwater algae species that were examined for chlorosulfolipids, all but one contained quantities of these novel lipids. For example, the phytoflagellate *Ochromonas danica* produces 12 such compounds, some of which contain six chlorines, e.g., **29**.[52]

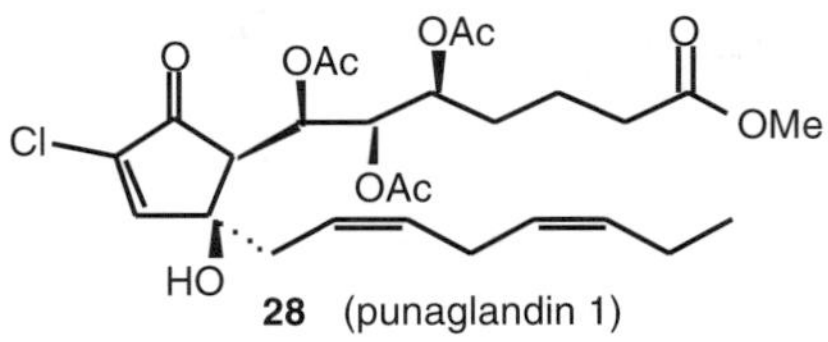

28 (punaglandin 1)

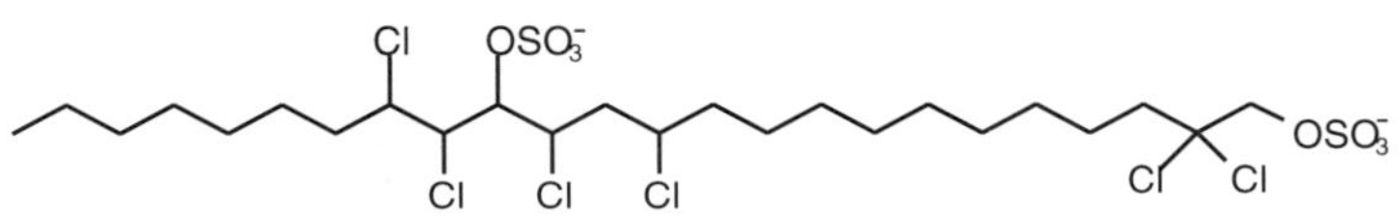

29

Pyrroles

The enormous reactivity of pyrroles in electrophilic substitution reactions endorses the observation that chlorinated and brominated pyrroles are found widely in nature. Following the isolation in 1958 of pyoluteorin (**30**), an antibiotic from *Pseudomonas aeruginosa,*[53] numerous additional natural chlorinated and brominated pyrroles have been discovered. The soil microbe *Streptomyces* sp. produces pyrrolomycin B (**31**),[54] and a soil *Streptomyces* sp. produces the novel neopyrrolomycin (**32**), a metabolite which has biphenyl-type optical activity.[55] More recently, a 3-chloropyrrole cytoprotective substance, rumbrin (**33**), was isolated from the fungus *Auxarthron umbrinum.*[56]

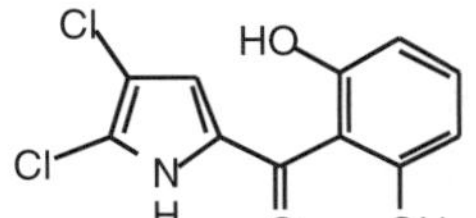

30 (pyoluteorin)

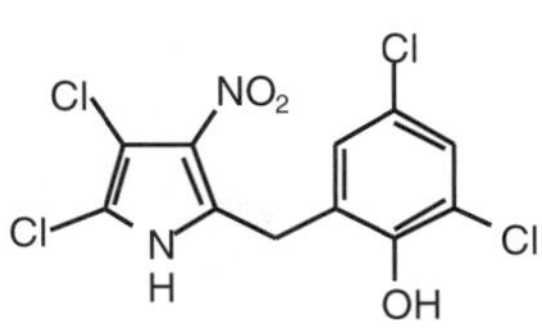

31 (pyrrolomycin B)

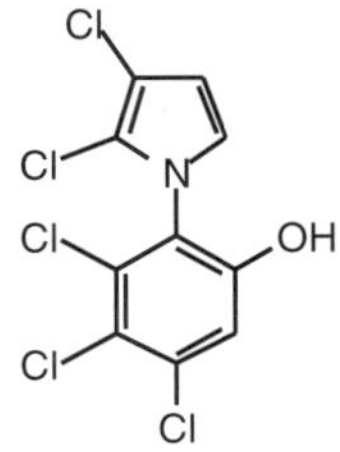

32 (neopyrrolomycin)

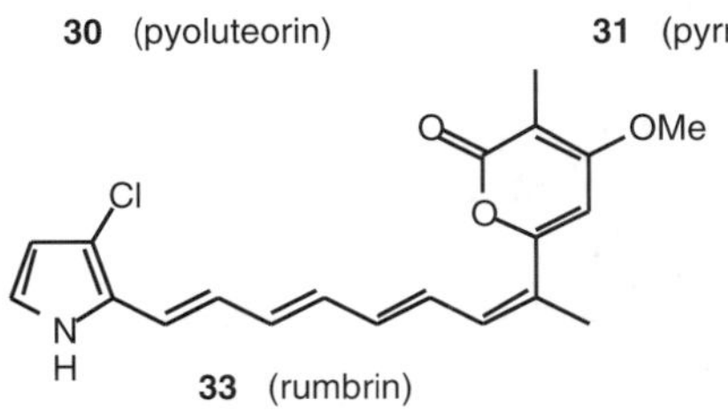

33 (rumbrin)

Indoles

The indole ring system has the distinction of being embodied in the indigo derivative Tyrian purple (**34**), the ancient Egyptian dye extracted from mollusk shells and the focus of a significant industry in Mediterranean countries in ancient times.[57] The marine acorn worm *Ptychodera flava laysanica* produces the unstable 3-chloroindole (**35**), which is in part responsible for the peculiar "iodoform-like" odor of this animal.[58] The New Zealand red seaweed *Rhodophyllis membranacea* contains ten novel halogenated indoles (e.g., **36**).[59] The 4-chloroindole ester **37** and the corresponding carboxylic acid are found in *Pisum sativum* and several other terrestrial plants and vegetables (lentil, sweet pea, sea pea, vetch).[60] These compounds appear to function as powerful plant growth hormones. A blue-green alga, *Hapalosiphon fontinalis*, contains 12 chlorinated isonitriles, one of which is hapalindole A (**38**).[61]

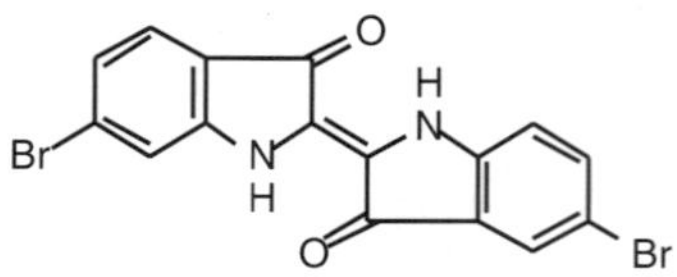

34 (Tyrian purple)

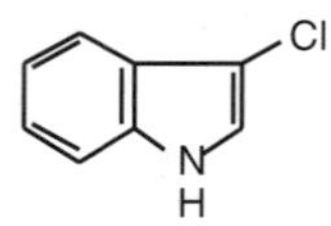

35
(3-chloroindole)

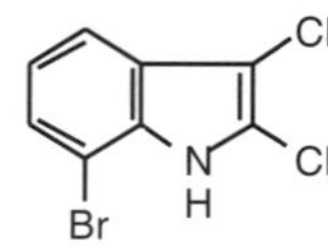

36

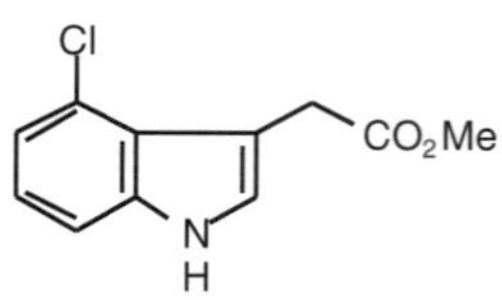

37

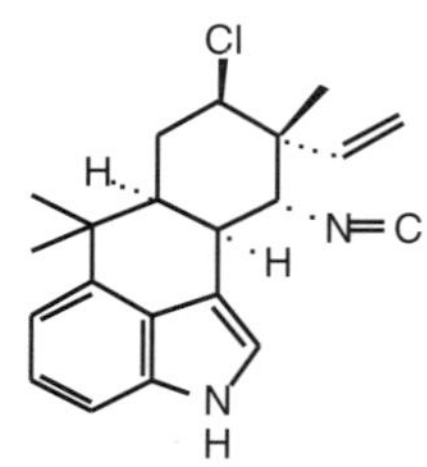

38 (hapalindole A)

The deepwater Bahamas sponge *Batzella* sp. produces six chlorinated indoles, e.g., batzelline A (**39**),[62] and nature's virtuosity in organic synthesis is further revealed by the marine bryozoan *Chartella papyracea*, which constructs a complement of five halogenated indoles, such as the structurally fantastic chartelline A (**40**).[63] The microbe *Penicillium crustosum* produces the three 6-chloroindole metabolites penitrems A (**41**), C, and F, of almost unbelievable molecular complexity.[64]

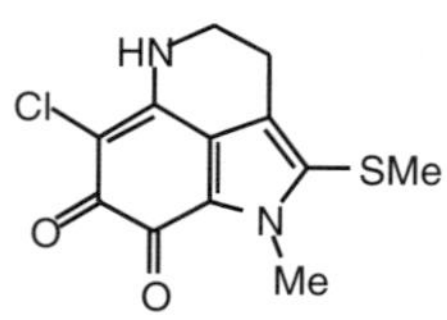

39 (batzelline A)

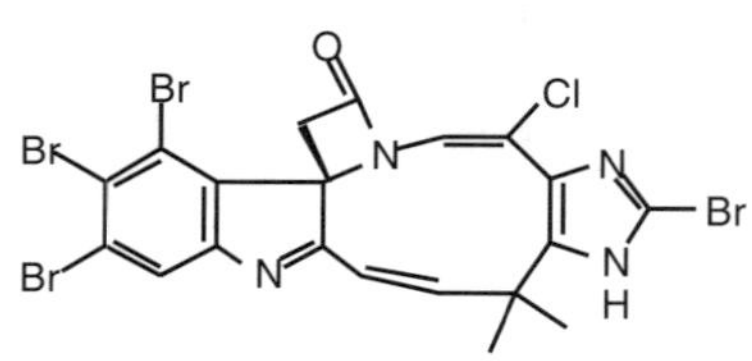

40 (chartelline A)

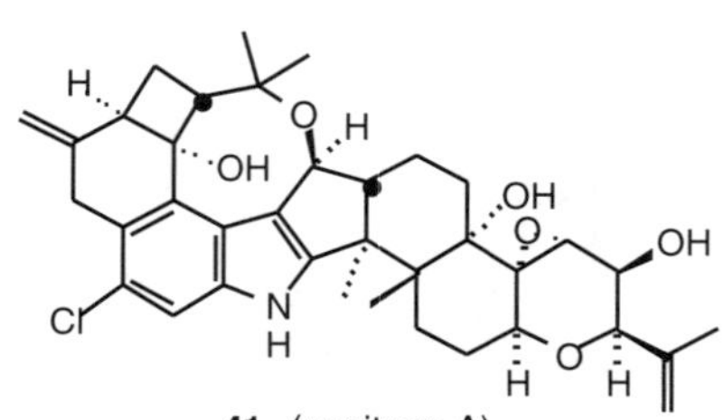

41 (penitrem A)

Miscellaneous Heterocycles

In contrast to pyrroles and indoles, chlorinated carbazoles are rare in nature. Chlorohyellazole (**42**) is produced by the blue-green alga *Hyella caespitosa,*[65] and 3-chlorocarbazole (**43**), which has potent biological activity, has been isolated from bovine urine.[66] This is the first carbazole of any type to be isolated from mammals. The several chlorinated examples of indolo[2,3-*a*]carbazoles include the anticancer rebeccamycin (**44**),[67] from *Nocardia aerocolonigenes*, a microbe from a Panamanian soil sample.

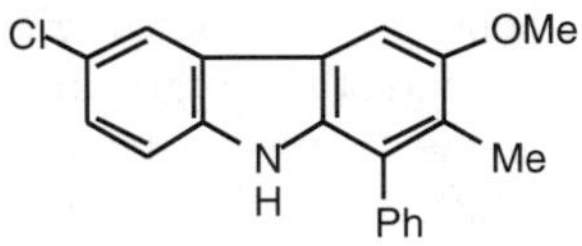

42
(chlorohyellazole)

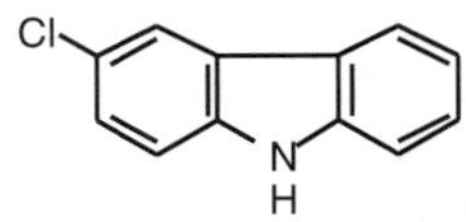

43

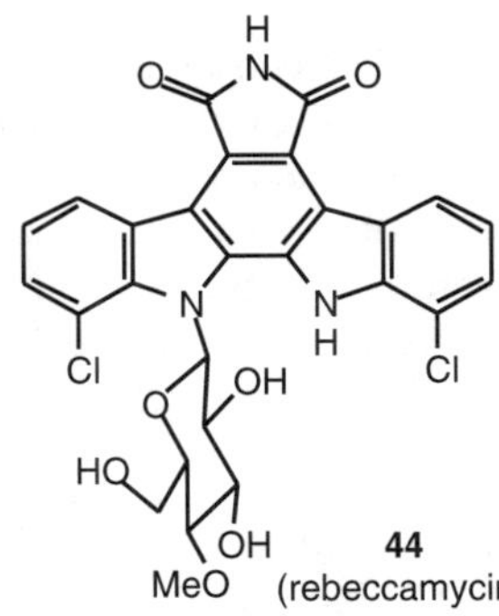

44
(rebeccamycin)

Only a few examples of nucleic acid bases are known to contain halogen. An *Actinomyces* sp. produces 2'-chloropentostatin (**45**),[68a] and kumusine, a novel chloroadenine riboside, is produced by an Indonesian *Theonella* sp. sponge.[68b] The first example of a chlorinated porphyrin was recently isolated in trace quantities from cultures of two *Pseudomonas* spp.,[69] and 2-chloropyridine is embodied both in the potent enediyne antitumor agent kedarcidin[70] and in the novel Ecuadoran poison frog metabolite epibatidine (**46**), which is secreted by several species of *Epipedobates* and which is 500–1000 times more potent than morphine as an analgesic.[71]

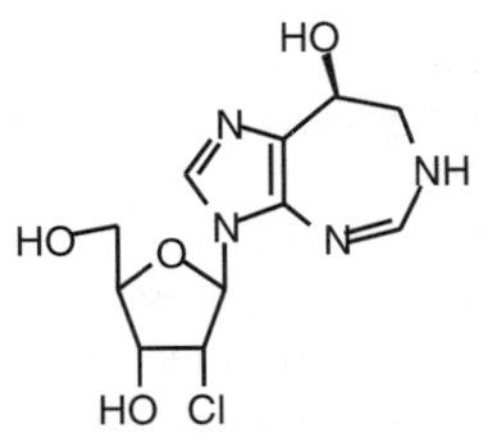

45 (2'-chloropentostatin)

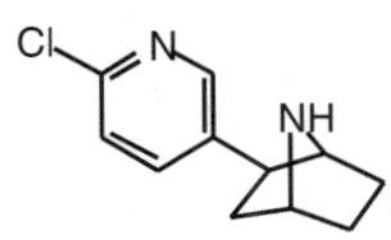

46 (epibatidine)

Aromatic Compounds

Although most of nature's chlorinated aromatic rings are phenolic, a few examples of simple chlorinated aromatics are known. The Mississippi salt marsh "needlerush" (*Juncus roemerianus*) contains 1,2,3,4-tetrachlorobenzene (**47**).[72] A study of the ash from the 1980 eruption of Mount Saint Helens has revealed the presence of chloroaromatics **48** and **49**, and, more surprisingly, three previously unknown isomers of pentachlorobiphenyls (PCBs),[73] although this latter observation needs to be corroborated. Another striking discovery from an organic chemistry standpoint is the discovery of azulene (**50**) from an unidentified deep sea gorgonian, collected at a depth of 350 m.[74] This was one of three halogenated azulenes isolated from this organism. Chlorobenzene is found in halite-containing minerals and 3-chloroanthranilic acid is produced by *Pseudomonas aureofaciens*.[4]

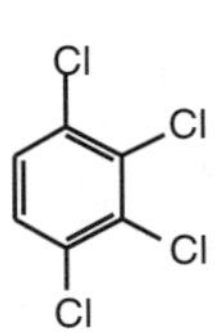

47

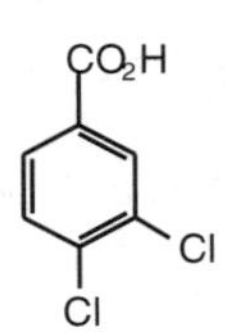

48

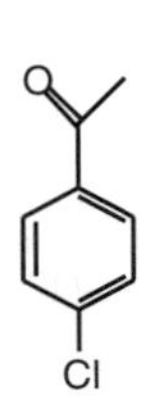

49

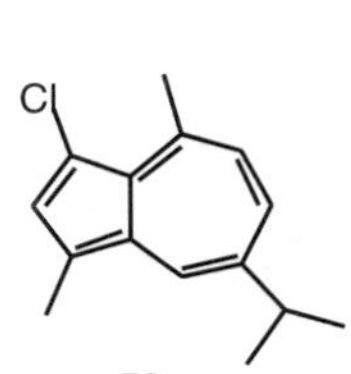

50

Phenols and Phenolic Ethers

Not surprising, in view of the susceptibility of the phenol ring to electrophilic halogenation, is the fact that nature has chosen to biosynthesize many halogenated phenols. A number of chlorinated tyrosine derivatives have been isolated from marine sponges, sea fans, and other gorgonians.[75] These amino acids, when incorporated into structural proteins, may stabilize the proteins in the organism by improving the adhesion between protein fibers and sheets.[75] The mollusk *Buccinum undatum* utilizes 3-chloro-5-bromotyrosine (**51**) in its structural proteins.[75] Isolations of 3,4-dichlorotyrosine, as well as **51**, from *Limulus polyphemus,*[76] and 3-chlorotyrosine from the cuticle of locusts (*Schistocerca gregaria*)[77] have been reported.

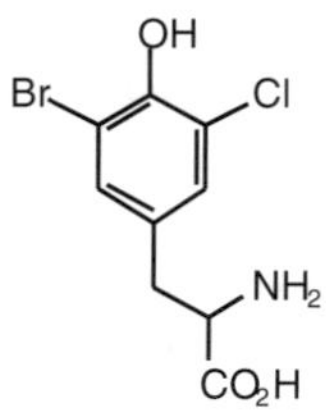

51

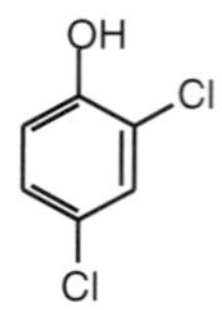

52

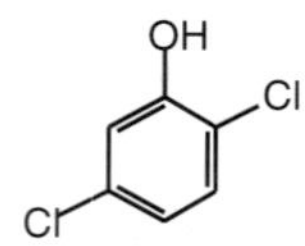

53

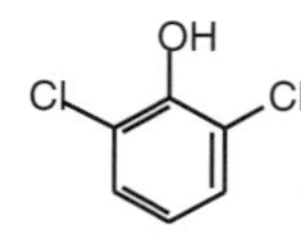

54

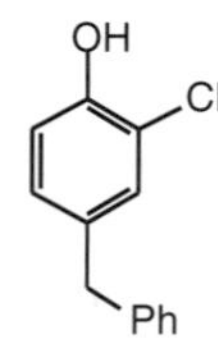

55

Several of the chlorophenols that are on the EPA's Priority Pollutant List are produced naturally. The soil fungus *Penicillium* sp. produces 2,4-dichlorophenol (**52**),[78] the industrial chemical used to manufacture the herbicide 2,4-D. Grasshoppers secrete 2,5-dichlorophenol (**53**), which is repellent to ants,[79] and the sex pheromone of the female Lone Star tick (*Amblyomma americanum*) and a dozen other hard ticks is 2,6-dichlorophenol (**54**).[80] The blue-green alga *Anacystis marina* contains chlorophenol **55**.[81]

A detailed examination of the white-rot fungus *Bjerkandera adusta* has revealed the presence of 3-chloro-4-methoxybenzaldehyde (**56**), 3-chloro-4-methoxybenzyl alcohol (**57**), 3,5-dichloro-4-methoxybenzaldehyde (**58**), and 3,5-dichloro-4-benzyl alcohol (**59**).[82] This fungus also produces diiodomethane and chloroiodomethane in the presence of added iodide. The production of these halogenated alkanes is related to the formation of chlorinated aromatic compounds.

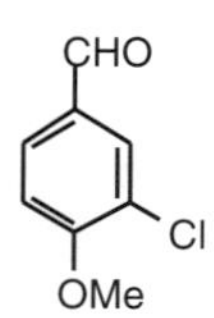

56

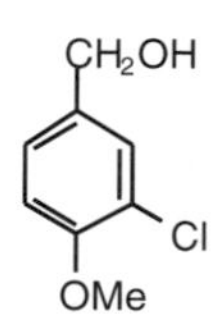

57

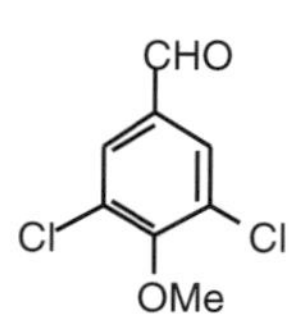

58

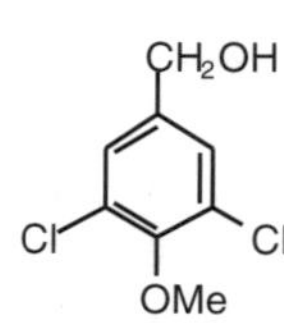

59

The lichen *Nephroma laevigatum* produces the new anthraquinone **60** as well as the novel **61** and **62**, in addition to some known chlorinated anthraquinones.[83]

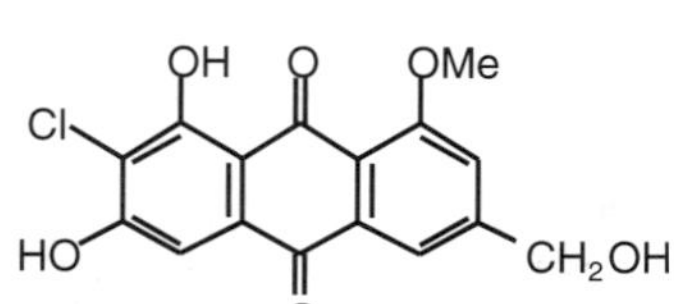

60

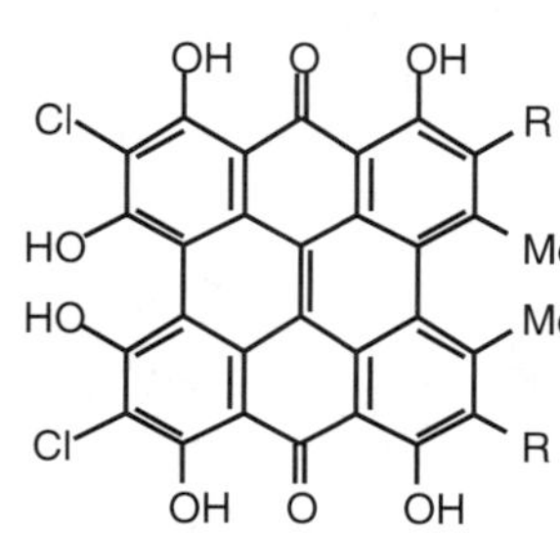

61 R = H
62 R = Cl

Vancomycin is one of more than 100 chlorine-containing glycopeptides with important antibiotic activity,[4] and vancomycin has been used for nearly 40 years to treat methicillin-resistant *Staphylococcus aureus* infections, particularly those in hospital patients. The role of the phenolic chlorine is to contribute both to the stability and to the specificity of the peptide-binding site.

Dioxins and Related Compounds

"Dioxin" — 2,3,7,8-tetrachlorodibenzo-*p*-dioxin (**63**) — was once considered to be the most toxic man-made chemical known and the "doomsday chemical" of the 20th century, but, after many years of study, the only documented human toxicity is the skin disease chloracne in heavily exposed people.[84,85] Nevertheless, the high toxicity of some polychlorinated dioxins (PCDDs) and related compounds, such as the polychlorinated dibenzofurans (PCDFs) (e.g., **64**), in some animals is sufficient reason for the continued study of these compounds, especially since it is now recognized that PCDDs and PCDFs are produced by natural combustion sources.

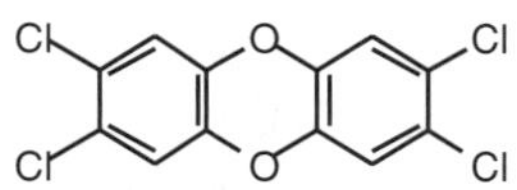

63 ("dioxin")

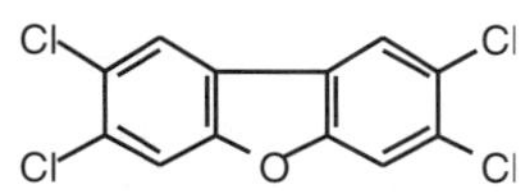

64

It has been found that PCDDs form during most combustion processes.[86] Laboratory studies have revealed that PCDDs, including "dioxin" (**63**), and PCDFs form in ppb amounts during the combustion of wood (treated or untreated).[87] The relatively poor efficiency and incomplete oxidation when damp vegetation and wood are burned in the presence of high chloride concentrations (up to 2,100 ppm in wood pulp)[6] are conditions conducive to PCDD formation, and forest and brush fires may be a major source of PCDDs and PCDFs in the environment.[88] It is estimated that some 130 pounds of PCDDs are produced in Canadian forest fires annually.[88a] Since most forest fires are lightning-caused, and there are 200,000 forest fires annually worldwide, burning 27,000 square miles,[6] it is logical to assume that PCDDs have been present in the environment for eons, and an 1877 soil sample was found to contain PCDDs and PCDFs.[89] Due to the obvious sampling problems, only one field study has shown that PCDDs and PCDFs are formed in forest fires.[88b]

Another surprising observation is the enzymatic conversion of chlorophenols into both PCDDs and PCDFs in the ppm range by horseradish peroxidase enzyme (HRP) and other peroxidases.[90] This extraordinary revelation opens the door to the possibility that a source of environmental PCDDs and PCDFs may be their completely natural formation from (natural) chlorophenols by soil and water microbes. In fact, recent evidence indicates that the natural chlorination of humic acids by soil chloroperoxidase enzymes does occur.[4,91] Equally unexpected is the observation that fresh garden compost forms PCDDs and PCDFs.[91a]

The discovery of the first chlorine-containing molecule — hydrogen chloride — in interstellar space[92] opens the window to the possible detection of organochlorine molecules and other organohalogens outside of our solar system, since nearly 100 stable organic molecules (alkynes, alcohols, amines, amides, alkenes, aldehydes) have been detected in interstellar space.

Mechanisms of Biogenic Chlorination

Marine organisms, terrestrial plants, fungi, microorganisms, and mammals all contain haloperoxidase enzymes that can chlorinate, brominate, and iodinate organic compounds in the presence of chloride, bromide, or iodide ions, respectively.[93] Chloroperoxidase, which was first isolated from the mold *Caldariomyces fumago*, has been extensively studied.[94] Soil extracts have been shown to have chlorinating ability, and more than 80 species of Death Valley fungi displayed chloroperoxidase activity.[93] Peroxidases are present in the thyroid gland of mammals.[95] The enzymes lactoperoxidase, chloroperoxidase, HRP, and vanadium peroxidase are capable of chlorinating phenols and phenol ethers (anisole) in the presence of chloride.[96] Chloroperoxidase appears to catalyze the chlorination step in the biosynthesis of the antibiotic chloramphenicol by *Streptomyces venezuelae*.[97] Lactoperoxidase, an enzyme present in mammalian milk, is capable of chlorinating organic substrates.[98]

An extensive investigation of the production of chlorinated aromatic compounds by fungi has revealed that this biogenesis is common to many types of common wood- and forest litter–degrading fungi.[99] These compounds (**56–59**) are found at concentrations of 75 mg/kg of wood in the environment. Not only are these compounds found in the fungi, but they are also present in the surrounding environment (leaves, twigs, branches, nut husks, rotting logs). However, these chlorinated aromatics were not found in fresh forest litter or intact wood where fungi are not present. Indeed, rotting wood or leaf litter adjacent to the fruiting bodies of these fungi contain chlorinated aromatics in "concentrations high enough to be considered hazardous," according to Dutch environmental regulations concerning chlorophenols in soil (≥ 10 mg/kg).[99] Likewise, these chlorinated anisyl metabolites were found in the edible mushroom *Lepista nuda* at a level of 0.8 mg/kg. The authors conclude that these chlorinated aromatic–producing fungi are widespread in the environment and the resulting organochlorines are likely transformed into chlorine-containing humic substances, which are known to be present in pristine surface waters and groundwaters.[100]

An electrifying development is the utilization of the halide/peroxidase/hydrogen peroxide chemical system by humans and other mammals to generate active halogen (HOCl, HOBr) in order to destroy microorganisms. Thus, human white blood cells (eosinophils and neutrophils) contain myeloperoxidase, which, in the presence of chloride, bromide, or iodide, and hydrogen peroxide, rapidly forms active halogen, resulting in the death of bacteria and fungi by as yet unidentified halogenation reactions.[101] Intact cells show a greater preference for bromide. It would appear that biohalogenation is an integral component of our immune system!

Metabolism and Biodegradation

Contrary to common perception, organohalogen compounds, both natural and unnatural, are readily metabolized and biodegraded to the halide ion by assorted microorganisms. The simple organochlorines CH_2Cl_2, $CHCl_3$, CCl_4, $Cl_2C{=}CCl_2$, and others can be degraded to CO_2 (or other simple organics) and chloride by soil bacteria.[102] The biodegradation of chlorophenols has been thoroughly studied. For example, 2,4-dichlorophenol is rapidly metabolized in soils,[103] and an exhaustive mechanistic study of the degradation of this compound to chloride and CO_2 by the white-rot fungus *Phanerochaete chrysosporium* has been reported.[104] This versatile and remarkable fungus is capable of degrading other chlorinated phenols,[105] DDT,[106] and even pentachlorophenol.[107] PCBs are also degraded by this fungus[108] as well as by a *Pseudomonas* sp.[109]

The white-rot fungus *Trametes versicolor* is capable of degrading chlorinated lignin derivatives in a combined bleach plant effluent from sulfite pulping to the extent of 45% reduction of adsorbable organohalogens (AOX).[110] Several other fungi were also effective. Chlorophenols are degraded under anaerobic conditions by sulfate-reducing enrichment cultures.[111] The enzymatic oxidative coupling of chlorophenols has been found to result in the loss of chloride, both in the polymerization process and in the binding to humic substances.[112] The degradation of pentachlorophenol by polyurethane-immobilized *Flavobacterium* sp. cells has been studied and found to be an excellent method for the removal of pentachlorophenol from highly contaminated water.[113] The degradation of mono- and polyhalomethanes by *Methylosinus trichosporium* follows the reactivity order $CH_3X > CH_2X_2 \geq CHX_3$ (where X = Cl, Br).[114] In these biodegradation reactions, loss of halide follows the insertion of oxygen into the C-H bond, leading to the ultimate formation of carbon dioxide. The responsible enzyme in these rapid and relatively nonselective reactions is methane monooxygenase. Methyl bromide is degraded anaerobically in saltmarsh sediments after reaction with sulfide to yield methanethiol.[115] Further reaction affords dimethyl sulfide. Moreover, under these conditions methyl bromide reacts with chloride to form methyl chloride, which can also react with sulfide. Thus, sulfide-rich sediments, such as those in coastal saltmarshes, are likely to be a natural sink for atmospheric methyl bromide.

The mechanism of dehalogenation of dichloromethane and CH_2FCl with the enzymes dichloromethane dehalogenase from *Methylophilus* sp. and rat liver glutathione *S*-transferases involves nucleophilic attack of glutathione on the substrate to produce a halomethylthioether intermediate, which subsequently hydrolyzes to formaldehyde and HCl or HF.[116]

Natural vs. Anthropogenic Organohalogen Compounds

Chloromethane is overwhelmingly biogenic in origin. However, what about the relative quantities of the other natural organochlorines and organohalogens vis-à-vis the approximately 20 million tons/year of 150 industrial organohalogens currently in use worldwide?[117] Table 7.2 summarizes several estimates of the quantities of natural and anthropogenic organohalogens.

Several researchers have reported the identification of chlorinated phenols in ocean waters and in soil extracts.[118] The major source of most of these compounds, such as 2,4,6-trichlorophenol, is biogenic and from natural chlorination processes, such as reactions between humic acid and soil microbes (chloride and chloroperoxidase).[91,118] Humic acids are not only present in soil, but also in rivers, lakes, and the oceans. The yearly net production of environmental humic acid from the decomposition of organic material has been estimated at 63 billion tons.[119] This can be added to the estimated existing global 1.0–1.5 trillion tons of soil humic acids and one trillion tons of ocean humic acid.[119] Numerous studies have demonstrated the ease with which humic acid model compounds (phenolics) react with chlorine to produce chloroform and other simple organochlorines via chlorophenols.[91,118] As we have seen, chloride is ubiquitous in soil, plants, rivers, lakes, and, of course, oceans. These ingredients, in combination with natural chloroperoxidase and other enzymes, provide a plausible rationale for the natural formation of chlorinated phenols. In addition, the chloroform that is frequently found in groundwaters with no clear anthropogenic source is most likely due to the degradation of natural chlorinated humic acids.[91,118]

Although a large number of halogenated compounds have now been isolated from natural sources, especially marine, only a relatively few marine organisms have been involved in these studies. For example, whereas some 12,000 natural products of all types have been isolated from terrestrial plants, only 500 had been isolated from marine plants (algae) up to 1987.[120] This contrasts with the fact that there are known some 500,000 species of marine animals, plants, and bacteria.[121] For example, there are 80,000 living species of mollusks, 5,000 species of sponges, and 4,000 species of bryozoans (moss animals), but only six of the latter have been examined.[2] Given such diversity and magnitude of biological species, it is indisputable that a very large number of unique organochlorine compounds will be isolated and identified from ocean life, to say nothing about the novel chlorinated and other halogenated compounds waiting to be discovered in terrestrial plants, animals, bacteria, fungi, and perhaps even humans.

Indications are that organohalogen compounds have been with us for centuries. For example, halogenated fulvic acids have been isolated from groundwater samples that date back 1,300, 4,600, and 5,200 years![122] Organohalogen material was identified in sediments dating back to the 13th century.[123] The analysis of preindustrial glacial ice from Antarctica and northern Sweden has revealed organohalogen levels of 1–3 ppb, including measurable amounts of trichloroacetic acid. Moreover, the latter compound, as well as chloroform, is found to be ubiquitous in soil, results confirmed by $^{37}Cl^-$ labeling experiments. The analysis of 35,000-year old organic matter (^{14}C dating) shows the presence of organochlorine, organobromine, and organoiodine compounds.[91a] Organochlorines have also been found in 1,000-year old peat and 4,000-year old marine clay. Even more remarkable are the discoveries of organochlorines (200–300 ppm) in several-thousand-year old peat from the Holocene period, in two lignite samples (107 and 166 ppm) that date from the Tertiary era, 15 million years ago, and in a 300-million year old bituminous coal sample (74 ppm) from the Upper Carboniferous period.[91a] Whether these natural organochlorines in sediments result from the deposition of biogenic material from plants already containing organochlorines or arise by the formation of organochlorines within the sediments remains to be established, although most of the evidence to date would support the former theory. These studies of sediments — ancient and modern — should provide information about

Table 7.2. Estimates of Naturally Occurring Halomethanes from Various Sources.[4]

Compound	Source	Tons per Year
CH_3Cl	oceans	2,500,000–5,000,000
CH_3Cl	biomass burning	460,000–2,000,000
CH_3Cl	anthropogenic	30,000–300,000
CH_3Br	oceans	40,000–300,000
CH_3Br	anthropogenic	50,000–80,000
CH_3I	all biogenic sources	1,300,000
CH_3I	all sources	40,000,000
CH_3I	oceans	300,000–2,000,000
CCl_4	all sources	1,900,000
$CHBr_3$	oceans	1,000,000–2,000,000
$CHBr_3$	Arctic ice microalgae	4,700–70,000
$CHBr_3$	Antarctic ice microalgae	53,000–80,000
$CHBr_3$	macroalgae	40,000–200,000
$CHBr_3$	anthropogenic	600–6,000

the global quantities of natural organochlorines. In any event, these studies clearly demonstrate that organochlorine compounds predate the arrival of man on Earth! Microfossils in Precambrian rocks, which are a billion years old, are identical to the blue-green alga *Nostoc*, and other microfossils are morphologically indistinguishable from *Oscillatoria*, two present-day species rich in organohalogen compounds.[124]

Function of Natural Organohalogens

Why does nature create organochlorine compounds? Many marine and terrestrial organisms use organochlorine and other organohalogen compounds in chemical defense — feeding deterrents, irritants, or pesticides — or in food gathering.[125] A recent, careful study of the sea hare *Stylocheilus longicauda* clearly indicates a feeding-deterrent role for its organohalogen compounds.[126] Marine algae store organohalogen compounds in the vesicular cells for facile secretion to discourage predators, and the terrestrial blue-green alga *Fischerella ambigua* secretes into the medium two chlorinated phenols that show antibacterial, antifungal, and molluskicidal activity.[127] The marine algae metabolite telfairine (Figure 7.2) is 100% lethal to mosquito larvae at 10 ppm,[128] and the related chlorinated monoterpene plocamene B is three times more effective than the commercial pesticide lindane against mosquito larvae.[129] Recently, it was found that the German cockroach (*Blattella germanica*) utilizes two chlorinated steroid glycosides, blattellastanosides A and B, as aggregation pheromones.[130] The fungal production of chlorinated aromatics seems to be involved in the generation of hydrogen peroxide for the lignin-degrading peroxidases, since the chlorinated anisyl alcohols **57** and **59** are excellent substrates for extracellular aryl alcohol oxidases.[131] One very attractive hypothesis is that these natural organohalogen compounds play an essential role in the survival of the organism, and the ability of the organism to synthesize such compounds for chemical defense and food gathering has evolved over time under the stress of natural selection.[132]

CONCLUDING REMARKS

Close to 2,800 different chlorinated and other halogenated chemicals are discharged into the biosphere by plants, marine organisms, insects, bacteria, fungi, mammals, and by other natural processes. In a few cases, the quantities of these naturally occurring halogenated compounds far exceed the amounts of the same chemicals from anthropogenic sources. Enzymatic, thermal, and other natural processes are constantly occurring in the ocean, in the atmosphere, and in the soil that lead to the formation of chlorinated phenols and many other chlorinated chemicals, including dioxins and CFCs, that previously were thought only to result from the actions of humans. It is clear that these natural processes have been producing chlorinated compounds and have been a vital component of our ecosystem for eons. As our understanding of biohalogenation progresses, it will be possible to differentiate and quantify more accurately natural and anthropogenic sources of chlorinated and other halogenated chemicals.

REFERENCES

1. Fowden, L. 1968. The Occurrence and Metabolism of Carbon-Halogen Compounds. *Proc. Roy. Soc. B* 171, 5.
2. For reviews, see (a) Siuda, J.F. and J.F. DeBernardis. Naturally Occurring Halogenated Organic Compounds. *Lloydia* 36, 107 (1973). (b) Engvild, K.C. Chlorine-Containing Natural Compounds in Higher Plants. *Phytochem.* 25, 781 (1986). (c) Gribble, G.W. Naturally Occurring Organohalogen Compounds — A Survey. *J. Nat. Prod.* 55, 1353 (1992). (d) Naumann, K. Chlorchemie der Natur. *Chem. Zeit.* 27, 33 (1993). (e) Hoekstra, E.J. and E.W.B. De Leer. Organochlorines: The Natural Alternatives. *Chem. Brit.* 127 (1995). (f) Harper, D.B. and D. O'Hagan. The Fluorinated Natural Products. *Nat. Prod. Rep.* 11, 123 (1994). (g) Gribble, G.W. The Natural Production of Chlorinated Compounds. *Environ. Sci. Technol.* 28, 310A (1994). (h) "The Natural Chemistry of Chlorine in the Environment," Euro Chlor publication, 1995, Brussels.
3. Amato, I. 1993. The Crusade Against Chlorine. *Science* 261, 152.
4. Gribble, G.W. 1996. Naturally Occurring Organohalogen Compounds — A Comprehensive Survey. *Prog. Chem. Org. Nat. Prod.* 68, 1.
5. Harper, D.B. 1985. Halomethane from Halide Ion — A Highly Efficient Fungal Conversion of Environmental Significance. *Nature* 315, 55.
6. Isidorov, V.A. 1990. *Organic Chemistry of the Earth's Atmosphere*, Berlin, Springer-Verlag.
7. Palmer, T.Y. 1976. Combustion Sources of Atmospheric Chlorine. *Nature* 263, 44.
8. (a) Lovelock, J.E. 1975. Natural Halocarbons in the Air and in the Sea. *Nature* 256, 193. (b) Reinhardt, T.E. and D.E. Ward. 1995. Factors Affecting Methyl Chloride Emissions from Forest Biomass Combustion. *Environ. Sci. Technol.* 29, 825.
9. Stoiber, R.E., D.C. Leggett, T.F. Jenkins, R.P. Murrmann, and W.I. Rose, Jr. 1971. Organic Compounds in Volcanic Gas from Santiaguito Volcano, Guatemala. *Geol. Soc. Amer. Bull.* 82, 2299.
10. Rasmussen, R.A., M.A.K. Khalil, R.W. Dalluge, S.A. Penkett, and B. Jones. 1982. Carbonyl Sulfide and Carbon Disulfide from the Eruptions of Mount St. Helens. *Science* 215, 665.
11. Rasmussen, R.A., L.E. Rasmussen, M.A.K. Khalil, and R.W. Dalluge. 1980. Concentration Distribution of Methyl Chloride in the Atmosphere. *J. Geophys. Res.* 85, 7350.
12. Moore, R.E. 1977. Volatile Compounds from Marine Algae. *Acct. Chem. Res.* 10, 40.
13. Ohta, K. and M. Takagi. 1977. Antimicrobial Compounds of the Marine Red Alga *Marginisporum aberrans*. *Phytochem.* 16, 1085.
14. McConnell, O.J. and W. Fenical. 1980. Halogen Chemistry of the Red Alga *Bonnemaisonia*. *Phytochem.* 19, 233.
15. Abrahamsson, K., A. Ekdahl, J. Collén, and M. Pedersén. 1995. Marine Algae — A Source of Trichloroethylene and Perchloroethylene. *Limnol. and Oceanog.* 40, 1321.
16. Yamamura, S. and Y. Hirata. 1963. Structures of Aplysin and Aplysinol, Naturally Occurring Bromo-Compounds. *Tetrahedron* 19, 1485.
17. Kupchan, S.M., J.E. Kelsey, M. Maruyama, J.M. Cassady, J.C. Hemingway, and J.R. Knox. 1969. Tumor Inhibitors. XLI. Structural Elucidation of Tumor-Inhibitory Sesquiterpene Lactones from *Eupatorium rotundifolium*. *J. Org. Chem.* 34, 3876.
18. Naylor, S., F.J. Hanke, L.V. Manes, and P. Crews. 1983. Chemical and Biological Aspects of Marine Monoterpenes. *Prog. Chem. Org. Nat. Prod.* 44, 189.

19. Faulkner, D.J. 1993. Marine Natural Products. *Nat. Prod. Rep.* 10, 497.
20. Estrada, D.M., J.L. Ravelo, C. Ruiz-Pérez, J.D. Martín, and X. Solans. 1989. Dactylomelol, a New Class of Diterpene from the Sea Hare *Aplysia dactylomela. Tetrahedron Lett.* 30, 6219.
21. Rao, C.B., C. Satyanarayana, D.V. Rao, E. Fahy, and D.J. Faulkner. 1989. Metabolites of *Aplysia dactylomela* from the Indian Ocean. *Indian J. Chem. B* 28, 322.
22. Pordesimo, E.O., F.J. Schmitz, L.S. Ciereszko, M.B. Hossain, and D. van der Helm. 1991. New Briarein Diterpenes from the Caribbean Gorgonians *Erythropodium caribaeorum* and *Briareum* sp. *J. Org. Chem.* 56, 2344.
23. Murakami, T. and N. Tanaka. 1988. Occurrence, Structure and Taxonomic Implications of Fern Constituents. *Prog. Chem. Org. Nat. Prod.* 54, 1.
24. Henkel, T. and A. Zeeck. 1991. Secondary Metabolites by Chemical Screening. 15. Structure and Absolute Configuration of Naphthomevalin, a New Dihydro-Naphthoquinone Antibiotic from *Streptomyces* sp. *J. Antibiot.* 44, 665.
25. Demuth, H., S.R. Jensen, and B.J. Nielsen. 1989. Iridoid Glucosides from *Asystasia bella. Phytochem.* 28, 3361.
26. Poiner, A., V.J. Paul, and P.J. Scheuer. 1989. Kumepaloxane, a Rearranged Trisnor Sesquiterpene from the Bubble Shell *Haminoea cymbalum. Tetrahedron* 45, 617.
27. Wright, A.D., G.M. König, and O. Sticher. 1991. New Sesquiterpenes and C_{15} Acetogenins from the Marine Red Alga *Laurencia implicata. J. Nat. Prod.* 54, 1025.
28. Moore, R.E. 1977. Toxins from Blue-Green Algae. *Bioscience* 27, 797.
29. Chen, J.L., R.E. Moore, and G.M.L. Patterson. 1991. Structures of the Nostocyclophanes A-D. *J. Org. Chem.* 56, 4360.
30. Barrow, R.A., T. Hemscheidt, J. Liang, S. Paik, R.E. Moore, and M.A. Tius. 1995. Total Synthesis of Cryptophycins. Revision of the Structures of Cryptophycins A and C. *J. Am. Chem. Soc.* 117, 2479.
31. Kasai, R., M. Katagiri, K. Ohtani, K. Yamasaki, C.-R. Yang, and O. Tanaka. 1994. Iridoid Glycosides from *Phlomis younghusbandii* Roots. *Phytochem.* 36, 967.
32. Youssef, D. and A.W. Frahm. 1994. Constituents of the Egyptian *Centaurea scoparia*; Chlorinated Guaianolides of the Aerial Parts. *Planta Med.* 60, 267.
33. Öksüz, S. and G. Topcu. 1994. Guaianolides from *Centaurea glastifolia. Phytochem.* 37, 487.
34. Hamburger, M., J.-L. Wolfender, and K. Hostettmann. 1993. Search for Chlorinated Sesquiterpene Lactones in the Neurotoxic Thistle *Centaurea solstitialis* by Liquid Chromatography-Mass Spectrometry, and Model Studies on their Possible Artifactual Formation. *Natural Toxins* 1, 315.
35. Fukuyama, Y., N. Shida, Y. Hata, and M. Kodama. 1994. Prenylated C_6 - C_3 Compounds from *Illicium tashiroi. Phytochem.* 36, 1497.
36. Fukuyama, Y., K. Okamoto, Y. Kubo, N. Shida, and M. Kodama. 1994. New Chlorine-Containing Prenylated C_6 - C_3 Compounds Increasing Choline Acetyltransferase (ChAT) Activity in Culture of Postnatal Rat Septal Neurons from *Illicium tashiroi. Chem. Pharm. Bull.* 42, 2199.
37. Quaghebeur, K., J. Coosemans, S. Toppet, and F. Compernolle. 1994. Cannabiorci- and 8-Chlorocannabiorcichromenic Acid as Fungal Antagonists from *Cylindrocarpon olidum. Phytochem.* 37, 159.
38. Todd, J.S. and W.H. Gerwick. 1995. Isolation of a Cyclic Carbonate, a γ-Butyrolactone, and a new Indole Derivative from the Marine Cyanobacterium *Lyngbya majuscula. J. Nat. Prod.* 58, 586.
39. Pcolinski, M.J., R.W. Doskotch, A.Y. Lee, and J. Clardy. 1994. Chlorosilphanol A and Silphanepoxol, Labdane Diterpenes from *Silphium perfoliatum. J. Nat. Prod.* 57, 776.
40. Abaul, J., E. Philogene, P. Bourgeois, C. Poupat, A. Ahond, and P. Potier. 1994. Contribution à la Connaissance des Rutacées Américaines: Études des Feuilles de *Triphasia trifolia. J. Nat. Prod.* 57, 846.
41. Biskupiak, J.E. and C.M. Ireland. 1984. Revised Absolute Configuration of Dysidenin and Isodysidenin. *Tetrahedron Lett.* 25, 2935.
42. Narayanan, S., M.R.S. Iyengar, P.L. Ganju, S. Rengaraju, T. Shomura, T. Tsuruoka, S. Inouye, and T. Niida. 1980. γ-Chloronorvaline, a Leucine Analog from *Streptomyces. J. Antibiot.* 33, 1249.
43. Fukuchi, N., A. Isogai, S. Yamashita, K. Suyama, J.Y. Takemoto, and A. Suzuki. 1990. Structure of Phytotoxin Syringomycin Produced by a Sugar Cane Isolate of *Pseudomonas syringae* pv. *syringae. Tetrahedron Lett.* 31, 1589.
44. Moore, R.E., V. Bornemann, W.P. Niemczura, J.M. Gregson, J.-L. Chen, T.R. Norton, G.M.L. Patterson, and G.L. Helms. 1989. Puwainaphycin C, a Cardioactive Cyclic Peptide from the Blue-Green Alga *Anabaena* BQ-16-1. Use of Two-Dimensional ^{13}C-^{13}C and ^{13}C-^{15}N Correlation Spectroscopy in Sequencing the Amino Acid Units. *J. Am. Chem. Soc.* 111, 6128.
45. Morita, H., S. Nagashima, K. Takeya, and H. Itokawa. 1994. A Novel Cyclic Pentapeptide with a β-Hydroxy-γ-chloroproline from *Aster tataricus. Chem. Lett.*, 2009.

46. Fajardo, V., F. Podesta, M. Shamma, and A.J. Freyer. 1991. New Withanolides from *Jaborosa magellanica. J. Nat. Prod.* 54, 554.
47. Carney, J.R., P.J. Scheuer, and M. Kelly-Borges. 1993. Three Unprecedented Chloro Steroids from the Maui Sponge *Strongylacidon* sp.: Kiheisterones C, D, and E. *J. Org. Chem.* 58, 3460.
48. Hall, R.J. 1972. The Distribution of Organic Fluorine in Some Toxic Tropical Plants. *New Phytol.* 71, 855.
49. White, R.H. and L.P. Hager. 1977. Occurrence of Fatty Acid Chlorohydrins in Jellyfish Lipids. *Biochem.* 16, 4944.
50. Baker, B.J., R.K. Okuda, P.T.K. Yu, and P.J. Scheuer. 1985. Punaglandins: Halogenated Antitumor Eicosanoids from the Octocoral *Telesto riisei. J. Am. Chem. Soc.* 107, 2976.
51. Iguchi, K., S. Kaneta, K. Mori, Y. Yamada, A. Honda, and Y. Mori. 1985. Chlorovulones, New Halogenated Marine Prostanoids with an Antitumor Activity from the Stolonifer *Clavularia viridis* Quoy and Gaimard. *Tetrahedron Lett.* 26, 5787.
52. Haines, T.H. 1973. Halogen- and Sulfur-Containing Lipids of *Ochromonas. Ann. Rev. Microbiol.* 27, 403.
53. Takeda, R. 1958. Structure of a New Antibiotic, Pyoluteorin. *J. Am. Chem. Soc.* 80, 4749.
54. Kaneda, M., S. Nakamura, N. Ezaki, and Y. Iitaka. 1981. Structure of Pyrrolomycin B, a Chlorinated Nitro-Pyrrole Antibiotic. *J. Antibiot.* 34, 1366.
55. Nogami, T., Y. Shigihara, N. Matsuda, Y. Takahashi, H. Naganawa, H. Nakamura, M. Hamada, Y. Muraoka, T. Takita, Y. Iitaka, and T. Takeuchi. 1990. Neopyrrolomycin, a New Chlorinated Phenylpyrrole Antibiotic. *J. Antibiot.* 43, 1192.
56. Yamagishi, Y., M. Matsuoka, A. Odagawa, S. Kato, K. Shindo, and J. Mochizuki. 1993. Rumbrin, a New Cytoprotective Substance Produced by *Auxarthron umbrinum* I. Taxonomy, Production, Isolation and Biological Properties. *J. Antibiot.* 46, 884.
57. Baker, J.T. 1974. Tyrian Purple: An Ancient Dye, A Modern Problem. *Endeavour* 33, 11.
58. Higa, T. and P.J. Scheuer. 1975. 3-Chloroindole, Principal Odorous Constituent of the Hemichordate *Ptychodera flava laysanica. Naturwiss.* 62, 395.
59. Brennan, M.R. and K.L. Erickson. 1978. Polyhalogenated Indoles from the Marine Alga *Rhodophyllis membranacea* Harvey. *Tetrahedron Lett.*, 1637.
60. Engvild, K.C., H. Egsgaard, and E. Larsen. 1981. Determination of 4-Chloroindoleacetic Acid Methyl Ester in Vicieae Species by Gas Chromatography-Mass Spectrometry. *Physiol. Plant.* 53, 79.
61. Moore, R.E., C. Cheuk, X.-Q.G. Yang, G.M.L. Patterson, R. Bonjouklian, T.A. Smitka, J.S. Mynderse, R.S. Foster, N.D. Jones, J.K. Swartzendruber, and J.B. Deeter. 1987. Hapalindoles, Antibacterial and Antimycotic Alkaloids from the Cyanophyte *Hapalosiphon fontinalis. J. Org. Chem.* 52, 1036.
62. Sun, H.H., S. Sakemi, N. Burres, and P. McCarthy. 1990. Isobatzellines A, B, C, and D. Cytotoxic and Antifungal Pyrroloquinoline Alkaloids from the Marine Sponge *Batzella* sp. *J. Org. Chem.* 55, 4964.
63. Anthoni, U., P.H. Nielsen, M. Pereira, and C. Christophersen. 1990. Bryozoan Secondary Metabolites: A Chemotaxonomical Challenge. *Comp. Biochem. Physiol.* 96B, 431.
64. de Jesus, A.E., P.S. Steyn, F.R. Van Heerden, R. Vleggaar, P.L. Wessels, and W.E. Hull. 1983. Tremorgenic Mycotoxins from *Penicillium crustosum*: Isolation of Penitrems A-F and the Structure Elucidation and Absolute Configuration of Penitrem A. *J. Chem. Soc., Perkin Trans.* 1, 1847.
65. Cardellina, J.H., M.P. Kirkup, R.E. Moore, J.S. Mynderse, K. Seff, and C.J. Simmons. 1979. Hyellazole and Chlorohyellazole, Two Novel Carbazoles from the Blue-Green Alga *Hyella caespitosa.* Born. ET Flah. *Tetrahedron Lett.*, 4915.
66. Luk, K.-C., L. Stern, M. Weigele, R.A. O'Brien, and N. Spirt. 1983. Isolation and Identification of "Diazepam-Like" Compounds from Bovine Urine. *J. Nat. Prod.* 46, 852.
67. Bush, J.A., B.H. Long, J.J. Catino, W.T. Bradner, and K. Tomita. 1987. Production and Biological Activity of Rebeccamycin, a Novel Antitumor Agent. *J. Antibiot.* 40, 668.
68. (a) Schaumberg, J.P., G.C. Hokanson, J.C. French, E. Smal, and D.C. Baker. 1985. 2'-Chloropentostatin, a New Inhibitor of Adenosine Deaminase. *J. Org. Chem.* 50, 1651. (b) Ichiba, T., Y. Nakao, P.J. Scheuer, N.U. Sata, and M. Kelly-Borges. 1995. Kumusine, a Chloroadenine Riboside from a Sponge, *Theonella* sp. *Tetrahedron Lett.* 36, 3977.
69. Lin, W., R.S. Burkhalter, and R. Timkovich. 1993. Isolation of Novel Meso-Substituted Uroporphyrins from Cultures of *Pseudomonas. Heterocycles* 36, 2191.
70. Zein, N., A.M. Casazza, T.W. Doyle, J.E. Leet, D.R. Schroeder, W. Solomon, and S.G. Nadler. 1993. Selective Proteolytic Activity of the Antitumor Agent Kedarcidin. *Proc. Natl. Acad. Sci. USA* 90, 8009.
71. Spande, T.F., H.M. Garraffo, M.W. Edwards, H.J.C. Yeh, L. Pannell, and J.W. Daly: Epibatidine. 1992. A Novel (Chloropyridyl)azabicycloheptane with Potent Analgesic Activity from an Ecuadoran Poison Frog. *J. Am. Chem. Soc.* 114, 3475.

72. Miles, D.H., N.V. Mody, J.P. Minyard, and P.A. Hedin. 1973. Constituents of Marsh Grass: Survey of the Essential Oils in *Juncus roemerianus*. *Phytochem.* 12, 1399.
73. Pereira, W.E., C.E. Rostad, and H.E. Taylor. 1980. Mount St. Helens, Washington, 1980 Volcanic Eruption. *Geophys. Res. Lett.* 7, 953.
74. Li, M.K.W. and P.J. Scheuer. 1984. Halogenated Blue Pigments of a Deep Sea Gorgonian. *Tetrahedron Lett.* 25, 587.
75. Hunt, S. and S.W. Breuer. 1971. Isolation of a New Naturally Occurring Halogenated Amine Acid: Monochloromonobromotyrosine. *Biochim. Biophys. Acta* 252, 401.
76. Welinder, B.S. 1972. Halogenated Tyrosines from the Cuticle of *Limulus polyphemus* (L.). *Biochim. Biophys. Acta* 279, 491.
77. Andersen, S.O. 1972. 3-Chlorotyrosine in Insect Cuticular Proteins. *Acta Chem. Scand.* 26, 3097.
78. Ando, K., A. Kato, and S. Suzuki. 1970. Isolation of 2,4-Dichlorophenol from a Soil Fungus and Its Biological Significance. *Biochem. Biophys. Res. Commun.* 39, 1104.
79. Eisner, T., L.B. Hendry, D.B. Peakall, and J. Meinwald. 1971. 2,5-Dichlorophenol (from Ingested Herbicide?) in Defensive Secretion of Grasshopper. *Science* 172, 277.
80. Berger, R.S. 1983. Occurrence of 2,6-Dichlorophenol in *Dermacentor albipictus* and *Haemaphysalis leporispalustris* (Acari: Ixodidae). *J. Med. Entomol.* 20, 103.
81. Faulkner, D.J. 1980. Natural Organohalogen Compounds. In *The Handbook of Environmental Chemistry,* Vol. 1, Part A, (O. Hutzinger, Ed.), Berlin: Springer-Verlag. p. 229.
82. Spinnler, H.-E., E. de Jong, G. Mauvais, E. Semon, and J.-L. le Quere. 1994. Production of Halogenated Compounds by *Bjerkandera adusta*. *Appl. Microbiol. Biotechnol.* 42, 212.
83. Cohen, P.A. and G.H.N. Towers. 1995. Anthraquinones and Phenanthroperylenequinones from *Nephroma laevigatum*. *J. Nat. Prod.* 58, 520.
84. Tschirley, F.H. 1986. Dioxin. *Sci. Amer.* 254(2):29.
85. Gough, M. 1991. Human Health Effects: What the Data Indicate. *Sci. Total Environ.* 104, 129.
86. (a) Bumpus, R.R., W.B. Crummett, S.S. Cutie, J.R. Gledhill, R.H. Hummel, R.O. Kagel, L.L. Lamparski, E.V. Luoma, D.L. Miller, T.J. Nestrick, L.A. Shadoff, R.H. Stehl, and J.S. Woods. 1980. Trace Chemistries of Fire: A Source of Chlorinated Dioxins. *Science* 210, 385. (b) Addink, R. and K. Olie. 1995. Mechanisms of Formation and Destruction of Polychlorinated Dibenzo-*p*-dioxins and Dibenzofurans in Heterogenous Systems. *Environ. Sci. Tech.* 29, 1425.
87. (a) Clement, R.E., H.M. Tosine, and B. Ali. 1985. Levels of Polychlorinated Dibenzo-*p*-dioxins and Dibenzofurans in Wood Burning Stoves and Fireplaces. *Chemosphere* 14, 815. (b) Schatowitz, B., G. Brandt, F. Gafner, E. Schlumpf, R. Bühler, P. Hasler, and T. Nussbaumer. 1994. Dioxin Emissions from Wood Combustion. *Chemosphere* 29, 2005.
88. (a) Sheffield, A. 1985. Sources and Releases of PCDDs and PCDFs to the Canadian Environment. *Chemosphere* 14, 811. (b) Clement, R. and C. Tashiro. 1991. Forest Fires as a Source of PCDDs and PCDFs. Abstract Dioxin '91, Research Triangle Park, NC. p. 34.
89. Eduljee, G.H., D.H.F. Atkins, and A.E. Eggleton. 1986. Observations and Assessment Relating to Incineration of Chlorinated Chemical Wastes. *Chemosphere* 15, 1577.
90. (a) Svenson, A., L.-O. Kjeller, and C. Rappe. 1989. Enzyme-Mediated Formation of 2,3,7,8-Tetra-Substituted Chlorinated Dibenzodioxins and Dibenzofurans. *Environ. Sci. Technol.* 23, 900. (b) Öberg, L.G., B. Glas, S.E. Swanson, C. Rappe, and K.G. Paul. 1990. Peroxidase-Catalyzed Oxidation of Chlorophenols to Polychlorinated Dibenzo-*p*-dioxins and Dibenzofurans. *Arch. Environ. Contam. Toxicol.* 19, 930. (c) Öberg, L.G. and C. Rappe. 1992. Biochemical Formation of PCDD/Fs from Chlorophenols. *Chemosphere* 25, 49.
91. (a) Several authors in "The Natural Chemistry of Chlorine in the Environment," Euro Chlor publication, 1995, Brussels. (b) Carlsen, L. and P. Lassen. 1992. Enzymatically Mediated Formation of Chlorinated Humic Acids. *Org. Geochem.* 18, 477. (c) Grimvall, A. 1995. Evidence of Naturally Produced and Man-Made Organohalogens in Water and Sediments, in *Naturally-Produced Organohalogens,* de Leer, E.W.B., Ed. Kluwer Academic Publishers, The Netherlands. (d) Hoekstra, E.J. and E.W.B. de Leer. 1993. Natural Production of Chlorinated Organic Compounds in Soil, in *Contaminated Soil '93*, Arendt, F., G.J. Annokkée, R. Bosman, and W.J. van den Brink, Eds. Kluwer Academic Publishers, The Netherlands.
92. Blake, G.A., J. Keene, and T.G. Phillips. 1985. Chlorine in Dense Interstellar Clouds: The Abundance of HCl in OMC-1. *Astrophys. J.* 295, 501.
93. Neidleman, S.L. and J. Geigert. 1986. Biohalogenation: Principles, Basic Roles and Applications. New York: Halsted Press.
94. (a) Libby, R.D., J.A. Thomas, L.W. Kaiser, and L.P. Hager. 1982. Chloroperoxidase Halogenation Reactions. *J. Biol. Chem.* 257, 5030. (b) Zaks, A. and D.R. Dodds. 1995. Chloroperoxidase-Catalyzed Asymmetric Oxidations: Substrate Specificity and Mechanistic Study. *J. Am. Chem. Soc.* 117, 10419.

(c) Burd, W., O. Yourkevich, A.J. Voskoboev, and K.-H. van Pée. 1995. Purification and Properties of a Non-Haem Chloroperoxidase from *Serratia marcescens*. *FEMS Microbiol. Lett.* 129, 255.

95. Taurog, A. and E.M. Howells. 1966. Enzymatic Iodination of Tyrosine and Thyroglobulin with Chloroperoxidase. *J. Biol. Chem.* 241, 1329.
96. Walter, B. and K. Ballschmiter. 1991. Biohalogenation as a Source of Halogenated Anisoles in Air. *Chemosphere* 22, 557.
97. Knoch, M., K.-H. van Pée, L.C. Vining, and F. Lingens. 1989. Purification, Properties and Immunological Detection of a Bromoperoxidase-Catalase from *Streptomyces venezuelae* and from a Chloramphenicol-nonproducing Mutant. *J. Gen. Microbiol.* 135, 2493.
98. Geigert, J., S.L. Neidleman, D.J. Dalietos, and S.K. DeWitt. 1983. Haloperoxidases: Enzymatic Synthesis of α,β-Halohydrins from Gaseous Alkenes. *Appl. Environ. Microbiol.* 45, 366.
99. de Jong, E., J.A. Field, H.-E. Spinnler, J.B.P.A. Wijnberg, and J.A.M. de Bont. 1994. Significant Biogenesis of Chlorinated Aromatics by Fungi in Natural Environments. *Appl. Environ. Microbiol.* 60, 264.
100. Dahlman, O., R. Mörck, P. Ljungquist, A. Reimann, C. Johansson, H. Borén, and A. Grimvall. 1993. Chlorinated Structural Elements in High-Molecular-Weight Organic Matter from Unpolluted Waters and Bleached-Kraft Mill Effluents. *Environ. Sci. Technol.* 27, 1616.
101. Weiss, S.J., S.T. Test, C.M. Eckmann, D. Roos, and S. Regiani. 1986. Brominating Oxidants Generated by Human Eosinophils. *Science* 234, 200.
102. Bouwer, E.J. and P.L. McCarty. 1983. Transformations of 1- and 2-Carbon Halogenated Aliphatic Organic Compounds Under Metanogenic Conditions. *Appl. Environ. Microbiol.* 45, 1286.
103. Smith, A.E. and A.J. Aubin. 1991. Transformation of [^{14}C]-2,4-Dichlorophenol in Saskatchewan Soils. *J. Agric. Food Chem.* 39, 801.
104. Valli, K. and M.H. Gold. 1991. Degradation of 2,4-Dichlorophenol by the Lignin-Degrading Fungus *Phanerochaete chrysosporium*. *J. Bacteriol.* 173, 345.
105. Hammell, K.E. and P.J. Tardone. 1988. The Oxidative 4-Dechlorination of Polychlorinated Phenols is Catalyzed by Extracellular Fungal Lignin Peroxidases. *Biochem.* 27, 6563.
106. Bumpus, J.A. and S.D. Aust. 1987. Biodegradation of DDT [1,1,1-Trichloro-2,2-Bis(4-Chlorophenyl)Ethane] by the White Rot Fungus *Phanerochaete chrysosporium*. *Appl. Environ. Microbiol.* 53, 2001.
107. Mileski, G.J., J.A. Bumpus, M.A. Jurek, and S.D. Aust. 1988. Biodegradation of Pentachlorophenol by the White Rot fungus *Phanerochaete chrysosporium*. *Appl. Environ. Microbiol.* 54, 2885.
108. Bumpus, J.A., M. Tien, D. Wright, and S.D. Aust. 1985. Oxidation of Persistent Environmental Pollutants by a White Rot Fungus. *Science* 228, 1434.
109. (a) Abramowicz, D.A. 1990. Aerobic and Anaerobic Biodegradation of PCBs: A Review. *Crit. Rev. Biotechnol.* 10, 241. (b) Han, S., L.D. Eltis, K.N. Timmis, S.W. Muchmore, and J.T. Bolin. 1995. Crystal Structure of the Biphenyl-Cleaving Extradiol Dioxygenase from a PCB-Degrading Pseudomonad. *Science* 270, 976.
110. Bergbauer, M., C. Eggert, and G. Kraepelin. 1991. Degradation of Chlorinated Lignin Compounds in a Bleach Plant Effluent by the White-Rot Fungus *Trametes versicolor*. *Appl. Microbiol. Biotechnol.* 35, 105.
111. Häggblom, M.M. and L.Y. Young. 1995. Anaerobic Degradation of Halogenated Phenols by Sulfate-Reducing Consortia. *Appl. Environ. Microbiol.* 61, 1546.
112. (a) Dec, J. and J.-M. Bollag. 1994. Dehalogenation of Chlorinated Phenols During Oxidative Coupling. *Environ. Sci. Technol.* 28, 484. (b) Dec, J. and J.-M. Bollag. 1995. Effect of Various Factors on Dehalogenation of Chlorinated Phenols and Anilines during Oxidative Coupling. *Environ. Sci. Technol.* 29, 657.
113. Hu, Z.-C., R.A. Korus, W.E. Levinson, and R.L. Crawford. 1994. Adsorption and Biodegradation of Pentachlorophenol by Polyurethane-Immobilized *Flavobacterium*. *Environ. Sci. Technol.* 28, 491.
114. Bartnicki, E.W. and C.E. Castro: Biodehalogenation. 1994. Rapid Oxidative Metabolism of Mono- and Polyhalomethanes by *Methylosinus trichosporium* OB-3b. *Environ. Toxicol. Chem.* 13, 241.
115. Oremland, R.S., L.G. Miller, and F.E. Strohmaier. 1994. Degradation of Methyl Bromide in Anaerobic Sediments. *Environ. Sci. Technol.* 28, 514.
116. Blocki, F.A., M.S.P. Logan, C. Baoli, and L.P. Wackett. 1994. Reaction of Rat Liver Glutathione S-Transferases and Bacterial Dichloromethane Dehalogenase with Dihalomethanes. *J. Biol. Chem.* 269, 8826.
117. Leisinger, T. 1983. Microorganisms and Xenobiotic Compounds. *Experientia* 39, 1183.
118. (a) Asplund, G. and A. Grimvall. 1991. Organohalogens in Nature. *Environ. Sci. Technol.* 25, 1346. (b) Grimvall, A., K. Laniewski, H. Borén, S. Jonsson, and S. Kaugare. 1994. Organohalogens of Natural or Unknown Origin in Surface Water and Precipitation. *Toxicol. Environ. Chem.* 46, 183. (c) Hoekstra, E.J.

and E.W.B. de Leer. 1994. AOX Levels in the River Rhine: Fifty Percent of Natural Origin?! *Wat. Sci. Tech.* 29, 133.

119. Humic Substances II, In Search of Structure. M.H.B. Hayes, P. MacCarthy, R.I. Malcolm, and R.W. Swift, Eds. New York: Wiley-Interscience. 1989.
120. Paul, V.J. 1987. Feeding Deterrent Effects of Algal Natural Products. *Bull. Mar. Sci.* 41, 514.
121. Barbier, M. In *Marine Natural Products,* Vol. IV; P.J. Scheuer, Ed., Academic Press, New York, 1981.
122. Asplund, G., A. Grimvall, and C. Pettersson. 1989. Naturally Produced Adsorbable Organic Halogens (AOX) in Humic Substances from Soil and Water. *Sci. Total Environ.* 81/82, 239.
123. Paasivirta, J., J. Knuutinen, P. Maatela, R. Paukku, J. Soikkeli, and J. Särkkä. 1988. Organic Chlorine Compounds in Lake Sediments and the Role of the Chlorobleaching Effluents. *Chemosphere* 17, 137.
124. Fox, S.W. and K. Dose. 1972. *Molecular Evolution and the Origin of Life*; Freeman: San Francisco, p. 294.
125. Scheuer, P.J. 1990. Some Marine Ecological Phenomena: Chemical Basis and Biomedical Potential. *Science* 248, 173.
126. Paul, V.J. and S.C. Pennings. 1991. Diet-Derived Chemical Defenses in the Sea Hare *Stylocheilus longicauda* (Quoy et Gaimard 1824). *J. Exp. Mar. Biol. Ecol.* 151, 227.
127. Falch, B.S., G.M. König, A.D. Wright, O. Sticher, H. Rüegger, and G. Bernardinelli. 1993. Ambigol A and B: New Biologically Active Polychlorinated Aromatic Compounds from the Terrestrial Blue-Green Alga *Fischerella ambigua. J. Org. Chem.* 58, 6570.
128. Watanabe, K., M. Miyakado, N. Ohno, A. Okada, K. Yanagi, and K. Moriguchi. 1989. A Polyhalogenated Insecticidal Monoterpene from the Red Alga *Plocamium telfairiae. Phytochem.* 28, 77.
129. Crews, P., B.L. Myers, S. Naylor, E.L. Clason, R.S. Jacobs, and G.B. Staal. 1984. Bio-Active Monoterpenes from Red Seaweeds. *Phytochem.* 23, 1449.
130. Sakuma, M. and H. Fukami. 1993. Novel Steroid Glycosides as Aggregation Pheromone of the German Cockroach. *Tetrahedron Lett.* 34, 6059.
131. de Jong, E., A.E. Cazemier, J.A. Field, and J.A.M. de Bont. 1994. Physiological Role of Chlorinated Aryl Alcohols Biosynthesized De Novo by the White Rot Fungus *Bjerkandera* sp. Strain BOS55. *Appl. Environ. Microbiol.* 60, 271.
132. Williams, D.H., M.J. Stone, P.R. Hauck, and S.K. Rahman. 1989. Why are Secondary Metabolites (Natural Products) Biosynthesized? *J. Nat. Prod.* 52, 1189.

Chapter 8

Environmental Fate of Originated Organochloro Compounds from Pulp Mills

J. Paasivirta, H. Palm, T. Rantio, J. Koistinen,
P. Maatela, and R. Lammi

ABSTRACT

Organochlorine compound (OCC) emissions from bleaching of pulp with chlorine chemicals to recipient watercourses have significantly decreased due to process improvements in modern mills. The majority of the OCC discharges are nontoxic high-molecular-weight substances, which are slowly degraded in recipient waters and in shallow Finnish watercourses. They are in great part buried in sediments as persistent residues. The history of pulp production is readable in dated sediment layers by analyses of OCCs. From the low-molecular-weight OCCs, free (instantly extractable) polychlorophenols (FPCPs) are degradable in the water column, but they — especially polychloroguaiacols — accumulate in water biota. However, they metabolize out of the fish rather fast, and thus do not pose a long-term hazard. Microbial metabolites of chlorophenols — PCAs and PCVs — have been shown to be the main cause of bad taste in the recipient fish. Bound (extractable after treatment with strong alkali) polychlorophenols (BPCPs) are less degradable and are found in abundance in the recipient sediment. Persistent alkylaromatic chlorohydrocarbons and ethers formed in bleaching are also found in sediments. Their levels in biota are far below those of the common environmentally hazardous chlorohydrocarbons from other sources.

OCC EMISSIONS

Anthropogenic organochlorine compounds (OCCs) may enter the natural environment through accidents, but more commonly they do so from industrial discharges, pesticide uses, urban wastes, and especially from combustion and chlorination processes.[1] The latter include bleaching of pulp with chlorine chemicals, which in Nordic countries has been a major source of OCCs in forest industry recipient watercourses.

Today these discharges have decreased considerably due to a combination of external treatment and process improvements like modified cooking, extended delignification, substituting chlorine with chlorine dioxide or complete substitution of chlorine chemicals with oxygen, ozone, and hydrogen peroxide.

An overview of the recent OCC emissions from two modern pulp mills in Finland (see Figure 8.1) is shown in Table 8.1. Since 1985, these mills had secondary activated sludge treatment, and in their bleaching process elemental chlorine was gradually replaced with chlorine dioxide. A majority of the total organically bound chlorine (TOCl) in discharge was in high-molecular-weight organic matter ("chlorolignin"). In addition, low-molecular-weight chlorinated acids, phenols, and various neutral organochlorine compounds were emitted.[2] TOCl was often measured as AOX (adsorbable organic halogen). Part of the AOX was soluble in organic solvent (diethyl ether, ethyl acetate, 2-propanol, etc.) and could be determined as EOX (extractable organic halogen).

The best known low-molecular-weight toxic OCCs in discharge waters are chlorophenolics. One type, free polychlorophenols (FPCPs) are hexane-extractable (after acetylation). The other type, bound polychlorophenols (BPCPs), are extractable only after treatment with strong alkali. BPCPs are formed together with FPCPs in bleaching process stages.[3] Although chlorophenolics form the most significant toxic OCC load in discharge, the neutral OCCs are of concern since they are more persistent and in many cases more bioavailable micropollutants.

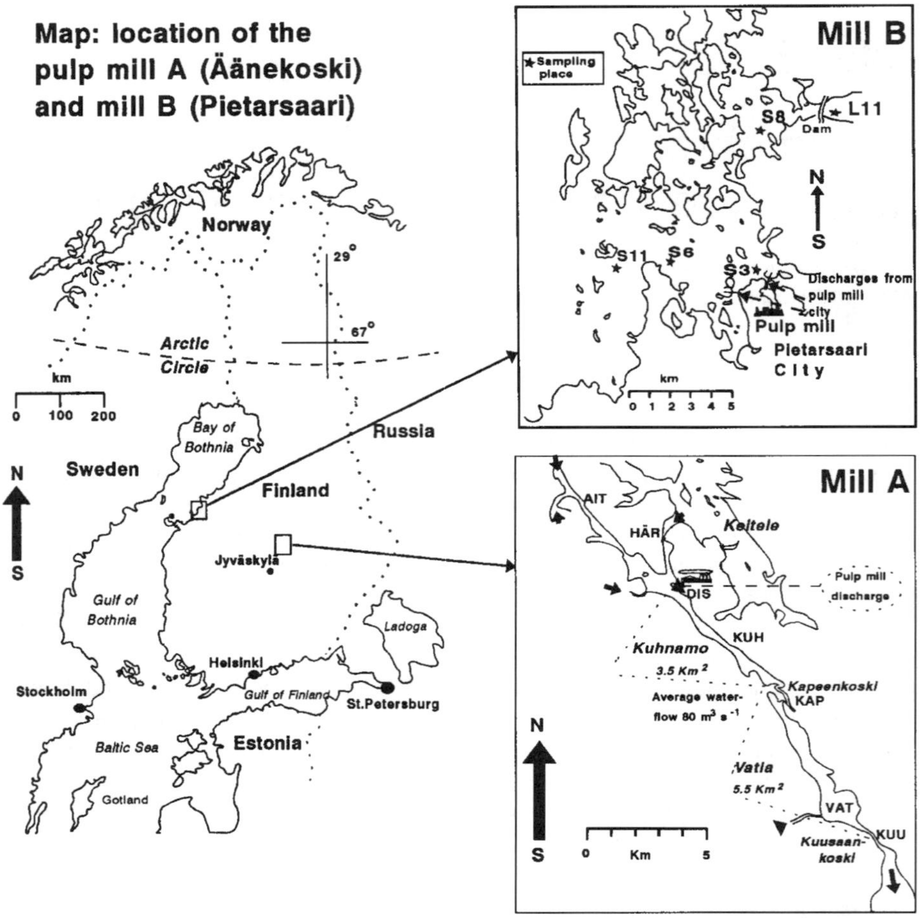

Figure 8.1. Location of the pulp mill A (Äänekoski) and mill B (Pietarsaari).

Table 8.1. Specific Emissions Grams Per Ton of Bleached Pulp (g/t) in Discharge to Recipient Watercourse from Two Kraft Mills in Finland. $$$ = Amounts in Removed Biosludge from Secondary Treatment in Mill B, September 1992.

	Mill A (Äänekoski, Central Finland)			Mill B (Pietarsaari, West Coast of Finland)			
	Oct 1983	Aug 1986	Aug 1989	Sep 1992	$$$	Apr 1993	Sep 1993
TOCl	2100	2900	1320	850	1.60	1250	110
EOCl				190		120	
Chloroform	15.1	4.9					
FPCPs	73.2	7.2		32.0	0.01	51.7	4.27
BPCPs				33.7	1.46	14.0	8.11

TOCl = total organically bound chlorine.
EOCl = extractable organically bound chlorine.
FPCPs = sum of the free chlorophenols, chloroguaiacols and chlorocatechols.
BPCPs = sum of the bound chlorophenols, chloroguaiacols and chlorocatechols.

Chlorobleaching produces very volatile halohydrocarbons, e.g., 1,2-dichloroethane, dichloromethane, chloroform, carbon tetrachloride, and 1,1,2-trichloroethene, which in large part are evaporated to atmosphere in the secondary treatment. Their annual global emission from pulp mills can be

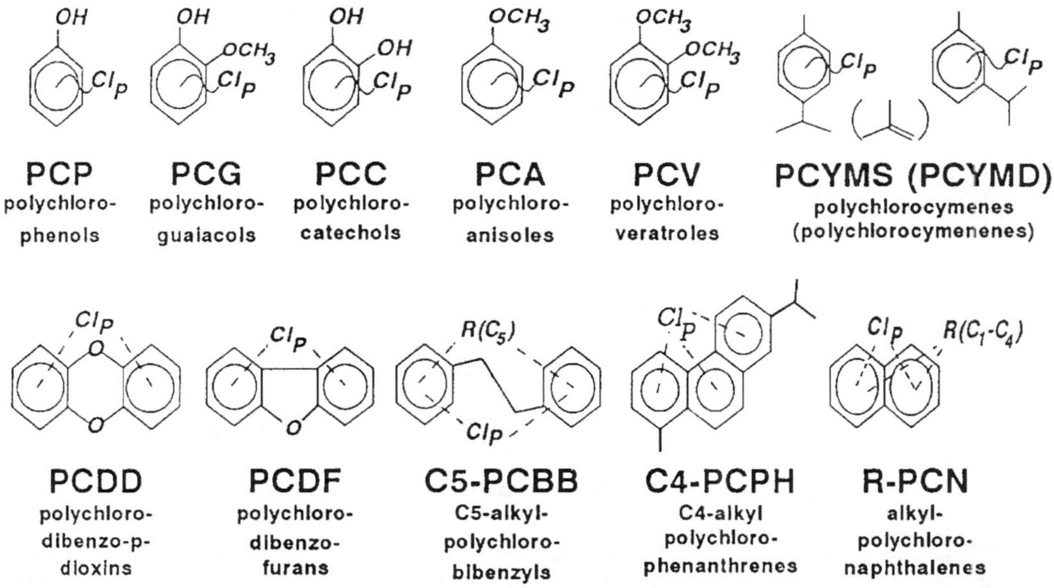

Figure 8.2. Environmentally interesting phenolic and neutral organochlorine compounds in pulp bleaching discharges.

estimated to be 5 kilotons. However, synthetic production and solvent use of these substances causes three orders of magnitude higher global emissions.[2]

Numerous semivolatile neutral OCCs in discharge waters include the most suspect long-term environmental toxicants, but only a small number of them have been identified thus far. Examples of the structure types of semivolatile phenolic and neutral OCCs in discharges are illustrated in Figure 8.2. PCDDs and PCDFs are well known as minor components emitted. Most of them are relatively nontoxic, but trace amounts of notoriously toxic 2,3,7,8-TCDD and 2,3,7,8-TCDF were also formed in measurable amounts when elemental chlorine was used in excess.[4–6]

Chlorocymenes and chlorocymenenes (PCYMS and PCYMD; Figure 8.2) are the major chlorohydrocarbons formed in bleaching.[7–9] Their emitted amounts were measured quantitatively only recently, after process changes had already significantly reduced them. From mill A, total specific emissions of PCYMS+PCYMD during softwood bleaching in October 1990, October 1991, and June 1993 were 230, 86.4, and 1.9 μg per ton of bleached pulp, respectively.[9]

Other alkylaromatic chlorohydrocarbon classes identified in chlorobleaching wastes are C5-PCBBs, C4-PCPHs, and R-PCNs (Figure 8.2).[10,11] In addition, trace amounts of alkylated polychlorodibenzofurans (R-PCDF)[10,12,13] and nonalkyl/alkyl polychlorofluorenes (C0-C2-PCFL)[14] have been identified in pulp mill products and discharges.

Environmental fate of these OCCs emitted from the example mills A and B (Table 8.1) has been studied by analyzing the recipient waters, sediments (also dated layers), and biota.

TOCl IN RECIPIENT WATERCOURSES

TOCl in recipient water consists mainly of chlorine bound to the high-molecular-weight dissolved organic matter (DOM). TOCl occurs at measurable levels also upstream from discharge and at other background areas. Typical TOCl concentrations are shown in Figure 8.3. Steep downward gradients from the discharge point to distances of a few kilometers were measured at both mill A (shallow sweetwater) and mill B (shallow brackish water) recipients.[15] The analyses of dated sediment layers[16,17] revealed that major removal of TOCl from the water column took place by precipitation of DOM together with particles. An example of TOCl deposition to sediment is shown in Figure 8.4. Excess of this deposition is demonstrated by mass balance estimations. TOCl emission from the mill

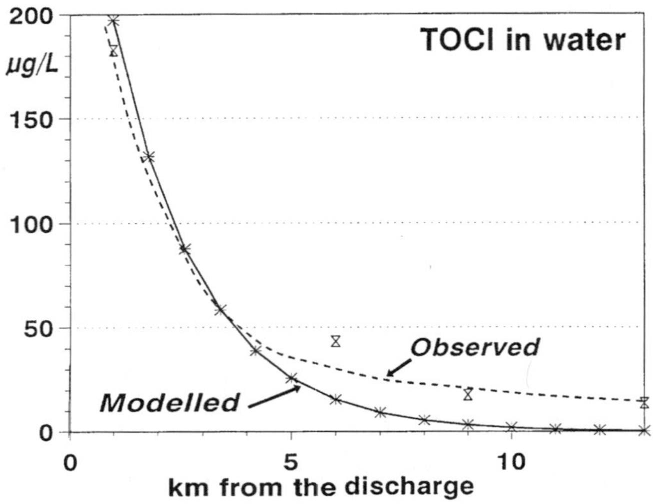

Figure 8.3. Observed and modeled (estimated removal rate 0.335 d^{-1}) concentrations of TOCl in water of the brackish water recipient area (westward stream) of the mill B in June 1993. At > 10 km distance the difference between observed and modeled concentration represents background TOCl at the area.

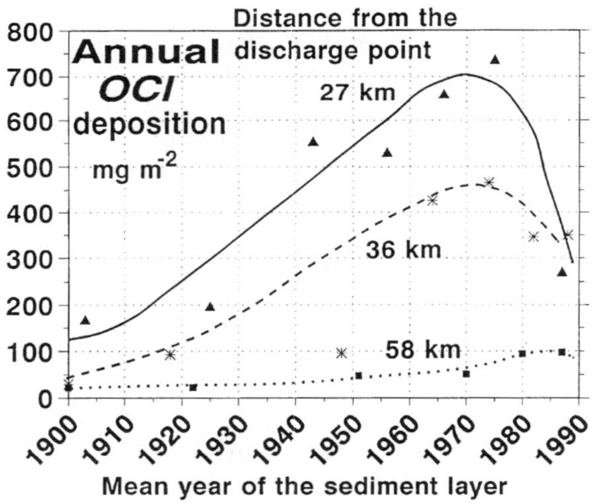

Figure 8.4. Total organically bound chlorine in sediment cores expressed as sedimentation rates in Jämsä watercourse in Central Finland where chlorobleaching of pulp had been stopped in 1981.

A (Äänekoski) 1974–1985 was approximately 3322 t, and 1835 t (55%) of this was found in recipient watercourse bottoms 70 km from the discharge point.[18] TOCl in the water layer is degraded (mainly biologically and by photolysis) rather slowly[19] and the hydraulic residence time within 70 km of the shallow Äänekoski watercourse is about one year. Therefore, the high rate of deposition of TOCl there is understandable. A quite different recipient is the deep Lake Vättern in Sweden, where only less than 15% of AOX emitted from pulp mills was found in sediments. There, the hydraulic residence time is 56 years, and, accordingly, pulp TOCl (AOX) from pulp mills has more time to degrade in the water column.[19,20] The concentrations of TOCl in dry sediments near pulp bleacheries vary quite a lot (depending on dilution), but based on organic matter (OM) they are characteristically more similar (Figure 8.5).[17,21,22]

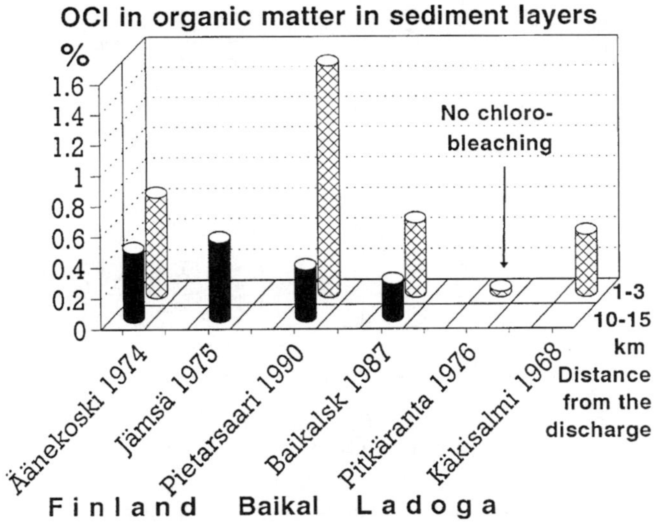

Figure 8.5. Total organically bound chlorine in organic matter at pulp mill influenced sediments in Äänekoski and Jämsä watercourses, Pietarsaari Archipelago, Baikal, and in the Lake Ladoga sediments near nonbleaching mill at Pitkäranta and near Käkisalmi, where pulping including chlorobleaching was stopped in 1986.

Extractable organic chlorine measured as EOCl or EOX in pulp mill recipient sediments is a fraction of AOX, which increases from 0.08 to 0.4 downward in sediment cores 1 km from mill B.[23] This is explained by biodegradation of high-molecular-weight organochlorines to low-molecular-weight (extractable) ones. At background sediment, the ratio of EOX/AOX remained constant (0.06) in layers from the year 1987 back to 1800. The general conclusion was that EOX persists for long periods in sediment.[23] EOCl in pulp mill-exposed fish is manifold related to the background fish, but its composition is totally different than that of the pulp bleachery EOCl.[24] Instead of the dominating chlorophenolics, chlorohydrocarbons, and ethers in pulp mill discharge EOCl, the fish EOCl seems to consist mainly of chlorinated fatty acids.[24]

CHLOROPHENOLICS

Free chlorophenols (FPCPs) in the recipient water column are degraded by photolysis and by biotransformation more rapidly than AOX.[19,20,25] Biodegradation rates in Nordic conditions seemed to depend more on seasonal variation of the microbial community than on the actual temperature of the water, as demonstrated for tetrachloroguaiacol in laboratory microcosms with natural water and sediment.[25] The biotransformation products PCAs and PCVs were nondetectable in recipient waters of mill A, but readily analyzable in fish and incubated mussels there. They were proven to be the major source of bad taste in these fish and mussels.[26,27]

The dominating chlorophenols in chlorobleachery recipient sediments are BPCPs (Figure 8.6).[28] They reflect quite well the history of bleached pulp production. The composition of BPCPs in nearby sediments is the same as in the BPCP fraction of the discharge: the main compounds in both are 3,4,5-trichlorocatechol, tetrachlorocatechol, 4,5-dichloroguaiacol, 3,4,5-trichloroguaiacol and tetrachloroguaiacol.[28] Free polychlorophenols (FPCPs), in contrast to the bound ones, are found in sediments also in background areas and in deep layers formed before any local industrial effluents occurred (Figure 8.7).[18,28] From FPCPs, some PCPs (see Figure 8.2 for structures) are found in fish and incubated mussels both in pulp mill recipient and in background areas. In freshwater fish in Finland, 2,3,4,6-tetrachlorophenol and pentachlorophenol especially originate in great part from combustion processes and from leakages of wood preservation. PCGs bioaccumulate at a higher rate than PCPs and PCCs in recipient fish and mussels, and are clear indicators of chlorobleaching. The major

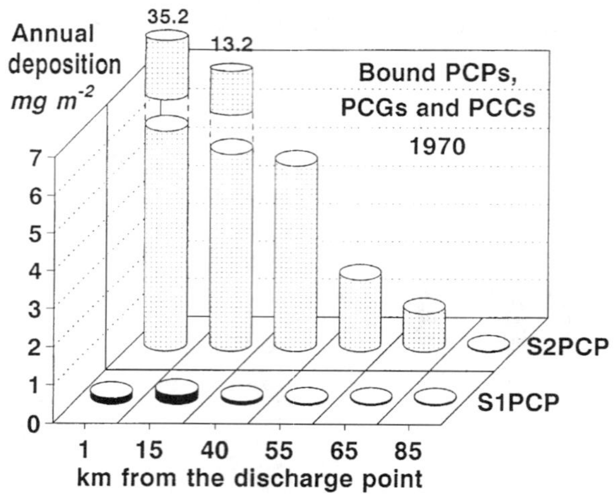

Figure 8.6. Annual deposition rates of bound chlorophenols (BPCPs) from sediment layers, 1970, in mill A recipient watercourse. S2PCPs are from bleaching only. S1PCPs could be also from wood preservation and combustions.

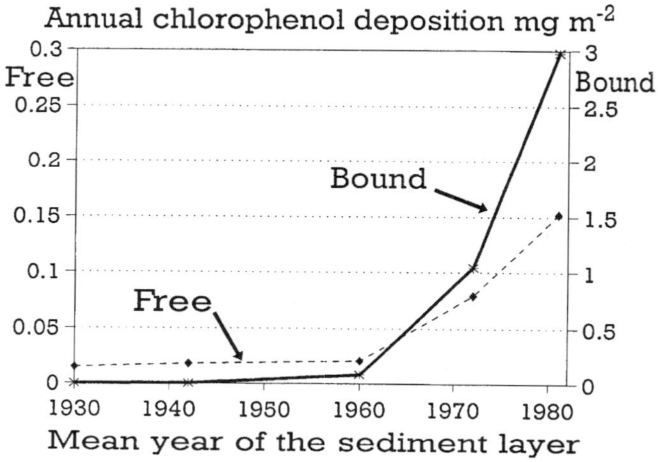

Figure 8.7. Chlorophenols in sediment layers of Ristinselkä (Päijänne) 65 km downstream of Äänekoski pulp mill (A), where chlorobleaching had started in 1956. Another small mill upstream had started bleaching in 1927.

chlorophenolics from bleaching, PCCs, are found only occasionally in fish, and not at all in incubated mussels. The present chlorophenolic contamination of biota at recipient watercourses of Finnish mills with ECF bleaching and activated sludge treatment has decreased to nearly background levels (Figure 8.8).[29,30]

SEMIVOLATILE NEUTRAL OCCs

Similarly, semivolatile neutral polychloro compounds from pulp bleaching can be found in sediments (Figure 8.9).[18,31] The toxic dioxin congeners formed measurably in elemental chlorine bleaching (mostly 2,3,7,8-TCDF and 2,3,7,8-TCDD) accumulated in recipient biota but not significantly in sediments.[6,31] Higher-chlorinated toxic dioxins and furans found in sediments originate mainly from sources (thermic reactions, chlorophenolic wood preservatives, or unknown origin) other than bleach-

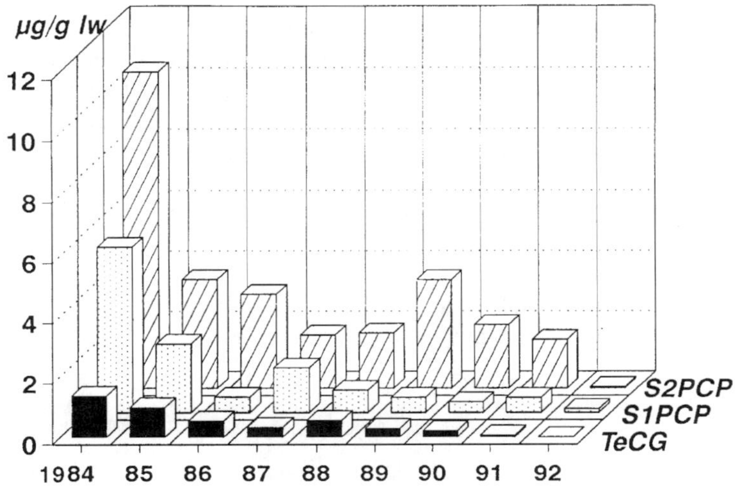

Figure 8.8. Concentrations in mussels incubated annually four weeks in August 18 km downstream of the mill A. TeCG = tetrachloroguaiacol, S1PCP = 2,4,6-trichlorophenol + 2,3,4,6-tetrachlorophenol + pentachlorophenol. S2PCP = sum of the other, mainly pulp bleaching-originated chlorophenolics.

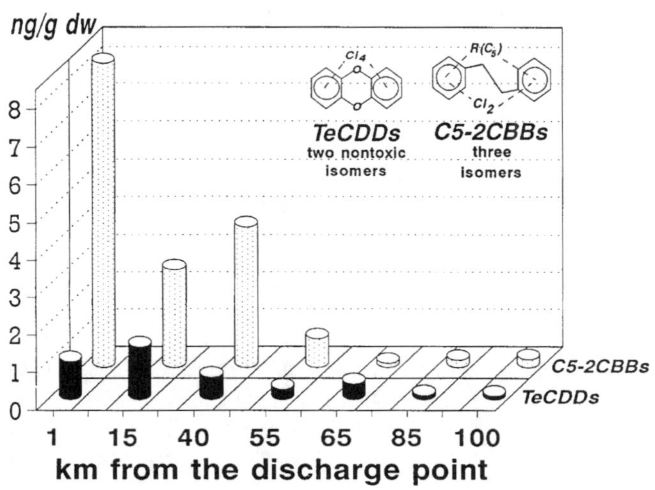

Figure 8.9. Concentrations of two nontoxic TeCDDs (1,4,6,8- and 1,4,7,9-TeCDDs) and three C5-alkyl-dichlorobibenzyls in surface sediments downstream of the mill A.

ing.[10,18,31,32] The minimal role of present modern pulp mills as a source of the total toxic dioxin load is supported by the U.S. EPA. According to the recent EPA estimation, total discharges of TEQs (dioxin toxic equivalents) in 1993 to watercourses from U.S. pulp mills (112 g TEQ/yr) is about 0.7% of the estimated emissions from other U.S. sources to atmosphere (16140 g TEQ/yr).[33]

The pulp mill–originated alkylaromatic chlorohydrocarbons PCYMSs, PCYMDs, C5-PCBBs, C4-PCPHs, and R-PCNs (Figure 8.2) are found in recipient sediment, fish, and mussels. Most abundant (up to 100–200 ng/g dw) among them in exposed sediment are dichlorocymenes (23DiCYMS and 25DiCYMS) and monochlororetene (C4-MonoCPH), which are very good indicators of the history of chlorobleaching.[22] Chlororetenes (C4-PCPH) and RPCNs were measured at elevated levels in spent biosludge of mill A.[11] C5-PCBBs and PCYMSs showed high bioaccumulation rates from water to biota.[34] Their measured levels in mussels incubated in 1988 at the Äänekoski watercourse were about 20 and 300 ng/g lw, respectively. Similar levels of PCYMSs were found in fish.[8] In Baltic Sea fish, the levels of common (mainly airborne) persistent chlorohydrocarbon pollutants — HCHs, hexachlorobenzene, chlordanes, DDT residues, and PCBs — are mainly two to three orders of magnitude higher.[35]

Due to the small number of analytical and toxicity studies done thus far, the environmental significance of alkylaromatic chlorohydrocarbons as potentially bioaccumulating toxicants cannot be estimated. Decrease of emissions due to recent process changes probably makes them nonsignificant as ecotoxicants. However, nonchlorinated retene (C4-PH) derivatives have been shown to form the highest MFO-inducer load in nonbleaching pulp mill discharge.[36]

SUMMARY

High-molecular-weight OCCs from bleaching are nontoxic. They degrade slowly in the water column. After sediment burial, they seem to persist. Bound chlorophenols are also persistent residues in sediments. Free chlorophenolics are degradable in the water column, but some — especially PCGs — accumulate in water biota. Microbial metabolites of chlorophenols, PCAs, and PCVs occur in pulp mill recipient waters mainly at subdetectable levels, but their bioaccumulation factor is very high. They have been shown to be the main cause of bad taste in recipient fish. Toxic dioxins from modern pulp mills occur in the environment at levels that are insignificant compared to dioxins from other sources. Persistent alkylaromatic chlorohydrocarbons formed in bleaching are bioaccumulating, but their levels in biota seem to be far below those of the common environmentally hazardous chlorohydrocarbons from other sources.

REFERENCES

1. Paasivirta, J. 1988. *Water Sci. Tech.* 20:119–129.
2. Paasivirta, J. and J. Koistinen. 1990. Environmental significance of organochlorines in water ecosystems. Relative risk of damages from pulp chlorobleaching wastes. *Pacific Paper Expo*, Vancouver BC, Canada 6-8.11.
3. Paasivirta, J., H. Tenhola, H. Palm, and R. Lammi. 1992. *Chemosphere* 24:1253–1258.
4. Kuehl, D.W., B.C. Butterworth, W.M. De Vita, and C.P. Sauer. 1987. *Biomed. Environ. Mass. Spectr.* 14:443–447.
5. Beck, H., K. Eckart, W. Mathar, and R. Wittowski. 1988. *Chemosphere* 17:51–57.
6. Rappe, C., B. Glas, and K. Wiberg. 1990. *DIOXIN '90. Organohalogen Compounds* 3:287–290.
7. Kuokkanen, T. 1989. Chlorocymenes and Chlorocymenenes: Persistent Chlorocompounds in Spent Bleach Liquors of Kraft Pulp Mills. *Department of Chemistry, University of Jyväskylä, Research Report* 32.
8. Rantio, T. 1992. *Chemosphere* 25:505–516.
9. Rantio, T. 1995. *Chemosphere* 31:3413–3423.
10. Koistinen, J., T. Nevalainen, and J. Tarhanen. 1992. *Environ. Sci. Technol.* 26:2499–2507.
11. Koistinen, J., J. Paasivirta, T. Nevalainen, and M. Lahtiperä. 1994. *Chemosphere* 28:1261–1277.
12. Buser, H-R., L-O. Kjeller, S.E. Swanson, and C. Rappe. 1989. *Environ. Sci. Technol.* 23:1130–1137.
13. Beck, H., A. Dross, K. Eckart, W. Mathar, and W. Wittowski. 1989. *Chemosphere* 19:655–660.
14. Koistinen, J., J. Paasivirta, T. Nevalainen, and M. Lahtiperä. 1994. *Chemosphere* 28:2139–2150.
15. Maatela, P. 1995. Determination of Organically Bound Chlorine in Industrial and Environmental Samples. *Department of Chemistry, University of Jyväskylä, Research Report* 53.
16. Paasivirta, J., J. Knuutinen, P. Maatela, R. Paukku, J. Soikkeli, and J. Särkkä. 1988. *Chemosphere* 17:137–146.
17. Maatela, P., J. Paasivirta, J. Särkkä, and R. Paukku. 1990. *Chemosphere* 21:1343–1354.
18. Paasivirta, J. 1992. In: *Environmental fate and effects of bleached pulp mill effluents.* A. Södergren (Ed.). *Swedish Environmental Protection Agency Report* 4031:168–179.
19. Carlberg, G.E. and T.R. Stuthridge. 1996. In: *Fate and Effects of Pulp and Paper Mill Effluents.* Servos, M.R., K.R. Munkittrick, J.H. Carey, and G.J. Van Der Kraak (Eds.). St. Lucie Press, Boca Raton, FL, pp. 169–178.

20. Grimwall, A., H. Borén, S. Johnsson, U. Lundström, and R. Sävenhed. 1991. *Water Sci. Tech.* 24:373–383.
21. Maatela, P., J. Paasivirta, M.A. Crachev, and E.B. Karabanov. 1990. *Chemosphere* 21:1381–1334.
22. Särkkä, J., J. Paasivirta, E. Häsänen, J. Koistinen, P. Manninen, K. Mäntykoski, T. Rantio, and L. Welling. 1993. *Chemosphere* 26:2147–2160.
23. Palm, H. and R. Lammi. 1995. *Environ. Sci. Technol.* 29:1722–1727.
24. Wesen, C. 1995. *Chlorinated Fatty Acids in Fish.* Lund University, Sweden.
25. Mikkelson, P., S. Herve, J. Paasivirta, and P. Heinonen. 1995. *Chemosphere* 31:3857–3871.
26. Paasivirta, J., P. Klein, M. Knuutila, J. Knuutinen, M. Lahtiperä, R. Paukku, A. Veijanen, L. Welling, M. Vuorinen, and P.J. Vuorinen. 1987. *Chemosphere* 16:1231–1241.
27. Paasivirta, J., A-L. Rantalainen, L. Welling, S. Herve, and P. Heinonen. 1992. *Water Sci. Tech.* 25:105–113.
28. Paasivirta, J., H. Hakala, J. Knuutinen, T. Otollinen, J. Särkkä, L. Welling, R. Paukku, and R. Lammi. 1990. *Chemosphere* 21:1355–1370.
29. Palm, H., J. Paasivirta, and R. Lammi. 1995. *Chemosphere* 31:2839–2852.
30. Herve, S., P. Heinonen, and J. Paasivirta. 1996. In: *Fate and Effects of Pulp and Paper Mill Effluents.* Servos, M.R., K.R. Munkittrick, J.H. Carey, and G.J. Van Der Kraak (Eds.). St. Lucie Press, Boca Raton, FL. pp. 335–340
31. Koistinen, J., J. Paasivirta, and J. Särkkä. 1990. *Chemosphere* 21:1371–1379.
32. Koistinen, J., J. Paasivirta, M. Suonperä, and H. Hyvärinen. 1995. *Environ. Sci. Technol.* 29:2541–2547.
33. Schaum, J., D. Cleverly, M. Lorber, L. Phillips, and G. Schweer. 1994. *DIOXIN '94. Organohalogen Compounds* 20:237–243.
34. Rantio, T., J. Koistinen, and J. Paasivirta. 1996. In: *Fate and Effects of Pulp and Paper Mill Effluents.* Servos, M.R., K.R. Munkittrick, J.H. Carey, and G.J. Van Der Kraak (Eds.). St. Lucie Press, Boca Raton, FL. pp. 341–345.
35. Paasivirta, J., T. Rantio, J. Koistinen, and P.J. Vuorinen. 1993. *Chemosphere* 27:2011–2015.
36. Hodson, P.V. 1996. In: *Fate and Effects of Pulp and Paper Mill Effluents.* Servos, M.R., K.R. Munkittrick, J.H. Carey, and G.J. Van Der Kraak (Eds.). St. Lucie Press, Boca Raton, FL. pp. 349–358.

Chapter 9

Effect of Biological Treatment on Pulp Mill Effluent Chemical Characteristics

L.E. LaFleur, T.M. Bousquet, and D.L. Cook

INTRODUCTION

For a variety of reasons, the major focus of research targeting identification of compounds formed in the pulp manufacturing process has been on internal process streams. As a result, extensive lists of compounds identified have been compiled (see Kringstad and Lindström, 1984; and McKague et al., 1988 for reviews of compounds identified in bleaching effluents). One of the reasons for this trend is that the higher concentrations found in undiluted process streams make it easier to detect, and perhaps isolate, the compounds. Another is that research objectives often focus on one aspect of pulp manufacturing, with the goal of better understanding the chemistry through characterization of by-products. A third reason is that in the 1970s, biological treatment was predominantly a feature of North American mills, while many European mills (where much of the research was taking place) did not utilize secondary treatment. For these reasons, there has been much greater emphasis on characterization of in-process wastewaters.

That is not to imply that nothing has been done to either characterize the effectiveness of secondary treatment or to identify compounds found in biologically treated effluents. Quite the contrary, in recent years a great deal of attention has been directed toward building our knowledge in these areas. The objective of this chapter will be to assemble published information pertaining to aerobic biological treatment removal efficiencies, and to report additional information relating specifically to the effectiveness of biological treatment, as practiced in the United States, for removal of selected organic compounds. Further, information characterizing mass discharge rates of these compounds for U.S. mills is summarized.

CHLORINATED PHENOLICS

Chlorinated phenolics are probably the most extensively studied class of compounds found in pulp mill effluents, and do not need further detailed discussion. Data providing estimation of overall removal efficiencies of individual compounds across biological treatment systems have been reported (Holmbom and Lehtinen, 1980; Leuenberger et al., 1985; and Lindström and Mohamed, 1988). However, the sampling regimes were somewhat limited, or were not described in detail, thereby limiting the value of specific quantitative estimates of treatment efficiency.

Gergov et al., 1988 and Stuthridge et al., 1991 summarized treatment efficiencies by summing compound classes (e.g., total chlorinated guaiacols, etc.). This presentation of the data tends to obscure the details, but does add to our general understanding of how the treatment efficiencies of different classes of chlorinated compounds compare. Recently, aquatic toxicity equivalency factors for chlorinated phenolics based on the degree of chlorination were published (Kovacs et al., 1993). This has caused a trend toward reporting chlorinated phenolics by degree of chlorination. One such report includes estimates of chlorophenolic removal efficiencies summarized in this fashion (Strömberg et al., 1995). These group removal efficiency estimates are summarized in Table 9.1.

Recently, the National Council of the Paper Industry for Air and Stream Improvement (NCASI) compiled and reported data pertaining to individual chlorinated phenolic compound treatment efficiencies (Wiegand et al., 1994). These results are summarized in Table 9.2.

Information relative to the mechanism of removal was also summarized (Wiegand et al., 1994). It was observed that partitioning to solids accounted for a small portion of the overall removal, generally less than 10%. This concurs with observations reported by other researchers (Saunamäki et al.,

Table 9.1. Removal Efficiencies of Chlorinated Phenolics by Compound Class and by Degree of Chlorination Reported in the Literature.

Removal Efficiencies (Percent)

	Gergov et al., 1988					Stuthridge et al., 1991	
	Mill 1 Pine	Mill 1 Pine	Mill 1 Birch	Mill 1 Birch	Mill 2	System	System
Treatment Type	AS[a]	AS	AS	AS	ASB[b]	ASB	Lagoon
Compound Class							
Chlorophenols	82	87	82	72	43		77
Chloroguaiacols	73	82	75	73	70		85
Chlorocatechols	97	99	96	86	15	80	
Chlorovanillins	65	83	14	53	89	84	91
Chlorosyringols			67				

Removal Efficiency (Percent)

	Stromberg et al., 1995				
	SW-2	SW-3	SW-4	HW-1	HW-2
Treatment Type	ASB	AS	ASB	AS	ASB
Degree of Chlorination					
Total Monochloro	89	45	75	57	83
Total Dichloro	66	93	73	24	79
Total Trichloro	21			36	67
Total Tetrachloro	0			400	

[a] AS = Activated Sludge; [b] ASB = Aerated Stabilization Basin.

1991; Gergov et al., 1988; and Leuenberger et al., 1985). The NCASI studies also found volatilization to be a minor component of the overall removal process. Biodegradation appears to be the predominant mechanism of removal.

MONOTERPENES

Summary of Literature Data

The earliest studies examining the fate of monoterpenes across biological treatment systems suggested that some monoterpenes can be converted to camphor or fenchone (Wilson and Hrutfiord, 1971). However, this conclusion was not well-supported by the data presented. It was observed that the levels of monoterpenes were unaffected by the bleaching operations. Monoterpene alcohols were more degradable than the hydrocarbons, and both of these were much more degradable than ketones. Specifically, the average removal for alcohols was 99%, for hydrocarbons it was 75%, and for ketones it was 12%. Results from another, smaller study were reported shortly thereafter (Hrutfiord et al., 1975). In that study, the data presented showed a slight increase in fenchone and a 25% reduction in camphor. The other monoterpenes were removed with 80 to 99% efficiency.

The removal efficiency of *p*-cymene (formed from monoterpenes in sulfite cooks) was reported to be >99% after activated sludge treatment (Leuenberger et al., 1985). Monochloro-*p*-cymenes (2 isomers) and dichloro-*p*-cymenes (2 isomers) had removal rates of >86% to >99%, and 69% to 89%, respectively. The removal mechanisms for *p*-cymeme and the monochloro-*p*-cymene isomers were thought to be due to slight biodegradation, but mostly due to volatilization. Volatilization alone was assumed to be responsible for the removal of the dichloro-*p*-cymene isomers.

In New Zealand, chlorinated monoterpenes and monoterpene alcohols were identified in two kraft bleach plant filtrates (Stuthridge et al., 1990). In a later study at one of the mills, it was determined

Table 9.2. Average Chlorinated Phenolic Removal Efficiencies at Twelve U.S. Bleached Kraft Mills.

	Removal Efficiencies Percentage											
Mill Code	A	B	C	D	E	F	G	H	I	J	K	L
Treatment Type	ASB[a]	ASB	AS[b]	AS	ASB	AS	ASB	AS	ASB	ASB	AS	ASB
Compound												
2,4-Dichlorophenol								68	42			
2,4,6-Trichlorophenol	61			5	35			68	24			
2,3,4,6-Tetrachlorophenol								62	43			
4,5-Dichloroguaiacol	98		>93	91	83		>97	>80	>91	>95	67	
3,4,5-Trichloroguaiacol				46			>88	39	31	>94	19	
4,5,6-Trichloroguaiacol								20	23			
Tetrachloroguaiacol								65	18			
4,5-Dichlorocatechol						93	58	56	>70	>92		
3,4,5-Trichlorocatechol	>56		>55	46			34	56	>57	84	63	
Tetrachlorocatechol								51	–41		59	
5-Chlorovanillin	>83		>85									
6-Chlorovanillin	99	>96	>90	93			>98			>93	>83	>94
5,6-Dichlorovanillin	>90			>77	>77	>95	>89					
2-Chlorosyringaldehyde	>91											
2,6-Dichlorosyringaldehyde	>80			58						>86		
Trichlorosyringol										>92		

[a] ASB = Aerated Stabilization Basin; [b] AS = Activated Sludge.

that the overall treatment system was capable of removing the chloromonoterpene hydrocarbons by 33% and the chloromonoterpene alcohols by 71% (Stuthridge et al., 1991). The improved treatability of the alcohols follows the same general trend observed by Wilson and Hrutfiord (1971). The treatment system investigated was comprised of two parallel systems, each treating different portions of mill wastewater prior to being combined for discharge into the receiving environment. Stuthridge et al., (1991), reports removal efficiencies across each of the parallel systems, as well as for the final combined outfall. It is interesting to note that the system with the highest removal efficiency for the chloromonoterpenes was the one with the more aggressive mechanical aeration. This may suggest that the main mechanism of removal is volatilization, a conclusion similar to that made by Leuenberger et al. (1985) for monochloro-*p*-cymenes.

Summary of Monoterpene Removal Efficiency in U.S. Mills

In order to provide an estimate of the removal efficiencies of monoterpene compounds typical of U.S. mill biological treatment systems, a limited number of mills were studied. The analyte list was developed in part from the compounds reported in effluents in the literature (Wilson and Hrutfiord, 1971; Hrutfiord et al., 1975; and Voss, 1984), in part by our previous experience in characterization of U.S. mill effluents, and in part by considerations related to the reliability of the analytical methodology employed for the work. The results of this survey are summarized in Table 9.3.

Generally speaking, the treatment efficiencies observed in this study were quite high for the monoterpenes and monoterpene alcohols, and are thus consistent with previous work. The ketones, camphor, and pipertone (where they were observed) were also readily removed, and there is no evidence of formation across treatment apparent in the data. The only compound that showed poor removal was verbenone.

Monoterpene-Related Compound Mass Discharge Rates in U.S. Mills

In order to further characterize the discharge of monoterpenes in biologically treated effluents from U.S. mills, a survey of mills was conducted. Analytes included in this survey were *p*-cymene, fenchyl alcohol, terpinene-4-ol, *p*-cymene-8-ol, α-terpineol, camphor and verbenone. The most commonly detected compound was α-terpineol, which was found in four of the 15 mills tested, with mass discharge rates ranging from 0.52 to 17.5 lb × 10^{-3}/T. Camphor was found in four of the mills, and discharge rates ranged from 0.28 to 3.70 lb × 10^{-3}/T. The other compounds were only occasionally detected. Generally, depending on the individual mill's flow and production, the analytical methodology utilized would have been capable of estimating mass emission rates as low as 0.016 to about 2.1 lb × 10^{-3} lb/T. Clearly, these compounds are a minor component of mill effluents.

FATTY AND RESIN ACIDS

Summary of Literature Data

Fatty and resin acids have been extensively studied since the mid-1970s. A number of reports in the literature provide either direct estimates of fatty and/or resin acid removal efficiencies or data that can be utilized for this purpose (Easty et al., 1978; CPAR, 1978; Holmbom and Lehtinen, 1980; Leach and Chung, 1980; Voss and Rapsomatiotis, 1985; NCASI, 1990; and Zender et al., 1993). These estimates are summarized in Table 9.4.

Additional estimates of overall resin acid treatment efficiencies (by group rather than by individual compound) are also available (Stuthridge et al., 1991; Dahlman et al., 1993; and Strömberg et al., 1995). The group removal estimates show that removal efficiencies for fatty acids range from 42% to 100%, for resin acids they range from 68% to 98%, and for chlorinated resin acids they range from 94% to 99%. Thus, they are similar to those observed for the individual compounds. Overall, the information available in the literature indicates a high degree of removal efficiency, but there are some exceptions.

Table 9.3. Summary of Monoterpene Removal Efficiencies in U.S. Mills.

	Estimated Removal Efficiencies[a] (Percent)							
	Activated Sludge Treatment				Aerated Stabilization Basins		Overall	
Mill Code[b]	8	9	10	14	11	12	Mean	RSD (%)
Hydrocarbons								
p-Cymene	>100	>100			>100	>98	99	1
Alcohols								
Fenchyl Alcohol	>100	>100	>100	80	>100	>100	97	8
Terpinen-4-ol	>100	>100			>100	>100	100	0
p-Cymene-8-ol		>97	>100	>100	>100		99	2
α-Terpineol	100	>100	>99	>99	>100	>100	100	0
Ketones								
Camphor	>97	>97	>99	>100	>100	>100	99	2
Verbenone		>100	>18	>100		>45	66	41
Pipertone						>99	99	

[a] The computational guidelines used in estimating treatment efficiencies are described in Appendix A of this chapter.
[b] Descriptions of selected facts about the mills are summarized in Appendix B of this chapter.

Since a number of the studies summarized in Table 9.4 had objectives other than estimating removal efficiencies, the sampling plans were not ideal for that purpose. In some cases, the data summarized in Table 9.4 were derived from a single analysis in and out of the treatment system. In these cases, normal process variability may give rise to estimates of treatment efficiencies that are not representative of the treatment technology. This, and the lack of information characterizing the type of treatment systems studied, makes it hard to examine the information for trends or indications of system performance.

Summary of Additional Fatty and Resin Removal Efficiency Estimates for U.S. Mills

In an effort to expand the available information base of estimates of treatment efficiency for U.S. mills, existing NCASI data were reviewed to identify situations where sampling schemes appropriate to the objective of estimating treatment efficiency existed. The treatment efficiency estimates derived from the data identified in this review are summarized in Table 9.5.

The data were examined to determine if any significant differences were apparent. In general, the chlorinated resin acids have slightly lower average removal efficiencies, although the differences were not statistically significant (using a t-test at the 95% confidence interval). However, it should be noted that, as a group, the chlorinated resin acid estimates had a higher frequency of nondetects in the data sets and the treatment estimates tend to be biased low. This may be a significant contribution to the lower than average removal efficiencies.

A comparison of the average performance of activated sludge (AS) systems versus that of aerated stabilization basins (ASB) failed to reveal any statistically significant differences. The activated sludge systems tended to have somewhat lower and more variable fatty acid removal efficiencies, whereas aerated stabilization basins had lower and more variable removal rates for resin acids. But again, it should be emphasized that these apparent differences are not statistically significant.

We attempted to correlate resin acid removal efficiency with BOD reduction where data were available. It was apparent that if a BOD removal efficiency of greater than about 75% was obtained, the total resin acid removal efficiency was in the 80% to 100% range.

Table 9.4. Removal Efficiencies of Individual Fatty and Resin Acids Reported in the Literature.

	Removal Efficiencies (Percent)																
	Easty et al. 1978	CPAR 408-3, 1978 Mill Code			Holmbom & Lehtinen 1980		Leach and Chung, 1980 Mill Number								Voss et al. 1985	NCASI 1993	Zender et al. 1993
		A	B	C	Pine	Birch	A1	A2	A3	A4	A5	A6	A7	A8			
Treatment Type		ASB[a]	ASB	ASB	ASB	ASB									ASB	AS[b]	ASB
BOD Removal		74	78	58	46	54										94	1990
Fatty Acids																	
Oleic Acid	80						90		75	81	>84	>82	>85	91	99	93	
Linoleic Acid	94	100	98	86	86	59	89	33	69	52	81	77	76	93	100	96	
Linolenic Acid	100				94	81											
9,10-Dichlorostearic Acid														100			
Resin Acids																	
Pimaric Acid	47	99	91	56	61	14	>99		89	90			99	>99	96	88	94
Sandracopimaric Acid														94	85	95	
Isopimaric Acid	52	99	93	67	65	62	96	70	92	78	87	>98	97	99	96	89	94
Palustric Acid					20	11									96	95	>99
Neoabietic Acid					8	50	>97		>99	>93			>99	>99	97	92	99
Abietic Acid	64	98	89	63	59	18	97	88	94	79	86	>95	98	>99	97	92	94
Dehydroabietic Acid	72	97	86	–29	18	30	97	81	94	84	94	>99	98	99	97	94	87
Levopimaric Acid		97	94	76			98	>99	98	90		>99	99	>99	92		
Chlorinated Resin Acids																	
12-Chlorodehydroabietic Acid																	
14-Chlorodehydroabietic Acid		84	6	–6	75[c]	33[c]									81[c]		
Dichlorodehydroabietic Acid		0	50	44											44	40	

[a] ASB = Aerated Stabilization Basin; [b] AS = Activated Sludge; [c] Sum of both monochloro isomers.

">" Signs signify that the analyte was reported as nondetect in one or more of the effluent samples. The detection limit (if available) or zero was substituted for these samples for the purposes of computing the treatment efficiency. This substitution makes the resulting treatment efficiency estimate a conservative underestimation of the true removal efficiency.

"–" Signs signify that the average concentration in the effluent was greater than the average concentration in the influent.

Table 9.5. Summary of Fatty and Resin Acid Removal Efficiencies in U.S. Mills.

	Removal Efficiencies[a] Percentage																			
	Kraft Mills													Sulfite Mills						
	Western Mills							Central Mill	Northeastern Mills		Southern Mills			Western Mills				Central Mills		
Mill Code[b]	4	6	38	2	3	5	39	33	31	34	1	7	40	28	32	41	45	35	36	37
Treatment	AS[c]	AS	AS	ASB[d]	ASB	ASB	ASB	AS	AS	AS	ASB	ASB	ASB	AS	AS	AS	AS	AS	AS	AS
BOD Removal Efficiency			90				90	97				91						92	97	98
Fatty Acids																				
Linoleic Acid	>91	80	>81	>95	83	76	>99		>99	78	>98	87	>99					92	99	>99
Oleic Acid	>47	90	>70	94	90	77	87	>99	79	31	>98	84	>99	80	95	18	23	82	>96	60
Dichlorostearic Acid			>95	>98			>99	>99	>99		>92	>92	>88		>99			>98		
Resin Acids																				
Pimaric Acid	>89	74	>94	88	38	>94	>98		>97	>98	93	49	86					76	>96	>97
Sandracopimaric Acid	>85	>82		98	46	>92					>94	61	89							
Isopimaric Acid	95	79	>94	91	37	>98	>97	>86	98	89	93	59	85	>73	>97	>77	>83	79	>97	93
Palustric Acid	>84	84	>98	90	27	>97	>97		100		>86	64	88					59		>99
Dehydroabietic Acid	92	86	93	92	47	>98	98	>97	98	90	95	81	87	>97	100	94	>96	90	100	96
Abietic Acid	93	82	>95	88	42	>98	>97		98	90	>94	74	84	>89	>88		>77	79	>99	95
Neoabietic Acid	>69	86	>97	>89	48	>95	97		98	>95	>96	31	90					79	86	96
7-Oxodehydroabietic Acid												>96								
Chlorinated Resin Acids																				
14-Chlorodehydroabietic Acid			>86	>89	35		>93	>97	>90	>63		69	94	>85	99	>68	>96	66		>88
12-Chlorodehydroabietic Acid			>87	>90	15		>92	>98	>95	>86	>89	57	88	>92	98	>84	>94	70	>67	>90
Dichlorodehydroabietic Acid			>84	>79	2		85	98	>87				77		80		>88	46		

[a] The computational guidelines used in estimating treatment efficiencies are described in Appendix A of this chapter.
[b] Descriptions of selected facts about the mills are summarized in Appendix B of this chapter.

Resin Acid Removal/Biodegradation Mechanisms

In addition to microbial degradation, physiochemical processes, such as adsorption, serve to remove compounds during biological treatment of effluents. Also, as was noted in an earlier section, air stripping can contribute to the removal process. To date, no systematic studies have been undertaken to specifically quantify which of these mechanisms predominate over the biological treatment of fatty and resin acids. Given the low volatility of fatty and resin acids, it stands to reason that this will be a relatively minor contributor. This leaves, essentially, biodegradation and adsorption as the most likely mechanisms of removal.

Considering microbial degradation more closely, if the process goes to completion (i.e., complete conversion to CO_2 and water) the compound will be completely destroyed. However, in some cases, microbial degradation may not be complete, and the result will be the formation and discharge of altogether new compounds. Studies comparing concentrations of compounds going into the treatment system versus levels being discharged (such as those summarized in Tables 9.6 and 9.7) cannot differentiate complete versus partial degradation. More detailed work is required to gain insight into questions of this nature.

Researchers in New Zealand have been very active in the investigation of changes and/or transformations of resin acids across treatment systems. Recent studies report that aerobic treatment can lead to the formation of hydroxylated resin acids (Wilkins et al., 1988; and Zender et al., 1993). Other studies suggest that, under anaerobic conditions, resin acids can undergo reduction of double bonds, resulting in more saturated ring structures (McFarlane and Clark, 1988; and Zender et al., 1993).

While aerobic treatment is the technology of choice in the United States, many facilities have portions of their system that, either intentionally or unintentionally, have become anaerobic. To assess the significance of the New Zealand findings for typical U.S. treatment systems, U.S. effluents were screened for selected compounds that might be indicative of the processes observed by the New Zealand workers. Previous work had reported on the concentration and discharge of 13-abieten-18-oic acid (NCASI, 1989), which would be indicative of a degradation pathway involving double bond reduction. Additional screening of a larger number of mills indicated that 13-abieten-18-oic acid can be, but is not always, present. NCASI has also surveyed 15 different mill effluents for 7-isopimarenic acid, and found the compound present only once, at just above the detection limit (NCASI, 1993).

The only oxygenated resin acid for which there are U.S. data available is 7-oxodehydroabietic acid. As can be seen from the data for Mill 7 in Table 9.5, the overall removal efficiency for this compound in an ASB system was greater than 96%. However, since it was detectable in the influent to the treatment system, it is unlikely that the original source of the compound was microbial transformation. Presently, we are aware of no data quantifying hydroxylated resin acids in U.S. mill effluents.

Another possible anaerobic microbial degradation pathway for resin acids is decarboxylation and reduction or aromatization (Wilkins and Panadam, 1987). These workers found fichtelite, dehydroabietin, 1,2,3,4-tetrahydroretene and retene in mill discharges. The source was identified as a small overflow stream from a sludge lagoon. NCASI studies are currently underway to investigate the potential significance of these findings for U.S. mills.

Summary of Fatty and Resin Acid Mass Discharge Rates for U.S. Mills

The data in Table 9.6 summarize fatty and resin acid mass discharge rates for U.S. mills. The information is organized by pulping technology (kraft vs. sulfite), and by geographic location. Since it is difficult to observe patterns or trends when the detailed data are arrayed in this manner, overall averages for kraft and sulfite and by different regions (where sufficient data existed to allow such a computation) were made, and are summarized in Table 9.7.

Unfortunately, the only sulfite data available were from western mills. However, due to pitch problems associated with sulfite pulping of softwoods, most mills in the United States use hardwoods. Although hardwoods do have fatty acids, they have no resin acids. This makes estimates of resin acid discharge rates for most sulfite mills meaningless. In the Pacific Northwest, a number of softwood species with low, or relatively low, extractive content (e.g., western hemlock, Sitka spruce,

Table 9.6. Estimated Fatty and Resin Acid Mass Discharge Rates for U.S. Mills.

Mill	Treatment	Estimated Mass Discharge Rates for U.S. Kraft Mills[a] (lb x 10^{-3}/T)												
Code[b]	Type	C18:2[c]	C18:1	DCS	Pim	Sand	Iso	Pal	DHA	Ab	Neo	14-CDHA	12-CDHA	DCDHA
Northeastern U.S.														
46	AS[d]	6.69	6.91		1.29	2.36	6.82	1.54	5.65	4.46	1.30			1.72
47	AS	5.81	1.37	0.347	0.734	0.247	1.51		0.258	0.384			0.479	0.372
Southern U.S.														
7	ASB[e]	22.0	25.4	2.48	20.4	4.91	17.7	22.1	54.4	66.0	9.40	2.27	6.77	
15	ASB	0.503	1.51		1.52	0.243	3.07	0.548	2.21	2.58	1.01		0.148	
17	ASB		0.511		1.84	0.350	1.35	0.832	5.11	4.67	0.69		0.311	
18	AS	3.33	8.10		25.2	3.73	11.8	7.63	43.5	54.0	6.34			
19	ASB			5.24	89.7	23.1	79.0	104	304	293	83.0	0.594	0.874	
23	ASB	0.818	3.58		4.20	0.854	4.92	1.80	13.8	12.9	2.97			
25	ASB	0.912	4.11	0.786	11.6	2.72	13.2	2.44	22.3	32.7	4.30	0.318	1.04	0.741
26	AS	0.256	0.277		0.199		0.402	0.118	0.551	0.67	0.15			
Western U.S.														
3	AS	10.9	5.00		20.1	5.08	32.7	18.6	36.7	58.1	13.5	2.94	6.74	9.36
4	AS						2.32		9.80	3.56				
5	ASB	10.5	4.14											
6	AS	10.3	6.94		4.28		11.2	3.33	8.94	9.70	2.09			
16	ASB	0.465	1.89		0.696	0.385	2.04	0.838	1.28	2.78	0.665	0.293		
22	ASB	0.065	0.371		0.044		0.067		0.115	0.057				
30	AS	7.11	12.0	2.30	6.52	5.72	40.0	22.5	27.9	43.4	9.09	0.909		
42	ASB		5.02											
44	ASB		8.27				8.82		12.7	8.82				

Mill	Treatment	Estimated Mass Discharge Rates for U.S. Sulfite Mills[a] (lb x 10^{-3}/T)												
Code[b]	Type	C18:2	C18:1	DCS	Pim	Sand	Iso	Pal	DHA	Ab	Neo	14-CDHA	12-CDHA	DCDHA
Western U.S.														
27	ASB		12.6										4.95	
28	AS		10.9											
29	AS	16.9	8.93						0.922				1.38	0.615
41	ASB		13.3						0.436				0.453	
43	AS		7.77						3.33					
45	AS								0.891				0.394	

[a] The computational guidelines used in estimating treatment efficiencies are described in Appendix A of this chapter; [b] Descriptions of selected facts about the mills are summarized in Appendix B of this chapter; [c] Compound abbreviations: C18:2 = linoleic acid; C18:1 = oleic acid; DCS = 9,10-dichlorostearic acid; Pim = pimaric acid; Sand = sandracopimaric acid; Iso = isopimaric acid; Pal = palustric acid; DHA = dehydroabietic acid; Ab = abietic acid; Neo = neoabietic acid; 14-CDHA = 14-chlorodehydroabietic acid, 12-CDHA = 12-chlorodehydroabietic acid; DCDHA = dichlorodehydroabietic acid; [d] AS = Activated Sludge; [e] ASB = Aerated Stabilization Basin.

Table 9.7. Overall and Regional Average Fatty and Resin Acid Mass Emission Rates.

	Estimated Mass Discharge Rates for U.S. Kraft Mills (16 x 10^{-3}/T)								
	Kraft Mills							Sulfite Mills	
	All U.S. Mills		N.E. Mill	Southern Mills		Western Mills		Western Mills	
	Mean	s		Mean	s	Mean	s	Mean	s
Fatty Acids									
Linoleic Acid	5.69	6.3	6.25	4.64	8.58	6.55	5.05	16.9	
Oleic Acid	5.61	6.1	4.14	6.21	8.88	5.46	3.67	10.7	2.34
9,10-Dichlorostearic Acid	2.23	1.9	0.347	2.84	2.25	2.30			
Resin Acids									
Pimaric Acid	12.6	22.9	1.01	19.3	29.9	6.32	8.13		
Sandracopimaric Acid	4.14	6.32	1.30	5.13	8.13	3.73	2.91		
Isopimaric Acid	13.9	20.2	4.16	16.4	26.0	13.9	16.0		
Palustric Acid	14.3	28.2	1.54	17.4	35.7	11.3	10.8		
Dehydroabietic Acid	32.3	72.0	2.95	55.8	102	13.9	13.6	1.39	1.31
Abietic Acid	35.1	70.1	2.42	58.3	98	18.1	23.0		
Neoabietic Acid	10.3	22.2	1.30	13.5	28.3	6.33	6.02		
Chlorinated Resin Acids									
14-Chlorodehydroabietic Acid	1.53	1.27		1.06	1.06	2.94			
12-Chlorodehydroabietic Acid	1.95	2.74	0.479	1.83	2.79	2.65	3.56	1.79	2.15
Dichlorodehydroabietic Acid	3.05	4.25	1.05	0.741		9.36		0.615	

western red cedar, and Douglas fir) are available, and are frequently used by sulfite mills. Since these species have inherently low resin acid content to begin with, it is hardly surprising that resin acids are usually not detectable in mill effluents. This is quite clearly illustrated in Table 9.7 by the low frequency of mass discharge rate estimates (due to frequent nondetects), and the low estimated discharge rate for dehydroabietic acid.

At first, the observations in the above paragraph might seem to be contradicted by the discharge rates for the chlorinated resin acids from sulfite mills. These rates are not all that different from the overall kraft mill average values. A plausible explanation is as follows: the low pH conditions of the sulfite cook do not efficiently solubilize the resin acids, which tend to partition onto the pulp, resulting in a higher amount of resin acid material being carried over into the bleach plant, thereby increasing the potential for formation of chlorinated resin acids.

The last observation from Table 9.7 is that fatty acid discharge rates are higher for sulfite mills. At the present time, we have no explanation for this observation.

The fatty acid discharge rates for kraft mills seem to be comparable between the different regions of the country. Due to the low frequency of detection and the limited number of data sets for northeastern mills, we would not consider the differences in 9,10-dichlorostearic acid discharge rates noteworthy.

Otherwise, the general trend is for southern mills to have the highest discharge rates, followed by western mills, and, finally, by northeastern mills. The fact that the southern mills have the highest discharge rates is hardly surprising. Southern softwoods (predominantly pines) have long been known for their high extractives content. As noted earlier, many softwood species in the Pacific Northwest have low extractives content and many northwest kraft mills do not attempt to recover tall oil soap.

Since data were available for only two northeast mills, and they pulp both softwoods and hardwoods, it is unlikely that this is a good representation of discharge rates indicative of the region. Thus, it would not be prudent to attempt to interpret this data further.

STEROLS

Summary of Literature Data

Several researchers have reported data that allow estimation of the treatment efficiency of sterols characteristic of woody plants. Holmbom and Lehtinen (1980) and Dahlman et al. (1993) report influent and effluent data or removal efficiencies for individual sterols. However, these data represent either limited sampling, or no description of the sampling, making it difficult to determine if they provide reliable estimates of treatment efficiency. Dahlman et al. (1993) and Strömberg et al. (1995) report total sterol removal across treatment systems, for a total of eight mills, that range from 53% to 99%. Given the small amount of information available, little can be concluded, except that the removal efficiencies can be highly variable.

Summary of Sterol Removal Efficiencies for U.S. Mills

In order to improve upon the existing database, and to access sterol treatment efficiency in the United States, a limited study was undertaken. The results are summarized in Table 9.8.

Overall, it appears that campesterol and ß-sitosterol are fairly well removed in the treatment systems for most of the mills studied, except mill 10, which had poor removal efficiency for ß-sitosterol. Stigmastanol removal was high in two of the seven mills surveyed, but for the remainder of the mills the removal was modest (54% to 69%) to low (13%).

The stigmasterol data are interesting in that, for the mills with activated sludge systems, the removals were comparable with the other sterols, but for the two mills with aerated stabilization basins, the data suggest that stigmasterol may actually be formed. Additional, more detailed examinations are required to confirm and explain this observation.

Table 9.8. Summary of Sterol Removal Efficiencies in U.S. Mills.

	Removal Efficiencies[a] (Percent)							Overall	
	Activated Sludge Treatment					Aerated Stabilization Basin		Mean	RSD (%)
Mill Code[b]	8	9	10	13	14	11	12		
Compounds									
Campesterol	71	93	>85		88	65	81	79	14
Stigmasterol	66	>95		>61	88	−377	−192	−69	299
ß-Sitosterol	71	95	>38	89	86	62	81	73	29
Stigmastanol	69	94	>13	96	90	55	54	64	48
Total Sterols	70	95	38	89	88	56	78	71	30

[a] The computational guidelines used in estimating treatment efficiencies are described in Appendix A of this chapter.
[b] Descriptions of selected facts about the mills are summarized in Appendix B of this chapter.

Table 9.9. Estimated Sterol Mass Discharge Rates for U.S. Mills.

Mill Code[b]	Treatment Type	Estimated Mass Discharge Rates for U.S. Kraft Mills[a] (lb x 10^{-3}/T)			
		Campesterol	Stigmasterol	ß-Sitosterol	Stigmastanol
Northeastern					
8	AS[c]	1.78	6.80	22.9	10.9
48	AS			0.438	
Southern					
10	AS			0.833	
11	ASB[d]	0.803	1.73	9.17	3.49
12	ASB	2.12	2.08	26.6	6.23
15	ASB	0.523	2.42	9.45	2.24
17	ASB	1.52	2.28	26.4	2.46
18	ASB	0.866	1.12	11.4	1.24
23	ASB	1.22	0.314	4.49	0.98
24	ASB	2.25	1.82	28.3	6.67
25	ASB	1.80	2.34	23.5	3.41
Western					
9	AS	3.82		7.45	4.04
16	ASB	3.00	3.08	16.0	3.59
20	ASB	4.82	6.70	23.5	5.24
22	ASB	0.145	0.174	1.01	0.189
Mill Code[b]	**Treatment Type**	**Estimated Mass Discharge Rates for U.S. Sulfite Mills[a] (lb x 10^{-3}/T)**			
		Campesterol	**Stigmasterol**	**ß-Sitosterol**	**Stigmastanol**
Central					
13	AS		6.06	5.81	3.17

[a] The computational guidelines used in estimating treatment efficiencies are described in Appendix A of this chapter.
[b] Descriptions of selected facts about the mills are summarized in Appendix B of this chapter.
[c] AS = Activated Sludge.
[d] ASB = Aerated Stabilization Basin.

Summary of Sterol Mass Discharge Rates for U.S. Mills

Table 9.9 summarizes sterol mass discharge rate estimates for 17 mills. The limited amount of sulfite data (one mill) precludes drawing any meaningful conclusions about the effect of pulping technology. Examination of the kraft mill data indicates that there is a wide range of discharge rate observed between the different facilities.

In general, the levels of campesterol tend to be higher in the western mills, relative to mills in the rest of the country. Stigmasterol discharge rate for the northeastern and central mills are among the highest observed, but at least one western mill (mill 20) had comparable levels. The same observations pertain to stigmastanol, except that the next highest discharge rate was observed at a southern mill (Mill 12). However, given the small number of samples for the central and northeastern mills, it would be prudent to consider this only a tentative observation. The ß-sitosterol discharge rates had the largest range, spanning from a low of 0.83×10^{-3} lb/T to a high of 26.65×10^{-3} lb/T. Given the variability observed, it is apparent that there are factors beyond geographic location and/or treatment system type which influence the discharge of ß-sitosterol. Although the database is quite small, it is interesting to note that the highest discharge rates for ß-sitosterol were observed at two of the three unbleached kraft mills included in the survey.

The data summarized in Table 9.9 were also examined with respect to treatment system type. Reviewed in this manner, there were no obvious correlations. This offers no support for the concept that stigmasterol may be generated in ASB systems (mentioned in an earlier section).

NEUTRAL CELLULOSE BREAKDOWN PRODUCTS

Summary of Literature Data

Although 3-methyl-2-cyclopentenone and 2,3,4-trimethyl-2-cyclopentenone were tentatively identified in treated effluent extracts by a GC/MS library search (Turoski et al., 1983), these compounds and a broader class of alkyl substituted cyclopentenones were not confirmed by comparison with authentic standards until 1984 (Voss, 1984). Although Voss (1984) reported qualitatively that the levels were significantly lower in treated effluents relative to untreated effluents, it was not possible to determine the specific removal efficiencies. Acetyl- and propionylthiophenes have also been reported in treated effluents (Voss, 1984). Again, qualitative observations suggested efficient removal across the treatment system, but quantitative estimates were not made. To date, we are not aware of any work that has been done to provide quantitative estimates of removal rates of either the alkyl cyclopentenones or the substituted thiophenes.

Studies have shown that carbohydrates will react with base to form a variety of alkyl cyclopentenones (Voss, 1984) and acetyl- and formylthiophenes (Lunde et al., 1991). Thus, the compounds discussed in this section presumably have common origins in the base degradation of cellulose.

Summary of Alkylcyclopentenone and Acetyl- and Propionyl-Thiophene Treatment Efficiencies in U.S. Mills

In order to establish a quantitative database of alkylcyclopentenone and acetyl- and propionylthiophene treatment efficiency in the United States, a limited study was undertaken. The results are summarized in Table 9.10.

There appears to be a general trend in the alkyl cyclopentenone removal rates suggesting relative removal efficiencies of 2,3-dimethylcyclopentenone > 2,3,5-trimethylcyclopentenone > 2,3,4,5-tetramethylcyclopentenone. In analytical method development studies, it was observed that the compounds with fewer methyl groups were more prone to volatilization losses in the analytical workup. This could suggest that the pattern observed for higher removal efficiency for dimethyl-substituted cyclopentenone supports the hypothesis that volatilization might be an important removal mechanism. Further study would be required to determine this.

Table 9.10. Summary of Alkylcyclopentenone and Acetyl- and Propionyl-Thiophene Removal Efficiencies in U.S. Mills.

	Removal Efficiencies[a] (Percent)							Overall	
	Activated Sludge Treatment					Aerated Stabilization Basin		Mean	RSD %
Mill Code[b]	**8**	**9**	**10**	**13**	**14**	**11**	**12**		
Alkyl Cyclopentenones									
2,3-Dimethylcyclopentenone	100	>100	95	>99	96	>100	76	95	10
2,3,5-Trimethylcyclopentenone	21	>100	32		99	>95	45	59	6
2,3,4,5-Tetramethylcyclopentenone	16	59				36	48	40	46
Acetyl- and Propionylthiophenes									
3-Acetylthiophene	>97	>100	42		39	65	85	78	31
2-Acetylthiophene	98	>99	42		57	68	82	78	31
2-Propionylthiophene	64		43			>90	>96	73	33

[a] The computational guidelines used in estimating treatment efficiencies are described in Appendix A of this chapter.
[b] Descriptions of selected facts about the mills are summarized in Appendix B of this chapter.

Table 9.11. Alkylcyclopentenone, Acetyl- and Propionyl-Thiophene Mass Discharge Rates for U.S. Kraft Mills.

Mill Treatment		Mass Discharge Rates[a] (lb x 10^{-3}/T)					
Code[b]	Type	2,3-DMCP[c]	2,3,5-TMC	2,3,4,5-TMCP	3-AT	2-AT	2-PT
Northeastern Bleached Mills							
8	AS[d]		29.0	3.13		0.923	24.2
48	AS			0.036			
Southern Bleached Mills							
10	AS	0.793	1.64	0.190	1.25	3.31	2.05
11	ASB[e]		0.543	1.20	1.34	4.27	1.06
14	ASB						
15	ASB	0.542	3.01	1.10	1.37	3.55	
17	ASB			1.85			
24	ASB	14.8	8.16	0.94	3.09	6.45	8.62
25	ASB	0.800	0.410	1.13	0.700	2.67	1.84
Southern Unbleached Mills							
12	ASB	6.52	6.22	0.635	0.229	0.583	0.124
18	ASB	32.7	11.3	0.514	2.41	4.67	1.48
23	ASB	27.1	14.6	0.786	2.35	4.97	2.38
Overall Southern Mean		11.8	6.41	0.928	1.59	3.81	2.38
Bleached Southern Mean		4.23	2.75	1.07	1.55	4.05	3.04
Unbleached Southern Mean		22.1	10.7	0.645	1.66	3.41	1.33
Western Bleached Mills							
9	AS			1.80			
16	ASB			5.67	0.307	0.804	
Western Unbleached Mills							
20	ASB			2.57	0.760	1.44	
22	ASB			0.080	0.230	0.450	0.220
Overall Western Mean				2.53	0.432	0.898	0.220
Bleached Western Mean				3.73	0.307	0.804	
Unbleached Western Mean				1.33	0.495	0.945	0.220
Overall Mean		11.42	8.35	1.48	1.09	2.51	4.17
Overall S.D.		14.62	9.84	1.52	0.81	1.73	8.13

[a] The computational guidelines used in estimating mass emission rates are described in Appendix A of this chapter.
[b] Descriptions of selected facts about the mills are summarized in Appendix B of this chapter.
[c] Compound abbreviations: 2,3-DMCP = 2,3-dimethylcyclopentenone; 2,3,5-TMCP = 2,3,5-trimethylcyclopentenone; 2,3,4,5-TMCP = 2,3,4,5-tetramethylcyclopentenone; 3-AT = 3-acetylthiophene; 2-AT = 2-acetylthiophene; 3-PT = 3-proprionylthiophene.
[d] AS = Activated Sludge.
[e] ASB = Aerated Stabilization Basin.

The variability in removal efficiencies for the acetyl- and propionylthiophenes makes it difficult to make comparisons or establish any trends. Thus, based on the information available to date, we can only observe that some systems are quite effective at removing these compounds, while others are less effective.

Summary of Alkylcyclopentenone and Acetyl- and Propionyl-Thiophene Mass Discharge Rates for U.S. Mills

The information summarized in Table 9.11 was compiled to broaden our understanding of typical mass discharge rates in the United States for these compounds. In addition to the data presented for kraft mills in Table 9.11, one sulfite mill from the central United States was analyzed. None of the alkylcyclopentenones or acetyl- or propionylthiophenes was detected. This is consistent with the presumption that these compounds arise from alkaline breakdown of cellulose, and the sulfite process is not alkaline.

There were significantly higher discharge rates for 2,3-dimethyl- and 2,3,5-trimethylcyclopentenones in the southern unbleached mills versus the southern bleached mills. Presumably, all the southern mills were utilizing similar softwood species, and all but one bleached mill utilized aerated stabilization basin treatment. Since the alkylcyclopentenones come from pulping, it would be logical to consider the differences in pulping practices. The most obvious is that unbleached mills usually pulp to much higher kappa numbers. Further controlled experiments would be required to determine if pulping to higher kappa numbers promotes greater formation of the di- and tricyclopentenones. Interestingly, this hypothesis could not be considered using the western mills since the di- and tricyclopentenones were never detected.

Acetyl- and propionylthiophene discharge rates for bleached and unbleached mills were quite similar when comparing southern mills together and western mills together. In general, the southern

Table 9.12. Compilation of Biological Treatment Removal Efficiencies Reported in the Literature and Determined by NCASI.

Chlorinated Compounds	Range of Removal Efficiencies Reported[a]		References
Chloroform	51–100	(x=94)	NCASI, 1977
	88–98		Easty et al., 1978
	65	(n=1)	Leuenberger et al., 1985
	85–94	(n=2)	Lindström & Mohamed, 1988
	63–100	(n=2)	Gergov et al., 1988
	67–88	(n=2)	Dahlman et al., 1993
1,1-Dichlorodimethyl Sulfone	8–17	(n=2)	Lindström & Mohamed, 1988
	2–22	(n=4)	this chapter
Chloroacetones	91–100	(n=2)	Gergov et al., 1988
Tetrachloroacetone	0–64	(n=2)	Holmbom & Lehtinen, 1980
Chloroacetic Acids	82–99	(n=2)	Dahlman et al., 1993
Chloroacetic Acid	70–100	(x=92)	Strömberg et al., 1995
Dichloroacetic Acid	92–95	(n=2)	Lindström & Mohamed, 1988
	59	(n=1)	Stuthridge et al, 1991
	93–99	(x=97)	Strömberg et al., 1995
Trichloroacetic Acid	22–99	(n=2)	Lindström & Mohamed, 1988
	39	(n=1)	Stuthridge et al., 1991
	0–99	(x=64)	Stromberg et al., 1995
MX (3-Chloro-4-(dichloromethyl)-5-hydroxy-2-(5H)-furanone)	0–83	(n=2)	Smeds et al., 1990
E-MX ((E)-2-chloro-3-(dichloromethyl)-4-oxobutenoic acid)	89–98	(n=2)	Smeds et al., 1990
Tetrachlorofuranone	96	(n=1)	Stuthridge et al., 1991
Nonchlorinated Compounds			
Acetophenone	73–100	(x=87)	this chapter
Furfurol	>99	(n=1)	Leuenberger et al., 1985
Pimaral	54–78	(n=2)	Holmbom & Lehtinen, 1980
Isopimaral	20–67	(n=2)	Holmbom & Lehtinen, 1980
Calamenene	61	(n=1)	Leuenberger et al., 1985
Juvabione/Todomatuic Acid	97–100	(x=99)	Leach & Chung, 1980
Betuligenol	35–60	(n=2)	Holmbom & Lehtinen, 1980
Lupenol	50–58	(n=2)	Holmbom & Lehtinen, 1980
Betulinol	40–58	(n=2)	Holmbom & Lehtinen, 1980

[a] If more than one estimate is presented, a range is given. If two estimates are given, n=2 is noted. If more than two estimates are given, the mean is noted (e.g., x = mean).

mills were higher than the western mills. The single mill in the Northeast where these compounds were detected had the highest emission factor for 2-propionylthiophene.

The small number of activated sludge treatment systems, along with the high frequency of non-detects for both the alkylcyclopentones and the acetyl- and propionylthiophenes, prevented any meaningful comparison based on type of treatment system.

MISCELLANEOUS TREATMENT EFFICIENCY INFORMATION

Summary of Literature for Selected Compounds

Throughout the literature, there are various studies of more focused scope (either in terms of a limited number of compounds considered, or in the number of sites investigated) which, when compiled and summarized, help establish the general efficacy of biological treatment systems in reducing the discharge of organic compounds. Table 9.12 attempts to summarize the majority of the information pertaining to this subject available in the literature.

Table 9.13. Estimated Acetophenone and 1,1-Dichlorodimethylsulfone Mass Discharge Rates for U.S. Mills.

Mill Code[b]	Treatment Type	Estimated Mass Discharge Rates for U.S. Kraft Mills[a] (lb x 10^{-3}/T)	
		Acetophenone	1,1-Dichlorodimethylsulfone[c]
Northeastern Bleached			
8	AS[d]		18.5
48	AS		10.9
Southern Bleached			
10	AS	0.610	19.7
11	ASB[e]	0.166	19.9
15	ASB	0.412	21.4
17	ASB		19.6
24	ASB	0.532	0.509
25	ASB	0.291	0.319
Southern Unbleached			
12	ASB	0.103	
18	ASB	0.231	
23	ASB	0.178	
Western Bleached			
9	AS		20.70
16	ASB		24.93
Western Unbleached			
20	ASB	0.28	
22	ASB	0.063	
Mill Code[b]	**Treatment Type**	**Estimated Mass Discharge Rates for U.S. Sulfite Mills[a] (lb x 10^{-3}/T)**	
		Acetophenone	**1,1-Dichlorodimethylsulfone[c]**
Central			
13	AS		1.61

[a] The computational guidelines used in estimating treatment efficiencies are described in Appendix A of this chapter.
[b] Descriptions of selected facts about the mills are summarized in Appendix B of this chapter.
[c] This compound's significance should only be considered in relationship to samples for mills using bleaching.
[d] AS = Activated Sludge.
[e] ASB = Activated Stabilization Basin.

Table 9.14. Summary of Treatment Efficiency Estimates Reported.

Compound Class	Range of Removal Efficiencies (%)
Monoterpenes	
hydrocarbon	98–100
alcohols	97–100
ketones	18–100
Fatty Acids	18–>99
Resin Acids	27–100
Chlorinated Resin Acids	2–99
Chlorinated Phenolics	0–97
Sterols	13–95[a]
Alkylcyclopentenones	16–100
Acetyl- and Propionylthiophenes	42–100
Acetophenone	73–100
1,1-Dichlorodimethylsulfone	0–22

[a] There are indications of increases of stigmasterol across two ASB systems.

U.S. Mill Mass Discharge Rates for Acetophenone and 1,1-Dichlorodimethylsulfone

In addition to the various compound information in the literature, NCASI has further expanded its effluent characterization studies to include acetophenone and 1,1-dichlorodimethylsulfone. The mass discharge rates resulting from these efforts are summarized in Table 9.13.

Treatment Efficiencies for Acetophenone and 1,1-Dichlorodimethylsulfone

Treatment efficiency of acetophenone was evaluated at six kraft mills (half with AS and half with ASB treatment) and one sulfite mill. The removal efficiencies ranged from 73% to 100%, with the average being 87%. Thus, consistently high removal efficiencies were observed for all mills and treatment types.

1,1-Dichlorodimethylsulfone treatment was accessed at five kraft mills (three with AS and two with ASB treatment). The removal efficiencies observed were a net increase of 1%, 2%, 2%, 13%, and 22% removal. Thus, most systems were ineffective at reducing the discharge of 1,1-dichlorodimethylsulfone.

SUMMARY

Although more highly effective for some groups of compounds than others, aerobic biological treatment systems almost uniformly reduce the amounts of organic compounds discharged into the receiving environment. Ranges of treatment efficiencies for U.S. mills, based on new information described in this chapter, are summarized in Table 9.14. 1,1-Dichlorodimethylsulfone was the only compound that uniformly responded poorly to biological treatment. All other groups of compounds generally showed moderate to high removal efficiencies. For all practical purposes, with one exception, activated sludge systems and aerated stabilization basin systems were equally capable of reducing discharge levels of the compounds investigated. The exception was a possible increase of stigmasterol across ASB systems, which is in need of confirmation. In general, one can anticipate significant benefit, beyond BOD removal, through utilization of well-operated, efficient, biological treatment systems.

Appendix A

Guidelines Used for the Computation of Treatment Removal Estimates and Mass Discharge Rates

I. Treatment Removal Estimate Guidelines

In order to ensure that all estimates of treatment efficiency were both meaningful and conservative, the following series of computational guidelines were established:

A. Sampling Considerations

1. Estimates based on a single influent and effluent composite sample were only made for mills with activated sludge treatment systems.
2. All estimates for aerated stabilization basin treatment systems and some activated sludge systems were based on multiple sampling episodes.

B. Calculation Considerations

1. Where sampling involved multiple episodes, the treatment efficiency was calculated based on the average influent versus the average effluent concentrations.
2. No attempt was made to estimate treatment efficiency where the analytes were never detected in the influent, or for situations where more than half the determinations in a multi-episode sampling were non-detect(s). This prevented estimates based solely on the assumptions of how non-detect(s) were handled.
3. The treatment efficiency was only estimated if the concentration in the influent was greater than two times the method detection limit or the lower limit of calibration. Again, this prevented estimates based on how non-detect(s) were handled.
4. The Method Detection Limit (MDL) is a method-specific performance parameter and is more representative of actual method performance than a Lower Calibration Limit (LCL). However, estimates for the MDL were not available for all the analytical procedures that were utilized. Thus, wherever a method-specific MDL was available, this concentration was substituted for any non-detect(s). When a MDL was not available, the LCL was used to substitute for non-detect(s).
5. Whenever a non-detect(s) was encountered in the effluent sample, the substitution process utilized would result in a conservative underestimation of the treatment efficiency. To signify where such situations occurred, the treatment efficiency given in the summary table was flagged with a ">" sign.

Taken collectively, these guidelines assure that only meaningful estimates are summarized in the summary tables and that the treatment efficiencies are not overstated.

II. Calculation of Mass Discharge Rates

A. The mass emission rates were determined only when the analyte was detected in the effluent according to guidelines described below. Where only a single analysis was available, substitution of a MDL or LCL for a non-detect(s) would not provide a representative estimate of mass discharge rate. Therefore, no mass discharge rates were calculated where the analyte was never detected.

B. Where multiple data points were available the following guidelines were applied to the data:
   1. The average concentration, average flow, and average production were determined for the sampling episodes. These averages were then used to compute a mass discharge rate.
   2. If more than half the effluent data points were not detected, a mass discharge rate was not estimated due to the fact that the estimate would be largely dependent on the detection limit substitution values. If less than half of the effluent data points had non-detected concentrations, the non-detect(s) data points were substituted with the MDL or LCL and the average concentration was calculated. The data presented, therefore, would provide a high biased or upper limit estimate of mass discharges.

Appendix B

Descriptions of U.S. Mills Studied

Mill Code	Pulping Type	Bleaching	Treatment Type	Geographic Location	Weed Species
1	Kraft	yes	ASB[a]	S	SW[c]/HW[d]
2	Kraft	yes	ASB	W	SW
3	Kraft	yes	AS	W	SW
4	Kraft	yes	AS	W	SW
5	Kraft	no	ASB	W	SW
6	Kraft	yes	AS	W	SW
7	Kraft	yes	ASB	S	SW/HW
8	Kraft	yes	AS	NE	HW/SW
9	Kraft	yes	ASB	W	SW
10	Kraft	yes	AS	S	SW/HW
11	Kraft	yes	ASB	S	SW
12	Kraft	no	ASB	S	SW
13	Sulfite	yes	AS	C	SW
14	Kraft	yes	AS	S	HW
15	Kraft	yes	ASB	S	SW
16	Kraft	yes	ASB	W	SW
17	Kraft	yes	ASB	NS	SW/HW
18	Kraft	no	ASB	NE	SW
19	Kraft	yes	ASB	S	SW
20	Kraft	no	ASB	W	SW
22	Kraft	no	ASB	W	SW
23	Kraft	no	ASB	W	SW
24	Kraft	yes	ASB	S	SW/HW
25	Kraft	yes	ASB	S	SW/HW
26	Kraft	yes	AS	S	SW/HW
27	Sulfite	yes	AS	W	SW
28	Sulfite	yes	AS	W	SW
29	Sulfite	yes	AS	W	SW
30	Kraft	yes	AS	W	SW
31	Kraft	yes	AS	NE	SW/HW
32	Sulfite	yes	AS	W	SW
33	Kraft	yes	AS	C	SW/HW
34	Kraft	yes	AS	NE	SW/HW
35	Sulfite	yes	AS	C	HW
36	Sulfite	yes	AS	C	HW
37	Sulfite	yes	AS	C	SW
38	Kraft	yes	AS	W	SW/HW
39	Kraft	yes	ASB	W	SW
40	Kraft	yes	ASB	S	SW
41	Sulfite	yes	AS	W	SW
42	Kraft	yes	ASB	W	SW
43	Sulfite	yes	ASB	W	SW
44	Kraft	yes	ASB	W	SW
45	Sulfite	yes	AS	W	SW
46	Kraft	yes	AS	NE	SW/HW
47	Kraft	yes	AS	NE	SW/HW
48	Kraft	yes	AS	NE	HW

[a] ASB = Aerated Stabilization; [b] AS = Activated Sludge; [c] SW = Softwood; [d] HW = Hardwood.

REFERENCES

CPAR (Cooperative Pollution Abatement Research). 1978. CPAR Project Report 408-3, Biodegradability of Various Toxic Compounds in Pulp Mill Effluents. CPAR Secretariat, Environmental Protection Service, Environment Canada, Ottawa, Ontario, K1A 1C8.

Dahlman, O., R. Mörck, L. Johansson, and F. De Sousa. November 1991. Chemical Composition of Modern Bleached Kraft Mill Effluents. Environmental Fate and Effects of Bleached Pulp Mill Effluents. Presented at the SEPA Conference, Stockholm, Sweden.

Dahlman, O., I. Haglind, R. Mörck, F. De Sousa, and L. Strömberg. April 1993. Chemical Composition of Effluents From Chlorine-Dioxide Bleaching of Kraft Pulps Before and After Secondary Treatment. Presented at the International Environmental Symposium, EUCEPA, Paris, France.

Easty, D., L. Borchardt, and B. Wabers. 1978. Wood Derived Toxic Compounds. Removal from Mill Effluents by Waste Treatment Processes. *Tappi J.,* 61(10):57–60.

Gergov, M., M. Priha, E. Talka, O. Välttilä, A. Kangas, and K. Kukkonen. 1988. Chlorinated Organic Compounds in Effluent Treatment at Kraft Mills. *Tappi J.,* 71(12):175–184.

Holmbom, B. and K.-J. Lehtinen. 1980. Acute Toxicity to Fish of Kraft Pulp Mill Waste Waters. *Paperi ja Puu,* 62(11):673–684.

Hrutfiord, B., T. Friberg, D. Wilson, and J. Wilson. 1975. Organic Compounds in Aerated Stabilization Basin Discharges. *Tappi J.,* 58(10):98–100.

Kovacs, T., P. Martel, R. Voss, P. Wrist, and R. Willes. 1993. *Environmental Toxicology and Chemistry,* 12:281–289.

Kringstad, K. and K. Lindström. 1984. Spent Liquors from Pulp Bleaching. *Env. Sci. Technol.,* 18(8):236A–248A.

Leach, J. and L. Chung. 1980. "Development of a Chemical Toxicity Assay for Pulp Mill Effluents." National Technical Information Service, Springfield, VA, Report No. EPA-600-2-80-206.

Leuenberger, C., W. Giger, R. Coney, J. Graydon, and E. Molnar-Kubica. 1985. Persistent Chemicals in Pulp Mill Effluents, Occurrence and Behaviour in an Activated Sludge Treatment Plant. *Water Res.,* 19(7):885–894.

Lindström, K. and M. Mohamed. 1988. Selective Removal of Chlorinated Organics from Kraft Mill Total Effluents in Aerated Lagoons. *Nordic Pulp and Paper J.,* 1:26–33.

Lunde, A., J. Skramstad, and G. Carlberg. 1991. Identification, Mutagenicity and Origin of Chlorinated Thiophenes in Kraft Bleaching Effluent. *Paperi ja Puu,* 73(6):522–526.

McFarlane, P. and T. Clark. 1988. Metabolism of Resin Acids in Anaerobic Systems. *Wat. Sci. Tech.,* 1:273–276.

McKague, B., M. Jarl, and K. Kringstad. 1988. An Up-to-Date List of Compounds Identified in Bleachery Effluent as of January 1989. SSVL-Miljö 90, Project Bleaching, Stockholm. Report available from AF-IPK, Box 8309, 104-20, Stockholm, Sweden.

NCASI. 1977. "Analysis of Volatile Halogenated Organic Compounds in Bleached Pulp Mill Effluent." NCASI Technical Bulletin No. 298; NCASI, 260 Madison Ave., New York, NY 10016.

NCASI. 1989. "Effects of Biologically Treated Kraft Mill Effluent on Cold Water Stream Productivity in Experimental Streams — Fifth Progress Report." NCASI Technical Bulletin No. 566; NCASI, P.O. Box 13318, Research Triangle Park, NC 27709.

NCASI. 1990. "Field Validation of the NCASI Predictive Model for Organic Compound Removal by Biological Wastewater Treatment Processes." NCASI Technical Bulletin No. 582; NCASI, P.O. Box 13318, Research Triangle Park, NC 27709.

NCASI. 1993. "Aquatic Community Effects of Biologically Treated Bleached Kraft Mill Effluent Before and After Conversion to Increased Chlorine Dioxide Substitution: Results from an Experimental Streams Study." NCASI Technical Bulletin No. 653; NCASI, P.O. Box 13318, Research Triangle Park, NC 27709.

Saunamäki, R., K. Jokinen, R. Järvinen, and M. Savolainen. 1991. Factors Affecting the Removal and Discharge of Organic Chlorine Compounds at Activated Sludge Treatment Plants. *Wat. Sci. Tech.,* 24(3/4):295–307.

Smeds, A., B. Holmbom, and L. Tikkanen. 1990. Formation and Degradation of Mutagens in Kraft Pulp Mill Sewers. *Nordic Pulp and Paper Research Journal,* 3:142–147.

Strömberg, L., R. Mörck, F. De Sousa, and O. Dahlman. 1995. Effects of Process Changes and External Treatment on Effluent Chemistry. *Environmental Fate and Effects of Pulp and Paper Mill Effluents,* St. Lucie Press: Delray Beach, FL, pp. 3–19.

Stuthridge, T., A. Wilkins, A. Langdon, K. Mackie, and P. McFarlane. 1990. Identification of Novel Chlorinated Monoterpenes Formed During Kraft Pulp Bleaching of Pinus Radiata. *Env. Sci. Tech.,* 24(6):903–908.

Stuthridge, T., D. Campin, A. Langdon, K. Mackie, P. McFarlane, and A. Wilkins. 1991. Treatability of Bleached Kraft Pulp and Paper Mill Wastewaters in a New Zealand Aerated Lagoon Treatment System. *Wat. Sci. Tech.,* 24(3/4):309–317.

Turoski, V., D. Woltman, and F. Vincent. 1983. Determination of Organic Priority Pollutants in the Paper Industry by GC/MS. *Tappi J.,* 66(4):89–90.

Voss, R. 1984. Neutral Organic Compounds in Biologically Treated Bleached Kraft Mill Effluents. *Env. Sci. Tech.,* 18(12):938–946.

Voss, R. and A. Rapsomatiotis. 1985. An Improved Solvent-Extraction Based Procedure for the Gas Chromatographic Analysis of Resin and Fatty Acids in Pulp Mill Effluents. *J. Chromatography,* 346:205–214.

Wiegand, P., D. Barton, and L. LaFleur. 1994. "Factors Influencing the Generation and Treatability of Chlorinated Phenolic Compounds and AOX in Bleached Chemical Pulp Mill Waste Waters." NCASI Technical Workshop — Effects of Alternative Pulping and Bleaching Processes on Production and Biotreatment of Chlorinated Organics. NCASI Special Report No. 94-01; NCASI, P.O. Box 13318, Research Triangle Park, NC 27709.

Wilkins, A. and S. Panadam. 1987. Extractable Organic Substances from the Discharges of a New Zealand Pulp and Paper Mill. *Appita,* 40(3):208–212.

Wilkins, A., A. Langdon, G. Mills, S. Panadam, and T. Stuthridge. 1988. Kinleithic Acid: A New Hydroxylated Resin Acid From the Biological Treatment System of a New Zealand Kraft Pulp and Paper Mill. *Aust. J. Chem.,* 42:983–986.

Wilson, D. and B. Hrutfiord. 1971. Sekor IV. Formation of Volatile Organic Compounds in the Kraft Pulping Process. *Tappi J.,* 54(7):1094–1098.

Zender, J., T. Stuthridge, A. Langdon, A. Wilkins, K. Mackie, and P. McFarlane. 1993. Removal and Transformation of Resin Acids During Secondary Treatment at a New Zealand Bleached Kraft Pulp and Paper Mill. Presented at the 4th IAWQ Symposium on Forest Industry Wastewaters, Tampere, Finland.

Chapter 10

An Environmental Assessment Before and After Conversion of a Bleached Kraft Mill to Elemental Chlorine–Free Bleaching

T.L. Deardorff, J.J. Renard, and R.B. Phillips

ABSTRACT

As part of our continuing efforts to understand the potential impact of biologically treated effluents to the aquatic ecosystem, we critically examined several environmental parameters before and after the conversion to elemental chlorine–free (ECF) bleaching at a kraft pulp and paper mill. The study involved evaluation of filtrates, pulps, sludges, and wastewater treatment within the mill and an integrated field study of the receiving stream, which included comprehensive water and sediment toxicity testing, benthic community evaluations, fish health testing, and physicochemical and biological parameters of water quality. Following ECF conversion, no detectable levels of 2,3,7,8-dioxin or polychlorinated phenolic compounds at the bleach plant and no indications of detrimental impacts to the ecosystem from the effluent were found.

INTRODUCTION

One of the current debates surrounding the pulp and paper industry is whether bleaching with an elemental chlorine–free (ECF) or total chlorine-free (TCF) technology is better. While the economic and quality factors favor the ECF technology, many people believe that TCF offers superior environmental performance. Some environmental interest groups, regulators, and, indeed, some customers, have suggested the industry adopt TCF as a more "environmentally friendly" process for producing bleached pulp. In this chapter we demonstrate that the aquatic ecosystem adjacent to a bleached kraft pulp and paper mill using high chlorine dioxide substitution (>55% chlorine dioxide; see below for actual substitution rates) was healthy, and that the environmental impact of the industrial, ECF bleaching process is so low that TCF offers no measurable advantage.

In 1993, we converted our Moss Point, Mississippi, kraft pulp and paper mill to ECF bleaching. Prior to the ECF conversion (termed "base case"), the softwood and hardwood pulps were bleached to a final brightness target of 89% GE using ($D_{55\%}$)(Eo+p)PD and ($D_{65\%}$)(Eo+p)D sequences, respectively. The mill is equipped with directly steamed batch digesters and produces unbleached kraft pulps with conventional kappa numbers of 30–32 from softwood and 18–20 for hardwood pulps. Chlorine dioxide is generated using a methanol-based process. Wastewater is treated with primary clarification and secondary biological treatment in an aerated stabilization basin (ASB) with an 11-day retention time. Final effluent is discharged into the brackish waters of the Escatawpa River. The base case sampling schedule was initiated two weeks before conversion while the mill operated under the conditions mentioned above.

METHODOLOGY

To better monitor mill parameters during conversion, we evaluated conditions continuously for 1 year after conversion. A combination of methodologies was used to analyze for dioxins and furans. Sample extraction followed the U.S. Environmental Protection Agency (EPA) Method 1613A (U.S. EPA Method 1613) and the resulting extracts were processed for quantification following a National Council for Air and Stream Improvement (NCASI) protocol. Chlorinated phenolics were analyzed using NCASI Method CP-85.01 (NCASI, 1986). For AOX analysis, the adsorption-pyrolysis-titrimetric procedure, Standard Method 5320, was used (Clesceri et al., 1989). The chemical oxygen demand (COD) was measured using the closed reflux colorimetric procedure, Standard Method 5220 D (Clesceri et al., 1989). For true color analysis, each 30-mL sample was adjusted to pH 7.6, filtered through a 47-mm Metricel Membrane Filter (Gelman Sciences, Inc.) to remove suspended material, transferred to cuvettes, and light absorption was measured in a Hach DR/3000 spectrophotometer set at a wavelength of 465 nm. Units of color were converted to mg/L of platinum, in the form of chloroplatinate ion. Measurements presented in Table 10.4 are expressed as kilograms per air dried, unbleached metric tons (ADUBMT) of pulp produced.

When more than one data point was collected at 1 month ECF or 1 year ECF, the average value was presented. For AOX, the average for the entire sampling period is provided.

Inside the Mill

Throughout the conversion, we performed a comprehensive, continuous examination of numerous parameters inside the mill. A summary of the results from within the mill is presented in Tables 10.1, 10.2, and 10.4. Table 10.1 shows the levels of 2,3,7,8-tetrachlorodibenzo-*p*-dioxin (TCDD) and 2,3,7,8-tetrachlorodibenzofuran (TCDF) in final softwood and hardwood pulps and primary sludge during the base case (2 weeks prior to conversion), 1 month ECF, and 1 year ECF. TCDD was not detected for any solid vector at any time period. Minimal concentrations of TCDF were detected in hardwood for the base case but no detectable quantities were revealed following conversion. TCDF was significantly decreased in the primary sludge from 91 parts per trillion (ppt) to 3.9 ppt (96% reduction). Table 10.2 shows TCDD and TCDF levels, expressed as parts per quadrillion (ppq), in the bleach plant filtrate and wastewater treatment system before and after biological treatment in the ASB. TCDD was not detectable using EPA standards before or after conversion. TCDF was only detected in the base case. Thus, we conclude that the production of TCDD and TCDF remained at nondetectable concentrations in all bleach plant filtrates and in the final discharge from the ASB, without oxygen delignification.

Concerning the effect of ECF bleaching on the 12 polychlorinated phenolic compounds (PCPs) selected by the U.S. EPA for possible regulation (Table 10.3), all were below the level of quantification

Table 10.1. TCDD and TCDF in Pulps and Primary Sludge Before and After Conversion to Elemental Chlorine–Free Bleaching.[a]

Medium	Base Case		1 Month ECF		1 Year ECF	
	TCDD	TCDF	TCDD	TCDF	TCDD	TCDF
Softwood Pulp	ND[b]	ND	ND	ND	ND	ND
Hardwood Pulp	ND	5.1	ND	ND	ND	ND
Primary Sludge	ND	91	ND	8.5	ND	3.9

[a] All values are expressed in parts per trillion (ppt).
[b] ND indicates that the compound was not detected using EPA methodology.

Table 10.2. TCDD and TCDF in Bleach Plant Filtrates and the Wastewater Treatment System Before and After Conversion to Elemental Chlorine–Free Bleaching.[a]

Medium	Base Case		1 Month ECF		1 Year ECF	
	TCDD	TCDF	TCDD	TCDF	TCDD	TCDF
Bleach Plant	ND[b]	13.8	ND	ND	ND	ND
Inlet to ASB	ND	31.5	ND	ND	ND	ND
ASB Discharge	ND	ND	ND	ND	ND	ND

[a] All values are expressed in parts per quadrillion (ppq).
[b] ND indicates that the compound was not detected using EPA methodology.

Table 10.3. Polychlorinated Phenolic Compounds with Minimum Analytical Detection Limits Proposed for Regulation by the U.S. EPA.

Compound	Minimum Level, ppb (μg/L)[a]
2,4,5-Trichlorophenol	2.5
2,4,6-Trichlorophenol	2.5
2,3,4,6-Tetrachlorophenol	2.5
Pentachlorophenol	5
3,4,5-Trichloroguaiacol	2.5
3,4,6-Trichloroguaiacol	2.5
4,5,6-Trichloroguaiacol	2.5
3,4,5,6-Tetrachloroguaiacol	5
3,4,5-Trichlorocatechol	5
3,4,6-Trichlorocatechol	5
3,4,5,6-Tetrachlorocatechol	5
3,4,5-Trichlorosyringol	2.5

[a] The minimum level is the concentration at which the analytical system gives recognizable mass spectra (corrected for background) and acceptable calibration points, using EPA Method 1653.

in bleach plant filtrates, primary sludge, and in the final discharge. Increasing the percent substitution of chlorine dioxide for elemental chlorine in the first bleaching stage decreases the production of the 12 priority PCPs. In our conversion study, PCPs were not detected in the bleach plant, the inlet to the ASB, or the final treated discharge. Laboratory and field research (Tsai et al., 1994; Yin et al., 1994; Deardorff et al., 1994; Graves et al., 1995) have shown that discharge from a well-controlled bleach plant operating at 70% substitution or ECF (with an efficient ASB) has significantly reduced or undetectable levels of these PCPs.

Table 10.4 shows other effluent parameters identified in the proposed U.S. EPA Cluster Rules for the Pulp and Paper Industry. The production and discharge of adsorbable organic halides (AOX), a barometer of the chlorinated content of water, was decreased from 0.6 kg AOX/ADUBMT in the base case to 0.32 kg AOX/ADUBMT (or a 47% decrease) after one year. The average AOX discharge for the first year was 0.39 kg AOX/ADUBMT (a 35% decrease). These results occurred while the mill operated at conventional kappa number without oxygen delignification. (*Note:* AOX concentrations do not indicate potential toxicity and are not appropriate to evaluate the potential impacts on the receiving ecosystem.)

At 1 year ECF, effluent color and chemical oxygen demand (COD) in the final discharge showed no significant change. Because color and COD are predominantly influenced by lignin levels in the unbleached pulps, reductions in these parameters were not anticipated. Parameters such as TCDDs, TCDFs, PCPs, AOX, and chloroform are influenced by lignin and chlorine usage and were expected to show a marked decrease, which they did.

The Aquatic Ecosystem

In addition to testing within the mill, we initiated an integrated field study of the receiving river to document the general health of the aquatic ecosystem prior to conversion to ECF and subsequent to conversion. Concerning dioxin levels in fish, it is important to realize that an appreciable reduction of dioxin in the edible musculature of fish can be achieved using high substitution. For example, in 1988, International Paper's Georgetown, South Carolina, mill had 640 ppq of TCDD detected in its effluent. Since that time, the mill has voluntarily reduced dioxin formation within the bleach plant to current levels that are not detectable using state-of-the-art technology; correspondingly, the catfish adjacent to the mill's discharge eliminated the levels of TCDD in their tissues (Figure 10.1). The state regulatory agency has removed the limited-fish-consumption advisory that it posted on the Sampit River near the mill, acknowledging that the TCDD concentrations in the edible musculature of fish did not represent a level of human health concern.

Table 10.4. Adsorbable Organic Halides[a] (AOX), Polychlorinated Phenolic Compounds[b] (PCP), Color,[c] and Chemical Oxygen Demand (COD) in Bleach Plant Filtrates and the Wastewater Treatment System Before and After Conversion to Elemental Chlorine–Free Bleaching.

	Base Case				1 Month ECF				1 Year ECF			
Collection Site	AOX	PCP	Color	COD	AOX	PCP	Color	COD	AOX	PCP	Color	COD
Bleach Plant	1.8	273	48	80	1	ND	35	73	1.1	ND	20	41
Inlet to ASB	1	185	108	74	0.5	ND	95	79	0.67	ND	95	57
ASB Discharge	0.6	23	109	41	0.4	ND	104	50	0.32	ND	104	40

[a] Measured as m/L and converted to air-dried, unbleached metric tons of pulp.
[b] Values represent the sum of all individually tested PCPs listed in Table 10.3.
[c] Measured as mg/L of platinum in the form of chloroplatinate ion and converted to air-dried, unbleached metric tons of pulp.

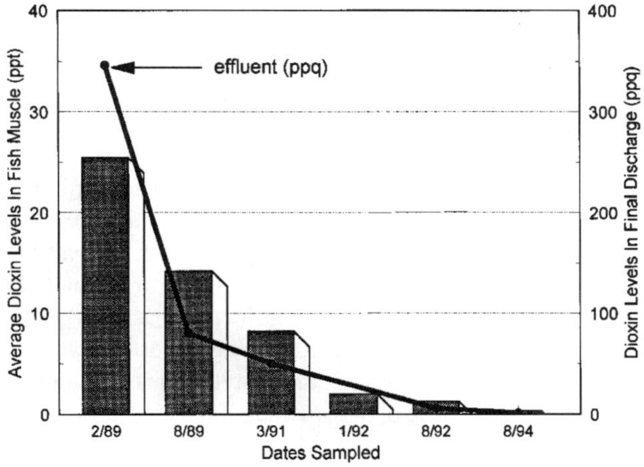

Figure 10.1. Showing the correlation between the reduction of TCDD and TCDF (as toxic equivalents) in the effluent from a bleached kraft paper mill (right side Y axis) and the concentration of TCDD in the edible musculature of catfish (left side Y axis) caught adjacent to the Georgetown mill's discharge. This mill converted to high-substitution bleaching during this time frame.

The Georgetown mill example demonstrates that fully treated discharge with nondetectable levels of TCDD according to the EPA guidelines into an estuarine environment show rapid TCDD reductions in adjacent native fish. Our Moss Point mill discharges into a similar estuarine environment; however, it is not the only potential source of TCDD or TCDF in the area. Nearly 500 fish comprise the Moss Point fish databank. The average dioxin values for each year of sampling near the Moss Point mill show that the fish have always been safe to eat. On average, the levels were always below FDA, EPA, and state guidelines (25 ppt, 7 ppt, and 5 ppt, respectively). Fish levels ranged as high as 12 ppt, 1 ppt, 28 ppt, 21 ppt, and 41 ppt for 1989, 1990, 1991, 1992, and 1993, respectively; however, the vast majority of fish have levels well below the federal and state guidelines or nondetectable. As expected, the highest levels of TCDD were found in older, larger, bottom-feeding fish. Although some fish had levels exceeding state guidelines, the majority of sampled fish had low or nondetectable levels. To put these data into proper perspective, it is important to note that the sample set for 1990 contained only four composite samples.

The levels of TCDD in fish from 1989 to 1993 showed, on average, a slight — but continual — decline which may be attributed to the mill's bleaching changes during that time frame. These changes included an increase in the percentage of chlorine dioxide substitution. The mill went to 35% substitution in 1989 and was at 60–65% substitution by 1991. At that time, dioxin levels in the mill's treated discharge were not detectable and have not been since. With several other sources of TCDD and TCDF in this area, such as a municipal waste incinerator, the rapid decline of dioxin in the edible musculature of fish seen at our Georgetown mill would not be expected.

The water quality and current toxicological status of the water and sediments in the river system near the mill showed no evidence of significant adverse impact from the mill's discharge either prior to conversion or after conversion to ECF. Water quality measurements always were within expected ranges and sufficient to support aquatic life in the study area. These measurements included color, turbidity, suspended solids, BOD_5, alkalinity, hardness, salinity, pH, dissolved oxygen, conductivity, temperature, and light extinction coefficients.

Chronic tests showed no apparent water toxicity problems using the 7-day waterflea *(Ceriodaphnia dubia)* and fathead minnow *(Pimephales promelas)* tests with full-strength effluent from the discharge canal. For this watershed, it was important to adjust water hardness. The Escatawpa River has a water hardness between 5.5 and 7.6 mg/L as $CaCO_3$ ("soft" water), which adversely affected the reproduction of the waterfleas. Reagent grade chemicals were used to amend the water to increase the hardness to "moderately hard" water (80–100 mg/L as $CaCO_3$). Without adjusting the water hardness, the *C. dubia* assay indicated that the river adversely impacts the reproductive success of the waterflea. This finding confirms our initial study of the river system in 1980 (unpublished report). In that study, because of the low hardness, four separate tests showed that no first instar waterfleas survived a 48-hr exposure to undiluted river water. Also, mortality and severe stress were suffered by bluegill fish *(Lepomis macrochirus)* exposed to the river water. No mortality or stress was observed in either test species exposed to river water adjusted to a hardness of 46 mg/L as $CaCO_3$ or greater. The Microtox® test has been used with varying success to assess the toxicity of pulp and paper mill effluents because it lacks sensitivity. We found that it lacked sensitivity but was predictive of acute toxicity.

The 10-day sediment toxicity tests, which use *Hyalella azteca* and *Chironomus tentans,* also indicated excellent conditions at all stations. The survival range for *H. azteca* was from 88 to 100% (>80% considered "passing"). For *C. tentans,* when not influenced by salinity due to high tides and onshore winds, the survival range was from 83 to 93% (>80% considered "passing"). No significant differences were observed between the treated discharge from ECF or high-substitution bleaching sequences.

Our investigation of the general health of fishes and invertebrate communities in the vicinity of the mill concluded that there was no indication of detrimental impacts resulting from effluent originating from the mill before or after conversion. This conclusion was based on the lack of chronic lesions in samples of small and large fish species and the high species richness and diversity of parasites and benthic macroinvertebrates found at or near the outflow canal.

Following histological examination of various tissues from over 2,000 fish (in addition to the nearly 500 fish caught and tested for TCDD levels in tissues) caught before and after conversion, we could not find any cancers or noncancerous toxicant-induced lesions that might be attributed to exposure to mill effluent. We even caught another group of small fish (collected above, below, at the mill, or from one site in another river system) and held them in aquaria for 6 or 12 months (grow-out period) and examined over 500 of them for lesions. No lesions were directly attributable to toxicant exposure.

Because early life-history stages of developing fish are sensitive to environmental changes, we joined with the EPA in evaluating the efficacy of a 7- to 10-day embryotoxic/teratogenic assay with the inland silverside, *Menidia beryllina.* Middaugh et al. (1987) provide a description of the methodology and additional information on this test. We compared the fish embryo assay data with the toxicity responses of the Microtox® 15-minute acute test and the *C. dubia* 7-day chronic test. Samples tested in these assays were: 1) river water prior to mill use; 2) wastewater at the inlet to the ASB following primary clarification; and, 3) biologically treated effluent collected from the aerated stabilization basin. Results (Figure 10.2) showed that this test may be a sensitive indicator of toxicity and, at the critical dilution to the receiving stream, no adverse effects would be anticipated to the developing embryo. No significant differences at this dilution were observed in the fish embryo test using final effluent prior to ECF or after conversion. The waterflea, fathead minnow, and Microtox® assays confirmed the inland silverside results.

Bottom-dwelling organisms reflect aquatic conditions due to their direct contact with the sediments that may contain pollutants; thus, they are excellent biomarkers of the health of the aquatic system. We studied macrobenthic organisms from several sites following the ECF conversion and concluded that overall species diversity near the outflow site indicated healthy conditions.

The above findings from the aquatic environment of the Escatawpa River were similar to the results reported by Overstreet et al. (1996) and Rakocinski et al. (1996) following a critical examination of the aquatic environment near our high-substitution bleached kraft mill in Texarkana, Texas. Overstreet et al. (1996) reported that the fish caught in the Sulphur River in Texas and Arkansas were free from lesions known to be associated with some other mills and demonstrated the utility of the western mosquitofish, *Gambusia affinis,* as a good sentinel or indicator species of the general health of fish associated with bleached kraft mill effluents and other contaminated waters. The use of parasites as a biomarker was shown. Rakocinski et al. (1996) reported that the macrobenthic assemblages showed relatively high species richness and typical taxonomic diversity at the discharge site.

CONCLUSIONS

Conversion to ECF improved the quality of the process streams at the bleach plant and changed the profile of the wastewater entering and exiting the biological treatment system. A summary of

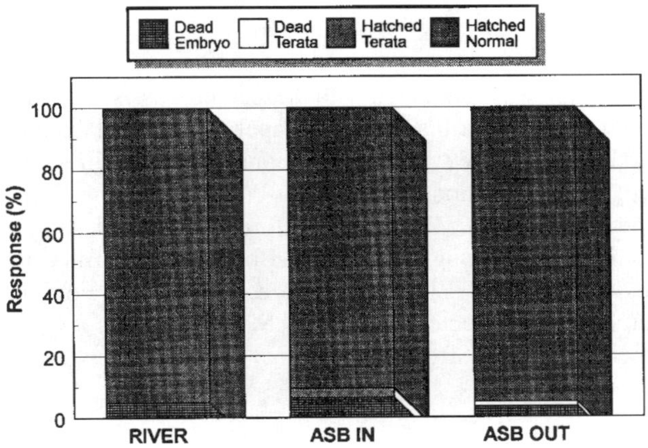

Figure 10.2. Showing results of a fish embryo assay indicating that Moss Point mill's discharge at the critical dilution to the receiving stream does not cause adverse effects in developing fish embryos. The "river" sample represented freshwater from the Escatawpa River collected approximately 12 miles upstream of the mill; the "ASB in" sample represents process water collected after primary treatment of the mill wastewater but prior to entering the secondary treatment; and the "ASB-out" sample was collected within the aerated stabilization basin prior to being discharged into the river.

findings from within the mill were: 1) TCDD and TCDF were at nondetectable concentrations in all bleach plant filtrates and in the final discharge from the ASB; 2) pulps and sludges contain no measurable TCDD; 3) PCPs were below the level of quantification in bleach plant filtrates, primary sludge, and final discharge; 4) AOX was decreased to 0.39 kg/ADUBMT, on average; and 5) color and COD in the final discharge show only a marginal effect following ECF conversion.

Regardless of whether the mill used high-substitution bleaching or ECF, the final effluent showed no significant effects on the surrounding aquatic environment. We were unable to find adverse impacts caused by our treated effluents to aquatic animals living on the bottom of the Escatawpa River (e.g., species richness, population density) or to fish (e.g., cancers, noncancerous lesions, deformities) swimming in it. Further, the treated discharge from the Moss Point mill was not toxic to any aquatic animals using approved EPA tests for water and sediments. The results from our study demonstrate that over the years, the Moss Point mill's pulping and bleaching processes have been well-controlled, its biological treatment system has been effective, and its discharge did not — and does not — adversely impact the associated aquatic ecosystem. Therefore, the ecosystem *remains* healthy.

The healthy condition of the aquatic environment adjacent to nearly all U.S. bleached kraft mills can be attributed to the biological treatment of the effluent prior to release into the receiving water body. Biological treatment oxygenates the water and breaks down many compounds that are toxic, persistent, and bioaccumulative. According to an international panel of expert scientists who reviewed pulping effluents in the aquatic environment, biologically treated effluent from well-operated North American pulp and paper mills is not likely to cause acute or chronic effects in fish in the receiving streams at dilutions below 10% (Mehrle et al., 1989). Adverse effects in fish resulting from U.S. kraft mills using high-substitution bleaching and secondary biological treatment of the wastewater are rare. Adverse effects in fish resulting from effluent discharged by U.S. mills using ECF bleaching and an ASB have not been reported.

The present study produced sound scientific evidence that demonstrated that our previous and current processes produced safe effluents. Going beyond ECF provides no incremental environmental benefit, and would cost our industry \$3.8 billion.

As for the ECF vs. TCF environmental debate, let's end it. We have clearly demonstrated with sound science that the environmental impact from the industrial, ECF bleaching process is so low that there is little room for TCF to be better. If effluent from kraft mills using TCF bleaching can equal the environmentally benign ECF discharge of U.S. mills, we are delighted. In that case, the ultimate winner in the ECF-TCF debate will be the environment.

ACKNOWLEDGMENTS

This extensive examination was completed with the technical assistance of several research teams. In particular, we acknowledge: from International Paper, Mobile, Alabama, Paul Stevenson, Tom Wiesemann, Dan Strizak, Craig Ratkey; from International Paper, Sterling Forest, New York, Dennis Iverson and June Lloyd; from International Paper, Moss Point, Mississippi, Vance Best, Tom Sweetser, and Ed Locke; from the University of Mississippi, Oxford, Mississippi, John Rodgers; from the Gulf Coast Research Laboratory, Ocean Springs, Mississippi, Bob Overstreet, Bill Hawkins, Chet Rakocinski, and Richard Heard; and from the U.S. EPA, Gulf Breeze, Florida, Jack Fournie and Doug Middaugh. Portions of this research were conducted with the U.S. EPA through a Cooperative Research and Development Agreement (#0100-93) from International Paper.

REFERENCES

Clesceri, L.S., A.E. Greenberg, and R.R. Trussell. 1989. *Standard Methods for the Examination of Water and Wastewater,* 17th ed. American Public Health Association, Washington, DC.

Deardorff, T.L., R.R. Willhelm, A.J. Nonni, J.J. Renard, and R.B. Phillips. 1994. Formation of polychlorinated phenolic compounds during high chlorine dioxide substitution bleaching. Part 3. Mill trial results. *Tappi Journal,* 77(8):163–168.

Graves, J., J. Jameel, and T.W. Joyce. 1995. Removal of chlorophenolics from an aerated lagoon. *Tappi Journal,* 78(5):99–108.

Mehrle, P.M., K. Dickson, R. Hartung, R. Huggett, D. McLeay, A. Oikari, and J. Sprague. 1989. Pulping effluents in the aquatic environment. A report prepared for the Proctor and Gamble Company, Cincinnati, OH.

Middaugh, D.P., M.J. Hemmer, and L.R. Goodman. 1987. Methods for spawning, culturing and conducting toxicity tests with early life stages of four Atherinid fishes: The inland silverside, *Menidia beryllina,* Atlantic silverside, *Menidia menidia,* tidewater silverside, *Menidia peninsulae,* and the California grunion, *Leuresthes tenuis.* EPA/600/8-87/004. U.S. Environmental Protection Agency, Office of Research and Development, Washington, DC.

National Council for Air and Stream Improvement. July 1986. NCASI methods for the analysis of chlorinated phenolics in pulp industry wastewaters. New York, Technical Bulletin No. 498.

Overstreet, R.M., W.E. Hawkins, and T.L. Deardorff. 1996. The western mosquitofish as an environmental sentinel: parasites and histological lesions, in *Environmental Fate and Effects of Pulp and Paper Mill Effluents.* Servos, M.R., K.R. Munkittrick, J.H. Carey, and G.J. Van Der Kraak, Eds., St. Lucie Press, Boca Raton, FL, pp. 495–509.

Rakocinski, C.F., M.R. Milligan, R.W. Heard, and T.L. Deardorff. 1996. Comparative evaluation of macrobenthic assemblages from the Sulphur River, Arkansas, in relation to pulp mill effluent, in *Environmental Fate and Effects of Pulp and Paper Mill Effluents.* Servos, M.R., K.R. Munkittrick, J.H. Carey, and G.J. Van Der Kraak, Eds., St. Lucie Press, Boca Raton, FL, pp. 533–547.

Tsai, T.Y., J.J. Renard, and R.B. Phillips. 1994. Formation of polychlorinated phenolic compounds during high chlorine dioxide substitution bleaching. Part 1: Laboratory investigations. *Tappi Journal,* 77(8):149–157.

U.S. EPA Method 1613. 1991. EPA Method 1613: Tetra- through octa-chlorinated dioxins and furans by isotope dilution HRGC/HRMS. 40 CFR Part 136, *Federal Register* 56(26), Thursday, February 7, 1991.

Yin, C., J.J. Renard, and R.B. Phillips. 1994. Formation of polychlorinated phenolic compounds during high chlorine dioxide substitution bleaching. Part 2: Biotreatment of combined bleach plant effluents. *Tappi Journal,* 77(8):158–162.

Chapter 11

Assessing the Aquatic Environmental Impact of ECF and TCF Pulping Operations Using Multispecies Mesocosms and Fish Biomarker Tests

K.-J. Lehtinen and J. Tana

INTRODUCTION

A primary objective of ecotoxicology is to develop test systems which predict real world environmental impacts of chemicals and effluents. The predictive capability of single-species toxicity tests is directly challenged by the complexity of natural ecosystems (Cairns, 1983). Furthermore, the limitations of short exposure duration and of waterborne exposure routes constrain single-species tests with persistent, bioaccumulative, or environmentally transformed chemicals. Recognizing these limitations, artificial mesocosms that mimic the intact ecosystem have been suggested as part of the testing requirements for the hazard assessment of industrial chemicals (Landner, 1988; Landner et al., 1989).

There are a number of factors complicating extrapolation of laboratory results to real ecosystems. Especially in arctic/boreal regions, the conditions in the laboratory do not greatly resemble those in nature, both in terms of test species and physicochemical conditions. Laboratory testing is based on strictly controlled experimental conditions (static temperature, static oxygen level, etc.). These strictly controlled conditions are at the same time a prerequisite for comparing data from tests with other toxicants. Such tests certainly can be defended if their objective is to rank the relative toxicity of different toxicants or wastewaters, but the scientific justification of using single-species tests to predict changes in competition, predation, community function, ecosystem energy flow, and nutrient cycling is questionable (Cairns, 1983). Single-species tests can only describe effects on individuals of one species.

One means of avoiding the limitations of the laboratory conditions is to study effects directly in the field. However, field studies also have their limitations. For example, in Scandinavia they are frequently performed adjacent to point sources close to the Baltic coast in a gradient from the discharge point. Process industries along this coast are, almost without exception, localized to areas with major freshwater sources, such as river estuaries. As a consequence, "ecological" gradients due to unstable physicochemical conditions are formed naturally, which in turn create a source of error, both when it comes to population studies as well as studies on the physiological status of individuals (Landner et al., 1994). An additional source of difficulties with field studies outside old established industrial plants, which are now applying improved environmental techniques, are the lingering impacts from previous operations and incomplete recovery. This may lead to unwarranted conclusions in investigations made several years after the discharge has been reduced or modified.

One way to avoid both the problems of strictly controlled laboratory conditions as well as those of uncontrolled field situations is to create a system, i.e., a mesocosm, that to a broad extent simulates the structural and functional mechanisms within a subsystem of the ecosystem of concern. This kind of approach has gained increasing support during the last decade. By using mesocosms, information is gained on how and when the species diversity and energy/material flow are affected. In ecotoxicological research, as well as in ecology, the structural (number and type of species) and functional (autotrophy, heterotrophy, nutrient cycling, energy flow) properties of the ecosystem are the two fundamental issues that must be understood in order to prevent environmental deterioration caused by human activities (Lehtinen, 1993).

Effects of effluents from the pulp and paper industry have been studied for several decades. The observed impacts have shifted from gross organic enrichment and suspended solids — effects which overwhelmed aquatic communities in the 1960s — to more subtle physiological and biochemical changes in the late 1980s (Owens, 1991). In the beginning of the 1980s, a systematic research was initiated of bleached kraft pulp mill effluents (BKME) in multispecies mesocosms simulating the littoral zone of the Baltic Sea (Rosemarin et al., 1990; Lehtinen et al., 1991). The main reason behind the application of the mesocosm technique was an increased awareness of the fact that predicting environmental impact is not possible by only using single-species tests. In order to obtain relevant and, for the environment, more realistic data, a broader scientific understanding of the structure and function of the ecosystem is needed. This includes an understanding of both direct effects of pollutants and indirect effects (responses) elicited by impacts on inherent ecosystem processes.

Pulp is produced by several methods. These production techniques yield different products and by-products and are usually grouped into mechanical and chemical pulping.

Mechanical pulping involves conversion of wood into fibers by physical or mechanical grinding. Chemical pulping is conducted by two processes — the sulfite process and the kraft (sulfate) process. The latter process dominates the industry because it more easily allows recovery of valuable chemicals and, as a result, minimizes discharges into the environment.

The main features of the kraft process involve the pulping, washing, and bleaching of the fibers. The pulping process removes the lignin from the wood fibers. Washing of the cooked pulp is done with water to remove dissolved organic material and the residual chemicals from the pulp. The remaining black liquor carried over to the bleach plant impacts the bleach plant chemicals consumption and the bleach plant effluent quality and the potential environmental effects. The handling technology of the pulp washing water (spills, open sea wash, etc.) will impact the quality of the mill effluent. The purpose of bleaching is to remove the residual lignin from the unbleached pulp and to make the pulp more white and bright according to quality specifications. Bleaching is carried out as a continuous process by alternating between delignifying stages (using chlorine, chlorine dioxide, oxygen, peroxide or ozone) and dissolved material extracting stages.

The techniques for production of chemical pulp and for treatment of the resulting effluents have developed considerably during the last two decades, to a large extent as a consequence of environmental pressure and concern. The focus was mainly on the discharges of chlorinated organic material. As a first step the kraft industry introduced low-chlorine multiple bleaching and an increased substitution of chlorine gas with chlorine dioxide in the first stage of the bleaching sequence. Subsequently, introduction of oxygen delignification, modified or extended cooking together with effective biological treatment, as well as total elimination of elemental chlorine in bleaching (ECF bleaching) occurred. Today several mills in Scandinavia have replaced chlorine dioxide with hydrogen peroxide, ozone, or peroxy acids (TCF bleaching) in their production of certain qualities of bleached pulp. These investments were generally accompanied by improved process control, good black liquor recovery and washing, and installation of spill recovery systems, all reducing also the discharges of unchlorinated organics from the mills. The modern mills have decreased their COD discharges from 100 kg/t pulp in 1970 to below 20 kg/t pulp today.

However, the uniqueness of each mill, both in design and operation, and its corresponding effluent must be considered in the assessment of environmental impacts.

The present study is a comparative investigation of biological effects of ECF (elemental chlorine–free) and TCF (totally chlorine-free) bleach plant and treated whole mill effluents from mills applying modern bleaching technologies. Effects on both ecosystem and individual levels have been studied using mesocosm techniques. In addition, the results reported here are compared with earlier results from mesocosm tests and an attempt has been made to relate biological variables to in-process conditions in order to evaluate the significance of process-related events on biological effects.

MATERIALS AND METHODS

Effluents

Effluents were sampled at a kraft mill with production of bleached softwood and hardwood pulp. Wastewater is treated with a secondary biological treatment in an activated sludge plant. The mill employs oxygen delignification and the kappa into the bleach plant was 14 during the time of effluent

Table 11.1. Effluents Tested in 400 and 2000 Times Dilution in Mesocosms, and Their Abbreviations. EHBU = ECF, Hardwood, Bleach Plant Effluent, Untreated; THBU = TCF, Hardwood, Bleach Plant Effluent, Untreated; THMT = TCF, Hardwood, Whole Mill Effluent, Treated.

Effluent	External Treatment	Flow (m^3/t)	COD[b] (kg/t)	AOX (kg/t)	Key
Bleach plant,ECF	no	25	40.7	0.8	EHBU*
Bleach plant,TCF	no	15	26.5	0.1	THBU
Whole mill,TCF/ECF 50% hardwood TCF 50% softwood ECF[a] paper mill	activated sludge	100	32,0	1.0	THMT

* Tested at one dilution (1:400) only.

[a] ECF softwood bleaching contained some chlorine gas (C-15) due to old-fashioned chlorine dioxide generation and occasional shortage of chlorine dioxide.

[b] Chemical oxygen demand.

sampling. Both externally treated whole mill effluent as well as untreated bleach plant effluents from hardwood production were tested. The effluents tested in model ecosystems and their abbreviations are presented in Table 11.1. The effluents were tested at two dilutions, 1:400 (high dose, HD) and 1:2000 (low dose, LD) based on water consumption of 50 m^3/t pulp. The treated TCF whole mill effluent is not a pure TCF effluent because of the parallel operations of the two bleach plant lines (Table 11.1).

The total amount of chlorinated phenolic compounds, resin acids, and neutral compounds in the effluents tested are presented in Table 11.2. Effluent samples were collected during the experimental period. The samples were analyzed using gas chromatography (Lehtinen et al., 1992).

The higher discharge of chlorinated phenolic compounds in THMT depends on several factors. The effluent contained discharges from a parallel ECF bleaching line, including some chlorine gas and old-fashioned chlorine dioxide generation. The level is, however, lower than levels normally found in untreated bleach plant effluents (Folke et al., 1993).

Experimental Setup

Ecosystem Level — Mesocosms

The mesocosm method used in the present study consists of outdoor land-based circular pools with an 8 m^3 water volume. Principally, the systems are established by transplanting part of the mother system — both plants and animals — into the respective pools. This procedure poses questions as to how and what is transplanted in order to avoid disturbances on the mother system, and comparative studies between the model ecosystem and the mother ecosystem are anticipated (Landner et al., 1989).

The main biotic component of the mesocosm is the bladder-wrack, *Fucus vesiculosus L.*, which is most significant for the Baltic ecosystem, where it constitutes about 90% of the biomass of the Baltic littoral zone. Studies on this community are made easier by its relatively low diversity compared with many other marine communities. Thus, the goal has been to build a community in the pools, based on bladder-wrack, as complete as possible. The main emphasis has been on studies of the benthic community, whereas less emphasis has been put on pelagic components like phyto- and zooplankton (see Lehtinen et al., 1992; Lindén et al., 1987).

The mother ecosystem, simulated in the mesocosm pools, is a typical small bladder-wrack-dominated sheltered bay with bottom substrate consisting of sand, stones, and gravel. The stones serve as substrate for the sessile bladder-wrack. Under natural conditions greater water volumes will pass through the benthic mother ecosystem than in the mesocosms. It is therefore important to adjust the plant biomass in order to maintain the similarity with the mother system, where the water current supplies the plants with necessary nutrients.

Table 11.2. The Total Amount of Chlorinated Phenolic Compounds, Resin Acids, and Neutral Compounds in Effluents Tested. The Amounts are Given as g/t Pulp.

Effluent	Chlorinated Phenolic Compounds	Resin Acids	Fatty Acids	Neutral Compounds
EHBU	0.3	10	145	35
THBU	0.005	2	50	35
THMT	10	5	40	75

The importance of the water flow for the growth of the bladder-wrack in the mesocosm was studied in detail by Landner et al. (1989). From those experiences the theoretical water residence time was set to a maximum of 2 days. Experiments with a dye showed that by adjusting the capillary of the incoming water obliquely toward the bottom, a current velocity was obtained, causing the water in a filled pool to circulate once every 3–4 minutes, giving a theoretical circulation of 360–480 times/day.

The outgoing water pipe was supplied with a mesh (1×1 mm) in order to prevent larger organisms from escaping the system. A small fraction of both incoming and outgoing water is drawn to a measuring unit for semicontinuous measurement of temperature, pH, and dissolved oxygen. The physicochemical characteristics of each pool were measured by a computer-operated valve system. The data were automatically stored in the computer and used for calculation of primary production in the model ecosystems in relation to incoming reference water. The calculation of primary production was made according to Gargas (1975). Accumulated light energy was calculated from the accumulated water temperature data during the part of the day when the system temperature was increasing as compared to incoming reference water. Insolation was also measured by a light meter. The experimental setup is schematically presented in Figure 11.1.

At the startup of the mesocosms, the bottoms of the pools were covered with a known volume of clean sand without organisms. After this, mature bladder-wrack specimens, attached to their stone substrate, were collected from a seashore and placed into the pools so that about 1/3 of the bottom was covered in each pool. The bladder-wrack specimens were collected by slipping a plastic bag (40×60 cm) over both the stone and the attached plant specimen to capture the plant and associated animal organisms. The bag contents were placed in a bucket and, after measurement of the plant volume using the displacement method, immediately introduced into the model ecosystem. The number of plants were counted and plant material introduced until a total plant volume of about 7 liters was reached in each pool. Plants were introduced randomly into the pools in order to avoid systematic errors regarding animal numbers, and substrates for quantitative calculations of periphytic growth were introduced to each pool.

Apart from the invertebrate organisms directly associated with the bladder-wrack, a number of other free-swimming animal species were introduced to the mesocosms, particularly broods of the three-spined stickleback, *Gasterosteus aculeatus L.* The model ecosystems were allowed to stabilize for 2–3 weeks before exposure to pollutants. The experiment lasted for 6 months, between May and November in 1993.

In order to avoid pseudoreplication on subsamples taken from each pool, invertebrate abundance and biomass from the different habitats were pulled together and expressed on a whole pool unit basis (Hurlbert, 1984). Statistical difference of bladder-wrack apical growth was established by Student's t test and Cochran's correction of variance of the means. Levels of significance were established at ***=$P<0.001$; **=$P<0.01$; *=$P<0.05$.

Individual Level — Fish Physiology

Physiological responses of rainbow trout, *Oncorhynchus mykiss*, were studied in separate 500-L pools connected to the outgoing water from the mesocosms; 20 individuals were placed in each test pool. The mean weight of the fishes was 130 g at the experimental start. The experiment lasted for 8 weeks from the beginning of September.

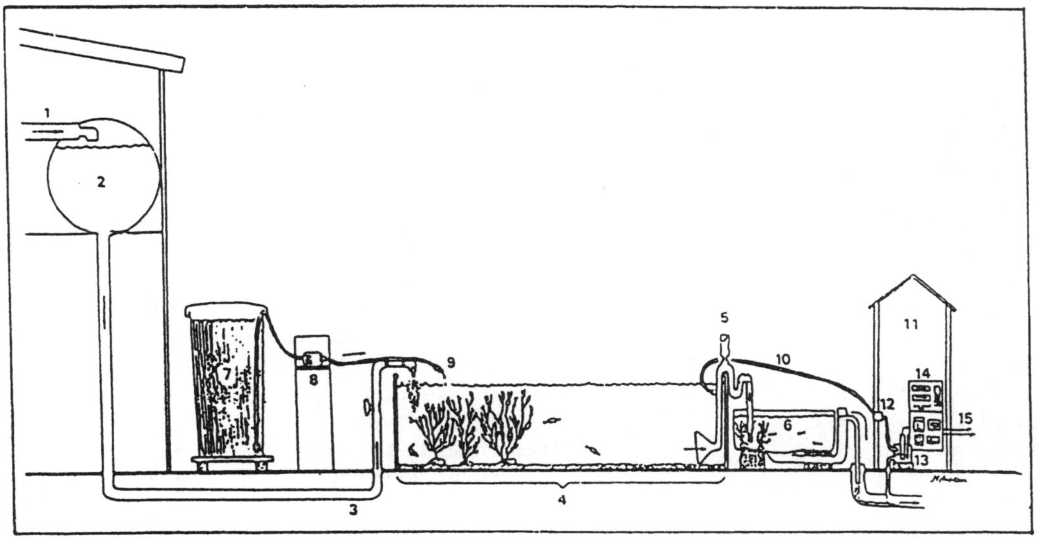

1. Incoming water
2. Stabilization tank
3. PEH-pipe to the pools
4. Model ecosystem pool
5. Siphon for out-going water
6. Rainbow trout pool
7. Effluent container
8. Membrane pump
9. Dosage of effluent
10. Tube to registration chamber
11. Registration chamber
12. Magnetic valve
13. Registration electrodes
14. Measuring units
15. Computer

Figure 11.1. Schematic diagram showing the model ecosystem setup.

At sampling the fish were caught with a dipnet and stunned with a blow on the head. Blood was collected from caudal vessels with a disposable syringe. After blood sampling the fish were weighed and their length measured and they were checked for external and internal lesions. The gallbladder was emptied using a 1-mL disposable syringe and the bile collected in a 5-mL glass septum bottle, with 3 samples per bottle. The samples were frozen for later analyses of conjugated chlorophenolics and extractives. The liver was then weighed and cut into pieces, which were stored in Eppendorf tubes in liquid nitrogen for later analysis of enzyme activities and different metabolic parameters. The analytical methods were the same as described by Lehtinen et al. (1992); altogether 27 different physiological and histological parameters were studied (Table 11.3). The physiological responses are presented as percent difference from the control values. The significance of differences between mean values of physiological parameters were established using Student's t test and Cochran's correction of variance of the means. Levels of significance were established at ***=$P<0.001$; **=$P<0.01$; *=$P<0.05$.

RESULTS AND DISCUSSION

The short-term toxicity of the effluents was determined using a set of single-species toxicity tests (Verta et al., 1996). The tests did not reveal any clear differences between untreated bleach plant effluents from ECF and TCF pulp production. Typically the EC_{50}-values were between 20–40% v/v. The treated whole mill effluent (THMT) did not show any short-term toxicity (Verta et al., 1996).

Table 11.3. Parameters Studied in the Rainbow Trout Experiment.

Morphometry	Hematology
1 = Length	9 = Hematocrit—packed red cell volume
2 = Weight, total	10 = Red blood cells, total amount
3 = Weight, somatic	11 = Immature red blood cells
4 = Weight of visceral mass	12 = Hemoglobin
5 = LSI—Liver somatic index	13 = Mean cellular hemoglobin concentration
6 = Condition factor	14 = Mean cellular hemoglobin
7 = Growth/day	15 = Mean cellular volume
8 = Food conversion	16 = Leucocrit—packed white cell volume
	17 = White blood cells, total volume
LIVER	18 = Lymphocytes
	19 = Granulocytes
23 = Liver glycogen	20 = Thrombocytes
24 = Liver lipids	21 = Aspartate-aminotransferase—ASAT[c]
25 = EROD[a]	22 = Aspartate-alaninetransferase—ALAT[c]
26 = UDP[b]	
27 = Liver structure	
BILE	
Conjugated chlorinated phenols	
Conjugated resin acids	

[a] Mixed-function oxidase enzyme, 7-ethoxyresorufin *o*-deethylase.
[b] Conjugation enzyme, UDP glucuronosyltransferase.
[c] Tissue damage enzymes (levels in plasma).

Effects on Ecosystem Level

Effects on both structural and functional levels were studied. Functional effects such as carbon binding capacity at different trophic levels describe changes in energy flow, material circulation, and energy conversion into biotic structures. Structural effects are described, for example, by bladder-wrack growth and the abundance and biomass of animals in the ecosystem. The apical growth of bladder-wrack exposed in model ecosystems to effluents from ECF and TCF pulp production is presented in Table 11.4.

Due to nutrient deficiency the apical growth of bladder-wrack in Control 2 was inhibited in the middle of the experimental period. Growth was later resumed, but this group consequently did not show the normal growth of the bladder-wrack (Lehtinen et al., 1992; Lehtinen et al., 1993). Bladder-wrack growth in the exposed pools, presented in Table 11.1, is thus compared to Control 1, where the growth was normal.

A clear inhibition in bladder-wrack growth as compared to control was observed at exposure to EFC bleach plant effluent. This may be related to chlorate present in the effluent, together with low water nitrate levels in the summer. Chlorate content depends on the amount of chlorine dioxide used in bleaching. The inhibiting effects on algae, mainly on the keystone species *Fucus vesiculosus*, have been previously observed to be attributable to chlorate (Rosemarin et al., 1986; Lehtinen et al., 1988). At pulp mills discharging to marine environments chlorate is nowadays removed by either direct chemical treatment or proper biological treatment systems. Effluents from TCF bleaching had no inhibitive effects on the growth of bladder-wrack.

Total invertebrate abundance in bladder-wrack, sediment, and pool walls as difference (%) from the two controls in mesocosms exposed to effluents from ECF and TCF pulp production is presented in Figure 11.2. The results show that effluents from TCF bleaching seem to have an inhibitive effect on the invertebrate abundance in model ecosystems, except for the low-dose THMT.

The circulation and synthesis of organic carbon is an important functional property of ecosystems. Disturbed ecosystems often show a decreased capacity for converting energy into organic structure

Table 11.4. Final Length of Apical Fronds (mm; Mean ± s.d.) of Bladder-Wrack and Growth as mm/wk. Statistical Significance According to Student's t Test: **=P<0.01: *=P<0.05. N is the Number of Apical Fronds Measured.

Effluent	Final Length of Apical Fronds (mm)	Growth/Week (mm)	N
Control 1	61.0±13.0	2.80	50
EHBU	30.1±10.0 **	0.57	9
THBU, low dose	66.0±15.5	4.03	50
THBU, high dose	60.0±15.0	2.30	50
THMT, low dose	62.5±15.5	4.16	50
THMT, high dose	67.0±12.5 *	4.20	50

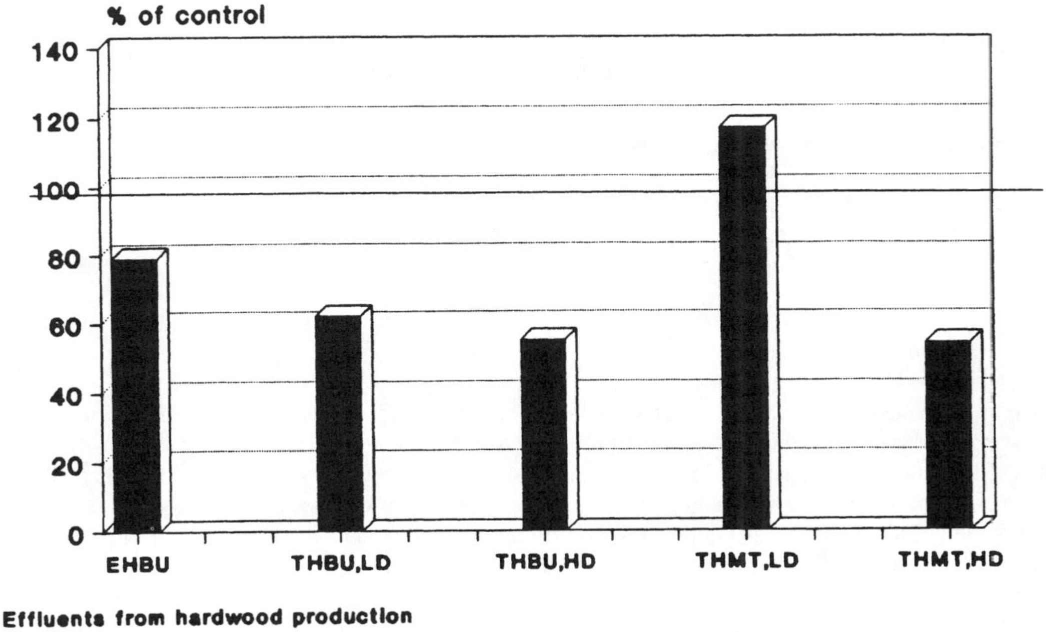

Figure 11.2. Total invertebrate abundance in bladder-wrack, sediment and pool walls as difference from the two controls in mesocosms exposed to effluents from hardwood ECF and TCF pulp production. EHBU = ECF, hardwood, bleach plant effluent, untreated; THBU = TCF, hardwood, bleach plant effluent, untreated; THMT = TCF/ECF, hardwood, whole mill effluent, treated. LD = low dose (1:2000); HD = high dose (1:400).

and will consequently lose carbon from the system (Odum, 1985). One way to study events connected to mechanisms of this kind is to measure input and output of organic carbon and how much is bound into primary producers and animals. The amount of particulate organic matter as well as total organic carbon are important parts of the carbon cycle in ecosystems. In the present study, incoming and outgoing total organic carbon (TOC) was measured at intervals (twice every month) (Figure 11.3). Difference between incoming and outgoing total organic carbon describe the carbon binding capacity of mesocosms.

The results show that clean control systems retained part of the total incoming TOC, whereas mesocosms exposed to effluents were leaking organic carbon in most cases. The mesocosm exposed to treated whole mill effluent (THMT) at 2000 times dilution retained a small net amount of organic

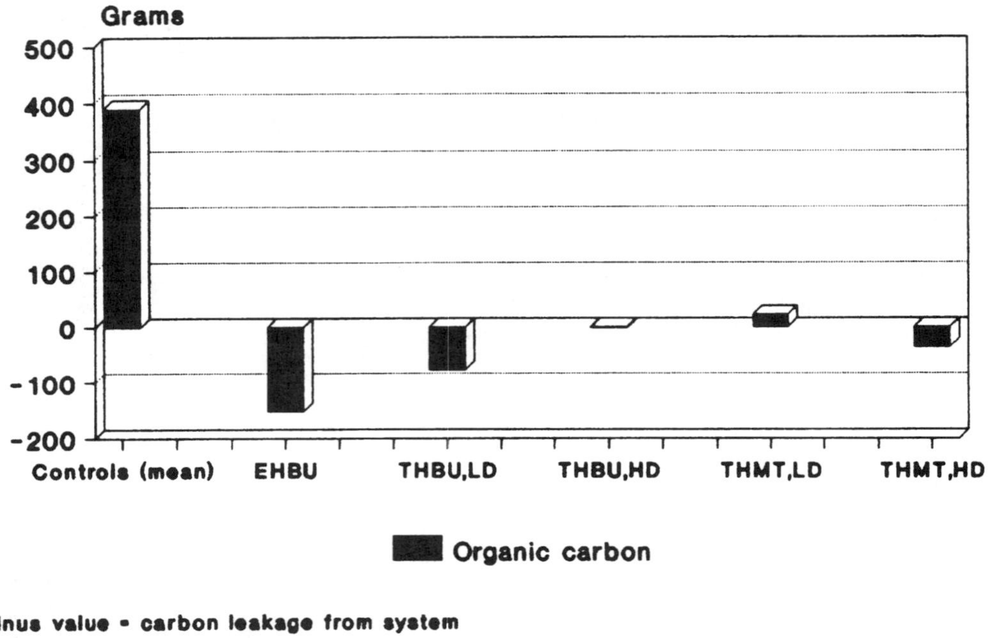

Figure 11.3. Difference between incoming and outgoing organic carbon in mesocosms exposed to effluents from ECF and TCF pulp production. Negative value indicates carbon leakage from system. For figure legends see Figure 11.2.

carbon. The leakage of TOC is a clear sign of a lower efficiency of retaining and converting energy into organic structure. However, it was not possible to detect any clear differences between effects of effluents from ECF and TCF pulp production. Despite several systems leaking carbon, there was a net oxygen production in all systems over the experimental period, meaning that the systems were autotrophic. This in turn means that the plants within the system were able to supply the animals with oxygen. Regarding effluent from the ECF production, the major part of the carbon leakage most probably was due to chlorate effects on bladder-wrack alga, which was not accompanied by any prominent increase of filter feeding invertebrates having the capacity of binding organic carbon.

The effects on abundance and biomass results reflected complicated events that occurred in each system during the experiment. The amount of energy assimilated by the system is crucial for the production of the system. Pollution, on the other hand, is also one of many driving forces. One important feature of mesocosms is that natural variabilities in physicochemical conditions are maintained, but kept under control. Due to external variability, such as wind directions, the position of the mesocosm pool, etc., it was not possible to maintain temperatures of each pool exactly the same. However, it is possible to measure any differences. A way to measure the incoming energy absorbed by each model ecosystem is to document the temperature increase during the light period. The present study included two controls, and it was calculated that Control 1 received less energy than Control 2 (Figure 11.4), resulting in high system metabolic rate in Control 2, followed by nutrient deficiency. The resulting difference in accumulated energy was reflected by a distribution of organic carbon from the bladder-wrack to the animal component, illustrated by an increased invertebrate abundance in Control 2.

The results (Figure 11.4) show a correlation between energy and total invertebrate abundance for the controls and mesocosms exposed to bleach plant effluent from ECF production. However, the response of animal abundance to accumulated energy was lower than anticipated in model ecosystems exposed to effluents from TCF production. The reasons for this inhibition in relation to available energy are not known presently.

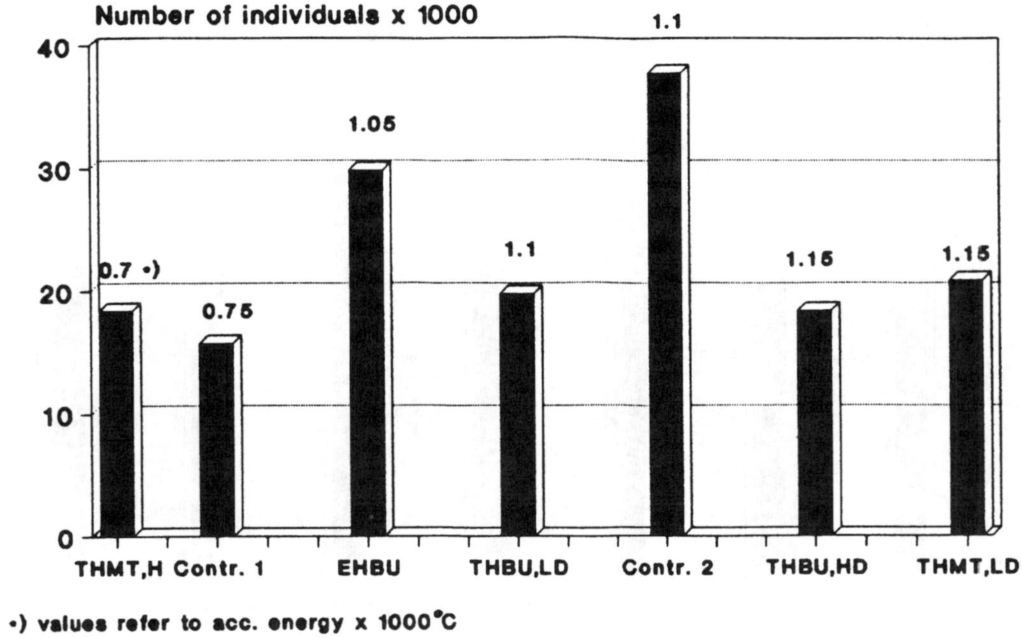

Figure 11.4. Total invertebrate abundance and accumulated energy in mesocosms exposed to effluents from ECF and TCF pulp production. Effluents are illustrated in order of increasing accumulated energy measured as accumulated temperature °C (values above the columns) in each mesocosm pool during the whole experimental period. For figure legends see Figure 11.2.

Earlier studies (Lehtinen, 1990; Lehtinen et al., 1992) have demonstrated both stimulatory and inhibitory effects on the growth of sticklebacks. It may be expected that water temperature will affect growth of these poikilothermic organisms. As the incoming energy was calculated, the relationship of temperature to stickleback mean individual weight could be plotted (Figure 11.5), to see if temperature was the only factor causing differences in fish production. Assuming that the only difference between the two controls was the temperature, stickleback exposed to effluents from production of TCF pulp had lower mean weights, indicating mechanisms affecting growth and metabolism.

Effects on Individual Level

The concentrations of conjugated chlorinated phenols and extractives in bile of fish exposed to high doses of the tested effluents are presented in Table 11.5.

The concentrations of chlorinated phenols in the bile of rainbow trout were low in all groups. Elevated levels as compared to control values could be detected in groups exposed to bleach plant effluent from ECF production and to treated whole mill effluent (THMT) containing 50% TCF hardwood and 50% ECF softwood effluent. The softwood pulp production line contained bleaching with some chlorine gas due to old-fashioned chlorine dioxide generation and occasional shortage of chlorine dioxide. The lowest concentration of conjugated extractives (fatty acids) was found in fish exposed to effluent EHBU — bleach plant effluent from ECF production. This group, on the other hand, showed some elevated concentrations of conjugated resin acids (Table 11.5).

Compared to background concentrations found in fish bile from water bodies uncontaminated by pulp mill effluents, the present values are low (Oikari and Kunnamo-Ojala, 1987; Oikari et al., 1985). The concentrations of conjugated chlorinated phenols and resin acids were also low and similar to levels in control fish in previous studies with pulp mill effluents in mesocosms (Lehtinen et al., 1992;

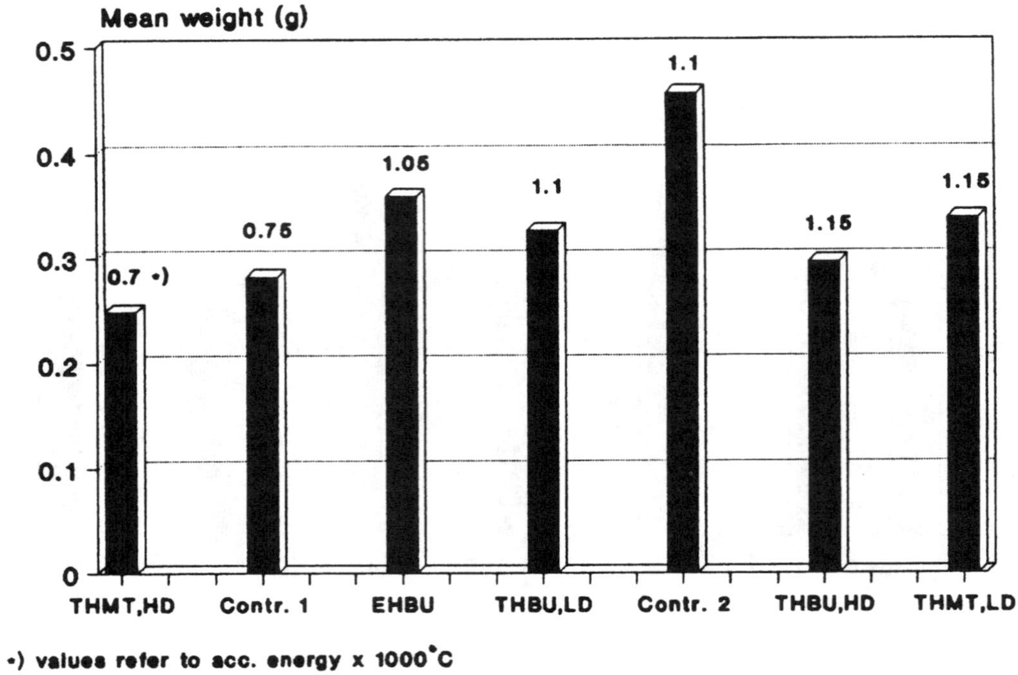

Figure 11.5. Stickle-back, *Gasterosteus aculeatus,* mean weight (g) and accumulated energy in mesocosms exposed to effluents from ECF and TCF pulp production. Effluents are illustrated in order of incoming accumulated energy measured as accumulated temperature °C (values above the columns) in each mesocosm pool during the whole experimental period. For figure legends see Figure 11.2.

Table 11.5. The Concentrations of Organochlorines and Extractives in Bile of Fish Exposed to Effluents from ECF and TCF Pulp Production in 400 Times Dilution (High Dose). The Concentrations are Given as μg/g Dry Weight.

	Cont.	EHBU	THBU	THMT
Chlorinated phenols	2.1	6.5	0.5	16
Resin acids	56	166	71	55
Fatty acids	1122	840	883	1234
Cholesterol	1672	1211	2031	1975
Unidentified	46	57	47	88

Lehtinen et al., 1993). The chlorinated phenolic and resin acid concentrations found in bile in this study were also lower than those found in the Grand Prairie study for a river environment downstream of a mill that uses 100% in bleaching (Swanson et al., 1994). The levels of conjugated resin acids in fish bile from different exposure groups in the present study correlated with corresponding concentrations of resin acids in the effluents.

The physiological responses in the rainbow trout exposed to discharges from the mesocosm pools are presented in Figures 11.6–11.8. Parameters differing by more than 20% from control values or showing statistical significance to control according to Student's t test are illustrated. The parameters described in Figures 11.6–11.8 are coded by numbers as listed for corresponding parameters in Table 11.3.

In the group exposed to untreated bleach plant effluent from ECF production, changes were mainly observed in hematology, but also some responses in liver function and increase in fish growth could

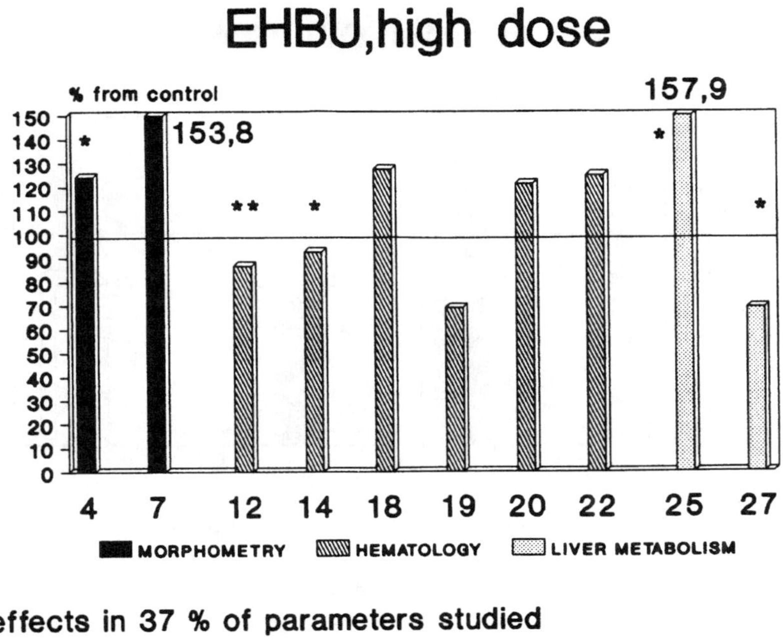

Figure 11.6. Fish physiological parameters (see Table 11.3) showing those with more than 20% difference or statistical significance to control in fish exposed to untreated bleach plant effluent from ECF pulp production. Statistical significance according to Student's t-test: $**=P<0.01$; $*=P<0.05$.

be seen (Figure 11.6). The activity of the MFO enzyme EROD was significantly elevated in this exposure group. In this group a statistically significant difference or a deviation of more than 20% from the control could be observed in 10 out of 27 parameters studied. This was more than in the other exposure groups.

The exposure to effluents from production of TCF pulp caused fewer responses in fish than the exposure to ECF bleach plant effluent (Figures 11.7 and 11.8). This is indicated by the lower number of parameters with statistically significant or over 20% difference from control. In the group exposed to TCF bleach plant effluent in a 2000 times dilution (THBU, low dose) there were only 4 parameters out of 27 showing over 20% difference (Figure 11.7). There were changes in energy allocation and fish growth, but no indications of decreased condition of fish or starvation. The oxygen carrying capacity seemed to be undisturbed, but structural changes in liver tissue occurred. This, however, did not affect the detoxification capacity of the liver. The treated whole mill effluent (THMT) containing more chlorinated phenolic compounds than the ECF bleach plant effluent (Table 11.2), did not, however, cause a statistically significant increase in the activity of the MFO enzyme EROD.

The significant responses observed in this experiment occurred in the same parameters in all exposure groups. The most significant responses were detected in energy metabolism appearing as changes in growth and food conversion factors. Previous model ecosystem studies (Lehtinen et al., 1993) usually identified a stimulation in fish growth. It is interesting that in the present study, there was a difference in the direction of these responses between low-dose groups and high-dose groups. Also, the same parameters were not always affected in each exposure group. The responses also varied between the different effluents. The reason for the opposite negative dose responses in growth are still unknown, but indicate changes in energy allocation.

Red blood cell parameters showed statistically significant changes as compared to control, mainly due to very low variation between exposure and control groups. The changes however, are so small that they are unlikely to disturb the oxygen carrying capacity of exposed fishes. The increased oxygen demand in the groups that showed increased growth occurred without any significant changes in the number of red blood cells or hemoglobin concentrations. The absence of significant changes in hematological parameters (Hct, MCV) also indicate that swelling of red blood cells was not prevalent in exposed fishes.

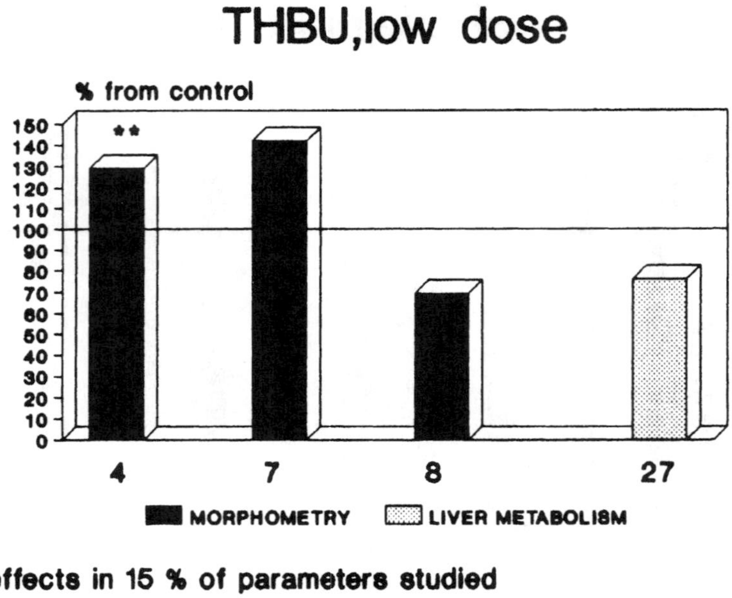

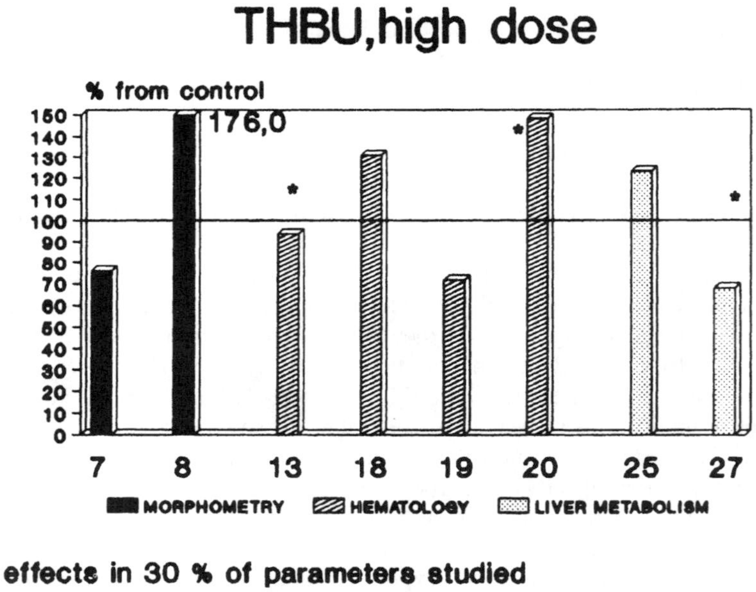

Figure 11.7. Fish physiological parameters (see Table 11.3) showing those with more than 20% difference or statistical significance to control in fish exposed to untreated bleach plant effluent from TCF pulp production. Statistical significance according to Student's t-test: **=P<0.01; *=P<0.05.

No changes or specific responses were detected in parameters describing liver metabolism (glycogen, lipids). The activity of MFO enzyme was higher in comparison to control in all exposure groups except in high-dose of THMT and in low-dose of THBU. In THBU a slight inhibition was observed. A more pronounced increase in the activity was found in the group exposed to ECF bleach plant effluent. However, responses in EROD did not correlate to substances analyzed in the effluent or to bile conjugates analyzed. MFO induction in fish has been considered as an informative biomarker of exposure to pulp mill effluents (Södergren, 1988; Andersson et al., 1988; Lindström-Seppä, 1990) but the role of environmental, physiological and toxicological factors on

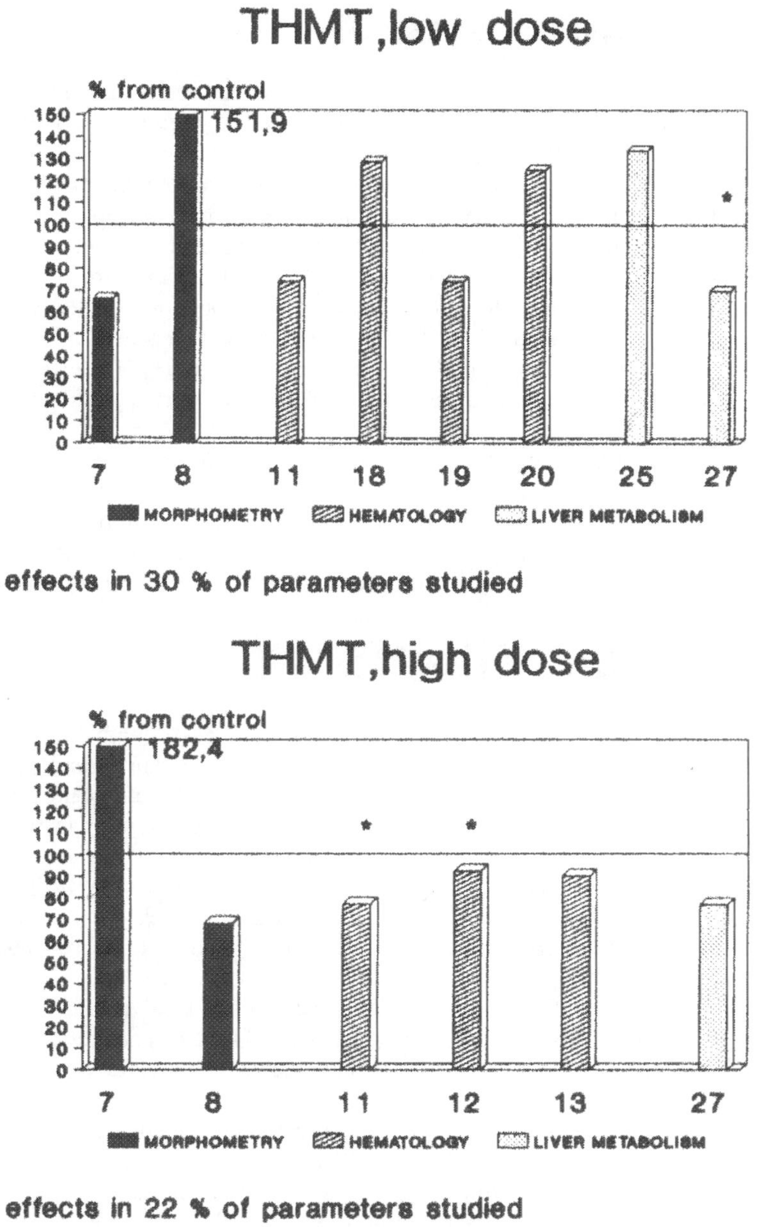

Figure 11.8. Fish physiological parameters (see Table 11.3) showing those with more than 20% difference or statistical significance to control in fish exposed to treated whole mill effluent from parallel ECF/TCF pulp production. Statistical significance according to Student's t-test: *=P<0.05.

the direction of MFO responses must be understood in order to properly interpret the biomarker response (Jimenez et al., 1990).

Changes in liver structure were also observed without any clear correlation to responses found in other liver parameters. The most pronounced change in liver structure was the increase of the amounts of necrotic cells in all exposure groups as compared to control.

In general, the physiological responses were detected in all exposure groups studied, without any clear difference between exposure to ECF and TCF. The responses were in general small and only in 9 of the 27 parameters studied could a difference of more than 20% be detected.

COMPARISON TO PREVIOUS STUDIES

The effects of total pulp mill effluents with different bleaching procedures and technical developments have been systematically studied in mesocosms since the beginning of the 1980s (Rosemarin et al., 1986; Rosemarin et al., 1990; Lehtinen et al., 1991; Tana et al., 1994; Lehtinen et al., 1994). Effects on both community and population level as well as whole-organism and within-organism levels have been studied in these works. At the community level, the main emphasis in the beginning was on community structure. From 1991 onward, functional mechanisms such as carbon binding, consumption of carbon dioxide, oxygen production, etc., have also been studied.

To date, both whole mill and bleach plant effluents from production of softwood and hardwood pulp have been studied in mesocosms, as well as effluents from unbleached and thermomechanical pulp (TMP) production. So far no paper mill effluents have been studied. The effluents from chemical kraft pulp production cover the spectrum from conventional chlorine gas bleaching to ECF and TCF bleaching.

All effluents have been normalized to water consumption of 50 m^3 per ton of pulp. This normalization has been considered necessary as different processes use different amounts of water and a high water use may lead to dilution of the effluent substances. The effluent dilutions were 400 and 2000 times. Originally, the mesocosm studies were designed to evaluate the ecosystem impact of whole mill effluents from mills employing new bleaching technologies, i.e., turning from conventional chlorine bleaching to oxygen delignification, higher use of chlorine dioxide instead of chlorine, low-chlorine multiple bleaching, etc. However, this approach does not permit comparison of the effect of contributions from the different compartments of the mill, especially when it has not always been possible to compare the same mill at every experiment.

In order to compare the different effluents originating from processes with different technological levels, some effects on ecosystem structure in mesocosms are compared in Figure 11.9; the responses illustrated are given as a percentage of the respective controls for different years, on bladder-wrack growth, sediment invertebrate abundance, bladder-wrack-associated invertebrate abundance, and total system abundance (including animals on the walls). The abundance of the animals has been considered more important than biomass since the abundance of individuals better reflects the total genetic information available within the community. Effects on bladder-wrack growth are mostly due to chlorate. The functional effects cannot be compared because they have only been studied in the last few experiments.

From Figure 11.9 it is seen that Effluents 1 and 2 distinctly inhibited all parameters. The effect was very similar despite differences in TOCl content of 6.5 and 2.4 kg/t, respectively. Interestingly, phytosterols alone also (No. 17 in Figure 11.9) affected bladder-wrack invertebrate abundance. Moreover, tertiary treated effluents from an integrated TMP newsprint mill (No. 16) affected the structural parameters in the same order as Effluents 1 and 2. Untreated bleach plant or untreated whole mill effluents from ECF and TCF pulp bleaching (19–21) inhibited invertebrate abundances. Activated sludge treatment of mixed ECF/TCF mill effluent (No. 22) was stimulatory on the abundance of invertebrates in mesocosms, as was the untreated effluent (No. 9) from a bleach plant with oxygen delignification and 85% chlorine gas in the first bleaching stage. Treated whole mill effluent (No. 11) with 80% chlorine dioxide in the first bleaching stage was the effluent that stimulated the invertebrate abundance most, and in an aerated lagoon treated effluent from low-chlorine multiple bleaching was also stimulatory for the invertebrates.

In summary, there was no relation between bleaching process and the outcome of the exposure on the invertebrate community. When it comes to the knowledge so far gained by the mesocosms, or different components such as fish physiology, it was shown in the late 1980s that AOX at 2.0–2.5 kg/t pulp did not correlate with observed effects (Lehtinen et al., 1992). Later studies indicated that a relationship between COD and effects seems to exist for various untreated and treated whole mill effluents. Bleach plant effluents have been studied to a lesser extent.

Sangfors et al. (1994) correlated response indices versus COD based on whole ecosystem responses (Figure 11.10); a correlation between lower COD discharge and lower response indices is observed. The quality of COD, particularly in less biodegradable substances, will influence heterotrophic activity, nutrients, and organic carbon for both autotrophic and heterotrophic processes

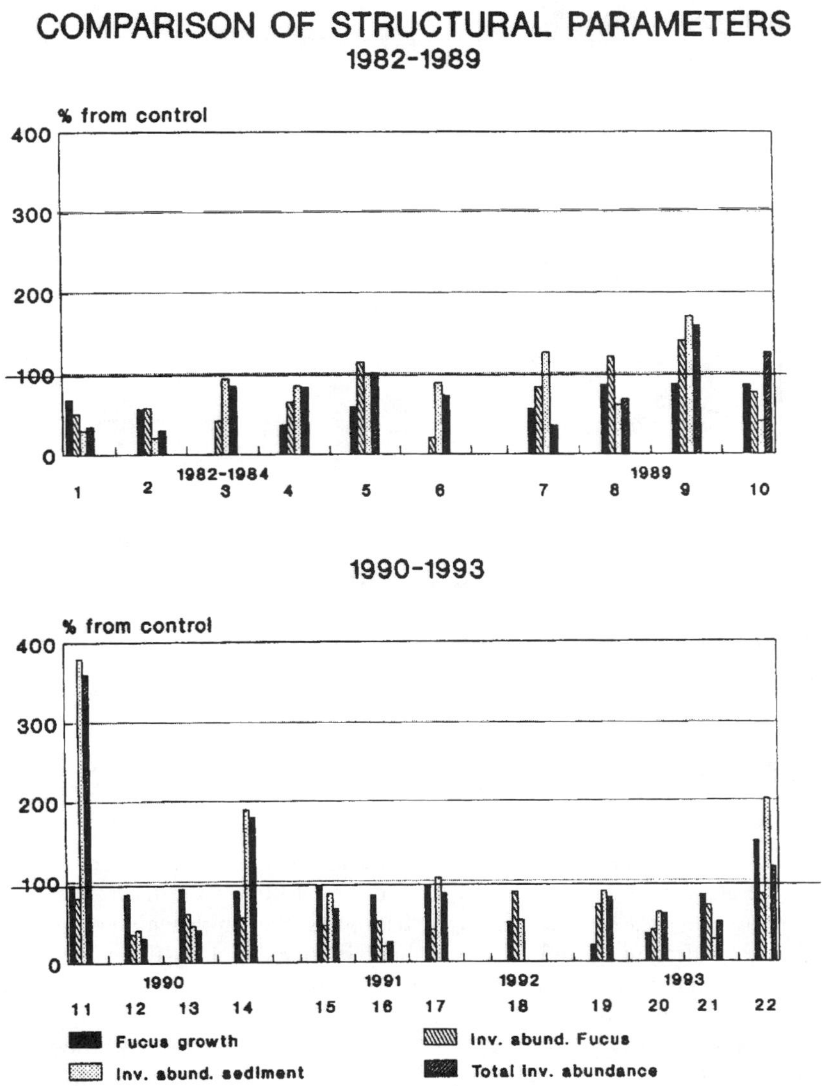

1. (C95+D5)EHDED	untreated	2. (C87+D13)EDED	aerated lagoon
3. O(C83+D17)EDED	untreated	4. O(C84+D16)EDED	partial aeration
5. O(C84+D16)EDED	aerated lagoon	6. O(C52+D48)EDED	untreated
7. O(C60+D40)(Eo)DED*	untreated	8. O(C60+D40)(Eo)DED	untreated
9. O(C85+D15)(Eo)DED*	untreated	10. OPD*	untreated
11. (D80+C20)(Eop)DED	untreated	12. (D80+C20)(Eop)DED	activated sludge
13. O(D27,C68+D5)EopEoD	untreated	14. O(D27,C68+D5)EopEoD	aerated lagoon
15. (C84+D16)(Eo)DED	activated sludge	16. TMP+newsprint	aerated lagoon
17. Phytosterols		18. O(D91+C9)(Eo)DED	activated sludge
19. DEoDED*	untreated	20. DEoDED	untreated
21. QeZEopAzEPA*	untreated	22. QeZEopAzEPA/DEoDED	activated sludge

* bleach plant effluent

Figure 11.9. Comparison of structural responses in mesocosms exposed in 400 times dilution to various pulp mill effluents. The responses illustrated are given as a percentage of the respective controls for different years.

(Figure 11.4). In any case, the mesocosm technique indicates that mills applying technology of the 1990s produce effluents with very low environmental impact regardless of whether bleaching is done with chlorine-containing chemicals or not.

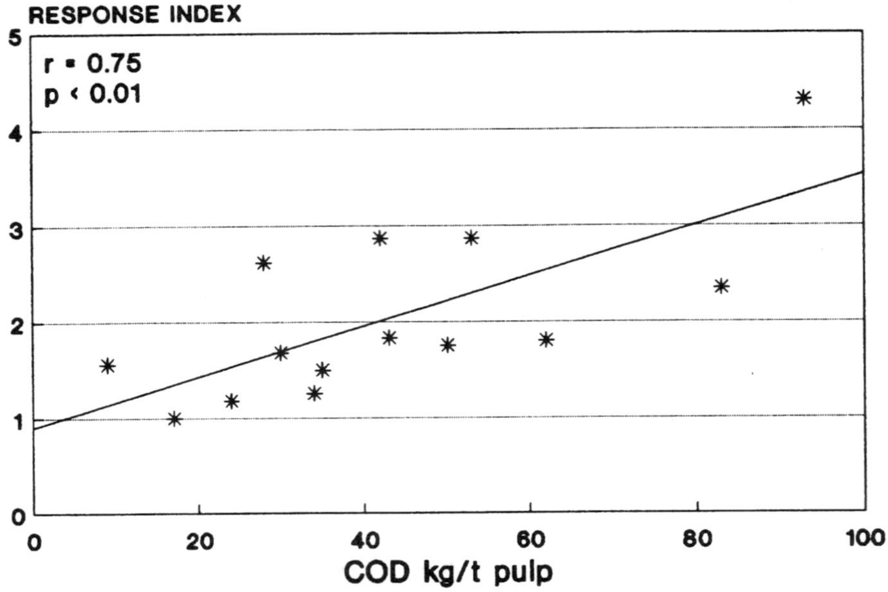

Figure 11.10. Ecosystem responses, including whole-organism responses, versus COD (kg/t pulp) after exposure to various pulp mill effluents (Sangfors et al., 1994).

CONCLUSIONS

The data imply that pre-bleach plant process practices may greatly modify the biological response of mesocosms. It should be kept in mind that existing data are incomplete, preventing full comparison, since the effluents tested originated from different mills with different in-mill solutions and bleaching sequences. The results also imply that one particular bleaching sequence is not uniquely associated with environmental effects versus alternative technologies. One will also need information about pulping operation, spill control, washing loss, and process control. These views are also supported by recent North American field studies outside mills which employ different pulping and bleaching technologies (Swanson et al., 1994; Hodson et al., 1992; Munkittrick et al., 1994). In the case of low-AOX bleaching sequences of today, comparison of bleaching technology is not environmentally relevant. Although far from complete from the toxicity point of view, the results suggest that it is not a question of whether to use ECF or TCF as the choice of bleaching, but rather to improve the pulping operation itself.

ACKNOWLEDGMENTS

The authors wish to thank the staff at the Baltic Sea Laboratory for skillful performance. Thanks are also due to Dr. J. William Owens and Dr. John Chapman for critical review of the manuscript. A major part of the work was funded by Kymmene Corporation as part of an evaluation of new bleaching technologies.

REFERENCES

Andersson T., L. Förlin, J. Härdig, and Å. Larsson. 1988. Physiological disturbances in fish living in coastal water polluted with bleached kraft pulp mill effluents. *Canadian Journal of Fisheries and Aquatic Sciences.* 45:1525–1536.

Cairns, J. 1983. Are single species toxicity tests alone adequate for estimating environmental hazard? *Environmental Monitoring and Assessment.* 4:259–273.

Folke, J., L. Landner, K-J. Lehtinen, H. Männistö, E. Männistö, and N. McCubbin. 1993. Chlorine dioxide in pulp bleaching — Technical aspects and environmental effects. Literature study. *CEFIC, Chlorate Group*, Brussels, Belgium.

Gargas, E. 1975. *A Manual for Phytoplankton Primary Production in the Baltic*. BMB Publication 2. Danish Environmental Protection Agency.

Hodson, P.V., M. McWhirter, K. Ralph, B. Gray, D. Thivierge, J.H. Carey, G. Van Der Kraak, D.M. Whittle, and M-C. Levesque. 1992. Effects of bleached kraft mill effluent on fish in the St. Maurice river, Quebec. *Environmental Toxicology and Chemistry*. 11 pp. 1635–1651.

Hurlbert, S.H. 1984. Pseudoreplication and the design of ecological field experiments. *Ecological Monographs*. 54(2):187–211.

Jimenez, B.J., A. Oikari, S.M. Adams, D.E. Hinton, and J.F. McCarthy. 1990. Hepatic enzymes as biomarkers: Interpreting the effects of environmental physiological and toxicological variables, in *Biomarkers of Environmental Contamination*. McCarthy, J.F, and L.R. Shugart, Eds. Lewis Publishers, Boca Raton, FL.

Landner. L. 1988. Hazardous chemicals in the environment — some new approaches to advanced assessment. *Ambio*. 17(6):663–673.

Landner, L., H. Blanck, U. Heyman, A. Lundgren, M. Notini, A. Rosemarin, and B. Sundelin. 1989. Community testing and mesocosm experiments: Ecotoxicological tools with high ecological realism, in *Chemicals in the Aquatic Environment — Advanced Hazard Assessment*, Landner, L., Ed, Springer Verlag, Berlin, Heidelberg, New York, pp. 216–254.

Landner, L., O. Grahn, J. Härdig, K-J. Lehtinen, C. Monfelt, and J. Tana. 1994. A field study of environmental impacts at a bleached kraft pulp mill site on the Baltic Sea coast. *Ecotoxicology and Environmental Safety* 27:128–157.

Lehtinen, K.-J. 1990. Mixed-function oxygenase enzyme responses and physiological disorders in fish exposed to kraft pulp mill effluents: A hypothetical model. *Ambio* 19(5):259–265.

Lehtinen, K.-J. 1993. Structure and function of brackish water littoral mesocosms, in *Proceedings of the International Workshop on Marine Mesocosms — Design and Applications, Perth, Australia, 1993.* CSIRO Marine Laboratories, North Beach, Western Australia, Australia, pp. 77–101.

Lehtinen, K.-J., M. Notini, J. Mattsson, and L. Landner. 1988. Disappearance of bladder-wrack (*Fucus vesiculosus L.*) in the Baltic Sea: Relation to pulp mill chlorate. *Ambio*. 17:387–393.

Lehtinen, K.-J., B. Axelsson, K. Kringstad, and L. Strömberg. 1991. Characterization of pulp mill effluents by the model ecosystem technique. *Nordic Pulp and Paper Research Journal*. 6(2):81–88.

Lehtinen, K.-J., J. Tana, P. Karlsson, C. Engström, K. Mattsson, S. Hemming, J. Hemming, and A-L. Fugleberg. 1992. Chemical characterization and effects in mesocosms of effluents from bleached kraft pulp production. *Publications of the Water and Environment Administration — Series A 105*, p. 146. ISBN 951-47-6321-1.

Lehtinen. K.-J., J. Tana, K. Mattsson, J. Härdig, P. Karlsson, C. Grotell, S. Hemming, C. Engström, and J. Hemming. 1993. Ecological impact of pulp mill effluents. *Publications of the Water and Environment Administration — Series A 133*. p. 139. ISBN 951-47-7137-0.

Lehtinen, K-J., K. Mattsson, C. Grotell, and J. Tana. 1994. Miljökunskap för processingenjörer: Ekologiska aspekter på ekotoxikologi. Technical review prepared for the process engineers. (Manuscript in Swedish).

Lindén, O., A. Rosemarin, A. Lindskog, C. Höglund, and S. Johansson. 1987. Effects of oil and oil dispersant on an enclosed marine ecosystem. *Environmental Science and Technology*. 21:374–382.

Lindström-Seppä, P. 1990. Biotransformation in fish. Monitoring inland water pollution caused by pulp and paper mill effluents. PhD thesis. *Publications of the University of Kuopio*. 8/1990 Kuopio, Finland.

Munkittrick, K.R., G.J. Van der Kraak, M.E. McMaster, C.B. Portt, M.R. van den Heuvel, and M.R. Servos. 1994. Survey of receiving-water environmental impacts associated with discharges from pulp mills. 2. Gonad size, hepatic EROD activity and plasma sex steroid levels in white sucker. *Environmental Toxicology and Chemistry*. 13(7):1089–1101.

Odum, E.P. 1985. Trends expected in stressed ecosystems. *Bioscience*. 35:419–422.

Oikari, A., B. Holmbom, E. Ånäs, M. Miilunpalo, G. Kruzynski, and M. Castrén. 1985. Ecotoxicological aspects of pulp and paper mill effluents discharged in water, and toxicant residues and physiological effects in caged fish (Salmo gairdneri). *Aquatic Toxicology*. 6:219–239.

Oikari, A. and T. Kunnamo-Ojala. 1987. Tracing of xenobiotic contamination in water with the aid of fish bile metabolites: A field study with caged rainbow trout (*Salmo gairdneri*). *Aquatic Toxicology*. 9:327–341.

Owens, J.W. 1991. The hazard assessment of pulp and paper effluents in the aquatic environment: A review. *Environmental Toxicology and Chemistry*. 10:1511–1540.

Rosemarin, A., J. Mattsson, K-J. Lehtinen, M. Notini, and E. Nylén. 1986. Effects of pulp mill chlorate on Fucus vesiculosus. A summary of projects. *Ophelia* 4:219–224.

Rosemarin, A., K.-J. Lehtinen, and M. Notini. 1990. Effects of treated and untreated softwood pulp mill effluents on Baltic Sea algae and invertebrates in model ecosystems. *Nordic Pulp and Paper Research Journal*. 2, pp. 83-82 and 103.

Rosemarin, A., K-J. Lehtinen, M. Notini, and J. Mattsson. 1994. Effects of pulp mill chlorate on Baltic Sea algae. *Environmental Pollution.* 85:3–13.

Sangfors, O., J. Tana, J. Härdig, and C. Grotell. 1994. Model ecosystem studies with pilot treated bleachery effluents from production of ECF and TCF pulp, in *Environmental Effects of ECF- and TCF-Bleached Pulp Mill Effluents,* Verta, M., Ed., Publications of the Water and Environment Administration — Series A 189. ISBN 951-47-9700-0 (in Finnish).

Södergren, A. 1988. Biological effects of bleached pulp mill effluents. *Final Report from Environment/ Cellulose, Project I. Report 3558, p. 134.* Swedish National Environmental Protection Board, Solna, Sweden.

Swanson, S.M., R. Schryer, R. Shelast, P.J. Kloepper-Sams, and J.W. Owens. 1994. Exposure of fish to biologically treated bleached-kraft mill effluent. 3. Fish habitat and population assessment. *Environmental Toxicology and Chemistry.* 13(9):1497–1507.

Tana, J., A. Rosemarin, K-J. Lehtinen, J. Härdig, O. Grahn, and L. Landner. 1994. Assessing impacts on baltic coastal ecosystem with mesocosm and fish biomarker tests. A comparison of new and old wood pulp bleaching technologies. *The Science of the Total Environment.* 145:213–234.

Verta, M., J. Ahtianinen, T. Nakari, A. Langi, and E. Talka. 1996. The effects of waste constituents on the toxicity of TCF and ECF pulp bleaching effluents, in *Environmental Fate and Effects of Pulp and Paper Mill Effluents.* Servos, M.R., K.M. Munkittrick, J.H. Carey, and G.J. van der Kraak, Eds., St. Lucie Press, Delray Beach, FL. pp. 41–51.

Chapter 12

Summary of Progress of the United States Pulp and Paper Industry in Reducing the TCDD and TCDF Content of Effluents, Pulps, and Sludges

R.P. Fisher and W.J. Gillespie

INTRODUCTION

Prior to the mid- to late 1980s, much scientific thought supported the position that the presence of 2,3,7,8-tetrachlorodibenzo-*p*-dioxin (TCDD) and 2,3,7,8-tetrachlorodibenzofuran (TCDF) in environmental matrices was a result of the formation of these substances in natural and anthropogenic combustion processes or inadvertently in certain chemical manufacturing processes. The presence of these substances was reported, however, in results of analyses of fish tissue samples collected downstream of bleached pulp mills in the National Bioaccumulation Study of 1985–1988 conducted by the United States Environmental Protection Agency (U.S. EPA; "the Agency"). This reporting prompted the conducting of three landmark studies of the association of pulp bleaching with TCDD and TCDF (TCDD/F) releases:

1. 1986–88 — The EPA/Industry Five Mill Screening Study, undertaken by the paper industry and the Agency to validate the association of pulp and paper manufacture with TCDD/F releases[1];
2. 1988–90 — The EPA/Industry 104 Mill Study, undertaken by the industry and the Agency to determine the magnitude of TCDD/F releases at all 104 bleaching mills in the United States[2];
3. 1988–90 — The Industry Intensive 25 Mill Study, undertaken by the pulp and paper industry to determine relationships of pulping and bleaching operating variables to the magnitudes and fates of TCDD/F formed in bleaching processes.[3]

Significant findings of these studies included the following:

1. TCDD and TCDF were the dioxin and furan congeners that were important in export vectors, constituting from 97 to 100% of the dioxin toxicity equivalents (TEQs, determined by use of toxicity equivalence factors, TEFs[4]) contained in pulp, sludge, and effluent.
2. Pulp bleaching was the source of the detectable TCDD/F.
3. The first stage or chlorination stage of bleaching was responsible for most of the TCDD/F formed.
4. The extent of molecular chlorine use correlated positively with TCDD/F levels.
5. Greater amounts of TCDD/F were formed from kraft pulp bleaching than from sulfite pulp bleaching.
6. Greater amounts of TCDD/F were formed from softwood pulp bleaching than from hardwood pulp bleaching.
7. Unchlorinated precursors to TCDF were present in unbleached pulp in amounts which could account for the TCDF formed and released via all three vectors, but measured precursors to TCDD were not present in high enough concentrations to account for the TCDD formed. (The mechanism by which TCDD is formed in pulp bleaching is still to be elucidated.)
8. Considering TCDD/F formation, there was no observable benefit from oxygen delignification with high substitution of chlorine dioxide for chlorine, as opposed to high substitution of chlorine dioxide for chlorine without oxygen delignification.

Subsequent studies by a number of researchers have concluded that TCDD/F and other congeners and isomers are ubiquitous in the environment and result from production by numerous sources not associated with pulp bleaching. Nonetheless, in the late 1980s and early 1990s, the pulp bleaching industry initiated pollution prevention process modifications to control, reduce, and eliminate the formation of TCDD/F. These modifications were based upon knowledge gained in the three studies reported above, plus other studies, and resulted in significant reductions in TCDD/F release in export vectors.

Information on the trends of these reductions has been compiled by the National Council of the Paper Industry for Air and Stream Improvement (NCASI), beginning with the original 104 mill study of 1988, and incorporating data on the status of releases in 1989, 1990, 1991, 1992, and 1993. An activity to collect data current as of the end of 1994 is ongoing.

The purpose of this summary is to compare TCDD/F data from the original 104 mill study (1986–1987 data) with more current TCDD/F release information.

TRENDS IN TCDD AND TCDF CONCENTRATIONS IN EFFLUENTS, PULPS, AND SLUDGES

Methodology

The base data collection activity for the trends information reported here is the U.S. EPA/paper industry cooperative study of 1988.[2] Sample collection and analyses were conducted according to strict protocols, with sample collection and data analyses overseen and reviewed by NCASI and U.S. EPA staff. At each of the 104 mills, samples of effluents, pulps, and wastewater treatment system sludges were composited over five days, generally to include eight aliquots of sample per day. Analyses for TCDD/F were performed by one of two laboratories using either EPA Method 1613,[5] or the method reported in NCASI Technical Bulletin No. 551.[6]

Following completion of the U.S. EPA/paper industry cooperative study of 1988, NCASI requested that representatives of bleached pulp mills provide current data on TCDD/F concentrations in effluents, pulps, and sludges, as of the ends of the years 1989, 1990, and 1991. The sample collection and analysis methodology employed for the generation of these later data were similar to those requested by U.S. EPA in the earlier study, differing primarily in that five-day composite samples were not collected. During this time period many bleached mills were collecting these data, for process improvement purposes, for National Pollutant Discharge Elimination System (NPDES) discharge monitoring report purposes, or both.

In anticipation of rulemaking and reevaluation activities, the U.S. EPA requested that the industry provide, via NCASI, current information on TCDD/F concentrations in effluents, pulps, and sludges, representative of conditions at the end of 1992. These data were collected via methodology as close as practicable to the methodology employed in the 1988 study, including the collection and analysis of five-day composite samples, and the simultaneous collection of effluent, sludge, and pulp samples. Results were provided to the Agency via the 1992 Industry-Wide Dioxin Profile.[7]

NCASI also requested that the industry provide updated TCDD/F release vector concentration information representative of conditions at the ends of years 1993 and 1994. Data from the 1993 update[8] (summarizing 1993 data) are available and are included here.

Results and Conclusions

Results are presented in Tables 12.1–12.2 and Figures 12.1–12.3. The tables and figures are based upon the 104 mill study data,[2] 1989, 1990, and 1991 NCASI data solicitation information,[7] the 1992 update for the EPA,[7] and 1993 data from an NCASI information solicitation.[8] TEQ values were computed by summing the TCDD release value and one-tenth of the TCDF release value.

Several of the data of Table 12.2 indicate the percentage of concentration data points below limits of detection, or below EPA's minimum level (ML) for Method 1613 of 10 ppq. Because of the tenfold lower TEQ for TCDF, data also show the percentage of TCDF concentration data points below ten times the ML.

Table 12.1. Summary of Progress in Reducing Releases of Chlorinated Dioxins and Furans.

	Total Releases from U.S. Pulp Bleaching, mg/day								
	Baseline 104 Mill Study Data[2]			1990 Data[7]			1993 Data[8]		
	TCDD	TCDF	TEQ	TCDD	TCDF	TEQ	TCDD	TCDF	TEQ
Effluent	551	4260	977	160	710	231	55	218	76
Pulp	718	6660	1384	276	1287	405	64	302	94
Sludge	577	3615	939	208	834	291	68	327	100
Sum	1846	14535	3300	644	2831	927	187	846	271

Table 12.2. Trends in Pulp Bleaching Export Vector Concentration Data.

	Baseline 104 Mill Study Data[2]					
	Effluent		Sludge		Pulp	
	TCDD ppq	TCDF ppq	TCDD ppt	TCDF ppt	TCDD ppt	TCDF ppt
Maximum	640	8400	1390	17100	116	2620
90th %	110	900	161	1300	22	157
Mean	51	443	72	607	8.7	91.5
Median	21	68	16	78	4.4	18
(% < ML[2])	38				19	
(% < 10×ML)		63		22		39
1993 Data, 97 Mills Responding[8]						
	Effluent		Sludge		Pulp	
	TCDD ppq	TCDF ppq	TCDD ppt	TCDF ppt	TCDD ppt	TCDF ppt
Maximum	80	510	133	735	10	323
90th %	6	33	31	95	2	9
Mean	5	23	11	49	0.8	5
Median	ND(5)	3	1	8	ND(0.3)	ND(0.1)
(% < ML)	93		45		81	
(% < 10×ML)		97		53		91

Refer to text.

These data document the significant improvements in the quality of effluents, pulps, and sludges which have been brought about by major research efforts and in many cases by significant capital expenditures by the industry in developing and implementing process modifications. The following conclusions are important:

1. Considering the period from 1988 through 1993, the industry has achieved an overall reduction in TCDD/F TEQs from all vectors of 92%.
2. Current TCDD levels in effluents are generally less than the minimum level of 10 ppq of Method 1613. (As of 1993, 93% of bleached mills reported less than 10 ppq.)
3. Demonstration of significant further progress in reducing TCDD/F levels may be difficult, given the detection limits of the methods employed for analyses, and the fact that most of the industry's effluents are at or below these values now.

REFERENCES

1. U.S. Environmental Protection Agency/Paper Industry Cooperative Dioxin Screening Study, NCASI Technical Bulletin No. 545, NCASI, P.O. Box 13318, Research Triangle Park, NC 27709, May, 1988.
2. U.S. EPA/Paper Industry Cooperative Dioxin Study: The 104 Mill Study, NCASI Technical Bulletin No. 590, May, 1990.

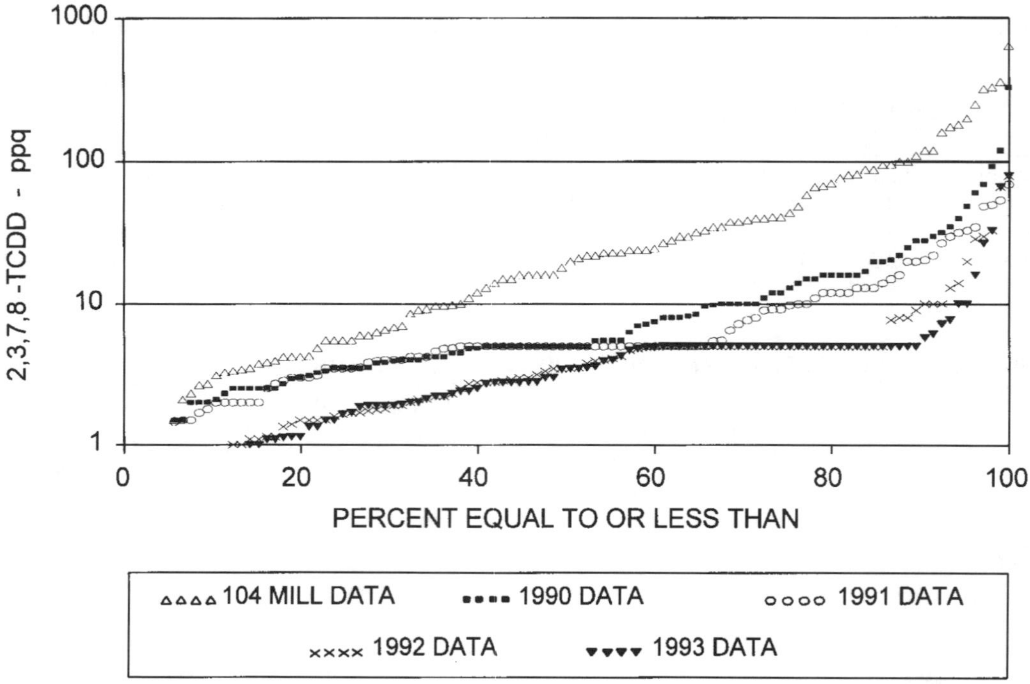

Figure 12.1. Log cumulative frequency distribution — effluent data — 2,3,7,8-TCDD.

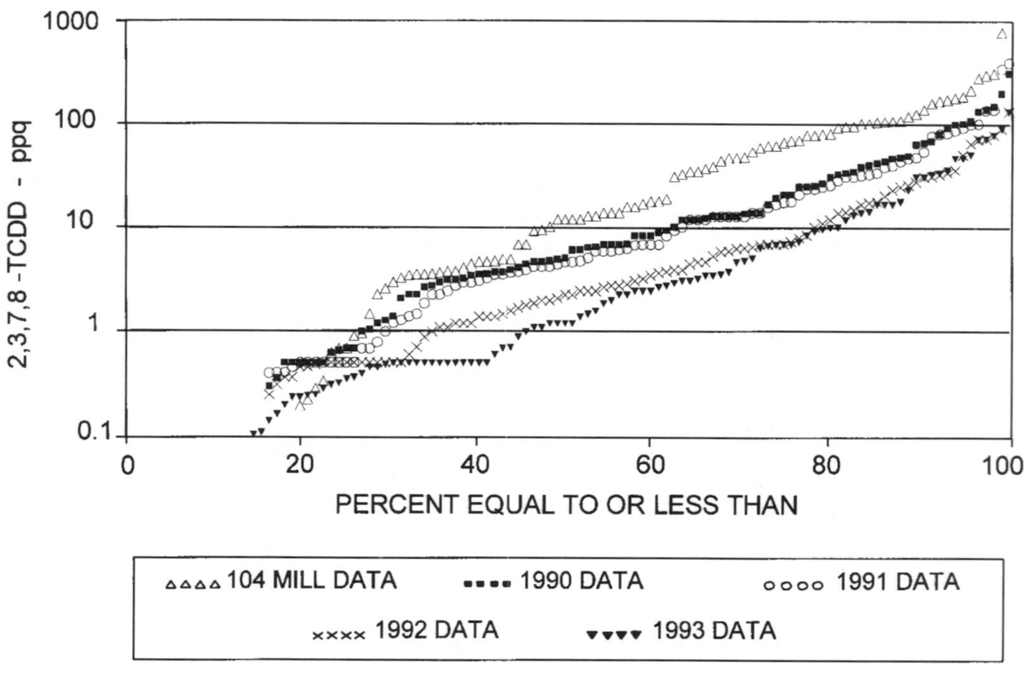

Figure 12.2. Log cumulative frequency distribution — sludge data — 2,3,7,8-TCDD.

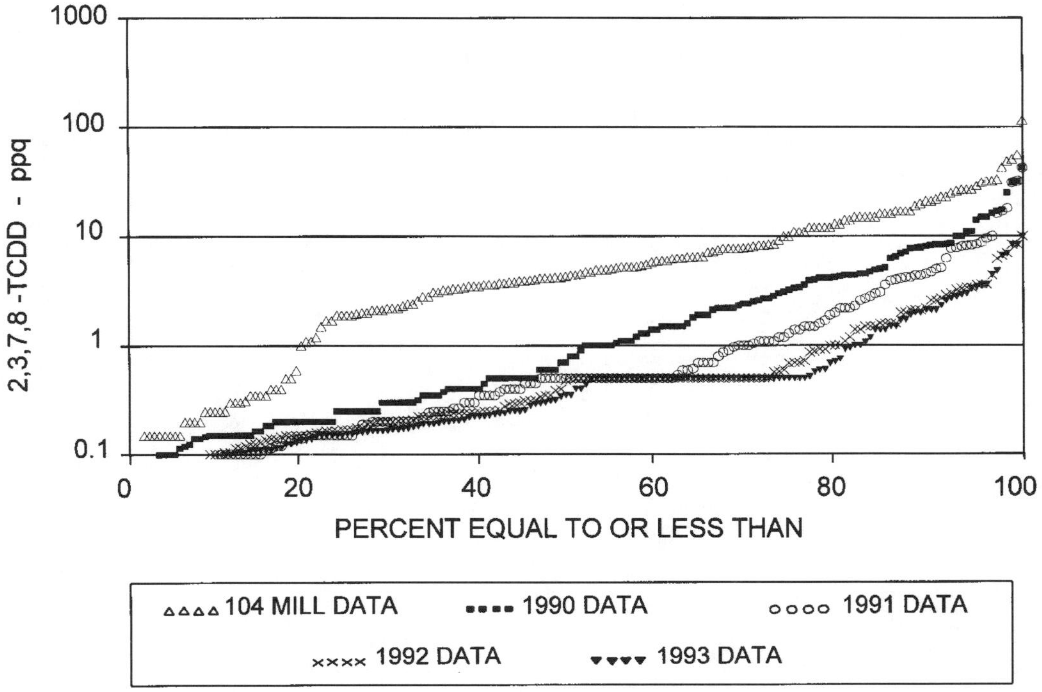

Figure 12.3. Log cumulative frequency distribution — pulp data — 2,3,7,8-TCDD.

3. An Intensive Study of the Formation and Distribution of 2,3,7,8-TCDD and 2,3,7,8-TCDF During the Bleaching of Kraft Pulps, NCASI Technical Bulletin No. 591, May, 1990.
4. Bellin, J.S. Interim Procedures for Estimating Risks Associated with Exposures to Mixtures of Chlorinated Dibenzo-*p*-dioxins and Dibenzofurans (CDDs and CDFs), U.S. EPA Risk Assessment Forum, U.S. EPA, Washington, DC, October 1986.
5. Method 1613: Tetra- through Octa-Chlorinated Dioxins and Furans by Isotope Dilution HRGC/HRMS, U.S. EPA, Office of Water Regulations and Standards, Industrial Technology Division, Revision A, April, 1990.
6. NCASI Procedures for the Preparation and Isomer Specific Analysis of Pulp and Paper Industry Samples for 2,3,7,8-TCDD and 2,3,7,8-TCDF, NCASI Technical Bulletin No. 551, May, 1989.
7. Summary of Data Reflective of Pulp and Paper Industry Progress in Reducing the TCDD/TCDF Content of Effluents, Pulps, and Wastewater Treatment Sludges, NCASI Special Report No. 93-08, August, 1993.
8. Progress in Reducing the TCDD/TCDF Content of Effluents, Pulps, and Wastewater Treatment Sludges, NCASI Special Report No. 94-08, July, 1994.

Part 3

Public Perceptions

Chapter 13

Chlorine: The Issue, the Reality, and the Solution

C.T. Howlett

Good afternoon. I'm very pleased to be here today to talk about the public policy debate surrounding chlorine — and in particular, the perceptions and misperceptions that are driving the issue in Washington, DC and state capitals around the country. Most importantly, I'd like to spend some time discussing the role that you, as scientists, can help play in setting the record straight. Mainly because chemists, more than any other group, understand completely the critical role chlorine plays in industrial manufacturing and chemical research.

I only wish the facts about chlorine chemistry would receive as much attention as the recent discovery about how to make fat mice thin. I sit in amazement at how excited Americans get about the *possibility* of a cure for obesity, but at the same time remain virtually ignorant about the innumerable, positive *documented contributions* that chlorine has made to our health and everyday lives.

Before I launch into chlorine's benefits, let me take a minute to tell you a little about the Chlorine Chemistry Council. The C-3, as we call it, is a council of the Chemical Manufacturers Association, created in 1993 to participate in the growing debate about chlorine chemistry. In this role, C-3 identifies and responds to health and environmental issues that involve chlorine. It also develops and implements programs that promote chlorine chemistry's benefits and the practice and understanding of responsible stewardship for chlorine chemistry and chlorine-based products. Headquartered in Rosslyn, Virginia, C-3's membership includes all U.S. leading chlorine producers.

Now let me turn to my official topic: "Chlorine: The Issue, The Reality, and The Solution." I won't be insulted if you find it odd that someone from Washington, DC is coming to talk to you about reality. Our nation's capital is not exactly a place where reality rules.

However, perception is reality in Washington, and it is perception, sprinkled with a strong dose of politics, that is driving the chlorine issue today.

And perhaps that is what is so vexing about the whole situation. Rather than being guided by what we *know* — the scientific facts — about chlorine and chlorinated compounds and the many benefits they have brought to society, the debate is revolving around what we don't know and the fears that spring from a lack of understanding and rush to judgment.

Environmental activists, such as Greenpeace, the Sierra Club, and others, are calling for a ban on all industrial uses of chlorine. They base their justification for this ban on alleged health and environmental issues that have been raised about a few chlorinated compounds, such as dioxins and PCBs. With hair-raising rhetoric, they tag chlorine and chlorinated compounds as the "root of all evil," responsible for a broad range of adverse health effects from cancer, to reproductive, to hormonal, to immune system problems.

No one — including myself and C-3 — would suggest for a moment that these are not serious health and environmental concerns that need addressing. We know that we don't have all the answers. We know that more information is needed. But we know the harm a rush to judgment can do, particularly when one of the fundamental chemical building blocks of modern life that has benefited society in thousands of ways and saved countless numbers of lives is at stake. That is why C-3 and our members are working to address these questions by sponsoring scientific research to fill in the information gaps and ensure that the chlorine industry follows the principles of Responsible Care in the manufacture, use, and disposal of the products of chlorine chemistry.

Parenthetically, I might add this concern and commitment is having results. Releases of chlorine and chlorine compounds in the Great Lakes states, one of the areas of greatest environmental concern, were down significantly between 1987 and 1993.

In that region, total air releases are down 56.2%, total water releases are down more than 81.6%, and total releases on land have dropped nearly 81.8%, for an overall reduction of chlorine releases of 56.5%.

Despite this progress and the fact that the use of chlorine and chlorine-related compounds as disinfectants and pharmaceuticals has *saved* more lives than any other chemical in the history of the human race, the war against chlorine continues to escalate.

Chlorine is "public enemy number one" to many environmental activists these days, which just goes to prove the lingering wisdom of Bertrand Russell's observation that, "The degree of one's emotion varies inversely with one's knowledge of the facts — the less you know, the hotter you get."

The problem with the charges being levied by chlorine's opponents is that while they might grab attention and make eye-catching headlines in the media, the scientific data to support a chlorine ban or restrictions on its uses are sketchy or nonexistent.

To understand how public policy is being made on chlorine, you need to understand how public perceptions about it are being fostered, and how these perceptions — or should I say, misperceptions — create an emotional environment that precludes responsible public policy making.

Let's start with a quick bit of history:

Public concerns about chlorine and chlorinated chemicals were first aroused by Rachel Carson in her 1962 book *Silent Spring,* which raised concerns about the effects of DDT and other chlorinated pesticides on human health, wildlife, and the environment. Attacks on chlorine and chlorinated chemicals have continued since then and intensified in the late 1980s with the launch of Greenpeace's "Chlorine Free" campaign.

In 1992, the International Joint Commission on the Great Lakes jumped on the bandwagon by recommending that the use of chlorine and chlorinated feedstocks be phased out in an effort to improve the Great Lakes ecosystem. Neither the United States nor Canada has accepted this proposal.

In February 1994, the Environmental Protection Agency announced its intention to "develop a national strategy for substituting, reducing or prohibiting the use of chlorine and chlorinated compounds" within the next three years. Rather than look at each chemical individually, the agency said it would look at the class of chlorine and chlorine chemicals comprehensively. This proposal died in the 103rd Congress.

Greenpeace would have you believe that chlorine is "absolute death." And by using sensationalist labels such as this, the organization is working to generate a knee-jerk reaction of fear of chlorine in the public. This rhetoric is then echoed in the media in stories with such headlines as, "Chlorine Attacked as a Health Hazard" (*The Times-Picayune*, February 21, 1994).

Lost in these stories, however, are the thousands of ways that chlorine and the products of chlorine chemistry have benefited society for more than 150 years. Moving the debate about chlorine from Greenpeace's slanderous characterization of "Absolute Death" to the scientific reality of its "Absolute Necessity" will therefore require two things: a strong dose of common sense, and a focus on scientific fact.

In other words, the debate over chlorine needs to get back to reality. And a realistic approach to public policy making must start with common sense — the voice that urges the greatest good for the greatest number; the voice that urges a solution that weaves us together as a society, not a solution that fragments us further.

Common sense about chlorine tells us a very different story than Greenpeace and others would have us believe.

Common sense tells us that chlorine is one of the eight fundamental building blocks for modern society. So important is chlorine to life that virtually every product manufactured today benefits in some way from chlorine. From plumbing pipes to X-ray film, medicines to seat belts, household bleach to rocket fuel, blood bags to computers, tap water to bullet-resistant vests, the products of chlorine chemistry have made life healthier, safer, more enjoyable, and more productive.

Common sense tells us that chlorine has saved millions of lives. Before chlorine was introduced into the United States' water supply, diseases like cholera and typhoid fever killed 25,000 Americans a year. Today — thanks to chlorination — these diseases have been virtually wiped out in this country. Just think of how many of the 25,000 people who die from waterborne diseases every day in developing nations could be saved, if they too had access to chlorinated water.

Common sense tells us that chlorine is a cornerstone of modern medicine. Eighty-five percent of all pharmaceuticals are made with chlorine, including many of the most commonly prescribed life-saving antibiotics, cancer treatment drugs, AIDS medications, heart and high blood pressure treatments, arthritis medications, cold and allergy medicines, and pain relievers like Tylenol. Without chlorine the war against AIDS would be even more difficult. Not only is it used to kill the AIDS virus in pediatric wards and in needle disinfection programs throughout the country, but it is also important in the diagnosis and treatment of the disease.

In addition, about 25% of all medial equipment is made from chlorine-based plastics, including I.V. and blood bags, X-ray and mammography film, electronic instruments, prosthetics, and coolers for organ transplants.

Common sense tells us that the economic costs of a chlorine ban in the United States would be tremendous. Up to 60% of commercially practiced chemistry depends on chlorine. If chlorine were banned, substitutes would have to be found for the products or processes now based on chlorine. New plants and equipment would have to be built to produce these substitutes. The entire process would require massive capital investments, mean job losses in plant communities, cause employment shifts and resulting economic disruption, and a ripple effect throughout the economy as indirect costs and hidden effects kick in.

According to a 1993 study by the economic consulting firm Charles River Associates, the net cost of these substitutes would be staggering for U.S. consumers — more than $90 billion a year, or $1,440 for a family of four.

Finally, common sense tells us that experience is the best guide to the future. Chlorine is a chemical that has been used for 150 years, studied for decades, is understood, and is routinely transported and handled safely. A chlorine ban would force reliance on a number of as yet unspecified substitutes, with unknown health, safety, and quality effects that could appear with time.

While common sense may tell you — as scientists — that government could not possibly outlaw the use of one of the most abundant elements in nature, especially one that provides so many important benefits to modern day living, common sense unfortunately often has nothing to do with public policy.

That is why the second element missing from the chlorine debate — the focus on scientific fact — is so important, and why your input as scientists is essential to a positive, productive outcome to the debate.

What does science tell us about chlorine and chlorinated compounds?

Three very important facts:

> First, that organochlorines are *not* a homogeneous family and do not all possess similar properties of toxicity, persistence or bioaccumulation. Rather, there is a broad spectrum of chlorinated compounds with a wide range of physical and chemical properties. At one end are compounds such as dioxins, like 2,3,7,8-TCDD, PCBs, and DDT that are persistent, toxic, and bioaccumulative. However, at the other end of the spectrum are many chlorinated organics that are used in pharmaceuticals and find applications as additives in many ordinary products, from foodstuffs to shaving cream.
>
> Second, because all chlorine compounds are not alike, it is inappropriate to lump all chlorinated products together as a class.
>
> And third, that the concept of banning an entire class of chemicals — let alone a basic element — to address the problems associated with a select number of compounds is not practicable, logical, or scientifically supportable.

The current debate over dioxin's health effects, one of the engines driving the chlorine issue, provides perhaps the best example of how public misperceptions can combine with uncertain science to pose a real challenge to common sense and responsible public policy making.

The term "dioxin" refers to a family of chemicals consisting of 75 dioxins and 135 related compounds called furans. Chemically, dioxins and furans are alike in that they all contain the elements carbon, chlorine, hydrogen, and oxygen. However, they differ from each other primarily by the location and number of chlorine atoms in the molecule.

No one makes dioxin on purpose. Rather, dioxin occurs as a natural — though unwanted — by-product of combustion. Automobiles, forest fires, incinerators, wood stoves, and even barbecues can be sources of dioxins and furans.

In September of 1994, the Environmental Protection Agency released a draft report of its reassessment of the potential human health risks arising from exposure to dioxin and related compounds such as furans and PCBs. Three years in the making, the reassessment evaluated possible health effects from exposure to dioxin, including cancer, reproductive, developmental, immune system, enzyme induction, hormone, and endocrine effects.

The EPA's conclusion? The agency said that dioxin is a "probable human carcinogen" based on the positive TCDD animal bioassay, and added that the class of dioxin, furan, and PCB compounds are capable of causing a broad spectrum of effects "at or near levels to which people in the general population are exposed."

Working with independent scientists, the EPA prepared and peer reviewed the reassessment's first eight chapters on the state of dioxin science. Unfortunately, the EPA drew its *own* conclusions in the ninth or final chapter, making default assumptions about dioxin's risk and, more importantly, the basis on which to determine science policy. In late 1994, many of these same independent scientists signed a letter to the editor of *Science* magazine urging the EPA to clearly distinguish regulatory policy from matters of scientific fact, noting that "otherwise, the press and public will surely misinterpret the hypothetical risk presented in the reassessment as real."

On May 16, 1995, the EPA's Scientific Advisory Board or SAB echoed the same concerns and asked the agency to rewrite its final chapter calling it "heavily dependent upon many unproved assumptions and hypotheses."

To its credit, the EPA is responding positively to the SAB, reopening the process and appearing determined to get it right the next time around. There is much from which to draw — the process has been an open one, with thousands of comments submitted.

We believe that once the EPA has revisited the weight of scientific evidence, it will agree that any health risk characterization about dioxins must include the following:

> There is simply no scientific evidence to show that cancers or other health effects in the general population are due to background levels of dioxin exposures.
>
> In addition, although dioxins have been shown to cause cancer in laboratory animals, there is no scientific consensus, based on the current 25 years of human evidence, to suggest that dioxins are a human carcinogen — even in groups of people exposed to very high levels of the compound through industrial accidents or workplace exposures.

Unfortunately, the public's perception about dioxin is something entirely different. To the public, dioxin is the most toxic chemical known to mankind. This belief persists despite a preponderance of scientific evidence that dioxin does not cause adverse human health effects other than chloracne, a condition that results only from extremely high levels of dioxin exposure.

Rather than advancing public knowledge about dioxin — and perhaps, calming some fears — the EPA's draft reassessment failed to differentiate its regulatory policy on dioxin from matters of scientific fact.

The pulp and paper industry knows the tremendous costs that can result from erroneous public perceptions about dioxins. Since the mid-1980s, when the first alarms sounded about high levels of dioxin in fish samples and river sediment downstream from pulp and paper mills and trace levels of dioxin in pulp products used to make a variety of consumer products, the industry has voluntarily spent more than $1 billion to eliminate the inadvertent presence of dioxin in its products.

While there was general agreement that no health hazard existed, the industry reacted very quickly with voluntary efforts to address public concerns. By changing its bleaching process from one that used elemental chlorine to one using chlorine dioxide, the pulp and paper industry has virtually eliminated dioxin in mill wastewater and pulp products.

As a result of this change, the U.S. EPA has recorded significant improvements in aquatic ecosystems. According to the agency, dioxin advisories on water bodies downstream of U.S. pulp mills remain on only 1.5% of the nation's 1,275 water bodies covered by fish consumption advisories.

What's more, the EPA predicts that all remaining dioxin advisories should be lifted following completion of the industry's conversion to elemental chlorine–free (ECF) bleaching.

Nevertheless, public concerns about dioxin in paper continue to this day. Greenpeace and others now want paper manufacturers to abandon chlorine dioxide bleaching and switch to totally chlorine–free (TCF) paper bleaching.

The costs of this switch would be enormous, requiring an investment of about $9.4 billion and the closing of about half the mills that could not be economically retrofitted.

Yet the environmental benefits of such a switch are highly uncertain and would not be without trade-offs of their own. For example:

Substitutes for chlorine dioxide bleaching will have undesirable by-products of their own.
TCF reduces paper's recyclability by damaging cellulose fibers.
Recent evidence indicates that TCF pulp manufacture increases wood consumption by about 10%. This means that if all North American pulp and paper mills were TCF, the annual increase in forest resources required would equal about *100 million mature trees.*

Tremendous economic costs for questionable environmental benefits. Is that a formula for a common sense solution? I think not. Fortunately, Environment Canada, the European Union, and, yes, even the U.S. EPA agree that industry's use of ECF represents the best available technology for dioxin reduction.

There is, however, another important lesson to be learned from the pulp and paper industry's experience with dioxin. And that is that chlorine chemistry can be part of the answer to improving our environment. Elemental chlorine–free bleaching represents the best of both worlds — and strikes a blow for common sense — by significantly improving the waters below paper mills, while still providing the highest quality recyclable white paper and saving precious forest resources.

Though society tends to quickly forget, chlorine chemistry over the years has provided other solutions that have allowed us to minimize or eliminate risks. For example, DDT, although now banned, helped rid malaria from much of the world decades ago. Today, two other products of chlorine chemistry — polyvinyl chloride (PVC) plumbing pipes and house paints based on titanium dioxide — are helping to reduce the public's exposure to lead.

Ironically, it may be the study of dioxin and its alleged health effects that hold the key to unlocking new pathways for treatment of breast cancer. This exciting new work is being carried out in several medical laboratories across the country, including the University of Massachusetts, Northwestern University and Texas A&M. Working on the identification and function of gene targets for the dioxin receptor in breast cells, researchers believe that depending on the level of dioxin exposure, chemoprotection against breast cancer may be observed. Let me be perfectly clear that no one is suggesting that there is a recommended "dose of dioxin" that is good for you. What is important about this work is that it further underscores the point that there is so much we still do not know in terms of dioxin's health effects. According to this new research, dioxin at certain exposure levels — even 2,3,7,8-TCDD — appears to provide a form of chemoprotection, similar to that which has been attributed to components in broccoli and other vegetables in fighting cancer. Interestingly, this theory is supported by information contained in the original rodent bioassay of the now famous Kociba dioxin study, as well as initial findings in the residents of Seveso, Italy, where in 1976 an accidental explosion in a chemical plant caused the highest documented exposure to TCDD. According to the Seveso study, published in the scientific journal *Epidemiology,* breast cancer incidence among the female residents was below expectations in most contaminated zones in the 10 years following the accident.

So what does this all mean in terms of chlorine? It means — to borrow an old expression — "we shouldn't throw the baby out with the bath water." In fact, common sense, scientific fact, and experience tell us that banning chlorine is not the answer. One only needs to look to the cholera epidemic in Peru that took thousands of lives in the late 1980s after the country stopped chlorinating its water out of concern over exposures to trihalomethanes (THMs) to see that in some instances, the real dangers we face without the benefits of chlorine chemistry far outweigh any hypothetical risks of chlorinated compounds.

This tragedy also illustrates the harm that can come when the public misunderstands EPA's policies. In a 1991 federal agency debriefing on the cholera outbreak, the concern was expressed that many Latin American countries had misunderstood the EPA's rationale for maximum contaminant levels of THMs in drinking water, and that widespread publicity and the large number of scientific articles about the potential health effects had caused many communities in Latin America to abandon chlorination. More importantly, new science on THMs says that drinking water contamination is insignificant and current exposure standards provide the public with as much as a 1,000-fold margin of safety.

So where does the solution to the debate lie? If the chlorine issue is to reach a common-sensical, scientifically sound resolution, we need scientists, such as yourselves, to speak out as advocates for chlorine chemistry.

Only scientific fact can cool the "hot emotion" of the debate. Richard Carpenter, a chemist and environmental consultant, came to the same conclusion in a recent article entitled "Environmental Science Uncertainties." Published in the journal *The Environmental Professional,* he urged scientists "to take extra efforts to translate, interpret and transfer their knowledge in as an objective manner as is possible. This means dealing explicitly with the uncertainties that are inherent in our work. We need to communicate clearly and patiently what we do know, don't know, could know and should know."

Each of you, because of your expertise, knowledge, and understanding of the chemical world, has the ability to credibly convey the overwhelming societal benefits that spring from chlorine chemistry and the importance in using scientific fact as the basis for any decisions regarding its future.

To shift the tide of public perception, we need your help communicating some clear and simple messages:

The multitude of societal benefits that spring from chlorine chemistry far outweigh any theoretical or hypothetical risks from chlorine's uses.

All chlorinated compounds are not equally hazardous and a broad-based ban of this class would be irresponsible and unscientific.

Regulatory decisions about chlorine and chlorinated compounds should be based on scientific fact and the weight of scientific evidence, not hypothesis and conjecture.

Substitutes for chlorine — if they can be found — may carry risks of their own.

The audiences for these messages are all around you — your elected representatives, federal and state regulatory officials, the media, and even opinion leaders in your own communities. You are a tremendous information resource for them on this issue; they will appreciate having the benefit of your knowledge and perspective. More importantly, they need it.

The issues surrounding chlorine and chlorine chemistry put our industry at the leading edge of the current discussion over how to balance continuing environmental progress and continued economic growth. This discussion is happening at a time during our political history when significant change is possible. We need to be smarter about how we tackle the environmental challenges facing us — and do it with common sense and a focus on the scientific facts.

I recently saw a bumper sticker with the saying, "Environment is one-tenth science and nine-tenths politics." And I thought — that is exactly the problem; that is exactly the opposite of how we should approach our common goal of environmental protection. With your help, I hope we will soon see the day when public policy decisions about our environment are nine-tenths science and only one-tenth politics. Thank you.

Chapter 14

Alternative Disinfectants for Drinking Water

S.D. Richardson, A.D. Thruston, Jr., T.W. Collette, T.V. Caughran,
K.S. Patterson, and B.W. Lykins, Jr.

ABSTRACT

Using a combination of spectral identification techniques — gas chromatography coupled with low- and high-resolution electron-impact mass spectrometry (GC/EI-MS), low- and high-resolution chemical ionization mass spectrometry (GC/CI-MS), and Fourier transform infrared spectroscopy (GC/FT-IR), we identified many drinking water disinfection by-products (DBPs) formed using alternative disinfectants. The alternative disinfectants included ozone, chlorine dioxide, chloramine, and titanium dioxide with ultraviolet irradiation. Many of these by-products are not present in any spectral libraries and are not on any regulatory list. Because some alternative disinfectants, such as ozone and TiO_2/UV, do not persist in water, full-scale drinking water treatment plants typically need to use a secondary disinfectant, such as chlorine or chloramine (in addition to the primary disinfectant) to maintain a disinfectant residual throughout the distribution system. Because of this, we also investigated the effect of secondary chlorine and chloramine treatment on the formation of DBPs. Secondary chlorination produced numerous chlorinated DBPs, among them some halomethanes. Chloramine also produced chlorinated DBPs, but they were fewer in number and lower in concentration than those produced with secondary chlorination.

INTRODUCTION

Because of concern over the presence of trihalomethanes (THMs) and other chlorinated compounds in chlorinated drinking water,[1] alternative disinfection methods are being explored. Those currently being investigated include ozone, chlorine dioxide, chloramine, and titanium dioxide/UV treatment. These alternative disinfectants are effective in killing harmful microorganisms in drinking water, and they do not produce appreciable levels of THMs. Although the work discussed in this chapter involves drinking water, it is also applicable to the paper industry, as some of these alternative disinfectants are also being used for the treatment and bleaching of wood pulp.

The U.S. Environmental Protection Agency's (EPA) National Risk Management Research Laboratory in Cincinnati, Ohio, and the EPA's National Exposure Research Laboratory in Athens, Georgia, are investigating these alternative disinfectants. TiO_2/UV treatment has not been used in the United States yet, but it may have potential for serving small treatment systems (less than 3300 people). Ozone, chlorine dioxide, and chloramine are currently the most popular alternative disinfectants being used. These three alternative disinfectants are already being used in cities around the country, an example being the city of Los Angeles, whose ozonation plant serves over 3 million people.[2] Although these disinfectants are currently in use, little is known about the chemical by-products that are produced. Because it is unknown whether these alternative disinfectants produce compounds as harmful or more harmful than those produced by chlorine, the EPA has set out to identify all potentially harmful by-products.

Because volatile organic by-products of these alternative disinfectants have been studied previously, the goal of our work was to identify the semivolatile organic DBPs produced. Rather than attempting to identify a few targeted by-products, our objective was to identify every compound that was detected in the sample extracts. Samples studied were produced using the following treatments: (1) ozone — treatment at a pilot plant in Jefferson Parish, Louisiana (that included the use of chlorine and chloramine as secondary disinfectants), and laboratory-scale ozonation reactions done in the presence of bromide; (2) chlorine dioxide — treatment done at a pilot plant in Evansville, Indiana (that included the use of chlorine as a secondary disinfectant); (3) TiO_2/UV treatment using a large laboratory-scale photoreactor (that included the use of chlorine as a secondary disinfectant).

Gas chromatography/mass spectrometry (GC/MS) is the common analysis tool for identifying compounds in environmental samples. Typically, GC/MS analyses use only one type of MS: low-resolution electron-impact (EI) mass spectrometry. This type of analysis is sufficient for regulated compounds that have been well-characterized, and whose spectra are in a library database. However, unknown pollutants, such as newly identified DBPs, are often impossible to identify from this type of data alone, as many of these compounds are not present in any library database. As a result, we use a combination of several mass spectrometry techniques, as well as infrared spectroscopy, to identify these unknown by-products.

EXPERIMENTAL

Treatment with Ozone

Ozonations were done on both a pilot plant and laboratory scale.[3,4] Pilot plant ozonations were performed with Mississippi River raw water (Jefferson Parish, Louisiana). There were two rounds of treatment: the first round occurred in January 1994 and the second in August 1994. The first round used ozone as the primary disinfectant followed by secondary disinfection with chlorine and chloramine. The second round included all of the sample types from the first round, plus samples ozonated in the presence of added bromide. The effect of bromide is of concern because it is present in the raw water of many coastal cities (due to saltwater intrusion) and also in the groundwater sources of inland cities that have a high mineral content. Bromide is rapidly oxidized by ozone to produce hypobromous acid,[5] which reacts with natural organic matter to form brominated DBPs. Typically, brominated compounds are more toxic than their chlorinated analogs.

In both study rounds, raw water was pumped to a full-scale water treatment plant, where it was clarified with diallyldimethyl ammonium chloride or dimethylamine-type cationic polymers. Following clarification, samples were taken before and after disinfection with ozone in the adjacent pilot facility. Ozone was generated from compressed, dry air using an electrically powered ozone generator with a maximum output capacity of 0.25 lb/d. The ozone contact chamber was 11 feet in height, with countercurrent operation consisting of water entering at the top of the contact chamber and ozone gas entering at the bottom. The water and ozone gas influent lines were oriented so that the influent water would be in contact with the ozone gas stream for approximately 30 min. Sufficient chlorine or chloramine then was added to achieve a residual ranging from 2 to 3 mg/L. The treated samples were collected following filtration through dual media (sand and anthracite). Samples that will be discussed in this chapter include: (1) raw water, (2) raw water + ozone, (3) raw water + NaBr + ozone, (4) raw water + ozone + chlorine + dual-media filtration, and (5) raw water + ozone + chloramine + dual-media filtration. Also, in order to compare these alternative disinfection technologies to chlorination, a separate treatment with chlorine was performed on the same raw water source.

Prior to the resin concentration (described below), 1 L of the treated water was removed for identification of polar by-products, which would not be extracted using the resins. Seven-hundred-and-fifty mL of this water was derivatized with *o*-(2,3,4,5,6-pentafluorobenzyl)hydroxylamine (PFBHA) and extracted with hexane, according to a method by Glaze and Weinberg.[6] This method allows low-molecular-weight aldehydes and ketones to be extracted and detected.

In addition to the raw water controls (with and without NaBr), four blanks were also analyzed: (1) ethyl acetate passed through the XAD resins and concentrated in the same manner as the treated samples; (2) deionized, distilled water passed through the XAD resins and concentrated; (3) deionized, distilled water treated with secondary chlorine and concentrated; and (4) deionized, distilled water treated with secondary chloramine and concentrated. The latter two blanks were analyzed to determine whether there were any artifacts due to reaction of residual disinfectants with the ethyl acetate or with resin impurities.

Laboratory-scale ozonations were performed on natural water collected from University Lake, Orange County, North Carolina. The ambient bromide level (0.02 mg/L) was enhanced to 8 mg/L by addition of sodium bromide. Ozonation of 2 L of this water was conducted in a semibatch, 70-L stainless steel reactor so that an absorbed ozone dose of 1.0 mg of O_3/mg of total organic carbon (TOC) was applied. The TOC of the water was 8.0 mg/L. After ozonation, active bromide was quenched with ammonium sulfate. The water was then extracted at neutral pH with 400 mL of methyl *tert*-butyl ether, and concentrated to 1 mL with a stream of nitrogen. Blanks were prepared by ozonating deionized, distilled water containing the same level of bromide, and extracting in the same manner as described above. Further details are provided elsewhere.[3,4]

Treatment with Chlorine Dioxide

At a pilot-scale drinking water treatment plant in Evansville, Indiana, raw river water (TOC=1.5 mg/L) was sampled before and after treatment with liquid and gaseous chlorine dioxide.[7] Chlorine dioxide residuals were 1.21 and 1.14 mg/L, respectively. Both treatment streams were clarified by the addition of alum and were treated with lime to maintain pH at 8.2–8.5. A reducing agent, $FeCl_2$ (0.1–1 mg/L), was added to control chlorite and chlorine dioxide concentrations. Sufficient chlorine then was introduced to achieve a free chlorine residual ranging from 2 to 3 mg/L. The last step in the treatment train was a final pH adjustment (pH 8.2–8.5) and filtration through dual media (sand and anthracite). Samples collected and analyzed in this study included the following: (1) raw water, (2) raw water + liquid chlorine dioxide, (3) raw water + gaseous chlorine dioxide, (4) raw water + liquid chlorine dioxide + $FeCl_2$ + chlorine + dual-media filtration, (5) raw water + gaseous chlorine dioxide + $FeCl_2$ + chlorine + dual-media filtration, and (6) raw water + chlorine (for comparison). In addition to the raw water control, a blank was performed with ethyl acetate passed through the XAD resins and concentrated in the same manner as the treated samples.

Treatment with TiO_2

For this study, water from a stream (Mill Creek, Cincinnati, Ohio) was treated using ultrafiltration and then recirculated through a TiO_2/UV photoreactor. The reactor consisted of a bundle of stainless steel tubes, each lined with a TiO_2-coated mesh with a UV light at its center.[8] Ultrafiltered water was recirculated through this reactor until a 40% reduction in the feed water TOC was obtained. This reduction in TOC was used as an indicator that the water had been adequately disinfected. A portion of this treated water was subsequently chlorinated. Samples collected and analyzed included the following: (1) raw water, (2) ultrafiltered raw water, (3) ultrafiltered raw water + TiO_2/UV, (4) ultrafiltered raw water + TiO_2/UV+ chlorine, (5) ultrafiltered raw water + chlorine, and (6) raw water + chlorine. Before resin concentration of most of the treated samples (described below), an aliquot of treated water was removed for separate analysis of the haloacetic acids (chloro-, bromo-, dichloro-, trichloro-, bromochloro-, and dibromo-acetic acid).[9] Blanks were done as described earlier. An additional blank was run using distilled water recirculated through the TiO_2/UV unit and concentrated in the same manner as the treated samples.

Sample Concentration

All water samples, except the laboratory-scale ozonation samples, were concentrated by adsorption on Amberlite XAD resins. Details about the preparation of these resins can be found elsewhere.[7] The water samples were acidified to pH 2 by the addition of HCl, prior to passage through columns containing a combination of XAD-8 resin over XAD-2 resin. The column was eluted with ethyl acetate. Residual water was removed from the ethyl acetate eluents using separatory funnels to drain off the water layers. Sodium sulfate was then added as a drying agent. After removal of an aliquot for mutagenicity testing, the ethyl acetate eluants (equivalent to approximately 90 L of treated water for ozone and chlorine dioxide treatment, and approximately 50 L for TiO_2/UV treatment) were shipped on cold packs to the National Exposure Research Laboratory in Athens, Georgia. At the laboratory, the samples were concentrated to 1 mL by rotary evaporation.

TOC and TOX Determinations

Total organic carbon (TOC) concentrations were determined using the persulfate-ultraviolet oxidation method, and the adsorption-pyrolysis-titrimetric method was used for total organic halide (TOX) analyses.[10]

GC/MS Analysis

High resolution GC/EI-MS and GC/CI-MS analyses were performed on a VG 70-SEQ high-resolution hybrid mass spectrometer, equipped with a Hewlett Packard model 5890A gas chromatograph. The mass spectrometer was operated at an accelerating voltage of 8 kV and at a resolution of 10,000. High-resolution accurate masses obtained were within 2 millimass units of theoretical masses calculated for molecular ions and fragment ions. Perfluorokerosene (pfk) was used as the internal mass calibrant over the mass range 33 to 500. Low-resolution GC/MS analyses were performed at a resolution of approximately 1000 on either a Finnigan 4500 mass spectrometer or the VG 70-SEQ mass spectrometer. Positive chemical ionization experiments were accomplished by using methane or 2% ammonia in methane gases. Injections of 1 to 2 μL of the extract were introduced via a split/splitless injector onto a J&W Scientific DB-5 chromatographic column (30-m, 0.25 mm i.d., 0.25 μm film thickness). The GC temperature program consisted of an initial temperature of 35°C, which was held for 4 min, followed by an increase at a rate of 9°C/min to 285°C, which was held for 30 min. Transfer lines were held at 280°C, and the injection port was controlled at 250°C. Semiquantitative concentrations of DBPs were determined by using an internal standard, 2,2'-difluorobiphenyl, and by using an external standard containing EPA's base-neutral "priority pollutants."[11] This external standard includes polyaromatic hydrocarbons (PAHs), phenols, amines, hydrocarbons, chlorinated aromatic and aliphatic compounds, phthalates, and ethers. Concentrations reported for DBPs identified in the resin extracts should be considered approximate, as standards for many of the DBPs identified are not available.

GC/FT-IR Analysis

GC/FT-IR analyses were performed on a Hewlett Packard Model 5890 Series II GC interfaced to a Hewlett Packard Model 5965B infrared detector (IRD). Spectra were generated at 8 cm^{-1} resolution with a useful range of 4000 to 700 cm^{-1}. Injections of 2 μL of the extracts were introduced onto a Restek Rtx-5 column (30-m, 0.32 mm i.d., 0.5 m film thickness) with a heated on-column injector (280°C). The GC temperature program consisted of an initial temperature of 35°C, which was held for 4 min, followed by an increase at a rate of 9°C/min to 280°C, which was held for 30 min. Transfer lines were held at 280°C, and the light pipe was controlled at 280°C.

RESULTS AND DISCUSSION

Overall, many DBPs were identified, several of which have never been reported previously. Many of the compounds were not present in any spectral library (NIST or Wiley), and many of the ones that were in the libraries did not give conclusive library matches. For many of the compounds, little information was provided in the mass spectra, due to the absence of molecular ions. As a result, we used CI-MS frequently to generate molecular ions. Methane gas was a good reagent gas in most instances, but 2% ammonia in methane was necessary to obtain molecular ions for some compounds (especially useful for brominated alcohols[3] and haloacids). High resolution EI-MS was an indispensable tool for determining structures, as it provided the necessary empirical formula information for the molecular ion and fragments. It also helped to limit the number of possible structures for each unknown DBP. GC/FT-IR was very useful for compounds where there was a choice of structures (based on mass spectral information). IR was definitive for the choice of functional group. When available, standards were purchased to confirm difficult identifications and to determine a particular isomer precisely, when spectra were not conclusive. In this way, both GC retention times and spectra of the standards were matched with those of the unknowns. It should be noted that, although we attempted to identify every DBP detected in the samples, several DBPs were present at such low concentrations (less than 1 ppt) that there was not sufficient spectral information to enable their identification.

Treatment with Ozone (Pilot Plant Study)

In the pilot plant study, we identified several of the same aldehydes (e.g., glyoxal, hexanal, heptanal, octanal, nonanal)[12] that have been identified previously at drinking water treatment plants and in laboratory-scale ozonation reactions with humic acid.[6,13–17] Table 14.1 lists the DBPs identified from treatment with ozone. Identifications should be considered tentative, as they have not yet been confirmed by analysis of standards. It should be noted that this work is ongoing, and as a consequence, the list of identified DBPs is likely to expand, and tentative identifications will be confirmed with standards. A complete list of DBPs will appear in a future publication. With the exception of benzeneacetonitrile, all of the ozonation by-products contained oxygen in their structures. Fourteen aldehydes, six carboxylic acids, two ketones, and one nitrile were identified as DBPs. These types of compounds are consistent with ozonation by-products found previously in laboratory studies.[6] The DBPs were present at concentrations ranging from approximately 5 ppt (ng/L) to 3 ppb (μg/L). Heptanal was the most concentrated ozonation DBP measured at approximately 3 ppb. Most of the aldehydes were extracted by the XAD resins; however, PFBHA derivatization was necessary for the detection of glyoxal, butanal, and 2-hexenal. These aldehydes were not detected in the resin extracts.

When chlorine and chloramine were used as secondary disinfectants (following treatment with ozone), many more by-products were obtained, most of which were halogenated (containing chlorine and bromine atoms). Among the DBPs identified were halo-aldehydes, halo-acetic acids, halo-alkanes, halo-acetonitriles, halo-propanones, halo-acetates, halo-ketones, halo-alcohols, and halo-nitro-alkanes. A full report of all of these by-products will appear in a subsequent publication. Although the compound classes of DBPs were similar for chlorine and chloramine, chloramine did produce fewer chlorinated DBPs, and at lower concentrations, than chlorine.

When NaBr was added to the raw water prior to ozonation, a few brominated DBPs were identified, namely, bromochloroacetonitrile, dibromoacetonitrile, bromoform, and bromocyclohexane. In addition, all of the same aldehydes and carboxylic acids produced by treatment with just ozone (Table 14.1) were formed. When chlorine and chloramine were used as secondary disinfectants, many more brominated by-products were formed.[18] Among these DBPs were bromo-alkanes, bromo-acetonitriles, bromo-propanones, bromo-acetic acids, bromo-nitro-alkanes, one bromo-acetaldehyde, bromo-ketones, and brominated alcohols. A thorough search was made for a compound class called bromohydrins, which are brominated alcohols that have a bromine atom and an OH group on adjacent carbon atoms. A number of these bromohydrins had been found in an earlier study (discussed below) involving laboratory-scale ozonation of water in the presence of bromide.[3,4] Although none of the same bromohydrins were identified in this study, two new bromohydrins were identified, 2-bromoethanol and 1,4-dibromo-4-methyl-3-pentanol. The latter compound is similar to a bromohydrin identified in the previous laboratory study, 1,4-dibromo-3-pentanol.

Treatment with Ozone (Laboratory-Scale Study)

For the laboratory-scale ozonation work, we identified 23 DBPs, among them a series of bromohydrins.[3,4] These identifications were confirmed by the analysis of purchased and synthesized standards. Most of the bromohydrins contained 5 carbons in their structures, and several of the bromohydrins exhibited stereoisomers (*threo* and *erythro* diastereomers). Almost none of these bromohydrins were in any spectral library, and they required much spectral information to solve their structures. In general, the EI mass spectra were very sparse. Molecular ions were generally not present, and few structurally distinguishing ions were present. CI-MS, and importantly, 2% ammonia in methane CI-MS, was key to determining the molecular weights. IR spectroscopy revealed the presence of the alcohol groups and also specified that each alcohol group must be on a carbon adjacent to the bromine group. High resolution EI-MS helped to determine the exact structures — how

Table 14.1. Ozonation Disinfection By-Products Identified.

Aldehydes	
1.	Glyoxal
2.	Butanal
3.	Hexanal
4.	2-Hexenal
5.	Heptanal
6.	Benzaldehyde
7.	Octanal
8.	Nonanal
9.	Decanal
10.	2-Methyl decanal
11.	Undecanal
12.	Dodecanal
13.	Tridecanal
14.	Tetradecanal
Carboxylic Acids	
1.	2-Methyl propanoic acid
2.	Pentanoic acid
3.	2-Methyl pentanoic acid
4.	*t*-Butyl maleic acid (identified as the anhydride)
5.	Benzoic acid
6.	Hexadecanoic acid
Other By-Products	
1.	Benzeneacetonitrile
2.	Heptadecadienone
3.	6-Methyl-5-hepten-2-one

the carbons were arranged in each molecule, and the location of the Br and OH groups. Finally, subtle differences in IR spectra allowed the assignment of specific structures to pairs of diastereomers that were present. Although diastereomers look very similar in structure, they often have different boiling points. As a consequence, the diastereomeric peaks were well-resolved by GC. However, because the EI and CI mass spectra were identical for each pair of diastereomers, additional information from IR spectroscopy was necessary to assign the specific structures.

Treatment with Chlorine Dioxide

Chlorine dioxide treatment, by itself, produced several DBPs that have not been reported previously.[7] Table 14.2 lists these by-products. Most of the DBPs contained oxygen in their structures: ketones, carboxylic acids, and maleic acids (identified in the anhydride form); only two of the by-products were chlorinated, and no halomethanes were observed. In all, there was a series of five maleic anhydrides identified — two were actual DBPs from chlorine dioxide treatment; the remaining three were in the chlorine dioxide–treated samples and in the raw water. The structures of the maleic anhydrides are shown below.

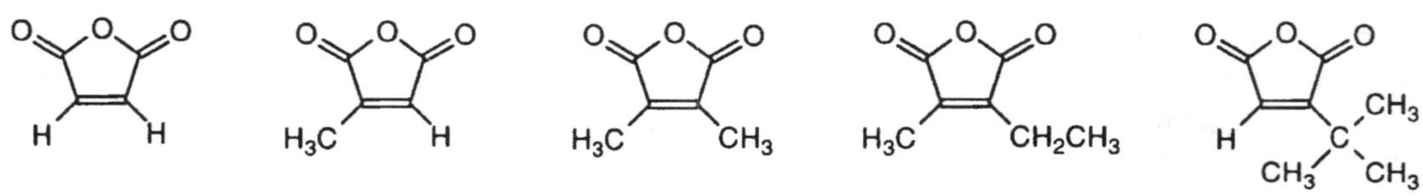

2-*tert*-Butylmaleic acid (identified as 2-*tert*-butylmaleic anhydride) was one of the most difficult-to-identify compounds we have ever studied. It is most likely that the original molecule was in the

Table 14.2. Chlorine Dioxide Disinfection By-Products Identified.

Carboxylic Acids

1. Butanoic acid
2. Pentanoic acid
3. Hexanoic acid
4. Heptanoic acid
5. 2-Ethylhexanoic acid
6. Octanoic acid
7. Nonanoic acid
8. Decanoic acid
9. Undecanoic acid
10. Tridecanoic acid
11. Tetradecanoic acid
12. Hexadecanoic acid
13. 2-*tert*-Butylmaleic acid[a]
14. 2-Ethyl-3-methylmaleic acid[a]
15. Benzoic acid

Chlorine-Containing Compounds

1. 1,1,3,3-Tetrachloro-2-propanone
2. (1-Chloroethyl)dimethylbenzene[b]

Ketones

1. 2,3,4-Trimethylcyclopent-2-en-1-one
2. 2,6,6-Trimethyl-2-cyclohexene-1,4-dione

Aromatic Compounds

1. 3-Ethyl styrene
2. 4-Ethyl styrene
3. Ethylbenzaldehyde[b]
4. Naphthalene
5. 2-Methylnaphthalene
6. 1-Methylnaphthalene

Esters

1. Hexanedioic acid, dioctyl ester

[a] Identified as the anhydride.
[b] Tentative identification.

diacid form, but converted to the anhydride during sample workup (extraction and concentration). Because no EI-MS ions were consistent and characteristic for a series of maleic anhydrides, and because this compound was not in any spectral library, mass spectrometry could not by itself provide an identification. Methane CI-MS provided the molecular weight, high-resolution EI-MS determined that there were two oxygens in the structure, and IR determined that the functional group must be an anhydride in a five-membered ring (furandione). With the furandione base structure, there only remained the placement of four methylene groups. Additional IR information was available to indicate that the maleic anhydride was mono-substituted (the four remaining methylene groups were attached to the same carbon atom on the ring). Finally, the exact structure was determined from an unusual chromatographic finding. This 4-carbon-substituted maleic anhydride eluted *before* a 3-carbon-substituted maleic anhydride. The only reasonable explanation for this elution order was that this unknown maleic anhydride must be heavily branched, as branched compounds typically have lower boiling points and elute before their unbranched counterparts. A *tert*-butyl branch would satisfy this finding. Also, a *tert*-butyl branch would be consistent with the lack of a significant molecular ion in the EI mass spectrum, as *tert*-butyl-substituted compounds typically do not exhibit molecular ions. All of this information led us to identify this compound as 2-*tert*-butylmaleic anhydride. Because this compound was so unusual, we set out to confirm the identification by synthesizing the standard. The

EI mass spectrum, IR spectrum, and GC retention time of the synthesized standard perfectly matched that of the unknown.

When chlorine was used as the secondary disinfectant (after treatment with chlorine dioxide), a greater variety of chlorinated and brominated DBPs were found, among them some halomethanes and halopropanones. Most of these compounds were by-products that have been identified in previous chlorination studies.[19–21] Because the halomethanes were only formed when secondary chlorination was used, it is possible that halomethanes can be eliminated by using a different secondary disinfectant, e.g., chloramine. Although some compounds identified are known to be toxic, and many others are of unknown toxicity, all of the semivolatile compounds identified appeared to be at low concentration (approximately 1 to 10 ppt — ng/L).

Liquid chlorine dioxide and gaseous chlorine dioxide produced very similar results. The types of by-products produced, relative concentrations of by-products produced, and the TOX were very similar for these treatments. Therefore, it appeared that the form of chlorine dioxide used was not important with regard to the formation of DBPs.

Treatment with TiO_2/UV

TiO_2/UV treatment has been the focus of intense research efforts for the remediation of contaminated waters,[22–24] as it photocatalytically degrades organic compounds in water.[25,26] Therefore, we expected that there would be few by-products produced (and in lower concentrations relative to chlorination) when this technology was used to treat drinking water. However, it was totally unexpected that we would detect only *one* DBP in the TiO_2/UV-treated drinking water samples. This by-product was tentatively identified as 3-methyl-2,4-hexanedione using the multispectral identification techniques mentioned earlier. Its concentration was approximately 50 ppt (ng/L). It is possible that some highly polar compounds, as well as thermally labile compounds, escaped detection. The ability of TiO_2/UV to degrade organic compounds was also evident in the reduction in TOC (40%) and the lowered background in the GC/MS chromatograms. In fact, TiO_2/UV has been shown previously to destroy DBP precursors.[27] The use of ultrafiltration pretreatment also reduced the amount of natural organic material present, as was evident in a lowered TOC (by 25%) and a lower background in the GC/MS chromatograms. Therefore, ultrafiltration may prove useful in reducing the amount of organic DBPs produced by removing precursor material.

Treatment with secondary chlorine produced many chlorinated and brominated DBPs, including some THMs, haloacetaldehydes, haloacetic acids, and halonitriles. The concentrations of these DBPs ranged from approximately 1 to 200 ppt (ng/L). Although most of the halonitriles had relatively simple structures, only one of them (dichloroacetonitrile) was in any spectral library. Some of the haloacetonitriles have been reported in previous chlorination studies.[19] The structures of the halonitriles identified are shown below.

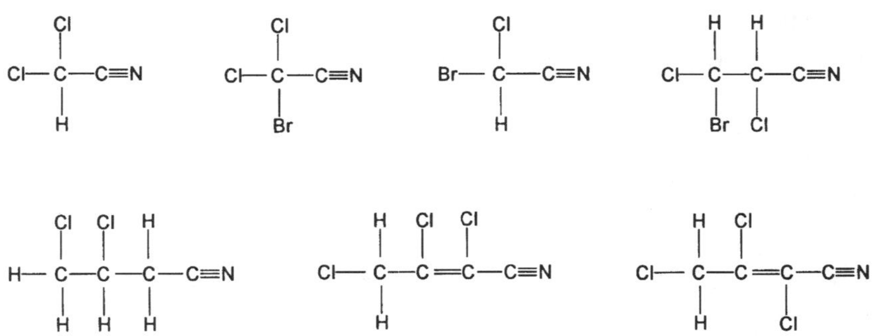

Most of the halogenated DBPs were traditional chlorination by-products that were also observed in the chlorinated control sample (raw water + chlorine). However, one of the more interesting compounds, tentatively identified as dihydro-4,5-dichloro-2(3H)furanone, was formed only by combination of TiO_2/UV treatment and secondary chlorination. A complete listing of all of these DBPs was reported in a recent publication.[28]

CONCLUSIONS

So far, the information we have obtained has been somewhat encouraging for the alternative disinfectants. Using the analytical methods employed, we detected fewer chemical by-products when the alternative disinfectants (ozone, chlorine dioxide, chloramine, and TiO_2/UV) were used to treat drinking water than when chlorine was used. Also, the concentrations of the by-products were lower. However, when chlorine was used as a secondary disinfectant (after initial treatment with an alternative disinfectant), many chlorinated by-products were produced, among them some halomethanes. Because some alternative disinfectants (ozone and TiO_2/UV) do not persist in water, drinking water treatment plants typically need to use a secondary disinfectant, such as chlorine or chloramine (in addition to the primary disinfectant), to maintain disinfection throughout the distribution system. Of the two secondary disinfectants studied (chlorine and chloramine), chloramine produced fewer chlorinated DBPs and at lower concentrations than chlorine.

ACKNOWLEDGMENTS

The authors wish to thank Wayne Koffskey, chief chemist, Jefferson Parish Department of Public Utilities, Jefferson Parish, Louisiana, and Mark Griese, manager of water quality and research for the Evansville Water and Sewer Utility, for their assistance. We would also like to acknowledge Robert Miller, John Glass, Sr., David Cmehil, Brad Smith and Betty Merriman for their contributions to this research; George Yager for assistance with GC/FT-IR analyses; and Terrance Floyd for his assistance with the chlorine dioxide identification work.

REFERENCES

1. *National Cancer Institute Report on Carcinogenesis Bioassay of Chloroform.* March 1976. Carcinogenesis Program, Division of Cancer Cause and Prevention: Bethesda, MD.
2. Georgeson, D.L. and A.A. Karimi. 1988. *Ozone Sci. Eng.* 10, 255.
3. Collette, T.W., S.D. Richardson, and A.D. Thruston, Jr. 1994. *Appl. Spectrosc.* 48(10):1181.
4. Cavanaugh, J.E., H.S. Weinberg, A. Gold, R. Sangaiah, D. Marbury, W.H. Glaze, T.W. Collette, S.D. Richardson, and A.D. Thruston, Jr. 1992. *Environ. Sci. Technol.* 26, 1658.
5. Haag, W.R. and J. Hoigne. 1983. *Environ. Sci. Technol.* 17, 261.
6. Glaze, W.H. and H.S. Weinberg. 1993. *Identification and Occurrence of Ozonation By-Products in Drinking Water*, American Water Works Association Research Foundation: Denver, CO, pp. 13–18.
7. Richardson, S.D., A.D. Thruston, Jr., T.W. Collette, K.S. Patterson, B.W. Lykins, Jr., G. Majetich, and Y. Zhang. 1994. *Environ. Sci. Technol.* 28, 592.
8. Ireland, J.C., P. Klostermann, E.W. Rice, and R.M. Clark. 1993. *Appl. Environ. Microbiol.* 59, 1668.
9. *Methods for the Determination of Organic Compounds in Drinking Water, Supplement 1*, July 1990. U.S. Environmental Protection Agency, Cincinnati, OH 45268, EPA/600/4-90, 020.
10. *Standard Methods for the Examination of Water and Wastewater*, 17th ed. 1989. American Public Health Association, Washington, DC.
11. Keith, L.H. and W.A. Telliard. 1979. *Environ. Sci. Technol.* 13, 416.
12. Richardson, S.D., A.D. Thruston, Jr., T.W. Collette, K.S. Patterson, and B.W. Lykins, Jr. *Multispectral Identification of Ozonation By-Products in Drinking Water*, Proceedings of the 43rd ASMS Conference on Mass Spectrometry and Allied Topics: Atlanta, GA, 1995.
13. Coleman, W.E., J.W. Munch, H.P. Ringhand, W.H. Kaylor, and D.E. Mitchell. *Ozone Sci. Eng.* 1992, 14, 51.
14. Glaze, W.H. 1986. *Environ. Health Perspec.* 69, 151.
15. Sievers, R.E., R.M. Barkley, G.A. Eiceman, R.H. Shapiro, H.F. Walton, K.J. Kolonko, and L.R. Field. 1977. *J. Chromatogr.* 142, 745.
16. Krasner, S.W., M.J. Sclimenti, and B.M. Coffey. 1993. *J. Am. Water Works Assoc.* 85(5):62.
17. Weinberg, H.S., W.H. Glaze, S.W. Krasner, and M.J. Sclimenti. 1993. *J. Am. Water Works Assoc.* 85(5):72.
18. Thruston, Jr., A.D., S.D. Richardson, T.W. Collette, K.S. Patterson, and B.W. Lykins, Jr. 1995. *Identification of Brominated Organics in Ozonated Drinking Water*, Proceedings of the 43rd ASMS Conference on Mass Spectrometry and Allied Topics: Atlanta, GA.

19. Coleman, W.E., J.W. Munch, W.H. Kaylor, R.P. Streicher, H.P. Ringhand, and J.R. Meier. 1984. *Environ. Sci. Technol.* 18, 674.
20. Oliver, B.G. 1983. *Environ. Sci. Technol.* 17, 80.
21. Trehy, M.L. and T.I. Bieber. 1981. In *Advances in the Identification and Analysis of Organic Pollutants in Water,* Vol. 2. Keith, L.H., Ed., Ann Arbor Science, Ann Arbor, MI, pp. 941–975.
22. Kenneke, J.F., J.L. Ferry, and W.H. Glaze. 1994. *The Origin of Chlorinated By-Products from the* TiO_2*-Mediated Photodegradation of Chloroalkenes in Water*, Proceedings of the 208th ACS National Meeting: Washington, DC.
23. Muszkat, L., M. Halmann, D. Raucher, and L. Bir. 1992. *J. Photochem. Photobiol. A: Chem.* 65, 409.
24. Matthews, R. 1990. *Water Res.* 24(5):653.
25. Ollis, D.F. 1995. *Environ. Sci. Technol.* 19(6):480.
26. Legrini, O., E. Oliveros, and A.M. Braun. 1993. *Chem. Rev.* 93, 671.
27. Hand, D.W., D.L. Parram, and J.C. Crittenden. 1995. *J. Am. Water Works Assoc.* 87(6):84.
28. Richardson, S.D., A.D. Thruston, Jr., T.W. Collette, K.S. Patterson, B.W. Lykins, Jr., and J.C. Ireland. 1996. *Environ. Sci. Technol.* 30, 3327.

Chapter 15

Cost-Benefit Analysis of Proposed Pulp and Paper Chlorinated Organic Discharge Standards

J. Festa

INTRODUCTION

In December 1993, the Environmental Protection Agency (EPA) proposed the most costly regulation the agency has ever issued affecting a single industry.

The so-called "cluster" proposed rule will revise effluent guidelines for the pulp and paper industry, and establish maximum achievable control technology standards (MACT) for certain chemicals listed as hazardous pollutants under the Clean Air Act. A final rule is expected in 1997.

This chapter will address the costs and benefits of the proposed standards for wastewater discharges of chlorinated organic compounds.

Before moving to this topic, some background on the U.S. pulp and paper industry will be helpful. The U.S. paper industry is recognized as a worldwide leader; it is globally competitive and an important contributor to the nation's economy. Employing nearly 700,000 people directly and producing products valued at more than $129 billion dollars per year, the industry ranks among our nation's top ten manufacturing industries.

On a global scale, the U.S. paper industry, with 5% of the world's population, has 12% of the world's paper and paperboard mills and 16% of its pulp mills. These mills produce 30% of the world's paper and paperboard and 35% of its pulp (Table 15.1).

Table 15.2 gives the number of paper and wood pulp mills in the United States. Currently there are nearly 600 paper mills and 260 wood pulp mills, with the kraft process accounting for 79% of total wood pulp production. Approximately 105 mills produce bleached chemical pulp. Although not represented in the figure, the number of mills producing recycled pulp has grown significantly in recent years. More will be said on this shortly.

ENVIRONMENT

Over the years, the paper industry has faced many, very complex environmental challenges. The industry has made major strides in meeting these challenges, and its record of accomplishments has been significant, as reviewed below.

As a major user of industrial process water, the paper industry strives continuously to reduce its water consumption. The industry has significantly reduced its effluent flow over the years through more efficient use of process water and through water recycling. Effluent flow is 70% lower today than it was back in 1960, and the trend continues downward.

Mill odor control is another important area of achievement. Years of research and investment in odor reduction technology have resulted in significant odor improvement at kraft mills. Average emissions of total reduced sulfur compounds have decreased by over 90% over the last two decades. Many kraft mills today have little perceptible odor.

The U.S. paper industry has also made great strides in improving the environmental compatibility of its wastewater discharges. The industry has led the way in biological wastewater treatment, now employed at virtually all U.S. mills. This has resulted in dramatic improvements in aquatic community health, as well as reduced river discoloration.

Table 15.1. U.S. Pulp and Paper in Perspective.

5% of the world's population
12% of the world's paper and paperboard mills
16% of the world's pulp mills
30% of the world's paper and paperboard production
35% of the world's pulp production

Table 15.2. U.S. Pulp and Paper Mills.

Paper Mills	599
Wood Pulp Mills	260
Kraft[a]	(122)
Sulfite	(15)
Semichemical	(29)
Mechanical and Chemimechanical	(94)

[a] Kraft pulp accounts for approximately 79% of total wood pulp production.

TOXICS

The U.S. paper industry has also made substantial progress in reducing releases of priority pollutants. U.S. pulp and paper mills have been a major supporter of the EPA's "33/50" voluntary pollution prevention initiative. The industry's level of participation in the program is nearly double the average of other industries, and most pulp and paper industry participants met the goal of 50% reduction in the use or release of 17 "priority chemicals"* by 1993, two years ahead of the EPA's timetable.

The industry's success in virtually eliminating dioxin from pulp bleaching is another pollution prevention accomplishment. As a result of process modifications, the industry has been able to reduce dioxin releases from pulp bleaching by 94% since 1988. Based on data compiled by the National Council of the Paper Industry for Air and Stream Improvement (NCASI), today the U.S. paper industry generates no more than 2.5 ounces annually of dioxin on a toxic equivalent basis (Figure 15.1), and dioxin is nonmeasurable in the effluent of 9 out of every 10 mills.

OTHER ENVIRONMENTAL ACHIEVEMENTS

Other important environmental accomplishments are in the areas of wastepaper recovery and efficient use of energy. Figure 15.2 shows the increasing paper recovery rate in the United States. In 1990, the industry implemented a national campaign to achieve a 40% recovery goal. This goal was achieved in 1993, two years ahead of schedule. The industry has now set a new goal of 50% recovery by the year 2000.

Efficient use of energy is also a primary component of the industry's pollution prevention efforts. Fossil fuel and purchased energy consumption per ton have declined by 41% in the last 20 years. Today the U.S. pulp and paper industry generates 56% of its own energy needs through combustion of waste wood, spent pulping liquors, and other wastes.

TREND OF ENVIRONMENTAL EXPENDITURES

The substantial investments made by the U.S. pulp and paper industry for environmental protection over the last 20+ years can be seen in Figure 15.3.

* Selected chemicals from Section 313 of the Superfund Amendments and Reauthorization Act.

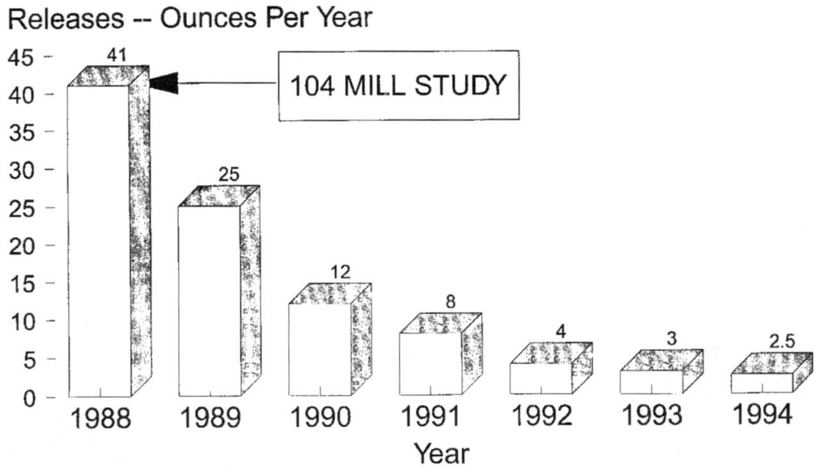

Figure 15.1. TEQ release from pulp bleaching.

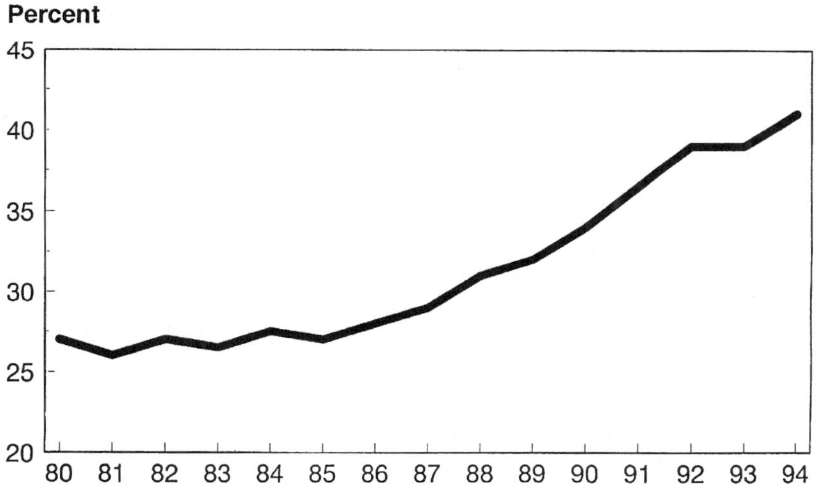

Figure 15.2. U.S. paper recovery rate.

The high rate of environmental expenditures during the decade of the '70s was followed by a smaller but still significant rate during the first half of the '80s, with another sharp rise beginning in '86–'87. The increase over the last 8 years reflects, in large part, industry expenditures for modifications and improvements in delignification and pulp bleaching processes.

Now, even with this brief summary, it is evident that the U.S. paper industry has shown a very strong commitment to the environment, with a record of accomplishment it can justly be proud of. This commitment is ongoing, and future environmental achievements by the industry can be expected.

CLUSTER PROPOSAL

As mentioned earlier, the purpose of the EPA's proposed cluster rule is to establish MACT standards for hazardous air pollutants, and to revise the industry's effluent guidelines.

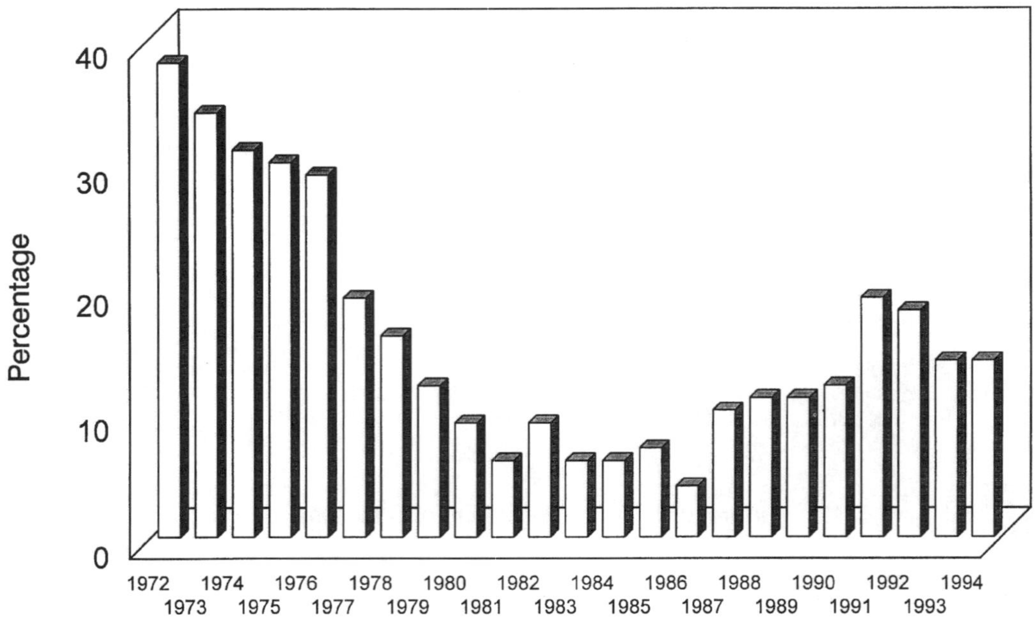

Figure 15.3. Percentage of total capital expenditures for environmental protection.

A summary of the compliance costs for the combined proposal is provided in Table 15.3. Both the EPA's and industry's* cost estimates are presented.

As seen, there is a substantial difference between the EPA's estimate of $600 million dollars in total annualized costs, versus industry's estimate of nearly $1.7 billion dollars.

The basis for industry's estimates has been provided in detail in the American Forest & Paper Association's (AF&PA) written comments to the EPA. The focus here will be on the costs and benefits of the proposed Best Available Technology (BAT) effluent standards for bleached paper grade kraft, and specifically on the chlorinated organic discharge standards.

PROPOSED STANDARDS

Table 15.4 gives the proposed effluent limitations for 2,3,7,8-tetrachlorodibenzo-*p*-dioxin (TCDD) and 2,3,7,8,-tetrachlorodibenzofuran (TCDF). The standards would apply to the bleach plant effluent rather than final wastewater treatment plant effluent, and are proposed to be nondetect for TCDD and 359 ng/kg (one-day maximum) for TCDF. The proposed effluent limitations for 12 chlorinated phenolic compounds are shown in Table 15.5. Most of the limitations would be set at nondetect levels, and again, they apply to the bleach plant effluent. The EPA has also proposed numerical effluent limitations for two additional organochlorine compounds: chloroform and methylene chloride.

Table 15.6 notes the proposed effluent limitations for AOX, or adsorbable organic halides, a measure of total organochlorine discharges.

The proposed limits are 0.267 kilograms per metric ton of pulp production (one-day maximum), and a monthly average of 0.156 kilograms per metric ton. The technology basis for all of these proposed BAT standards are complete substitution of chlorine dioxide, and oxygen delignification or extended delignification.

Table 15.7 summarizes the EPA's water-related benefits for the proposed standards for all subcategories, including bleached paper grade kraft. Ninety-nine percent of the quantified benefit estimates are based on dioxin reductions assumed to be achieved by the proposed standard.

* Industry cost estimates developed by the National Council of the Paper Industry for Air and Stream Improvement (NCASI).

Table 15.3. Compliance Cost Estimates for Proposed Air and Water Regulations. (Costs in millions, 1992 dollars.)

	Water Capital	O&M/yr	Air Capital	O&M/yr	Total Annualized
EPA	$3,100	$230	$750	$171	$600
NCASI	$7,100	$193	$2,300	$94	$1,694

Table 15.4. Bleached Paper Grade Kraft. Bleach Plant Effluent Limitation.

Pollutant	Maximum for Any One Day
TCDD	ND
TCDF	359 ng/kg

Table 15.5. Bleached Paper Grade Kraft. Bleach Plant Effluent Limitation.

Pollutant	Maximum for Any One Day
Trichlorosyringol	218 mg/kg
Trichlorocatecol (3,4,5-; 3,4,6-)	ND
Trichloroguaiacol (3,4,5-; 3,4,6-; 4,5,6-)	ND
2,4,5-Trichlorophenol	ND
2,4,6-Trichlorophenol	78.6
Tetrachlorocatecol	ND
Tetrachloroguaiacol	ND
2,3,4,6-Tetrachlorophenol	ND
Pentachlorophenol	ND

A major problem with the EPA's benefits analysis, however, is that the analysis is based on outdated dioxin discharge data, with much of the data having been collected prior to completion of industry's voluntary process changes.

Moreover, note that the benefits for human health range from $10 million to $350 million dollars. The EPA derives the upper end value of $350 million using an exposure model that assumes all dioxin in the water column is bioavailable to fish. The lower end estimate of $10 million dollars is based on a more realistic exposure model that takes into account the fact that much of the dioxin in the water column binds to particulates, thereby lowering bioavailability. The use of the former model, therefore, results in a highly inflated upper end benefits value.

The EPA's prediction of avoided sludge disposal costs is also based on a faulty assumption. Experience shows that most mills eligible for land application nonetheless choose other methods of sludge disposal. Almost all pulp and paper mills could switch to land application of sludge today if they so chose, without making any of the proposed rule's changes. This means the proposal will not lead to any sludge disposal benefits.

Benefits associated with recreational angling relate to dioxin fish advisories. Here again, however, the projected benefits are overstated, since a significant number of fish advisories have already been lifted as a result of dioxin reductions.

AOX

Although the EPA has proposed a very stringent AOX effluent limitation, the agency has not provided any evidence of a correlation between AOX and toxicity. In fact, the EPA concluded that

Table 15.6. Bleached Paper Grade Kraft. End-of-Pipe Effluent Limitation.

Pollutant	Maximum for Any One Day	Monthly Average
AOX	0.267 kg/kg	0.156 kg/kg

Table 15.7. EPA's Estimated Potential Water-Related Benefits.

Benefits	Millions of 1992 Dollars per Year
Human health	\$10.0–\$350.0
Recreational angling	\$5.2–\$24.1
Avoided costs of sludge disposal	\$56.3
Total	\$71.5–\$430.4

because AOX concentrations "do not provide information on the potential toxicity of the effluent,...[they] therefore are not appropriate to evaluate the potential impacts on the environment."[1] Thus, there appears to be no rational justification for an AOX limit.

Our overall conclusion is that the EPA's projected water-related benefits are unrealistically high. Moreover, these benefits bear little relation to the very significant costs associated with the proposed standards.

INDUSTRY'S ALTERNATIVE

The paper industry has presented an alternative to the EPA's BAT proposal. The alternative commits to meeting the EPA's proposed dioxin limits and standards for chlorinated phenolics with technology based on complete chlorine dioxide (ClO_2) substitution *without* oxygen delignification or extended delignification (Table 15.7).

While we see no justification for an AOX limit, if one is to be established, it should be based only on complete substitution.

We estimate this alternative will result in a \$3.6 billion dollar capital cost saving compared to what would be required under the EPA's proposal. And there would be no measurable difference in reductions in toxic discharges.

Data supporting the ability to meet the EPA's proposed standards for dioxin and chlorinated phenolic compounds with complete ClO_2 substitution alone have been published in a number of papers, including those by Deardorff, et al.[2] and Stinchfield and Woods.[3] Some of the data have been presented in this book.

I will close by noting recent trends in pulp bleaching chemical usage in the United States. AF&PA survey data show a sharp decline in the use of elemental chlorine, from 1.3 million tons in 1991 to a projected 500,000 tons in 1998 (a 65% reduction in usage over that period). Chlorine usage will continue to decline beyond 1998 as a result of BAT. At the same time, the consumption of sodium chlorate (used to generate chlorine dioxide) is growing rapidly, from 740,000 tons in 1991 to a projected 1,280,000 tons in 1998 (73% growth). The use of another bleaching chemical, hydrogen peroxide, has also trended upward. These chemical usage trends are indicative of the bleaching process changes that are occurring within the industry.

REFERENCES

1. U.S. EPA. 1993. Regulatory Impact Assessment of Proposed Effluent Guidelines and NESHAP for the Pulp, Paper, and Paperboard Industry (EPA-821-R-93-020).

Table 15.7. Industry's Alternative Proposal.

- Meet the EPA's proposed dioxin limits and standards for chlorinated phenolics with technology based on complete ClO_2 substitution without oxygen delignification or extended delignification.
- No justification for AOX limit; but if one is established, it should be based only on complete ClO_2 substitution.

Capital costs savings compared to the EPA's proposal = $3.6 billion.

2. Deardorff, T.L., R.R. Willhelm, A.J. Nonni, J.J. Renard, and R.B. Phillips. 1994. Formation of Polychlorinated Phenolic Compounds During High Chlorine Dioxide Substitution Bleaching. *TAPPI J.* 77(8):163–168.
3. Stinchfield, A.E. and M.G. Woods. 1995. Reducing Chlorinated Organic Compounds from Bleached Kraft Mills Through First Stage Substitution of Chlorine Dioxide for Chlorine. *TAPPI J.* 78(6):117–125.

Chapter 16

Establishing Safety for Paper Products

B.F. Vincent

As all of us know who have occasion to use the Code of Federal Regulations, the first section of a regulation usually contains definitions. Accordingly, during this discussion of safe products, let us at the outset define the concept conveyed by use of the adjective "safe." For this chapter, let us agree that it means "unlikely to cause significant harm." There are, of course, unreasonable actions one can take with essentially any item that will cause or lead to harm to the one engaging in the action. These situations are not in view in this discussion. Unless otherwise noted, it will be assumed in this chapter that products under discussion are used in the manner and for the purpose intended.

PRODUCTS DISCUSSED

Due to time constraints, only a few of the many items manufactured from wood pulp will be covered in this chapter. We'll begin with tissue and towel products and move on to some, but far from all, applications for paperboard. I'll also have some things to say about environmental considerations stemming from manufacture of these products.

USERS' EXPECTATIONS FOR TISSUES AND TOWELS

When purchasers use products, they expect that the needs to be met by these products will be fulfilled in a manner that is satisfying and that presents no unpleasant surprises. For example, use of bathroom tissue is not expected, in the United States, to lead to irritated mucous membrane areas. Thus, such an effect would likely be viewed as an unpleasant surprise. If use of the tissue leads to hives or a similar condition, even though transient, the customer's satisfaction with the product is likely to dwindle or even vanish. Thus, absence of irritancy potential for bathroom tissue is an important property. Similarly, other tissue products are expected to perform their function without leading to unexpected negative effects.

Another expected attribute of bathroom tissue, facial tissue, napkins, and kitchen towels is sanitation. Significant amounts of bacteria, yeasts, or molds at any level posing a risk of problems is another unexpected negative effect, the potential for which must be absent.

One of the major reasons that purchasers expect sanitation in napkins and towels is the use of these products in direct contact with food. Foods are often placed on these products during preparation, use, or serving and, perhaps, storage in the refrigerator. Contamination of these foods with microorganisms from the paper products could provide unpleasant surprises, particularly in the digestive tract.

FULFILLING THESE EXPECTATIONS

To minimize the likelihood of unpleasant surprises to users, appropriate testing of tissue products is carried out. The most useful tests to discover tissue irritation possibilities are those found at 16CFR 1500.41 and 42, often called the Draize tests after the physician who began to quantify the responses to irritants after deliberate, lengthy exposure of animals. Depending on the components and the extent of exposure, the finding of no irritation in these tests may be followed by a cumulative test on human volunteers carried out according to the guidelines of the appropriate reviewing body. Acceptably low scores from these latter tests normally mean users will not be unpleasantly surprised by unexpected irritation.

No evaluation of a product for irritation should be undertaken before a literature search for known effects of the components is carried out. As a sign in our library has noted, six months in the lab can often save two hours in the library. Usually, physiological effects of components are evaluated as much as possible prior to incorporation into the manufacturing procedure of a product that will be used in contact with sensitive areas of the body. Screening of new additions to a product's list of components can avert users' discovery of unpleasant surprises later on. This procedure is closely akin to that used to select materials intended to impart desirable physical properties to the finished product in that the effects of the material are as well known as possible before committing resources to producing product whose attributes rely on the properties of the material.

These types of tests are used rather than the more precise analytical ones as a consequence of the potentially wide array of substances that may be the cause of any effect found. Analytical testing requires knowledge of the substance of interest and the substance or substances causing the undesired effects may or may not be known. Further, even identification of a number of substances is still unlikely to provide the same degree of knowledge offered by the tests mentioned above. If, however, irritation is found in these practical tests, the analyst will likely be able to shed some light on the chemicals possibly involved. Reformulation without the suspects may be the most realistic route to the desired absence of irritation potentially caused by the product.

Assistance in maintaining a high sanitation level in white tissue and towel products accrues from bleaching the pulp from which such products are made. Oxidants are generally biocides, and their use provides pulp with low plate counts, which translate into production of sanitary tissues and towels. Clearly, other considerations must be taken into account to assure that the roll or box of product is, in fact, sanitary. The handling of the pulp, the condition of the water and additives at the paper machine, as well as the state of sanitation throughout the manufacturing, converting, and packaging operations must be audited regularly to assure that the desired level of product cleanliness is maintained.

TISSUE RECYCLE

The relatively recent advent of large quantities of consumer tissue products made largely or entirely from recycled fiber raises another interesting situation about safety. Since the control of raw material quality is far less with recycled than with virgin material, one could theorize that the potential for finding unwanted materials in the finished product is considerably greater. While the truth is similar in kind, it is far different in degree. The process of preparing the recycled material for conversion into tissues, napkins, and towels requires significant cleaning of the fiber. This cleaning is the primary objective of the repulping process to which the paper is subjected as it enters the recycle mill. A rule of thumb may help to put this situation into perspective: for each pound of fiber in the paper-forming process there are about 200 lb of water used to disperse that fiber during this operation. It is clear then that the fiber undergoes a significant amount of washing and, in most cases, detergents and separation processes are used to remove such things as inks, coatings, etc. Further, if we agree that "the proof is in the pudding," users have not shown dissatisfaction with the products so produced.

Sanitation of tissue from recycled fiber must be included as a consideration of safety. Bleaching products, most of which contain chlorine, are good sanitizing agents. Use of such products will usually reduce, often to undetectable levels, the microorganisms associated with post-consumer items, thus enhancing the sanitation of the finished product.

PAPERBOARD

Turning our attention to paperboard, let us look at safety with these products. Although cards are one product made from paperboard and from which the name "cardboard" came, these items are a small part of applications for these materials.

One of the more demanding uses for paperboard packaging is the use with food products. These applications are regulated by the U.S. Food and Drug Administration, and the board mill customer, who is usually a food company, will require assurance that the products they purchase meet these regulations. These food companies in turn must meet the requirements of their customers, who include members of the FDA's field offices.

Looking at a specific application, regular inspections by food safety officers can be expected by companies packaging milk and milk products. Thus, the supplier of paperboard milk cartons must provide assurances of compliance with FDA Food Packaging Regulations to his customers and also must maintain records demonstrating that he has maintained the required sanitation levels for his products. Sanitation testing includes determination of levels, if any, of microorganisms on both raw materials and finished product as set forth in the Interstate Milk Shippers Standards for the Fabrication of Containers and Closures for Milk and Milk Products. Recently ice cream cartons have been included in this system, and some carton manufacturers have been visited by food safety officers.

Where the paperboard is used to package meat products, an inspector from the Department of Agriculture's Food Safety and Inspection Service must be present and may wish to verify the FDA acceptability of the food package. To do so he obtains a copy of the letter guaranteeing compliance of the food package from the food company and sends this to the Product Assessment Division in Beltsville, Maryland. A letter goes from this division to the packaging supplier requesting on a confidential basis a list of all the components of the package. To permit USDA verification of FDA acceptability, the components must be listed by their chemical name as it appears in the FDA regulations. It is, of course, very helpful if the regulation number itself is indicated. For trade named materials the trade name is provided, and a letter from the packaging manufacturer to his supplier requests that supplier to provide the needed information to the Department of Agriculture. Thus, for meat and poultry product packaging, both FDA and USDA are involved in the assurance of safety of the packaging material.

Other applications for paperboard in contact with food would be paper cups, plates, and bowls, boxes used by bakeries for pastries, donuts, etc., and in some restaurants the box for taking home the food you could not eat. Although some of these items may not fully satisfy all of the requirements for "food packaging" as defined in FDA rules, you can be sure that if any of these products appear to contribute to food adulteration the FDA would immediately step in to solve the problem. This statement is not theoretical; the FDA has on a number of occasions done just that. Thus, the manufacturers of these products do well to assure that their food contact surfaces comply with FDA Food Packaging Regulations.

Accomplishing this assurance is not made easier by the many regulations which affect these situations. There is, as you may have heard, "more than one way to skin the cat." Most food packaging manufacturers have someone in the company who is responsible for assuring compliance with FDA regulations. One way for this individual to accomplish his job is to depend on outside counsel to do so, and there are a number of law firms around the country, including one which is well known here in Chicago, who specialize in assisting companies in this area. Another way is for the accountable individual to take the responsibility for assuring the compliance of the company's products. In some cases, the mill Technical Director may be responsible for such activity.

How, you may wonder, does the paperboard packaging manufacturer convey to his customer in a legally responsible fashion, without compromising confidentiality of the formulation, the assurance that his product complies with FDA regulations? It is probably not very surprising to learn that this issue was important during the enactment of the 1958 Food Additives Amendment to the 1938 Food, Drug and Cosmetic (FD&C) Act. Congress included a provision establishing what has come to be known as the FDA guarantee letter, which is a contract binding the issuer to convey to the recipient of the letter only materials that comply with FDA Food Packaging Regulations. The company issuing the letter is responsible, both civilly and criminally, for assuring that only such materials are included in his shipment of items named in the contract and is accountable if any of those items result in allegations of violation of the FD&C Act. Thus, the supplier must know for certain that his packaging material complies with FDA Food Packaging Regulations before issuing such a letter.

The food packer, on the other hand, having such a letter in his possession, should feel fairly comfortable that his food products are contained in a safe package and that this package will not cause food contamination problems.

PAPERBOARD RECYCLE

Paperboard can also be made from recycled fiber as well as virgin and, to be an effective means of recovering previously used paper and paperboard, these products must be able to serve many of the same functions, including some food packaging applications. Since the requirements for packaging are the same for virgin or recycled paperboard, namely no adulteration of the packaged food, there must be a reason to believe that this condition exists. We have carried out analyses of a number of lots of paperboard manufactured from recycled fiber and have been unable to find harmful amounts of a rather wide list of potential contaminants. It must also be remembered that paperboard cannot directly contact moist food for any length of time since it will rather quickly lose its packaging properties. Accordingly, it is coated with polyethylene, wax, or a similar material which will prevent moisture from getting to the paperboard. There is, therefore, no mechanism for migration even if there were undesirable amounts of material in the paperboard.

While not all applications have been studied, those in which room temperature and colder environments are the only ones encountered should be acceptable applications for this product. Thus, paperboard from recycled fiber is expected to be a safe food packaging material under the conditions described.

OTHER PRODUCTS

The pulping of wood and conversion of this pulp into various products, can affect water flowing from the mill, whether it be to a river, lake, or metropolitan sewage treatment system. For any of these receivers, the substances in the effluent from the mill should not have demonstrably adverse effects thereon.

As we have heard, one of the major surprises to the industry was the presence of chlorinated dioxins and furans in pulp mill effluent. We have heard how virtual elimination of those materials was accomplished.

Installation of biological treatment systems in the 1970s and early 1980s essentially eliminated demonstrable biological effects due to pulp and paper mill effluents. It seems unlikely that the actual versus theoretical benefits to be gained by current proposals purporting to "reduce biological impacts" would outweigh the very significant costs involved. Other chapters in this book have addressed these issues, so they will not receive further comment here.

Effluent testing now required must demonstrate no adverse effect on fathead minnow growth, ceriodaphnia reproduction, and in many cases algal growth as well. In nearly all cases, the organisms are not demonstrably affected by the chemical substances in 100% effluent. Comparison of COD, suspended solids, and other such effluent characteristics between the United States and Eastern Europe show that we have come a long way in this country in eliminating unwanted by-products of this very necessary commodity production. For example, Romanian chemical oxygen demand has recently been reported to be about 535 lb/ton of pulp while U.S. mills are about 70 to 90. Similarly, total suspended solids in Romanian mills are about 170 lb/ton with U.S. levels in the vicinity of 3, depending on several factors. Romanian mills generate nearly 3 lb of absorbable organic halides per ton of pulp produced, while U.S. mills are in the 0.3 to 1 range.

Solids produced by these processes usually go to landfills which must be demonstrably secure. Monitoring wells are the usual source for samples, the analysis of which demonstrates that leaching of unacceptable amounts of materials, such as those containing chlorine, does not occur. The placement of these wells is designed to uncover any downgradient migration of such materials.

Stormwater runoff is also monitored around pulp and paper manufacturing facilities to assure that loading and unloading areas, outdoor storage areas, and landfills do not contribute unacceptable levels of unwanted materials to rivers and lakes.

Discussions of the safety of paper products also includes the safety of those who work in the mills pulping the wood and converting it into the paper products we use.

In 1981 the industry decided that a study on worker health was needed, and after considerable study chose Johns Hopkins University Department of Public Health as the organization to collect and analyze the data. Since that time, data have been collected on over 100,000 workers from 50 of the largest pulp and paper mills in the United States, making it the largest and most powerful epidemiology study of the pulp and paper industry. The study, directed by the widely recognized epidemiologist, Professor Genevieve Matanoski, who chairs the Epidemiology Department at Johns Hopkins, has issued the first report of worker health showing that pulp and paper mill workers are at least as healthy as the average U.S. population, with little reason to be concerned about the effect of their employment on their health.

It is also worthwhile to note that the results just released covered the period of the 1970s and 80s when manufacturing processes were being upgraded and a great deal of change was taking place in the industry. Since many of the workers included in this initial report of results had been in the industry 30 or more years, there is indication that earlier manufacturing procedures had little, if any, negative impacts on the health of the workers in those times.

This chapter has shown that the safety of paper products encompasses not only the product itself and the intended use, but stretches back to selection of raw materials, the manufacturing process, conversion of intermediate forms to the product, and detoxification of manufacturing by-products. While there are still some issues under discussion, the manufacture of paper products provides items which many of us would agree are essential, and does so with no measurable adverse impact on health.

Chapter 17

Dioxin: The Magnitude of the Threat from Normal Daily Exposure

R.J. Scheuplein

Abstract

In September 1994 the EPA published its long awaited "Health Assessment Document for 2,3,7,8-tetrachlorodibenzo-*p*-dioxin (TCDD) and Related Compounds." Dioxin has generated concern because of its widespread occurrence in the environment, its persistence, its tendency to accumulate in human tissues, and its potency in eliciting a variety of adverse effects in experimental animals at very low exposures. The EPA document heightens that concern, in particular, by noting that subtle non-cancer effects have been reported in humans, by adding the toxicity of related compounds through toxic equivalent factors (TEF) and by considering higher-exposed population groups.

While everyone wants to reduce exposure to dioxin, there is disagreement about the magnitude of the threat to humans. Much of the chemical industry and significant parts of the scientific community have argued that the EPA's assessment overstates dioxin's human toxicity, overstates the human exposure and does not adequately describe the uncertainties in much of the dioxin database. This chapter will discuss many of the controversial issues involved in the EPA dioxin reassessment and present a history of the actions of federal agencies to control dioxin related risks.

INTRODUCTION

In September 1994 the EPA published its long awaited dioxin reassessment, "Health Assessment Document for 2,3,7,8-tetrachlorodibenzo-*p*-dioxin (TCDD) and Related Compounds." The central message received from the risk characterization part of the EPA report (Chapter 9) was that the U.S. public is at risk from a broad spectrum of adverse health effects produced by dioxins at or near normal background levels. These potential health effects include cancer, and damage to the reproductive, developmental, endocrine, and immune systems. However, if one reads the report carefully, it becomes apparent that the effects from near background levels are quite subtle, with no known clinical or functional consequences.

The perception created by the report is that all of us are in some danger. The sense of threat is heightened by two other findings that the EPA discusses. First, there appears to be more dioxin in the environment than the known sources produce. The mass balance does not add up. Second, the half-lives of the dioxins and furans in humans are approximately 6–10 years or longer, so whatever we do today to reduce body burdens will have little beneficial impact for many years. How did the EPA come to this position, and how do they justify it?

Dioxin Chronology

Dioxins have been of public concern for many years. In fact, even before we knew that dioxin itself was actually the active agent, U.S. federal regulatory bodies acted to reduce exposure to substances that contained it. Only later would the active principles be shown to be dioxins or furans. The chronology below is a brief history of the major events and federal regulatory actions concerning dioxin. It is certainly not complete (in particular, I've failed to include many dioxin-related incidents from other countries), but this limited chronology will serve to illustrate that dioxin is an old story and that its potential threat has not been neglected.

1949 — Nitro, West Virginia — An accident in the Monsanto chemical plant exposed more than 200 workers to dioxin. 122 workers developed chloracne within two weeks. (A follow-up epidemiological study in 1980 found no excess in total mortality or in deaths from malignant neoplasms.)

1950 — German researchers demonstrated the chloracnegenic potential of TCDD in man and on rabbit ear in the late 1950s (Kimmig and Schultz, 1957).

1957 — An outbreak of chick edema disease in the United States was traced to by-product fat used as a component of broiler feed. The toxic agents were later identified as TCDD and other polychlorinated dioxins.

1960 — "Chick edema factors" were found in food-grade oleic acid and derivatives. The FDA issued a Food Additive Regulation (21 CFR, 172.860) specifying that food additive fatty acids be free of chick edema factor.

1963–1971 — Vietnam exposure to dioxin-containing defoliants — Two major epidemiological studies were conducted by the U.S. The "Ranch Hand Study," completed in 1984, compared the 15–20 year health histories of 1,045 herbicide applicators with 19,000 other veterans believed to have had no herbicide exposure. No significant adverse effects were found and some borderline effects were judged "insufficient to support a causal relationship between herbicide exposure and adverse health." In a 1984 Centers for Disease Control (CDC) study, children with birth defects were examined to determine the impact of parents with Vietnam experience. No significant correlation of birth defects or changes in fertility were found to be associated with the children of Vietnam veterans.

1964 — Dow Chemical, Michigan — Plant employees were exposed to dioxin levels potentially reaching 10,000 ppm or 10 mg/kg body weight. Forty-nine of the exposed workers developed chloracne. (No excess in total deaths has occurred among these workers.)

1968 — The "Yusho Incident" — In the summer of 1968 approximately 1,000 people appeared at physicians' offices in Kyushu, Japan, complaining about acne-like rashes and other internal symptoms. It was determined that they had ingested PCB-contaminated rice oil. It was later found that the PCBs contained high levels of PCDFs, which were later determined to have produced the symptoms.

1969 — An outbreak of chick edema disease in North Carolina resulted in the death or destruction of more than a million chickens. The FDA's investigation traced the outbreak to the use of dioxin-contaminated vegetable oil soapstock as a feed ingredient. The vegetable oil industry was alerted to remove industrial chemical and pesticide manufacture from vegetable oil refinery sites.

1973 — FDA-promulgated regulations established temporary tolerances for unavoidable, indirect PCB contamination of fish, milk, other dairy products, animal feed, and eventually paper used in food packaging. The potential toxicity these regulations were designed to eliminate was probably caused by furans.

1975 — Levels of the higher-chlorinated dioxins up to 3.6 ppb were found in consumer packages of gelatin. The gelatin industry was alerted and this contamination was eliminated.

1976 — The "Yu-Cheng Incident" — An incident in Taiwan in many respects similar to the earlier "Yusho" incident, also probably due to PCDFs.

1976 — Seveso — A section of the community in Seveso, Italy was dusted with 1–4 pounds of TCDD following an explosion at the ICMESA chemical plant which was producing trichlorophenols. Some 37,000 people were potentially exposed and several hundred developed chloracne. A few individuals developed minor reversible nerve damage and presented abnormal liver function tests. Body fat burdens as high as 57,000 ppt were measured — the highest exposures on record. No reproductive effects were observed. No cancers definitively associated with the dioxin contamination have yet been definitely observed.

1981 — The FDA issued an advisory to the State of Michigan stating a level of concern for TCDD (>25 ppt) in the edible tissue of bottom fish from the Great Lakes.

1982 — PCP/dioxin contamination was found in market eggs in the Houston, Texas area. The source of the contamination was traced to feed ingredients made from protein meal and fat derived from PCP-treated sheep hides. Texas state officials were alerted and the use of PCP-contaminated ingredients in animal feeds was terminated.

1982 — Times Beach, Missouri — Dioxin-contaminated soil from the roadsides was dispersed widely throughout the town by floodwaters. The incident resulted in a federal buyout of the property and relocation of the residents. Studies completed in 1983 and 1986 by the CDC revealed no conclusive evidence of damaging health effects. (As of 1986 there was no evidence of increased liver disease or any general illness in the exposed group.)

1985 — A federal interagency group of scientists met to review data on TCDD levels in fish from the U.S. waterways and to evaluate information on bioaccumulation in fish.

1986 — The domestic fish monitoring program was substantially expanded to include collections and analyses of 255 samples of ocean, lake, bay, and river fish from locations throughout the U.S.

1987 — In January, the EPA Administrator formally made the EPA-TEF/87 procedure a part of official agency policy. It was emphasized that the TEF approach was an *interim* procedure.

1987 — An interagency work group with the FDA, EPA, and the Consumer Product Safety Commission (CPSC) was established to coordinate activities related to dioxin in various consumer paper products. The major focus was to obtain dioxin concentration and migration rate data so that appropriate risk assessments could be made.

1988 — The FDA analyzed nine brands of coffee filters for dioxin residues. The study found that dioxin (TCDD) was sometimes present at levels between 2–5 ppt. This translates to an average dioxin intake from brewed coffee of approximately 0.01–0.03 pg/kg/day.

1989 — The EPA issued an "Interim Strategy for the Regulation of Pulp and Paper Mill Discharges to the Waters of the United States." This was based on the EPA's National Bioaccumulation Study, which found dioxin in fish living in waters downstream from pulp and paper mills that used chlorine to bleach pulp.

1989 — The FDA surveyed dioxins and furans in milk packaged in bleached paperboard cartons. The mean dioxin intake level from milk was 0.28 pg/kg/day TEQ. The risk assessments for the average and 90th percentile consumer were slightly in excess of 10^{-6}. They have since been reduced well below this level.

1990 — An FDA-EPA-CPSC interagency committee completed work on a comprehensive risk assessment of dioxins leachable from all bleached paper products (including disposable paper plates, milk and juice cartons, oven trays, etc.). Current estimates of daily dioxin TEQ intake from these sources is less than 0.02 pg/kg/day, with associated risk levels less than 10^{-6}.

1992 — According to industry sampling, the average levels in milk carton paperboard stock collected in 1990 were 0.49 ppt for TCDD and 2.4 ppt for TCDF, below the 2.00 ppt TEQ goal. These levels assure virtually nondetectable levels in paperboard-cartoned milk.

1994 — The EPA published the "Dioxin Reassessment." The EPA claimed a variety of human health effects are likely to be produced at or near ordinary background levels of dioxin.

During the past 40 years, the policy at the FDA was to search out sources of dioxin that were seen to be both avoidable and that generated levels significantly larger than background; and then move to reduce or eliminate them. The FDA, through implementing regulations, forced the reduction of PCBs in food cartons and in some high-fat foods. The FDA also forced lower dioxin levels in milk packaged in chlorine bleached-paper cartons, and warned against consumption of Great Lakes fish with greater than 25 ppt TCDD residues. EPA regulations and revised industry practices have transformed the paper bleaching industry in this country. Former bleaching processes that used chlorine are rapidly being converted to chlorine dioxide processes that produce virtually no dioxin residues. Dioxin from pulp and paper effluent has been reduced by two orders of magnitude (Farland, 1995). The EPA has also issued regulations enforcing stricter incinerator emissions and wastewater effluent standards. Over the last ten years many state health regulatory agencies have issued warnings and guidelines concerning dioxin contamination in lakes and rivers.

These events over the last 40 years seem to me to be highlights of a modest success story. Government agencies did not ignore the dioxin problem. As soon as the toxicity of dioxin-containing substances was appreciated in the 1950s, government — both at the federal and state levels — moved to reduce exposures to them. Recent measurements of PCB levels in food and dioxins and furans in human fat tissue indicate a gradual decline in human tissue over the last 20 years (Stanley et al., 1990; Scheuplein, 1988). There do not appear to be any remaining foods in the United States where the contamination from dioxins is significantly larger than the natural background or where dioxin contamination can feasibly be reduced further. There is one significant exception to this — mother's milk. Because dioxins accumulate in the fat of human milk, a nursing child can receive a much larger dioxin dose than the normal daily adult intake. However, while noting this fact, the World Health Organization (WHO) concluded that the nutritive and protective value of mother's milk far outweighed any potential harm from dioxins.

Perhaps through more good luck than good sense, it has turned out that the occupational exposures and the accidental exposures to dioxin did not have serious human health consequences. We have been lucky because human exposures to dioxin have always been relatively small. Dioxin was never a desired end product; it was always a contaminant in either 2,4,5-trichlorophenol, 2,4,5-T, PCBs, hexachlorobenzene, hexachlorophene, Agent Orange, or some other chlorinated or brominated industrial chemical. Not one death has been reported in the literature that is definitively linked to exposure to dioxin. Compare this, for example, with the incident at Bhopal, India where from a single chemical spill there were between 2,500–5,000 people killed and up to 60,000 seriously injured. Approximately 40 tons of *methyl isocyanate* were dispersed at Bhopal. Approximately 2–4 pounds of dioxin were generated at Seveso.

The EPA Scientific Position

What are the EPA's reasons for heightening the public concern over dioxin and its related compounds at this time? The EPA contends that their policy on dioxins is soundly based on science and that this is described in their report (EPA, 1994). However, it would be far more accurate to say that the EPA dioxin policy relies in part on scientific judgment and in part on regulatory policy that is designed to fill gaps in scientific data.

The 1983 NAS Red Book (Risk Assessment in the Federal Government: Managing the Process) first called attention to "scientific inference options," i.e., necessary science policy choices that need to be made in order to formulate a scientific policy in absence of complete scientific information.

> The choices encountered in risk assessment rest, to various degrees, on a mixture of scientific fact and consensus, on informed scientific judgment, and on policy determinations (the appropriate degree of conservatism).... That a scientist makes the choices does not render the judgments devoid of policy implications. Scientists differ in their opinions of the validity of various options, even if they are not consciously choosing to be more or less conservative.... A scientist's weighting of these variables may not easily be expressed implicitly, and the result is a mixture of fact, experience (often called intuition), and personal values that cannot be disentangled easily. As a result, the choice may be perceived by the scientist as based primarily on informed scientific judgment. From a regulatory official's point of view, the same choice may appear to be a value decision as to how conservative regulatory policy should be, given the lack of a decisive empirical basis for choice. (NAS, 1983.)

The NAS Red Book also recommended that these policy choices be identified, and that a major contribution to the integrity of the risk assessment process would be to assure that the underlying rationale for such judgments are made explicit. A major difficulty with the EPA position is that the underlying scientific policy rationale is not clearly evident. Another problem is that the EPA scientific rationales, once discovered, are found to be inconsistent.

The EPA scientific position can be found in the 1994 Draft Dioxin Reassessment and also in the paper by Lucier et al. (1993). The Lucier et al. paper is quoted and liberally excerpted in the 1994 Draft Dioxin Reassessment. There appear to be two major scientifically based policy choices embedded *implicitly* in the EPA dioxin policy. One relates to low-dose extrapolation, and impacts cancer risk, the other relates to the definition of an adverse effect, and impacts non-cancer effects. These two EPA inference options are:

1. *Low-dose effects cannot be prudently modeled with either a threshold or a nonlinear dose-response model.*
2. *Subclinical effects are to be regarded as the equivalent of frank toxic effects for regulatory purposes.*

The first assures that low-dose extrapolated cancer risks for dioxins and related compounds will be worst-case (upper bound) and based on a linear dose-response model. The second assures that non-cancer risks for dioxins and related compounds will have comparatively higher margins of safety than other compounds with the same toxic endpoints. Neither of these policy choices appears to be notably consistent with the EPA's proposed new Cancer Guidelines (EPA, 1996). Dioxin and its related compounds seem to be regarded as a special case.

Low-Dose Extrapolation

The EPA starts with the proposition that TCDD is a member of a class of xenobiotics that is nonmutagenic, binds to a cellular receptor, and alters cell growth and development (Lucier et al., 1993). It is acknowledged that the biology of receptor-mediated events should be included in any risk assessment modeling for TCDD. However, the EPA emphasizes that just because an effect is receptor-mediated does not mean that it has a threshold. The EPA points out that the shape of the receptor occupancy–biological response dose-response curve is theoretically linear in the low-dose region (Lucier et al., 1993). (This is true for some models of receptor occupancy, but not all — notably not for those that include antagonistic effects.) They point out that CYP 1A1 and CYP 1A2 enzyme induction by dioxin, epidermal growth factor binding inhibition by dioxin, and other dioxin-mediated changes in gene expression all show linearity in the low-dose region. They state that a receptor-mediated model approach must include submodels *"of exposure, tissue distribution, tissue diffusion, cellular biochemistry, cellular action, tumor incidence, and cancer mortality"* (Lucier et al., 1993). At each stage, data must be collected and understood before the model can be accepted as valid for understanding the observed effects and for confidently predicting the effects of TCDD at low dose. The EPA admits that the required science is still in its infancy, implying very clearly that it will be some time before the modeling will be acceptable. This policy, although extraordinarily conservative, would at least be self-consistent, were it not for the inclusion of the TEFs. It's hard to see how the expressed desire to obtain the most detailed mechanistic information for TCDD in order to be confident in the prediction of risk can possibly be consistent with adding on the presumed low-dose cancer risk of a dozen or more dioxin, furan, and coplanar PCB congeners about which we know next to nothing.

The attention the EPA pays to receptor-based modeling is carried out with apparently little regard for the concept of "additive to background linearity" (Lucier et al., 1993). This concept, which the EPA also applies to dioxin, holds that if a compound's low-dose effects are additive to background biological process, then a linear response is predictable at low doses under very general conditions (Crump et al., 1976). If the EPA truly believes the "additive to background linearity" theory is applicable to dioxin, it is not clear why they would take modeling the complex sequence of molecular and biological events involved in dioxin-mediated toxicants so seriously. Such modeling could not possibly make any significant difference in the outcome.

The conclusion is inescapable — the EPA will continue to use a linear extrapolation model. The EPA model, used since 1980 and still in use today, is the LMS (linearized multistage), which yields 6 fg TCDD/kg/day for the one-in-a-million risk dose. All the other dioxin and furan congeners are referred to this primary TCDD potency through the TEF procedure, and thus additional conservatism is built in.

Along with Canadian and European regulators, the EPA acknowledges that TCDD does not induce DNA damage and is not mutagenic (Shu et al., 1987). The EPA also acknowledges that the dose-response relationships for TCDD's effects on cell proliferation or on altered hepatic foci or on tumor development itself are different from TCDD's effects on CYP 1A1 gene expression. These responses are not linear and do not occur at the same low doses (Lucier et al., 1992; Maronpot et al., 1993). Van den Heuvel et al. measured the TCDD induction of hepatic CYP 1A1 mRNA from 0.1–10,000 ng/kg with a PCR technique. The lowest significant response occurred at 1.0 ng. The observed low-dose response between TCDD and enzyme induction could have been linear, nonlinear, or shown a threshold. Despite the sensitivity of the PCR technique, the resolution at low doses is too poor to decide. The same paper reported that the induction of EROD by TCDD was approximately 100-fold less sensitive, with statistically significant increases observed only at 100 ng/kg (Van den Heuvel et al., 1994).

The interesting question is why these observations are not given any significant weight by the EPA. Data on the dose dependence of cellular proliferation or hepatic foci, being muticellular events, should be, if anything, better surrogates for cancer risk assessment than enzyme induction. One is forced to conclude that the EPA recognizes that risk assessments based on cellular proliferation will likely permit a considerably higher acceptable daily intake for TCDD, and this is an option they reject (Lucier et al., 1992). What the EPA has effectively done by their statements and actions is to add the class of receptor-mediated carcinogens to those substances for which no practical threshold can exist in principle or for which no practical threshold can be determined in the laboratory in our lifetime. The other two classes are genotoxic substances, and those substances that are additive to background biological processes — presumably, all substances.

Defining an "Adverse Effect"

The second conservative policy inference is the EPA's definition of an adverse health effect. The EPA does not appear willing to make a distinction between temporary biochemical changes, e.g., dioxin-induced changes in the level of a circulating hormone or enzyme, and any resulting alterations in the functional status of the corresponding endocrine or target organ. The EPA seems to define both biochemical perturbations as adverse effects worthy of the same concern.

In the 1994 Reassessment Document, the EPA discusses a "continuum of response to dioxin-like compounds" progressing from the earliest subtle effect, e.g., receptor binding, to gene regulation, to enzyme induction, through intermediate cellular level changes, then to a functional change and ultimately to frank toxicity. The EPA apparently believes any subsequent step in the "continuum" is an obligatory consequence of the preceding one. It is enough for the EPA to observe the biochemical effect and to assume that the functional and clinical effects will follow.

> Observations on effects from exposure to dioxin and related compounds in humans and other animals suggests that fundamental changes in cellular biochemistry and biology may be related to frankly adverse effects which can be more readily observed at higher levels of exposure. (EPA, Chapter 9, Dioxin Reassessment, 1994.)

The EPA acknowledges that this inference may be contentious, but they include it nonetheless as one of the key inferences in their policy (EPA, 1994). So they cite evidence of thyroid changes in circulating levels of T_4 and TSH and neglect the author's comment that these changes do not prove that the functional status of the thyroid has been affected (Pluim et al., 1993). And they cite evidence for a small decrease in serum testosterone in response to 3,000 ppt of serum dioxin (blood lipid) and ignore the fact that the dioxin levels are over 1,000 times background, and the apparent decline is still well within the normal range (Egeland et al., 1993).

When the EPA states that adverse impacts on human health may occur from exposure to dioxin-like compounds at or within an order of magnitude of current background body burdens, they either mean: (1) that the adverse outcomes are the kinds of nonclinical effect described above, or (2) that the adverse outcomes are frank toxic effects that are as yet unseen but are predictable (on the basis of the assumed continuous biological spectrum). The EPA argues that there are highly exposed populations and that there may be specially sensitive populations that must be protected. And they state that dioxin-induced symptoms may be lurking below the level of epidemiological detectability in these people. When one combines all these unproven concerns, it is not hard to understand why the EPA concludes that a significant number of people can be at risk.

In any event, their position seems to stretch our common understanding of adverse health impacts. This is not to say that such data are not worth exploring further, but such data cannot honestly be interpreted *per se* as implying an existing health threat. The point of view that even the most subtle biological effect is worth regulating, in effect, turns toxicological evaluation into a default testing system, with an outcome that depends entirely on the quality of the instrumentation and the sensitivity of the technique, and not at all upon scientific judgment. The EPA's position on noncancer effects very adroitly, if unhelpfully, answers one of the more difficult questions in toxicology: When does a detectable biological effect become an adverse biological effect? The EPA's answer is: Always!

This answer is, of course, not helpful, because many things we do to sustain life produce these same biological effects. Enzyme induction occurs in profuse amounts after every meal. It has been shown that a wide variety of natural substances from food also bind strongly to the Ah-receptor — in amounts far exceeding TCDD (Ames, 1989; Safe, 1995). In effect, by ignoring the natural background, the EPA interprets these experiments with dioxin in the absence of the necessary control groups for determining human risk.

The conclusion we can draw, given what the EPA has said and done, is that the EPA prefers to be conservative, both in the choice of the linear model, in their essentially empty application of receptor-based biology, in its concern over special populations, in its estimation of exposure, in the use of TEFs, and in use of worst-case risks. Why the EPA continues to insist on this degree of conservatism, in view of the fact that dioxin has never been demonstrated to result in any fatalities, and that chloracne is the only established human adverse clinical effect, is not clear. Unfortunately, the identification of the underlying scientific rationales helps explain only how the EPA rationalizes its policy, not why it holds it.

RISK ASSESSMENT OF DIOXIN

The EPA linearized estimate for the one-in-a-million cancer risk is about 6 fg/kg/day (6.0×10^{-15} g/d). Given that the detailed biology that the EPA contends is necessary for reliable receptor modeling is largely unknown at this time, is there another way we can decide whether or not there is a significant risk from low levels of dioxins? I think we can, and the risk assessment is similar to that conducted in the European countries.

Natural Background Exposure

Table 17.1 shows the sources and magnitude of the normal background exposures to dioxin. Most of the normal dioxin exposure is from fat-containing foods; or more accurately, food is the final exposure pathway to humans. Air, soil, and water contribute very little to the human body burden. The typical TCDD intake in the United States and Europe is approximately 0.23 pg/kg/d, or 16 pg/day. If one includes all the dioxin and furan congeners in the exposure, in terms of their TCDD equivalents, the total from this data is 86.7 pg TEQ/day (Table 17.1). To obtain these estimates, TEF values for each congener are collected, weighted by their concentration, and summed. The EPA's estimate is 119 pg TEQ/day. Thus the total dioxin TEQs are about 5–7 times greater than TCDD itself, without the consideration of the planar PCBs. Using a single-compartment model for TCDD, a half-life of 7 years, and the fact that most dioxin is stored in body lipids, the dioxin intake of 0.23pg/kg/day predicts a human blood lipid level of about 4 ppt, in good agreement with observation.

Direct human adipose tissue measurements were conducted by the EPA through the National Human Adipose Tissue Survey (NHATS). According to this survey, the mean TCDD fat concentration for the 15–44 age group has declined over the period 1971–1987 from 18 ppt to 4 ppt (Stanley et al., 1990). Table 17.2 shows the 1987 NHATS figures for dioxin and its congeners. A State of California survey is also given for comparison (California, 1989). It seems reasonably certain that TCDD blood lipid levels nationally average between 4–6 ppt and the total TEQs average between 20–100 pg TEQ/day.

What of the toxicology of dioxin? Do we know enough of the toxicity of dioxin to predict the range of its effects and the doses at which they occur? TCDD and its congeners produce a wide variety of effects in animals at doses significantly above human background. The important effects for human risk assessment in normal populations are, however, those that can occur at doses close to normal human exposure levels.

Animal Toxicology Data

A very extensive database exists on the animal toxicity of TCDD. Some of the major findings are shown in Table 17.3. One of the earliest and still most intriguing observations is the marked species variation observed in the acute toxicity of dioxin (Table 17.3). Even if one ignores the guinea pig, which is by far the most sensitive species, there still is a 200-fold difference between the most com-

Table 17.1. Major Sources of Exposure to Dioxins.[a]

Source	Intake 2,3,7,8-TCDD (pg/d)	Conc. 2,3,7,8-TCDD (ppt)	Intake TEQ (pg/d)	References
Media				
Air	0.075	—	3.0	Theelan (1991)
Soil	0.007	—	0.25	Theelan (1991)
Water	0.004	—	1.0	Travis and Fry (1991)
Food				
Meats	6.6	0.035	22[b]	Henry et al. (1992)
Fish	8.6	0.5	14[b]	Henry et al. (1992)
Milk	0.2	0.002	2.0	Henry et al. (1992)
Cheese	0.31	0.016	20[a]	Henry et al. (1992)
Butter	0.11	0.044	6.0[b]	Henry et al. (1992)
Eggs	—	—	4.0	EPA Exposure (1994)
Ice cream	0.04	0.006	0.4	Henry et al. (1992)
Oils	—	—	14	Theelan (1991)
Totals	16.0 pg/d		86.7 pg/d	
Totals	0.23 pg/kg/d	1.24 pk/kg/d		

[a] The major sources and levels of dioxin exposure are shown. Most of the background exposure to dioxins and furans comes from fat-containing foods, because of their very high solubility for dioxins.

[b] TEQ values from Theelan.

Table 17.2. NHATS and SCARB Mean Adipose Tissue Data for Dioxins and Furans.[a]

Congener (2,3,7,8-)	Concentration (ppt)		TEQ (NHATS) (pg/g)
Dioxins	**NHATS**	**SCARB**	
TCDD	5.38	4.58	5.38
PeCD	10.7	10.1	5.35
HxCD	86.8	9.91	8.68
HpCD	110	80.3	1.1
OCDD	724	519	0.72
Furans			
TCDF	1.88	1.98	0.18
1-PeCDF	0.31	0.56	0.02
2-PeCDF	9.7	6.01	4.85
HxCDF	14.2	13.6	1.42
HpCDF	16	9.48	0.16
OCDF	2.28	0.82	0.002
Totals			27.9 pg/g

[a] The human tissue levels of the major dioxins and furans (the 2,3,7,8 congeners) are shown. The central column compares the NHATS data with an independent survey from the State of California. The data shows that the average TCDD concentration in fat (or in blood lipid) is approximately 4–6 ppt and that the total TEQs are approximately 5–6 times higher.

mon rat strain and the hamster. Admittedly these are acute effects, but there appears to be some correlation between the acute toxicity of a chemical and its ability to produce a cancer response (Kimbrough, 1990). Rao et al. (1988) had to administer a total dose of 600 mg/kg of TCDD to hamsters to produce the tumor response requiring only 70 mg/kg in rats (Rao et al., 1988).

Table 17.3. Summary of TCDD Animal Toxicity — Acute. Acute Toxicity Studies Show Marked Species and Strain Differences.

Oral LD_{50} (µg/kg)	
Hamster	1,157–5,051
Bullfrog	500–1,000
Mouse	132–2,570
Dog	100–200
Rabbit	50–115
Rhesus monkey	~70
Chicken	< 25
Rat (L/E)	10–17.7
Rat (S-D)	25–60
Rat (Fischer)	340
Rat (H/W)	3,000–7,200
Guinea pig	0.6–2.0

The oral LD_{50}'s for TCDD in a variety of animal species and rat strains are shown. The table illustrates the large species-strain sensitivity for the acute effects of TCDD.

From: Pohjanvirta and Tuomisto (1994).

The chronic NOEL for TCDD (cancer) for the rat is near 1.0 ng/kg/day, about 4,000 times more than the human daily background (Table 17.4). This is also approximately the lowest effect dose observed for any frank toxicity. Regulatory officials in Canada and Europe, who worry about low levels of dioxins from food, are comforted by these data. In part because dioxin is not mutagenic, they believe that the dioxin cancer response is thresholdable. They typically apply a 100–1,000 fold safety factor to this NOEL to obtain an ADI for humans of from 1.0 to 10 pg/kg/day — 160 to 1,600 times higher than the EPA 6 fg/kg/day (10^{-6} risk level). The rat data, when linearly extrapolated to low doses under the EPA no-threshold assumption, give a human risk from background dioxin levels between 10^{-3}–10^{-4}, depending on the assumptions used for dose scaling. If one-in-a-million is an agreed virtually safe level, background dioxin exposures are therefore considered trivial in Europe but quite significant by the EPA.

Dioxin produces a variety of responses, ranging from frankly adverse effects to biological changes, such as induction of drug-metabolizing enzymes (Table 17.4). Because dioxin affects enzymes, growth factors, and hormones, virtually every tissue in the body and every class of biological reactions is modulated by dioxin at high enough doses. The most sensitive of these effects, at least for humans, appears to be cancer or possibly reproductive or developmental effects. And as long as cancer is the only effect in which a threshold response is questioned and linear extrapolation is used, cancer remains the most sensitive regulatory endpoint (Birnbaum, 1994).

There are essentially two human risk assessment issues: (1) Can we assume a threshold for cancer or not, and (2) Where does the human response fit within this spectrum of different species sensitivities? European and Canadian regulatory scientists say we can assume a threshold for dioxin's effects. They point to the nongenotoxicity of TCDD and the near toxic levels required for tumors to develop in the Kociba study. In more recent years they can also cite the dose-response data on cellular proliferation and the induction of hepatic foci, which is nonlinear (Lucier et al., 1992; Maronpot et al., 1993). The EPA, on the other hand, says we cannot assume a threshold. They point to the theoretical linearity of the receptor-based mechanism and the findings of linearity in TCDD's effects on enzyme induction and gene expression. Each group weighs the same experimental data differently. On the animal sensitivity issue, all regulatory agencies appear to accept the rat model as best reflective of human response, at least for now.

Table 17.4. Summary of TCDD Animal Toxicity — Mostly Chronic. NOAELs[a] for Longer-Term Studies and for the Most Sensitive Effects are Greater Than ~1.0 ng/kg/d.

Dose (ng/kg/d)	
Multigeneration study in rats, NOAEL	1.0
Lifetime chronic study in rats, NOAEL	1.0
Enzyme induction in rats, NOAEL	1.0
Teratogenicity in mice, NOAEL	100
TCDD promotion study with DEN with ~ NOEL at	3.5
Minimal effect acnegenic dose	100

[a] No Observed Adverse Effect Levels (NOAELs) are shown for the longer-term toxicity of TCDD and for some other effect. Cancer, reproductive effects, teratogenicity, and other effects all occur at levels higher than approximately 1.0 ng/kg/day. (See text.)

Human Toxicology Data

As a practical matter, relatively few individuals have been exposed to high levels of TCDD. Summarized information on the more important accidental and occupational exposures to TCDD are shown in Table 17.5. All of the exposures listed in the table were at least at 100 times normal background levels, and some, like Seveso, were over a thousand times higher. There are really two generic types of exposure: (1) a single large exposure, like the Seveso explosion, or (2) the more frequent day-to-day gradual accumulation, typified by occupational exposure from a chemical plant. Human blood levels anticipated from these two types of exposure are modeled in Figure 17.1 (Scheuplein and Bowers, 1995). They are actually data from Seveso and for the NIOSH–12 Plant study.

Chloracne is the universal toxic lesion, and it occurs, in at least some members of every exposed group, in virtually all of the occupational exposures. Investigators have also reported sporadic abnormalities in liver function tests, porphyria, reversible nerve damage, and fatigue. All of these have been reversible effects and have disappeared over time, although chloracne can hang on with discomfort for several years. Some of the epidemiological studies of these cohorts have reported significant standardized mortality rates for lung cancer, soft tissue sarcoma, non-Hodgkin's lymphoma and other cancers all allegedly caused by TCDD (Fingerhut et al., 1991; Manz et al., 1991; and Zober et al., 1990). All of these incidents are characterized by exposures orders of magnitude above the normal average background level of 0.2–0.4 pg TCDD/kg/day.

The two occupational studies having the largest cohort sizes, longest follow-up periods, and the highest measured dioxin blood levels are the NIOSH study (Fingerhut et al., 1991) and the Boehringer-Ingleheim, Hamburg FWG plant study (Manz et al., 1991). The two largest industrial accidental exposures occurred at the BASF chemical plant in Ludwigshafen FRG in 1953 and at the ICMESA trichlorophenol plant near Seveso, Italy in 1976. In each of these four studies, to a greater or lesser extent, blood TCDD levels were obtained for some exposed workers. It was therefore possible to base exposure estimates on actual experimental human data.

Table 17.6 shows the predicted risk from those exposures, assuming that people are similar to Sprague Dawley rats in their response. The expected number of cases is the product of the cohort size (N) and the cancer risk (R). Table 17.6 shows that the highest dioxin exposures combined with the largest exposed cohorts were the NIOSH 12-Plant study and the Seveso accident. In the case of the NIOSH study, for example, one would expect on average to observe 32 cases of lung or liver cancer in humans due to TCDD exposure of the U.S. Plant workers, if dioxin had the same potency in humans as in rats. Since the background risk of respiratory cancers in the 2,000-man cohort was about 40, the expected SMR for lung cancer was approximately 2; this is small but still detectable. According to Fingerhut et al., lung cancer was in fact observed just at this level. In contrast, neither of the other two remaining occupational cohorts had sufficient power to detect major TCDD cancers; if such cancers were present they would be undetectable above background.

Table 17.5. Significant Occupational Exposures to Dioxins[a] (Concentrations in ppt).

Ref.	Exposure Site	Number Exposed	Range	Mean	Median
1	Missouri A	9	42–750	245	122
1	Missouri B	16	5–557	145	24
2	Seveso	10	828–56,000	19,144	16,600
3	Ranch Hand	888	n.d.–618	—	12.4
4	New Jersey Plant	103	2.2–3,390	294	84
4	Missouri Plant	32	3.4–1,290	177	109
5	BASF	69	5.0–170	—	24.5
6	Boehringer Ingelheim	118	5.9–2,252	83–296	137
7	NIOSH	5,172	2–3,400	233–418	~100

[a] The major occupational exposures to dioxin are shown: 1. Missouri Dioxin Study 1986 (A) horse arena, (B) TCP production area (Andrews, 1989). 2. Zone A residents with chloracne (Facchetti et al., 1980). 3. Handlers of Agent Orange, all personnel (Needham et al., 1991). 4. Workers in New Jersey and Missouri chemical manufacturing plants (Sweeney et al., 1990). 5. Workers in BASF chemical manufacturing plant (Zober et al., 1990). 6. Workers in German chemical plant producing 2,4,5-T and 2,4,5-trichlorophenol. Blood levels are based on only 48 volunteers (Manz et al., 1991). 7. Composite study of 12 U.S. chemical plants making dioxin-contaminated products. The two means are for exposures less than and greater than one year (Fingerhut et al., 1991).

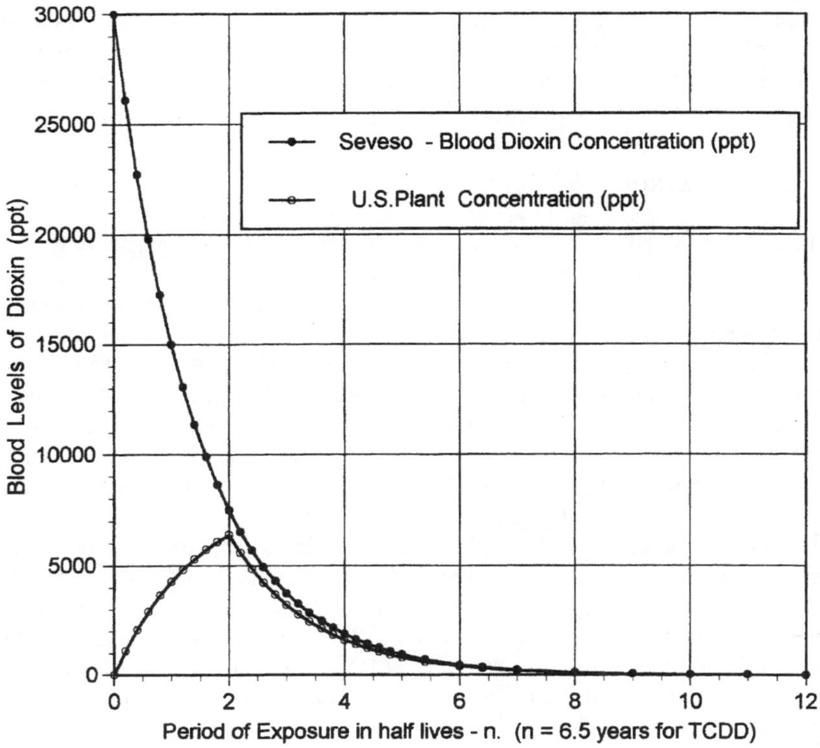

Figure 17.1. The upper curve (closed circles) shows the decline in dioxin blood levels from the Seveso incident. This is typical of a single exposure. The lower curve (open circles) shows a cohort from the NOSH 12 Plant study. This lower curve is typical of sustained occupational exposure, in this case for 13 years or two TCDD half-lives.

Table 17.6. Cancer Risks and Dioxin Blood Levels (Lipid Adjusted) for Four Major Exposures.[a]

Study	Median Exposure TCDD (ppt)		Cohort Size N	Cancer Risk R	Expected Cases
	C_{meas}	C_{avg}	N	R	N x R
NIOSH	333	2,000	2,000	0.02	32
B–I	137	1,220	118	0.01	1.6
BASF	24.5	247	114	0.003	0.35
Seveso	30,000	13,330	50	0.09	4.3

[a] The predicted cancer risks from the four major dioxin exposures are shown. The first column gives the median measured exposure. The second column is the corresponding extrapolated value. The extrapolated value is higher for occupational exposures, since the measured values represent blood levels taken years after the working period, and after considerable decline in the blood levels. In contrast, for Seveso the average blood level over the period of exposure is less than the measured value, since it was sampled immediately after exposure. The cancer risk was computed from the average human dose and the animal potency. (See Scheuplein and Bowers for details, 1995.)

The Seveso incident delivered the highest recorded exposures to individuals (57,000 ppt), although when back-extrapolated, some of the NIOSH exposures were in this same range. However, up to the present only a few of the blood samples have been analyzed for TCDD. Of the fifteen or so that have been, three samples were around 30,000 ppt. I used 30,000 ppt in the calculation and assumed that as many as 50 of the Seveso victims were exposed. This gives an SMR also of about 2 for dioxin-related cancers. There is still no definitive answer from Seveso. And the NIOSH study has been substantively challenged. (There has been recent criticism of the NIOSH study, in that respiratory cancers were primarily observed in only one of the 12 plants, and smoking may not have been adequately corrected for.) Nonetheless, if the NIOSH study is actually detecting dioxin-induced lung cancer, it is from an average sustained exposure to dioxin of 25 years, with dioxin levels averaging about 400 times background (Scheuplein and Bowers, 1995)

In Utero and Infant Exposure

Infants and developing children are exposed to dioxins *in utero* and particularly through breast feeding. Over a period of years the lipids in maternal milk concentrate fat-soluble substances, including dioxins, PCBs, and other fat-soluble contaminants. Dioxin levels in mother's milk are typically 10–100 times greater than in cow's milk or in typical fatty foods. This means that a child during breast feeding receives a substantially larger dose of dioxin than if the child were not breast-fed. During a typical 6 month nursing period, the blood level of dioxin TEQs in an infant can be an order of magnitude or two greater than in a adult. Therefore if low levels of dioxin and related compounds had any adverse effects in humans, one would expect them to show up first in infants. No adverse clinical effects of these higher dioxin exposures in infants have yet been reported. As the child grows, the blood concentration of dioxin decreases due to increasing body size and to the cessation of breast feeding, and the overall increment to the adult's total dioxin body burden due to breast feeding is only approximately 5% (WHO, 1988). The convincing evidence on the benefits of breast feeding to the overall health and development of the child, in the opinion of the World Health Organization (WHO) far outweigh any potential harm from dioxins, and in 1988, the WHO strongly endorsed breast feeding despite the higher dioxin exposures (WHO, 1988). Recent data have not altered this recommendation. The 1994 EPA comprehensive review of dioxin rather surprisingly made no specific mention of breast feeding.

SUMMARY AND CONCLUSIONS

In summary, the only reliable *human* adverse effect from modest dioxin exposures is chloracne, which occurs after exposure(s) to doses much higher than the normal background. With animal toxic responses disappearing at doses lower than 1.0 ng/kg/day and with human TCDD intakes slowly decreasing below 1.0 pg/kg/day; it seems we ought to be reasonably confident that normal background levels of dioxin are safe. We certainly should continue to investigate the unique biological properties of dioxin and reduce unnecessary human exposures, but on the basis of the present evidence, the threat to human health from current background exposures is negligible.

REFERENCES

Ames, B.N. 1989. What are the major carcinogens in the etiology of human cancer? Environmental pollution, natural carcinogens and the causes of human cancer: six errors. In: *Important Advances in Oncology 1989,* V.T. De Vita, Jr., S. Hellman, and S.A. Rosenberg, Eds. J.B. Lippincott Co., Philadelphia. pp. 237–247.

Andrews, J.S., Jr. 1989. "Summary of Missouri dioxin studies," Presented at the Fourth National Environmental Health Conference, June 23, San Antonio, TX.

Bertazzi, P.A., C. Zocchetti, A.C. Pesatori, S. Guercilena, M. Sanarico, and L. Radice. 1989. Ten-year mortality study of population involved in the Seveso incident in 1976. *Am J. Epidemiol.*, 129:1187–1199.

Bertazzi, P.A., A.C. Pesatori, D. Consonni, A. Tironi, M.T. Landi, and C. Zocchetti. 1993. Cancer incidence in a population accidently exposed to 2,3,7,8-tetrachlorodibenzo-*para*-dioxin. *Epidemiology,* 4(5):398–404.

Birnbaum, L.S. 1994. The mechanism of dioxin toxicity: relationship to risk assessment. *Environ. Health Perspect.,* 1023(Suppl 9):157–167.

Crump, K.S., D.G. Hoel, C.H. Langley, and R. Peto. 1976. Fundamental carcinogenic processes and their implications for low dose risk assessment. *Cancer Research,* 36:2973–2979.

Egeland, G.M., M.H. Sweeney, M.A. Fingerhut, K.K. Wille, T.M. Schnorr, and W.E. Halperin. 1993. Total serum testosterone and gonadotropins in dioxin exposed workers.

EPA. August 1994. "Health Assessment Document for 2,3,7,8-Tetrachlordibenzo-*p*-dioxin (TCDD) and Related Compounds," Volume III, External Review Draft, U.S. Environmental Protection Agency, Office of Health and Environmental Assessment.

EPA. 1996. Proposed Guidelines for Carcinogen Risk Assessment, *Federal Register*/Vol. 61, No. 79/April 23, 1996, 17960–18011.

Facchetti, S., A. Fornari, and Montagna. 1980. Distribution of 2,3,7,8-TCDD in the tissues of a person exposed to the toxic cloud at Seveso (Italy). *Adv. Mass Spectrom.* 8B, 1405–1414.

Farland, W. 1995. Presented at the ASPET Colloquium on Receptor-Acting Xenobiotics and Their Risk Assessment, Tyson's Corner, VA, October 13, 1995.

Fingerhut, M.A., W.E. Halperin, D.A. Marlow, M.S. Piacitelli, P.A. Honchar, M.E. Sweeney, A.L. Greife, P.A. Dill, K. Steenland, and A.J. Suruda. 1991. Cancer mortality in workers exposed to 2,3,7,8-tetrachlorodibenzo-*p*-dioxin. *New Engl. J. Med.* 324:212–218.

Henry, S., G. Cramer, M. Bolger, J. Springer, and R. Scheuplein. 1992. Exposures and risks of dioxin in the U.S. food supply. *Chemosphere,* 25(1-2):235–238.

Kimbrough, R. 1990. How toxic is 2,3,7,8-tetrachlorodibenzodioxin to humans? *J. Toxicol. Environ. Health,* 30:261–271.

Kimmig, J. and K.H. Schultz. 1957. Occupational chloracne caused by aromatic cyclic ethers. *Dermatologica,* 115:540–546 (in German).

Kociba, R.J., D.G. Keyes, J.E. Beyer, J.E. Carreion, R.M. Wade, C.E. Dittenber, D.A. Kalnins, R.P. Frauson, L.E. Park, C.N. Barnard, S.D. Hummel, and C.G. Humiston. 1978. Results of a two year chronic oncogenicity study of 2,3,7,8-tetrachlorodibenzo-*p*-dioxin in rats. *Toxicol. Appl. Pharamacol.* 46:279–303.

Lucier, G.W., G. Clark, A. Tritscher, J. Foley, and R. Maronpot. 1992. Mechanisms of dioxin tumor promotion: Implications for risk assessment. *Chemosphere,* 25:177–180.

Lucier, G.W., C.J. Portier, and M.A. Gallo. 1993. Receptor mechanisms and dose-response models for the effects of dioxins. *Environ. Health Perspect.* 101:(1)36–44.

Lucier, G.W., A.M. Tritscher, T. Goldworthy, J. Foley, G. Clark, J. Goldstein, and R. Maronpot. 1991. Ovarian hormones enhance 2,3,7,8-tetrachlorodibenzo-*p*-dioxin-mediated increases in cell proliferation and preneoplastic foci in a two-stage model for rat hepatocarcinogenesis. *Cancer Research,* 51:1391–1397.

Manz, A., J. Berger, J.H. Dwyer, D. Flesch-Janys, S. Nagle, and H. Walstgott. 1991. Cancer mortality among workers in chemical plant contaminated with dioxin. *Lancet,* 338:959–964.

Maronpot, R.R., J.F. Foley, K. Takahashi, T. Goldsworthy, G. Clark, A. Tritscher, C. Portier, and G. Lucier. 1993. Dose response for TCDD promotion of hepatocarcinogenesis in rats initiated with DEN: Histologic, biochemical, and cell proliferation endpoints. *Environ. Health Perspect.* 101(7):634–641.

Mocarelli, P., A. Marocchi, G.D. Patterson, P. Brambilla, P.M. Gerthoux, L. Meazza, and V. Carreri. 1991. Serum concentrations of 2,3,7,8-tetrachlorodibenzo-*p*-dioxin and test results from selected residents of Seveso, Italy. *J. Toxicol. Environ. Health* 32:357–366.

NAS Redbook. 1983. *Risk Assessment in the Federal Government: Managing the Process,* Commission on Life Sciences, National Research Council, National Academy Press, Washington, DC.

Needham L.L., G.D. Patterson, and V.V. Houk. 1991. Levels of TCDD in selected human populations and their relevance to human risk assessment, in M. Gallo, R. Scheuplein, and K.A. Van der Heijden (Eds.), *Banbury Report 35: Biological Basis for Risk Assessment of Dioxins and Related Compounds,* Cold Spring Harbor Laboratory Press, Plainview, New York, pp 229–257.

Pluim, H.J., J.J.M. de Vijlder, K. Olie, J.H. Kok, T. Vulsma, D.A. van Tijn, J.W. van de Slikke, and J.G. Koppe. 1993. Effects of pre- and postnatal exposure to chlorinated dioxins and furans on human neonatal thyroid hormone concentrations. *Environ Health Perspect.* 101:504–508.

Pohjanvirta, R. and J. Tuomisto. 1994. Short-term toxicity of 2,3,7,8-tetrachlorodibenzo-*p*-dioxin in laboratory animals: effects, mechanisms, and animal models. *Pharmacol Rev.* 46(4):483–549.

Poiger, H. and C. Schlatter. 1986. Pharmacokinetics of 2,3,7,8-TCDD in man, *Chemosphere,* 15:1489–1494.

Poland, A. and J.C. Knutson. 1982. 2,3,7,8-Tetrachlorodibenzo-*p*-dioxin and related halogenated aromatic hydrocarbons: Examination of the mechanisms of toxicity. *Ann. Review Pharmacol. Toxicol.* 22:517–554.

Rao M.S., V. Subbarao, J.D. Prasad, and D.G. Scrapelli. 1988. Carcinogenicity of 2,3,7,8-tetrachloro-*p*-benzodioxin in the Syrian golden hamster. *Carcinogenesis,* 9:1677–1679.

Safe, S.H. 1995. Environmental and dietary estrogens and human health. Is there a problem? *Environ. Health Perspect.* 103:346–351.

Scheuplein, R.J. 1988. Risk assessment and risk management of environmental food contaminants by FDA. In Cothern C.R. (Ed.) *Advances in Modern Environmental Toxicology.* Princeton Scientific Publishing Co. Inc., Vol. 16, pp. 109–122

Scheuplein R.J. and J.C. Bowers. 1995. Dioxin — An analysis of the Major Human Studies: Comparison with Animal-Based Cancer Risks. *Risk Analysis,* 15(3):319–333.

Shu, H.P., D.J. Paustenbach, and F.J. Murray. 1987. A critical evaluation of the use of mutagenesis, carcinogenesis, and tumor promotion data in a cancer risk assessment of 2,3,7,8-tetrachlorodibenzo-*p*-dioxin. *Regulatory Toxicol. Pharmacol.* 7:57–88.

Stanley, J.S., R.E. Ayling, and P.H. Cramer. 1990. Polychlorinated dibenzo-*p*-dioxin and dibenzofuran concentration levels in human adipose tissue samples from the continental United States collected from 1971 through 1987. *Chemosphere,* 20:859–901.

State of California Air Resources Board. 1989. MRI Report Determination of body burdens for poylchlorinated dibenzo-*p*-dioxins (PCDDs) and poylchlorinated dibenzofurans (PCDFs) in California residents. ARB Contract No. A6-195-33, MRI Project No. 8941-A.

Sweeney, M.H., M.A. Fingerhut, D.G. Patterson, and L.B. Connally, et al. 1990. Comparison of serum levels of TCDD in TCP production workers and in an unexposed comparison group. *Chemosphere,* 20:993–1000.

Theelan, R.M.C. 1991. Modeling of human exposures to TCDD and I-TEQ in the Netherlands: Background and occupational. In Gallo, M., R. Scheuplein, and K. Van der Heijden, Eds. *Biological Basis for Risk Assessment of Dioxins and Related Compounds.* Banbury Report #35, Cold Spring Harbor Laboratory Press, Plainview, NY.

Travis, C.C. and H.A. Hattemer-Frey. 1991. Human exposure to dioxin. *Sci. Total Environ.* 104:97–127.

U.S. Environmental Protection Agency. 1994. Health Assessment for 2,3,7,8-TCDD and related compounds. Washington, DC, Office of Research and Development. EPA 600/BP-92/001.

Van den Heuvel, J.P., G.C. Clark, M.C. Kohn, A.M. Tritscher, W.F. Greenlee, G.W. Lucier, and D.A. Bell. 1994. Dioxin-responsive genes: Examination of dose-response relationships using quantitative reverse transcriptase-polymerase chain reaction. *Cancer Res.* 54:62–68.

WHO. 1988. *Assessment of Health Risks In Infants Associated With Exposure To PCBs, PCDDs and PCDFs in Breast Milk,* Report on A WHO Working Group, WHO Regional Office for Europe, Copenhagen.

Wilson, J.D. 1987. A dose-response curve for Yusho Syndrome. *Reg. Toxicol. Pharmacol.* 7:364–369.

Zober, A., P. Messerer, and P. Huber. 1990. Thirty-four year mortality follow-up of BASF employees exposed to 2,3,7,8-TCDD after the 1953 accident, *Int. Arch. Occup. Environ. Health,* 62:139–157.

Chapter 18

Chlorinated Organics — Sorting Reality from Myth in Maine and Beyond

M.K. Hill

ABSTRACT

Peoples' beliefs about chlorine and chlorinated chemicals are often polarized. Some individuals do not admit to any problems with chlorinated organic chemicals, whereas others believe they are harmful at all levels and in all circumstances. Reality is more complex. On the one hand, chlorine disinfection of drinking water has served a tremendously beneficial function, although there are concerns related to chlorinated disinfection by-products. Conversely, some chlorinated organic chemicals, of which the pesticide DDT is a notorious example, have caused obvious environmental damage. Other serious concerns continue to surface — consider the accumulation of polychlorinated chemicals in the Arctic. One intractable issue centers around the dioxin family, which has become so disputatious that some cannot consider it dispassionately. In this presentation, Maine is used as an example of a state where even undetectable levels of dioxin are considered a major threat.

CHLOROORGANIC CHEMICAL CONCERNS

The perception that all chlorinated organic chemicals are dangerous has far-reaching implications. Consider the city of Lima, Peru, which made a decision to stop chlorinating its drinking water. Saving the cost of water disinfection probably played a role in this decision, but the reason given to justify the move was a U.S. Environmental Protection Agency (EPA) report which concluded that ingesting chlorinated chemical by-products in chlorine-disinfected drinking water was linked to an increase in human cancer. It took an epidemic of cholera to change that attitude. Before the outbreak, starting in January, 1991, cholera had been absent from Latin America for many years. Some cholera cases resulted from ingesting raw shellfish that were contaminated with cholera bacteria, but others resulted from contaminated drinking water. Since 1991, cholera has again become endemic in Latin America, millions have become ill, and thousands have died.[1]

At the other end of the spectrum are instances where concerns about chlorinated organic chemicals were only too fully justified.[2] An example is Florida's Lake Apopka, which received a large spill of the polychlorinated pesticide dicofol in 1980. The dicofol was heavily contaminated with *DDT* (dichlorodiphenyltrichloroethane).[3] Years later, it was observed that only 5% of the lake's alligator eggs were hatching, as compared to a normal hatch rate of 70–80%. Half of those that did hatch died within two weeks. Surviving male alligators had feminine characteristics that were found to be associated with *p,p′*-DDE, a degradation product of DDT, which blocks the action of androgens.[4] Lake Apopka is now on EPA's National Priority List of hazardous waste sites.

The problems at Lake Apopka can be blamed on a spill. More disturbing are instances where polychlorinated chemicals became widely dispersed in the environment over a period of many years. A familiar example is DDT, banned in 1972, which was associated with abnormal eggshell thinning in peregrine falcons and other birds. Another example is fish kills resulting from the polychlorinated cyclodiene insecticide endrin. Or, consider PCBs (polychlorinated biphenyls). Although manufacture of PCBs was banned in the mid-1970s, they were still found in some locales more than twenty years later at levels high enough to adversely affect a sensitive species like mink. In one study, Michigan State University investigator John Giesy examined laboratory-raised mink that were fed Great Lakes fish contaminated with PCBs. When the contaminated fish comprised 20% of their diet, the mink had a reduced number of kits, with a reduced survival rate. When the fish comprised 40% of their diet, the mink did not reproduce at all. The fish used in this experiment contained 8.4 parts per million of TCDD toxic equivalents (TEQ), of which PCBs constituted about 97% and TCDD 3%.[5] See the following for an explanation of TEQ.

> *Toxic equivalents.* There is not just one dioxin; there are 75. In addition, there are 135 closely related furans included in the dioxin category. Each dioxin and furan differs in the number and positions of the chlorine atoms that it contains, and each has a different toxicity. To account for toxicity in a mixture of dioxins, a toxic equivalent factor (TEF) is used. The most toxic dioxin is 2,3,7,8-tetrachlorodibenzo-*p*-dioxin (2,3,7,8-TCDD, *the* dioxin), which was given a TEF of 1. The toxicity of other dioxins and furans is rated relative to 2,3,7,8-TCDD. For example, the most toxic furan, one-tenth as toxic as 2,3,7,8-TCDD, has a TEF of 0.1. To determine the total toxicity of a mixture, the amount of each important dioxin and furan is determined and then the toxicity of each is factored in to give a final toxic equivalent (TEQ). To further complicate the situation, certain PCBs have dioxin-like toxicity and these are also counted as dioxins. Indeed, in two major examples used in this article (of TEQs in the flesh of Great Lakes and Baltic fish) all or almost all TEQs were from PCBs.

In 1988, greater than half the seal population in the North Atlantic and Baltic Seas, about 20,000 seals, died after being infected with a distemper virus. Suspicion arose that the animals were especially susceptible to infection as a result of exposure to chlorinated organic chemicals through the fish they consumed. A Dutch scientific team set out to test this hypothesis by capturing a group of seal pups. Control pups were fed fish from the relatively unpolluted North Atlantic Sea. The others received fish from the Baltic Sea that contained about ten times more chlorinated organic chemicals than North Atlantic fish. PCBs constituted about 93% of the TEQ in the fish and the remaining 7% were from dioxins and furans.[6] The Dutch investigators observed that pups fed contaminated fish showed a number of immune system changes not seen in control pups. Their natural killer cells, a population of immune cells, capable of destroying cells infected with virus, were 25–50% less active than those of control seals. The activity of the animals' T cells, another population of immune cells, was also suppressed. These were changes with the potential to interfere with the seals' ability to fight infection. Results were interpreted as being consistent with the hypothesis that exposure to polychlorinated chemicals may have played a role in the 1988 kill-off of seals. To confirm that the pups' immune systems were indeed weakened would require exposing them to a pathogen and comparing their infection rate to pups fed the less contaminated fish. No one would willingly carry out such a test. Questions were also raised as to nutritional differences between fish from the Baltic and the North Atlantic seas, which could also affect immune system function. Despite these and other caveats, these results raised serious concerns about the Baltic Sea's high PCB levels.[7]

Assuming that the seal study results were indeed due to high PCB levels, one could argue that the polluted, shallow, and almost enclosed Baltic Sea should not be compared to less polluted North American waters. One could also point out that concentrations of PCBs in the U.S. environment have been decreasing since the manufacture of these chemicals was banned in the mid-1970s. Since that time, PCB concentrations in Great Lakes fish have decreased as much as 90%.[8,9] However, there are troubling exceptions. As noted earlier, mink populations around the Great Lakes are still adversely impacted by PCBs. There are also numerous hot spots (locales containing high PCB concentrations) in lakes, rivers, and coastal areas. In addition, environmental concentrations of PCBs are now decreasing much more slowly than was the case in earlier years.[9,10]

Disturbingly, concentrations of some polychlorinated chemicals may be *increasing* in the Arctic environment. The toxaphene mixture of insecticides is an example. Its components are volatilized in warmer climates and blown toward the Arctic.[11] Once far enough north, the extreme cold prevents further volatilization, while also preventing their degradation. The result is a buildup in toxaphenes in locales where they have never been used. Consider that toxaphene was recently discovered in fish at concentrations above the Canadian health-based standard in a "pristine" Yukon lake. Although most polychlorinated pesticides, including toxaphene, have been banned in Western countries, many are still used in underdeveloped countries. The ability of volatile polychlorinated chemicals to "leapfrog" from warmer to colder locales makes them important far beyond the areas in which they are used, and an effort has been mounted to ban them worldwide.

Although it is easy to dismiss the importance of polychlorinated chemicals in countries such as the United States, where they are no longer used, there are several reasons not to do so. The "leapfrog" effect is a reminder of their continued use elsewhere. The existence of numerous hot spots in this country also makes them relevant. Past and present problems with these chemicals also serve as reminders of how important it is to prevent future environmental buildups of persistent chemicals. Consider "dioxin" (2,3,7,8-tetrachlorodibenzo-*p*-dioxin and related dioxins and furans). Even given the range of beliefs about dioxin toxicity, all would agree that it is desirable to keep environmental concentrations as low as possible. Indeed, major dioxin sources are being strictly controlled.

Another reason that already banned chemicals remain relevant is that they still strongly influence public opinion. To understand chlorinated chemical issues and to reach even tentative conclusions is an ongoing challenge even to those individuals willing to make the effort. Most don't make an effort and receive the bulk of their information from the media, information sometimes interpreted in a sensationalist manner or from a single viewpoint. It is thus not surprising that people become alarmed about *all* chlorinated chemicals at *all* concentrations.

DIOXIN

Dioxins are the most feared of the chlorinated organic chemicals. Even tiny amounts of the most toxic dioxin, 2,3,7,8-TCDD (TCDD), administered to laboratory animals, cause adverse effects. For example, rat dams given a one-time oral dose of TCDD on day fifteen of their pregnancy had male offspring with slowed sexual maturation and smaller than normal sex organs. Even the lowest dose administered to the dams (0.064 μg/kg body weight) consistently lowered sperm production.[12] In another study, TCDD displayed a one hundred times greater ability to suppress immune system function in mice than did corticosterone, a hormone known for its immune-suppressing effects.[13] Many other instances of dioxin's sometimes dramatic toxicity could be cited. However, a problem arises when individuals look at a laboratory study where TCDD is directly administered to animals at relatively high concentrations and then directly equate results to an effluent that contains undetectable or barely detectable amounts of TCDD. A similar problem arises when effects of a chlorinated organic chemical observed at hot spots are equated to average environmental levels.

The risk characterization section of EPA's 1994 draft risk reassessment concluded that there is no threshold for dioxin's adverse effects and that even everyday exposures are at or within an order of magnitude of a level that could adversely impact health.[14,15] This conclusion was criticized by a panel of scientists organized by EPA's own Science Advisory Panel.[16] However, if we assume that EPA's worst-case conclusions are correct, then it is especially important to obtain accurate information on dioxin sources, especially those resulting in the greatest human exposure. An EPA report noted that about 95% of all dioxin emissions to the environment come from combustion processes, especially medical waste and municipal solid waste combustion.[14] Another combustion process, wood burning, contributes about 3.5% of the 95%, or 3.9% if residential wood burning is included. Only about 0.03% of total dioxins released to the environment were directly discharged to water by pulp bleaching mills that used chlorine-containing chemicals. However, because dioxins bioaccumulate, this 0.03% is magnified to the point that about 6.6% of the average American's exposure to dioxins comes from fish consumption. Most of the remaining 93% of human exposure is from consumption of meat and dairy products.

How does human exposure to dioxins occur? Dioxin emissions into the atmosphere from combustion sources settle onto vegetation that is eaten by cattle and other animals. Humans are exposed to dioxins by eating the meat or eggs or drinking the milk of these animals. A different pathway of exposure occurs for dioxin discharged to waterways from a mill bleaching with a chlorine-containing chemical. Discharged dioxins bind tightly to sediment particles and are ingested by small aquatic organisms. These organisms are eaten by fish, which in turn are eaten by larger fish. At the top of the food chain, the larger fish are eaten by animals, which include humans. Ingested dioxins accumulate in body fat and become progressively more concentrated in animals higher on the food chain. In nursing women, dioxins concentrate in the fat of breast milk. Nursing infants can have unusually high exposures to dioxins — ten to twenty times greater than the average adult exposure. This fact troubles more individuals than those concerned about adult exposure to dioxins.

DIOXIN IN MAINE

In Maine the word dioxin is often equated with paper mills and fish. This is true even though Maine's mills have reduced dioxin discharges 95% as compared to the late 1980s. At that time, TCDD concentration in bleach plant effluents was about 100 parts per quadrillion. Now it is usually undetectable at a detection limit of 10 parts per quadrillion. The move from chlorine to chlorine dioxide bleaching is further decreasing this value. Nonetheless, despite the absence of dioxins at this tremendously low detection level, a common observation is: "Just because you can't measure it doesn't mean it's not there." Indeed, a common opinion may be summarized as follows. Dioxin comes from paper mills and contaminates fish. Fish is the major source of human exposure to dioxin, which is toxic in any amount. Thus, switching from chlorine to chlorine dioxide isn't enough. Mills must go to totally chlorine-free bleaching.

After Maine's mills made process changes that greatly lowered discharges, dioxin levels in fish caught below bleaching mills also fell. Decreases varied with fish species and size as well as locations in which they were caught. In twelve of fifteen samples analyzed in 1992, dioxin levels had fallen between 51% and 92% as compared to the same species caught at the same locations in 1988–1990. Of the three remaining samples, one showed a 13% decrease, one no decrease, and one an increase.[17] Since 1992, fish concentrations have not further decreased and may have increased in some fish. A Maine Department of Environmental Protection publication observed that "Increases seemed more related to the capture of larger fish or unexplained variability than to any increased discharge of dioxin."[17] One explanation for the absence of a continuing decline in fish dioxin levels is a possible reservoir of dioxin in sediments. Such a reservoir could prevent further decreases for years to come.

In 1993, for the first time, lobster flesh and lobster tomalley (hepatopancreas) were monitored for dioxin content. In lobster flesh, the dioxin TEQ was about 1.2 parts per trillion. Had the level been slightly higher, an advisory would have been issued advising people to limit their consumption of lobsters. The State of Maine issues advisories at dioxin levels of 1.5 parts per trillion, as compared to the U.S. Food and Drug Administration's level of 25.0 parts per trillion. However, the lobster tomalley contained up to 30 parts per trillion and the Maine Department of Health issued an advisory to limit consumption of this organ. The results upset the lobster industry and many members of the public. In a letter to the editor of the *Bangor Daily News,*[18] I pointed out that dioxins were also present in *human* blood fat at about 30 parts per trillion, a level that certainly did not come from eating lobster tomalley. I was later told that some individuals eat whole containers of tomalley; for these people, tomalley was potentially a significant source.

Like a number of other states, Maine has been working on an environmental priorities project. Reports of the technical committees evaluating issues came out in 1995.[19] Dioxin was rated a high priority contaminant in both air and water. One report contained the following comment on dioxin (italicization added): "Atmospheric deposition is one means of distribution, although not the primary one in Maine. *Discharges to waterways are more significant.*" Another comment on dioxin was "*Extremely toxic even at levels near analytic detection limits.*" No documentation was given to support these statements.

In Maine, are dioxin "*discharges to waterways more significant*" than atmospheric emissions? To answer this question, we need to compare dioxin sources in Maine as compared to the nation. Nationwide, the EPA estimated that about 95% of total dioxin emissions were from combustion processes, primarily from municipal solid waste, MSW, and medical waste incineration.[14] Nationwide, Americans burn 16% of their MSW, whereas in Maine 36% of MSW is burned.[20] This fact is consistent with combustion being an even greater source of dioxin in Maine than in the country as a whole. However, unlike many combusters that the EPA evaluated, two of Maine's four facilities have state-of-the-art control technology. Thus, even though Maine burns a higher than average percentage of MSW, dioxin emissions may not be greater than the national average. Nobody knows.

An environmental injustice. About 250 miles of Maine's more than 30,000 miles of rivers are under fish consumption advisories because of dioxin. Native Americans living along the rivers, traditionally heavy fish consumers, are most affected. The argument has been made that the risk assessment process used to set advisories was overly conservative and has resulted in a substantial overestimate of exposure.[21] Furthermore, Maine places advisories on fish consumption at 1.5 parts per trillion as compared to the U.S. Food and Drug Administration's higher level of 25 parts per trillion. There is also the argument that dioxin in fish may no longer be coming from bleaching plants, but rather from a sediment sink. Nonetheless, the advisories are in place and the net result is that Native Americans and others avoid eating fish caught in these sections of Maine's rivers. Not surprisingly, Native Americans are among those calling for totally chlorine-free pulping.

Wood combustion also produces dioxins. Nationwide, wood burning produces about 3.9% of total dioxin emissions. However, Maine burns much more wood than the national average. Among eleven Northeastern states, Maine generates about 450 megawatts a year by burning about four million tons of wood. The next highest wood burning state, New Hampshire, generates about 200 megawatts.[22] In addition, Maine industrial facilities, including pulp and paper mills, burn another two million tons of wood per year to meet internal energy needs. Most Maine homes also have at least one fireplace or stove and some homeowners obtain all their heat from wood. Overall, it can be hypothesized that dioxin emissions in Maine from wood burning are greater, perhaps much greater, than the national average.

No one knows the average profile of human exposure to dioxin in Maine. Nationwide, exposure to dioxins via fish consumption averages 6.6%. In Maine, one may assume that Indians and sports fishermen who eat fish caught beneath bleaching mills have a greater exposure from fish than is the case nationwide, perhaps much greater in some cases. Beyond these sub-populations, we have no information as to whether the majority of Maine's population has exposure patterns significantly different than average Americans, who experience their greatest dioxin exposure from consuming meat and dairy products. One explanation for the common perception that fish is *the* major source may relate to the observation that, "Most human exposure to dioxin comes through the food chain, yet fish is the only food regulated for dioxin content."[26] Indeed, many people seem unaware that meat, dairy products, and even eggs contain dioxins. Another possible source of dioxin exposure is indoor air. See inset.

Dioxin exposure in indoor air. It is worthwhile to consider dioxin contamination of *indoor* air, especially in a state like Maine where most homes have a wood burning stove or fireplace. A German study found that chimney soot in homes with oil central heating contained 147 parts per trillion dioxin TEQ. Soot in homes with wood central heating had 14,896 parts per trillion — a level 100 times greater.[23] Dioxin levels in the soot of wood burning American homes is unknown, but one can hypothesize that they are high relative to fossil fuel soot. Furthermore, soot is a substance to which wood burning homeowners are regularly exposed. As with other pollutants with indoor sources, one could hypothesize that dioxin levels may be higher indoors than out. Cigarette smoke is another indoor source of dioxins. One study, which examined the dioxins (as furans) in cigarette smoke, calculated that smoking one pack a day contributed an exposure of about 4.3 pg/kg body weight/day.[24] Since total dioxin intake has been calculated at 3–6 pg/kg/day,[25] cigarette smokers could have twice the intake of nonsmokers.

Trying to explain dioxin levels in fish or in mill effluents is complicated. Not only are there many dioxins and furans, but the EPA also includes the toxicity of several coplanar PCBs as part of the dioxin TEQ.[14] Thus, a citizen trying to understand dioxin exposure must first understand what dioxins are as well as the EPA's definitions of TEF and TEQ. That person also needs to know that PCBs are not formed by pulp bleaching. A Maine TV program on dioxin produced in 1995 included a film of Lake Michigan fish that showed adverse effects, reportedly due to high PCB levels. This was a Maine program with a major emphasis on dioxin discharges by Maine paper mills. Nonetheless, no attempt was made to distinguish between PCBs and dioxins or to point out that these two groups of chemicals come from different sources.

Are dioxins *"extremely toxic even at levels near analytic detection limits"?* Consider the following illustration. The detection limit for 2,3,7,8-TCDD is about 10 parts per quadrillion in water and about 100 parts per quadrillion in fish. In the example given earlier of an effect of dioxin at an extremely low level, male offspring of rat dams given a one-time oral dose of TCDD were adversely affected by a dose as low as 0.064 μg TCDD per kg body weight.[12] However, for a 1-kg rat dam to obtain even that lowest dose of 0.064 μg from water containing TCDD at 10 parts per quadrillion, it would have to ingest (in one drink) 6.4 kg water, a physical impossibility. If the TCDD was in fish, the rat would need to ingest 64% of its weight in fish at one sitting. In both cases, to obtain 0.064 μg, it would need to absorb 100% of the dioxin that was ingested.

In Maine — and probably in other states with bleaching mills — fish consumption *is* probably a major source of dioxin exposure for some Native Americans who eat fish caught beneath those mills. However, even before paper mills decreased dioxin discharges 95%, fish consumption was unlikely to have been a major source of exposure for the large majority of Maine's people. Conversion to chlorine dioxide bleaching will almost certainly result in a further decrease in dioxin discharges, discharges that are already typically below detection limits. Dr. Robert Huggett, U.S. Environmental Protection Agency Science Advisor, has made the following statement: "Chlorine dioxide is a solution to dioxin and other persistent, bio-accumulative, toxic substances in mill waste water."[28] Do misunderstandings about dioxin sources matter? To the extent that dioxins in the environment continue to be taken seriously, they should matter. Publicizing some sources while other sources are largely disregarded will not accomplish the goal of continuing to reduce dioxin exposure.

POSTSCRIPT

For dioxins, other chlorinated chemicals or any chemical, there is no precise level or dose at which "no danger" becomes "clear danger." This ambiguity, in itself, is taken as a reason to believe that no level of a chemical can be considered safe. The only fact that is clear is the need to go beyond the existing polarization of opinions on dioxins and other chlorinated chemicals and to sort out conflicting opinions in an atmosphere of goodwill. Maine's Independent Governor, Angus King, attempted one such effort. In early 1996, he brokered an agreement between the state and Maine's seven bleaching mills in which the companies agreed to a "long-term vision" of eliminating the discharge of dioxin and other pollutants. The only relatively short-term goal in the agreement is to eliminate fish advisories due to dioxin contamination by the year 2000. The agreement does not specify how this will be accomplished. Not surprisingly, it has not quieted criticism, and is referred to as a public relations gimmick, especially in the absence of specific steps and deadlines. Others believe that it does show a commitment to solve the problem and, from that perspective, should be viewed favorably. Clearly, polarization of opinion on the dioxin issue continues to exist, but one must applaud the Governor's effort to lessen the divide.

REFERENCES

1. Garrett, L. 1994. *The Coming Plague,* Farrar, Straus and Giroux, New York, p. 563.
2. LeBlanc, G.A. 1995. *Environmental Health Perspectives*, 103(10):888.
3. Raloff, J. 1994. *Science News*, 145(2):24.
4. Culotta, E. 1995. *Science,* 267, 330.
5. Anonymous. 1992. *Health & Environment Digest*, 5(12):1.
6. Ross, P.S., R.L. DeSwart, J. Reijnders, H. Van Loveren, J.G. Vos, and A. Osterhaus. 1995. *Environmental Health Perspectives,* 103(2):162.
7. Raloff, J. 1994. *Science News*, 146(1):8.
8. Anonymous. 1993. *Health & Environment Digest*, 7(2):6.
9. Fisher, B.E. 1994. *Environmental Health Perspectives*, 102(9):743.
10. Miller, S. 1994. *Environ. Sci. Technol.,* 28(2):92A.
11. Raloff, J. 1995. *Science News*, 148(3):38.

12. Mably, T.A., R.W. Moore, and R.E. Peterson. 1992. *Toxicol. Appl. Pharmacol.,* 114, 97 *and* Mably, T.A., R.W. Moore, R.W. Goy, and R.E. Peterson. 1992. *Toxicol. Appl. Pharmacol.,* 114, 108.
13. Schmidt, K.F. 1992. *Science News*, 141(2):24.
14. U.S. Environmental Protection Agency. *Estimating Exposure to Dioxin-Like Compounds*, Vol. 1: Executive Summary; external review draft; Office of Health and Environmental Assessment. Office of Research and Development. U.S. Government Printing Office: Washington, DC, June 1994; EPA/600/6-88/005Ca.
15. Farland, W.H. 1995. *Health & Environment Digest*, 9(2):10.
16. Stone, R. 1995. *Science*, 268, 1124.
17. Mower, B. 1995. *Dioxin Monitoring Program, State of Maine, 1994.* Department of Environmental Protection, Augusta, Maine.
18. Hill, M.K. 1994. *Bangor Daily News*, September 23.
19. Maine Environmental Priorities Project, Reports from the Technical Working Groups to the Steering Committee, July, 1995.
20. Lord, D., Maine Waste Management Agency, Personal communication, May, 1995.
21. Keenan, R.E., M.H. Henning, P.E. Goodrum, M.N. Gray, R.A. Sherer, and P.S. Price. 1993. Proceedings of the 13th International Symposium on Chlorinated Dioxins and Related Compounds, Vienna, Austria, September.
22. Connors, J. May 1995. Maine State Planning Office, Personal communication.
23. Thoma, H. 1988. *Chemosphere*, 17(7):1369.
24. Muto, H. and Y. Takizawa. 1989. *Archives Environ. Health*, 44(3):171.
25. Anonymous. 1995. *Environ. Sci. Technol.,* 29(1):26A.
26. Anonymous. 1994. *Environ. Sci. Technol.,* 28(9):406A.
27. Giesy, J.P., D.A. Verbrugge, R.A. Othout, W.W. Bowerman, and M.A. Mora. 1994. *Arch. Environ. Contam. Toxicol.,* 27:213.
28. Dr. Robert Huggett, testimony given at U.S. EPA's public hearing on the proposed Cluster Rule for the pulp and paper industry, Washington, DC, Feb. 10, 1994.

Part 4

Regulatory Aspects

Chapter 19

Common Sense and Chlorine in the Paper Industry

T.T. Davies

INTRODUCTION

This chapter discusses a new approach that the U.S. Environmental Protection Agency (EPA) is using to develop regulations that are based on sound scientific information and protect the environment and human health. An important component of the process is to ensure that the regulations the EPA develops are reasonable, effective, and cost-efficient.

One item discussed in this chapter is how the new approach was used in the development of the final pulp and paper rule that the EPA proposed in 1993. Also presented in this chapter is a brief overview of two other major efforts that are currently underway that are also of interest to the pulp and paper industry: (1) the EPA's dioxin reassessment and (2) the National Academy of Sciences study of health effects related to hormone-related environmental toxicants.

A significant portion of this chapter is also devoted to how the EPA's new approach to rule making was successfully used in the agency's recent landmark rule-making effort that will protect human health and the sensitive ecosystems that are unique to the entire Great Lakes basin.

This chapter will now describe the EPA's new approach to rule making. Following that, this chapter will discuss — as examples of current efforts — the Great Lakes Water Quality Initiative, the EPA's dioxin reassessment, the National Academy of Sciences study of health effects related to hormone-related environmental toxicants, and the development of the final pulp and paper rule.

EPA'S NEW APPROACH TO RULE MAKING

A major goal of the U.S. Environmental Protection Agency is to develop pollution control regulations that will reach the right balance so that the environment and human health are protected and the regulations are cost-effective. The new approach to rule making for the agency involves close cooperation among the EPA regulators and the state and local government agencies, Native American tribes, and the plant managers, scientists, and marketing experts in the industries that are being regulated. It also requires the active involvement of academia, environmental groups, and the general public to ensure that sensitive ecosystems will be adequately protected.

It is a "common sense" approach because it emphasizes greater involvement of all the stakeholders and is a holistic, process-oriented approach which considers total system impacts, with cost-effectiveness as well as environmental protection as a goal. The following discusses some of these aspects of this new approach.

One important aspect of the EPA's new approach to rule making is the development of "cluster rules." In the past, each EPA office writing a regulation would look almost exclusively at discharges to a single medium: air, water, or land. The reason for this was that the legislation that gave the agency the authority to develop the regulations was medium-specific. While our legislation remains medium-specific, we are changing the way we are writing our regulations and are increasingly considering discharges to all media (surface water, air, groundwater, and land disposal). An example of this is the way the EPA developed the final pulp and paper rule, which will be described later in this chapter.

"Cluster rules" have many advantages. For example, industries — and specific dischargers — will know, up front, what the total discharge limitations are to all media and can choose the most cost-effective means to achieve those limitations. This gives the discharger the opportunity to incorporate pollution prevention measures, alter processes, or revise treatment in a cost-efficient manner. The environment will be protected because we are better able to protect entire ecosystems in a more logical manner; and the EPA will benefit because, in the long run, we are more efficient and devote less time and money while developing better regulations.

Another important aspect of the agency's new approach to rule making is the involvement of stakeholders throughout the rule development process — from inception, through development, to implementation. This involvement has proved instrumental in reaching accord on the types of controls and methods of implementation because there is involvement by everyone affected throughout the process.

Another aspect of the changing approach to regulation is the focus on addressing the concept of ecosystem protection by stressing the development of basin-wide or regional pollution protection to protect a sensitive ecosystem. This involvement by the EPA, state, and local pollution control agencies along with industries, environmental groups, academia, the scientific community, and the general public will help to make our regulations protective of human health and the environment and will ensure that they are reasonable and cost-efficient for those who implement them. An example of this new approach to include basin-wide or regional protection of a sensitive ecosystem is the Great Lakes Water Quality Initiative.

THE GREAT LAKES WATER QUALITY INITIATIVE

The Great Lakes Initiative resulted in the development of basin-wide pollution reduction measures that were specifically designed to protect human health and sensitive ecosystems in the Great Lakes basin. As such, it is a good example of the EPA's new regulatory approach of basin-wide or regional protection of a sensitive ecosystem.

The Initiative stemmed from widespread concern over the extent and severity of pollution — particularly chlorinated compounds — that were seriously affecting (as described below) human health, aquatic life, and wildlife in the basin. The resulting Guidance sets minimum basin-wide water quality standards, antidegradation policies, and implementation procedures that are to be carried out by all of the states and Native American tribes in the basin.

Once adopted by the Great Lakes States, the final Guidance will improve consistency of water quality standards and permit requirements across the Great Lakes basin, with an emphasis on provisions for decreasing loadings of bioaccumulative chemicals of concern (BCCs). These are toxic chemicals, such as PCBs, dioxin, and other organochlorine compounds, that biomagnify in fish, leading to exposure for wildlife and humans as well.

Over the years, scientists have observed increases in the number and severity of the effects of BCCs on fish, bird, and mammal populations in the Great Lakes basin. These effects include cancer, premature death, eggshell thinning, population declines, reduced hatching success, abnormal behavior (such as abandonment of nests), infertility, birth defects (such as crossed beaks and club feet), and illnesses such as chick edema. They also include more subtle effects, including abnormalities in the thyroid, liver, and endocrine systems. For example, reproductive effects associated with exposure to organochlorine compounds in the Great Lakes area have been documented in populations of mink and the Foster's tern, and increased incidence of tumors and certain cancers have been documented in native fish populations (Kubiak, et al., 1989; Auerlich, et al., 1973; Baumann, et al., 1987).

With regard to effects of these bioaccumulative chemicals on human health, virtually the entire Great Lakes watershed is under fish advisories due to elevated levels of various chemicals in fish, including PCBs, dioxin/furans, mercury, toxaphene, chlordane, and others. Also, published Agency for Toxic Substances and Disease Registry (ATSDR) health studies have shown that, within the Great Lakes basin, populations continue to eat fish that contain PCBs and that there are significant health consequences associated with that fish consumption. (1996 EPA/ATSDR summary paper *Public Health Implications of PCB Exposures.*)

That many bioaccumulative chemicals of concern are present in the Great Lakes basin and have the potential to adversely affect human health, aquatic life, and wildlife through long-term exposure constitutes a major concern in the Great Lakes basin because of the extraordinarily long retention time of the waters in the basin. The water, and the bioaccumulative pollutants that have been discharged into it, may be retained in the basin for as long as two hundred years before it is flushed from the system.

From the start, one goal of the Great Lakes Initiative was to use open public forums, drawing upon the extensive expertise and interest of individuals and groups within the Great Lakes community. The idea was to reduce disparities between state water quality controls in the entire basin. The final Guidance represents a milestone in the 30 years of effort on the part of all the Great Lakes stakeholders to define and apply innovative, comprehensive environmental programs in protecting and restoring the Great Lakes.

In developing the *Water Quality Guidance for the Great Lakes System,* three committees were formed to facilitate the process and enhance the involvement of its many stakeholders: A Steering Committee, which was composed of directors of water programs from the Great Lakes States' environmental agencies and the EPA's National and Regional Offices; a Technical Work Group, which consisted of technical staff from the Great Lakes States' environmental agencies, the EPA, the U.S. Fish and Wildlife Service, and the National Park Service; and a Public Participation Group that was made up of representatives from environmental groups, municipalities, industry, and academia. The public participation group observed the deliberations of the other two committees, advised them of the public's concerns, and kept its various constituencies apprised of ongoing activities and issues.

The EPA published the proposed *Water Quality Guidance for the Great Lakes System* in early 1993 and the final Guidance in March 1995. Between the time it was proposed and the time it was issued as final, the EPA reviewed over 26,500 pages of comments, data, and information from over 6,000 respondents commenting on the proposal and from meetings with members of the public. To enhance intergovernmental partnership and to stay abreast of public expectations as the EPA developed the final Guidance, the Agency continued an open dialog with local, state, and tribal government officials, financial officials and coregulators, the regulated community, and environmental interests so that the agency could understand and openly discuss their concerns. During the post-proposal process, the EPA participated in several meetings with the eight Great Lakes States as coregulators of the water quality program and over 35 such meetings with over 500 other stakeholder representatives.

In addition, and critical to the success of the Agency's multimedia efforts to protect and restore the basin's ecosystem, was the adoption of numeric water quality criteria for bioaccumulative chemicals of concern and other toxic pollutants by the Great Lakes States and Native American tribes. These numeric criteria will serve as the water quality goals for other elements of the Great Lakes program, such as Remedial Action Plans and Lakewide Management Plans which address both point and nonpoint sources of pollution.

EPA'S DIOXIN REASSESSMENT

Another current activity at the EPA related to the regulation of chlorinated compounds is EPA's dioxin reassessment. In reviewing the mounting scientific evidence on the potential for adverse effects of chlorinated compounds on human health and the environment, the Agency became increasingly aware of the potential risk to human health and the environment that was presented by one chlorinated compound: dioxin. Over five years ago, the EPA initiated an extensive reassessment of the potential effects of dioxin. This reassessment, which is well underway, is considering both the cancer and noncancer risks of a broad class of compounds that includes dioxins, dibenzofurans, and dioxin-like polychlorinated biphenyls (PCBs).

In the fall of 1994, the EPA released a draft report that concluded that the general public may be affected by a broad range of ill effects (ranging from endometriosis to cancer) resulting from exposure to dioxin and related compounds in the environment. The draft report was reviewed by a 39-member panel of the agency's Science Advisory Board. In May 1995, the Board's findings were presented; and the agency will address them in developing the final reassessment. This final reassessment was planned for release in December 1996.

The implications of the dioxin reassessment on water programs could be significant. The EPA's policy has been to continue moving forward in implementing dioxin control programs while the reassessment is underway. Once the final reassessment is completed, a number of programs carried out by the Office of Water will be affected:

- development of revised dioxin water quality criteria for human health protection supporting water quality standards and permits.
- development of regulations for dioxin in sewage sludge (biosolids).
- potential revision of the human health criterion for dioxin in the *Water Quality Guidance for the Great Lakes System,* guidance on fish consumption advisories, and guidance on managing dioxin-contaminated dredged material.

NATIONAL ACADEMY OF SCIENCES STUDY ON HEALTH EFFECTS IN HUMANS AND WILDLIFE FROM HORMONE-RELATED ENVIRONMENTAL TOXICANTS

Another area currently receiving focused attention under the EPA's new approach to regulation are hormone-related environmental toxicants and their possible health effects. In late 1993 and early 1994, there was considerable attention concerning exposure to environmental contaminants that mimic or interfere with hormonal processes in humans and animals. Some of the effects attributed to this exposure include increased breast cancer in women, decreased sperm count in men, and reproductive impairment in wildlife. A congressional hearing on a possible relationship between estrogenic pesticides and breast cancer also was held to explore these concerns.

The EPA has initiated a significant in-house research program on the effects of endocrine disruption. The Department of the Interior, the Centers for Disease Control, and the EPA have asked the National Academy of Sciences to develop a report of the health effects in humans and wildlife from hormone-related environmental toxicants. Participants in this study include experts in the fields of ecology, comparative endocrinology, immunology, mammalogy, medicine, reproductive physiology, toxicology, wildlife management, tumor biology, and zoology. The objectives of this study are to evaluate findings from the diverse research on the health and ecological impacts associated with endocrine disrupters; identify the conclusions that can be drawn from existing data; and establish a research agenda that would help clarify remaining uncertainties. The recommendations developed by the National Academy of Sciences will enable the EPA to focus on several key activities under the Safe Drinking Water Act and the Clean Water Act.

The project began during the summer of 1995, and the report will become available in April 1997. The information from this study — and other related efforts — should enable us to reduce exposure to those hormone disrupters found to be significant threats to humans, aquatic organisms, and wildlife.

THE PULP AND PAPER RULE

An example of the EPA's new approach to rule making for an industry is the final pulp and paper rule. In 1993, the EPA requested comments on proposed regulations that would reduce air and water pollutant emissions from pulp and paper mills. Those proposed regulations were based on the information, data, and industry practices that were current at the time.

The pulp and paper rule making represented a monumental effort. It has been considered a landmark rule for the agency because it analyzed the manufacturing process as a whole and addressed multi-media discharges. It was the EPA's first "cluster rule."

Since the proposal, the Agency has aggressively continued to seek input from industry experts and the concerned public as we develop the final rule. During this time, the pulp and paper industry has committed to reducing dioxin formation in the bleach plant. Significant reduction of dioxin has been achieved through modification of the bleaching process. The agency is using the most recent information, while the industry is experimenting and implementing major changes in bleaching technology.

Since 1993 we have reviewed thousands of comments on the proposed rule. We have also collected new data on pollution prevention processes, treatment technologies, costs, and environmental effects. We have met with stakeholders, including industry and environmental leaders, to discuss new processes and technologies. I have seen first-hand where mills had been using conventional bleaching processes for many years but had recently invested in new processes that are cleaner and more efficient. The Assistant Administrator for Water, Bob Perciasepe, and I have toured mills together to see the differences between old processes that were in use for many years and alternative processes that have been suggested as better choices for the final rule.

As we move toward issuing a final rule, we have tried to incorporate flexibility and common sense into the regulatory framework. In July 1996, we announced in the *Federal Register* that the EPA's long-term environmental goals for the industry include not only regulations, but also incentives to reward those mills choosing advanced pollution prevention technologies. The EPA introduced this incentives program, in part, to encourage mills to invest in environmental improvements beyond those that might be required through regulation.

The incentives program stemmed from the EPA's extensive discussions with a range of stakeholders. The EPA's senior management in both the air and water programs have been meeting regularly with vice presidents and chief executive officers of some of the major pulp and paper companies in the country. We have also met with leaders in the environmental community. We have supported and encouraged dialog among all the stakeholders. The pulp and paper industry and its environmental issues are complex. The EPA plays an important role, addressing difficult scientific and policy issues as the agency responds to the environmental statutes and, at the same time, balances competing interests.

In all of our discussions and throughout our substantial data collection, we have seen real environmental progress. Mills are changing their processes, and we are encouraged with the environmental results that the industry has achieved since the pulp and paper rule was proposed. Among the more noteworthy improvements is the reduced discharge of chlorinated compounds, including dioxin. As an illustration, 6 of the 22 dioxin-related fish advisories (downstream from 26 bleached kraft mills) have been lifted based on reduced dioxin levels in fish.

Much of the controversy that followed the proposed rule focused on the technology basis for the effluent limitations. Based on the information available in the summer of 1996, the EPA determined that two technologies identified in the proposal merit careful consideration for the final rule. Although, as of this writing, the analysis is not yet complete and the decisions are not yet made, both options are being given equal weight. Discharges of dioxins and furans will be substantially reduced under either of the options.

CONCLUSION

The U.S. Environmental Protection Agency is, and always has been, a dynamic agency. We are changing our philosophy and we are changing the way we do things — in particular, the way we develop regulations. Our new approach involves "cluster rules," where they are appropriate; stakeholder involvement, to the greatest extent possible; and geographic coverage, wherever it is appropriate to protect sensitive ecosystems. Our goal is to develop pollution control regulations that are based on sound scientific information that will reach the right balance where the environment and human health are protected and the regulations that we develop are cost-efficient.

REFERENCES

1. U.S. EPA. June 1994 (draft). "Estimating Exposure to Dioxin-Like Compounds" (EPA/600/6-88/005Ca)
2. U.S. EPA. August 1994 (draft). "Health Assessment Document for 2,3,7,8-Tetrachlorodibenzo-*p*-Dioxin (TCDD) and Related Compounds" (EPA/600/BP-92/001c).
3. National Research Council (National Academy of Sciences), "Hormonally-Active Agents in the Environment," (to be published in summer of 1997).
4. *Federal Register,* Vol. 60, No. 56, March 23, 1995. Rules and Regulations (015366). "Final Water Quality Guidance for the Great Lakes System; Final Rule."
5. Kubiak, T.J., H.J. Harris, L.M. Smith, T.R. Schwartz, D.L. Stalling, J.A. Trick, L. Sileo, D.E. Docherty, and T.C. Erdman. 1989. Microcontaminants and Reproductive Impairment of the Foster's Tern on Green Bay, Lake Michigan — 1983. *Arch. Environ. Contam. Toxicol.,* 18:706–727.

6. Auerlich, R.J., R.K. Ringer, and S. Iwamoto. 1973. Reproductive Failure and Mortality in Mink Fed on Great Lakes Fish. *J. Reprod. Fertil. Suppl.,* 19:365.
7. Baumann, P.C., W.D. Smith, and W.K. Parland. 1987. Tumor Frequencies and Contaminant Concentrations in Brown Bullheads from an Industrialized River and a Recreational Lake. *Transactions of the American Fisheries Society.* 116:79–80.
8. U.S. EPA and the Agency for Toxic Substances and Disease Registry. 1996. "Public Health Implications of PCB Exposures."
9. *Federal Register,* Vol. 58, No. 241, December 17, 1993, Proposed Rules (066078). "Effluent Limitations Guidelines, Pretreatment Standards, and New Sources Performance Standards: Pulp, Paper, and Paperboard Category; National Emission Standards for Hazardous Air Pollutants for Source Category: Pulp and Paper Production; Proposed Rule."
10. *Federal Register,* Vol. 61, No. 136, July 15, 1996, Proposed Rules (036835). "Effluent Limitation Guidelines, Pretreatment Standards, and New Source Performance Standards: Pulp, Paper, and Paperboard Category; National Emission Standards for Hazardous Air Pollutants for Source Category: Pulp and Paper Production; Availability."

Chapter 20

Assuring the Environmental Safety of Kraft Pulp Bleaching: A Cost-Benefit Analysis

L.M. Lancaster, J.J. Renard, and R.B. Phillips

ABSTRACT

The goal of assuring the safety of bleached kraft mill effluent can be achieved without necessarily implementing oxygen delignification. The environmental and economic impacts of various process options are discussed from the point of view of implementation in the near future to meet the requirements of the Environmental Protection Agency's impending cluster rule, and in the perspective of long-term process development needs for the pulp and paper industry.

INTRODUCTION

The outstanding response of the pulp and paper industry to the discovery of minute quantities of dioxins in the effluents of bleached pulp mill remains an unmatched example of swift, voluntary action by an industry to correct a perceived environmental problem. In less than five years, the industry as a whole had decreased dioxin emissions by over 90%.[1] During this period, our own company decreased dioxin formation and release in its 11 bleached pulp mills by 98.5%. Both 2,3,7,8-tetrachlorodibenzodioxin and 2,3,7,8-tetrachlorodibenzofuran (2,3,7,8-TCDD and 2,3,7,8-TCDF) are now below the quantification level of 10 ppq in our mill effluents. These results were obtained by implementing a series of in-plant equipment and operations modifications. International Paper achieved this with a capital investment of over $120 million during one of the deepest and longest recessions in recent years.

The dioxin issue triggered concerns about the potential formation of other toxic, bioaccumulable organic compounds as by-products of chlorine bleaching. The EPA decided to address those concerns as part of their integrated effort to minimize the environmental impact of paper manufacturing by piloting the cluster rule program.[2] The agency's proposal to address the chlorinated organic issue would be to:

- reduce the concentration of TCDD and TCDF below quantification levels in the effluents leaving the bleach plant and, therefore, in the treated effluents discharged to the receiving bodies of water.
- decrease the concentrations of polychlorinated phenolic compounds (PCPC) commonly found in bleach plant effluents below their respective quantification limits (also at the bleach plant discharge).
- limit the total amount of chlorinated organic compounds discharged with the mill effluent to less than 0.156 kg/ton of pulp. The adsorbed organic halogen (AOX) tests were selected to measure the concentration of chlorinated organic compounds.

These proposed regulatory limits were based on a set of values determined at a newly built mill operating the Best Available Technology (BAT) selected by the agency. The BAT process sequence minimizes the concentration of residual lignin entering the bleach plant by either: maximizing the amount of lignin removed in the digester (extended delignification); or by delignifying the pulp with oxygen and alkali (oxygen delignification, or OD) ahead of bleaching. Final delignification and bleaching are effected without using any elemental chlorine (i.e., by using only chlorine dioxide, oxygen, hydrogen peroxide, and caustic), otherwise known as elemental chlorine–free (ECF) bleaching. We have found, however, that all compounds of potential biological significance can be reduced to similar levels without oxygen delignification.

ENVIRONMENTAL IMPACT OF CHLORINATED ORGANIC COMPOUNDS

As with any other class of compounds, chlorinated organics differ widely in their individual toxicity, in their propensity to persist in the environment, and in their ability to accumulate in various organisms. Bioaccumulability is particularly important in the case of bleach plant effluents which are typically discharged, after biological treatment, into nearby rivers, lakes, and estuaries. Accumulation of these compounds in fish and other aquatic organisms could lead to their propagation throughout the food chain to other organisms not exposed directly, such as birds and mammals (including humans). An instructive way of considering the ecological impact of chlorinated organic compounds is to look at their position in the toxicity-bioaccumulation grid (Figure 20.1). Polychlorinated dioxins and furans in the upper right quadrant are both highly toxic and bioaccumulative. Polychlorinated phenols in the lower right quadrant have moderate to low toxicity, but are highly bioaccumulative. Therefore, it is appropriate to minimize the formation and release of these classes of compounds as proposed by the cluster rule.

The other components of the AOX in bleach plant effluents tend to fall into the lower left quadrant, and have both low toxicity and bioaccumulativity. In fact, extensive field and laboratory studies have established that the AOX fraction of bleach plant effluents is not responsible for any known biological response in exposed aquatic organisms, including subtle effects such as induction of detoxifying enzymes, such as mixed function oxidases.[3] Moreover, numerous studies have shown that AOX is readily dechlorinated and degraded by sunlight[4] and microorganisms.[5] Therefore, the regulation of these compounds is not imperative and should be subjected to a rational cost-benefit analysis.

FORMATION OF CHLORINATED ORGANIC COMPOUNDS DURING ECF PULP BLEACHING

By specifying oxygen delignification as part of BAT, the EPA assumes that reducing the amount of residual lignin entering the bleach plant will decrease the amount of chlorine-bearing bleaching chemical

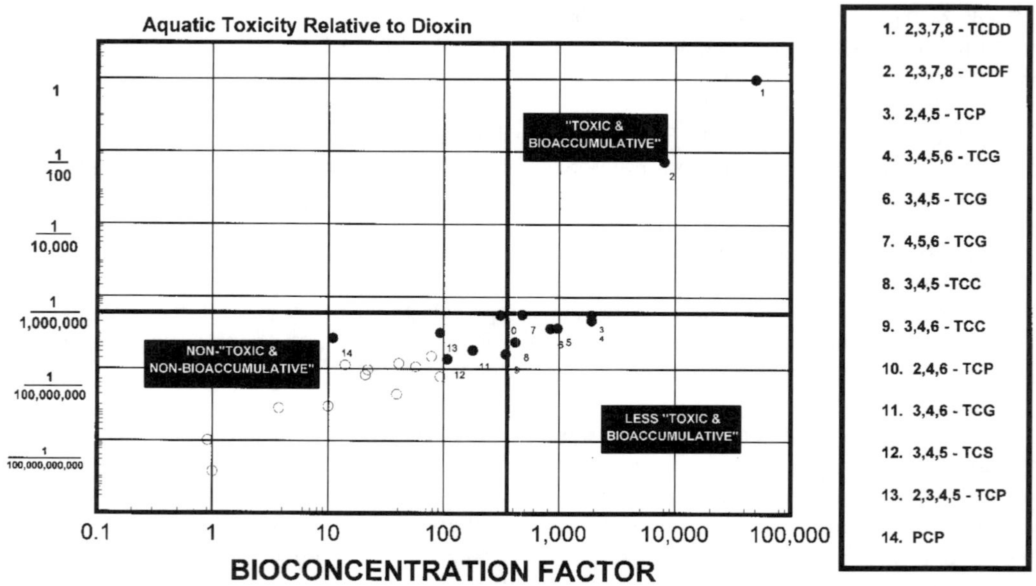

Figure 20.1. EPA assessment of aquatic toxicity and bioaccumulation.

used, and therefore the amount of chlorinated organic compounds formed. This is certainly true if chlorinated organic compounds are quantified through an aggregate measurement, such as AOX; however, this general logic does not necessarily hold true for specific chlorinated compounds such as polychlorinated dioxins and phenols.

The National Council for Air and Stream Improvement (NCASI) has recently surveyed several U.S. bleached pulp mills using ECF bleaching (Table 20.1).[6] When present, 2,3,7,8-TCDD was always below the level of detection whether the mill practiced oxygen delignification or not. Polychlorinated phenolic compounds were also below their level of detection in 99.8% of the cases as long as ECF bleaching was used; however, in a small number of instances, 2,3,7,8-TCDF was detected. The incidence of detection was surprisingly higher when oxygen delignification was used (12%) than when it was not (4%). The reason for this unexpected result is not known. Nevertheless, it suggests that although ECF bleaching, whether preceded by oxygen delignification or not, will virtually eliminate the formation of the potentially toxic, bioaccumulative chlorinated compounds. Demanding the inclusion of oxygen delignification will not add a measure of safety and might even be counterproductive.

This conclusion, based on actual mill data, is consistent with the results obtained in laboratory investigations under carefully controlled conditions. As shown in Figure 20.2, when elemental chlorine is present, the formation of PCPC increases with increasing lignin content.[7] However, when elemental chlorine is absent, the lignin content of the pulp has little or no impact on PCPC formation.

Table 20.1. Impact of Oxygen Delignification on Dioxins Discharged from ECF Mills.

Dioxins and Furans	ECF	OD + ECF
Number of mills	9	4
Number of analyses	114	51
Percent of TCDD nondetects	100%	100%
Percent of TCDF nondetects	96%	88%
Polychlorinated Phenolic Compounds	**ECF**	**OD + ECF**
Number of mills	9	6
Number of analyses	1704	876
Percent of PCPC nondetects	99.8%	99.8%

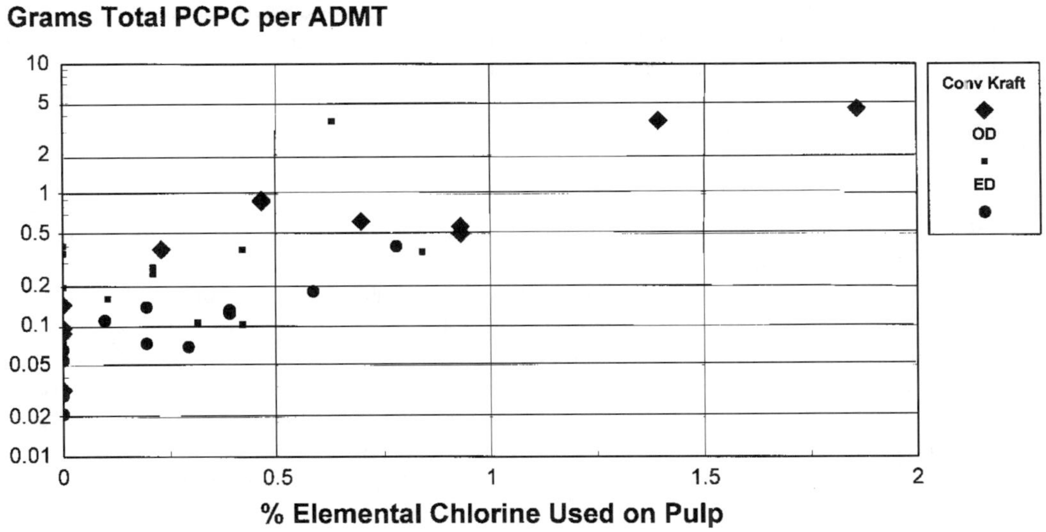

Figure 20.2. Pulp lignin content is not a factor in formation of PCPC at 100% sub. Unbleached Kappa Nos.: Conventional Kraft-31; OD-14; ED-13. *Source:* Tsai et al., *Tappi,* August, 1994.

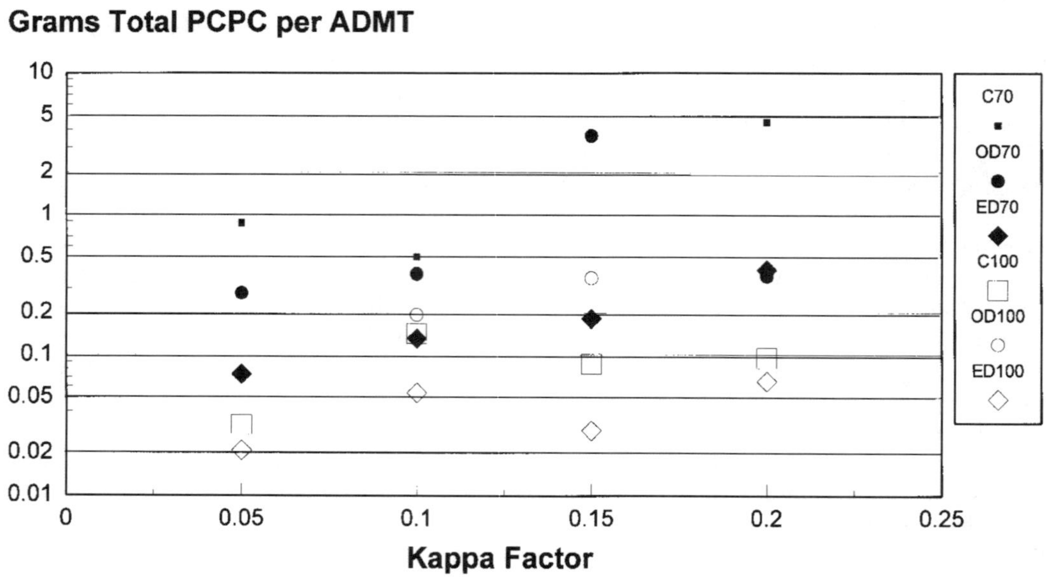

Figure 20.3. "Active chlorine factor" is more significant to PCPC formation than pulp lignin content. Unbleached Kappa Nos.: Conventional Kraft-31; OD-14; ED-13. *Source:* Tsai et al., *Tappi,* August, 1994.

Figure 20.3 shows that at 100% substitution, there is no evidence that OD produces fewer PCPCs than either ED or conventional cooking. In fact, examination of the individual PCPC concentrations (Figure 20.4) indicates that OD sequences tend to produce more polychlorinated catechols and guaiacols than the bleaching of either ED or conventionally delignified pulps. Recently, Graves et al.[8] found that oxygen-delignified pulps produce systematically more 4,5-dichloroguaiacol than the corresponding kraft pulp before OD, under the same bleaching conditions, from conventional chlorine bleaching to ECF (Figure 20.5).

Therefore, the evidence now available indicates that ECF bleaching alone will virtually eliminate the formation of TCDD, TCDF, and PCPC. Although OD will undoubtedly reduce the amount of AOX formed, it fails to further reduce the formation of TCDF and certain PCPCs and therefore does not provide an additional measure of safety over ECF.

COST-BENEFIT ANALYSIS

To fully assess the cost of reducing the discharge of chlorinated organic compounds in a bleached kraft mill operation, the overall capital and operating costs for the mill must be considered. It is not simply a matter of quantifying only the capital and operating cost impact of adding oxygen delignification. The impact on other parts of the mill, and the resulting effect on compliance with other regulatory issues, such as air pollutant emission control, have to be included as well.

For example, if only ECF bleaching is implemented to eliminate the formation of toxic, bioaccumulable chlorinated organic compounds, the quantity of black liquor solids sent to the recovery furnace is unaffected and amounts to about 3,600 lb of solids/ton of pine pulp (Figure 20.6). However, if oxygen delignification is implemented, the quantity of solids sent to the recovery furnace will rise by 120 lb/ton, a net increase of 3% (Figure 20.7). These other impacts are possible:

- To minimize the steam needed to reach the temperature required by the oxygen delignification process, the screen room water circuits must be closed, thus sending more solids to the recovery furnace.

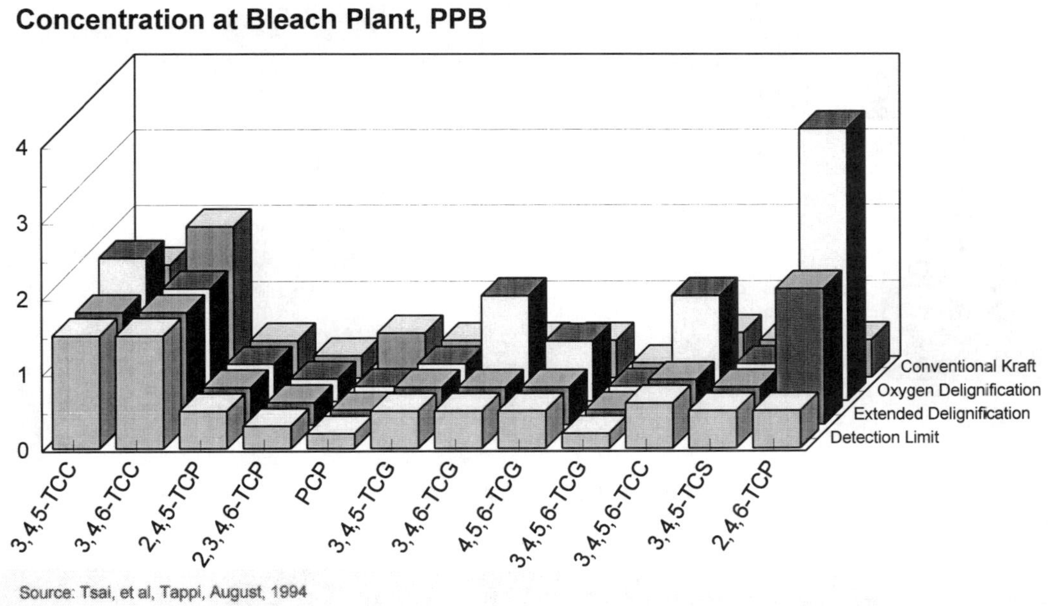

Figure 20.4. Production of polychlorinated phenolic compounds. *Source:* Tsai et al., *Tappi,* August, 1994.

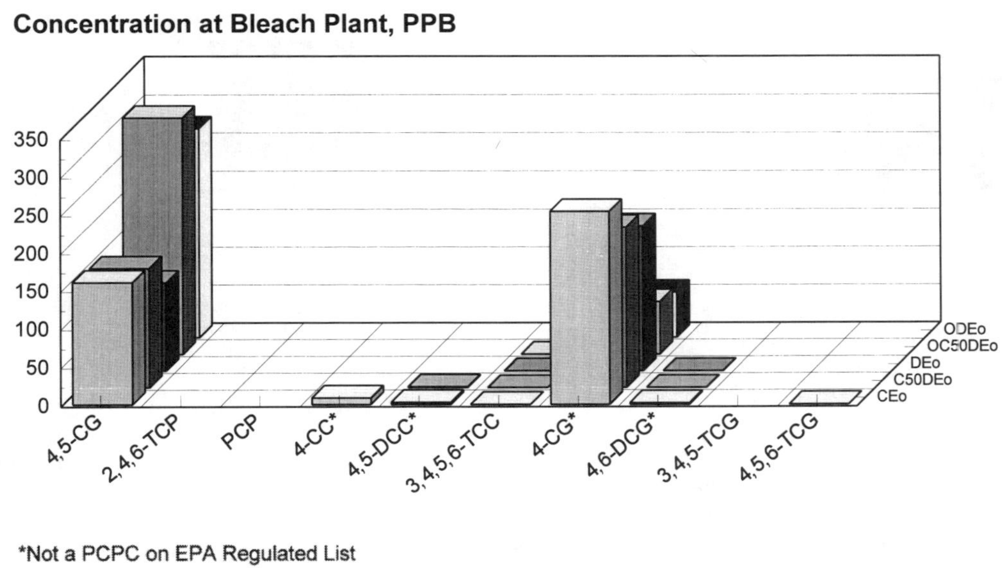

Figure 20.5. Formation of chlorinated phenolics in alternative bleaching processes. *Source:* Graves et al., to be published.

- The number and composition of the process vents to be collected and incinerated to meet the air pollution regulations of the cluster rule will change somewhat, resulting in a different cost picture.
- The oxygen delignification stage is more selective than conventional pulping, but less selective than ECF bleaching in removing lignin (Figure 20.8). Therefore, at constant bleached pulp production, more wood will be required for the OD+ECF option than for ECF alone.

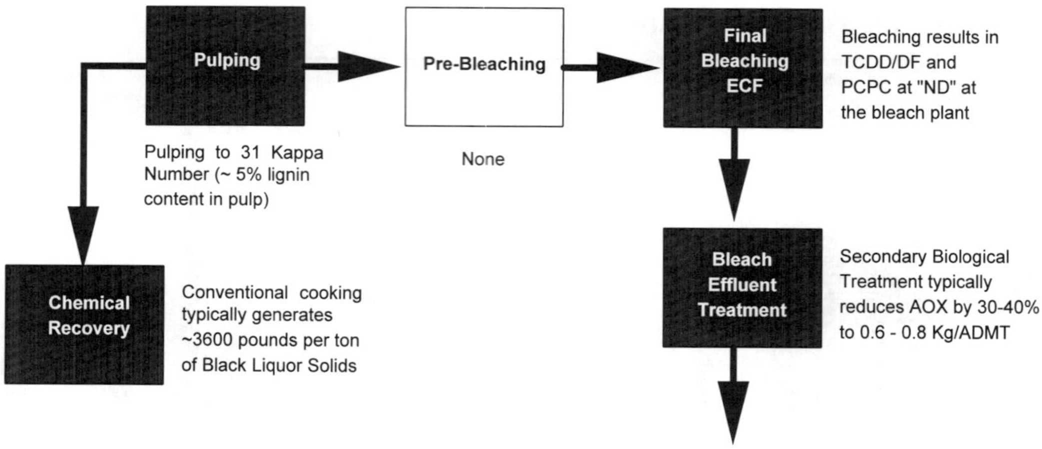

Figure 20.6. Industry proposal.

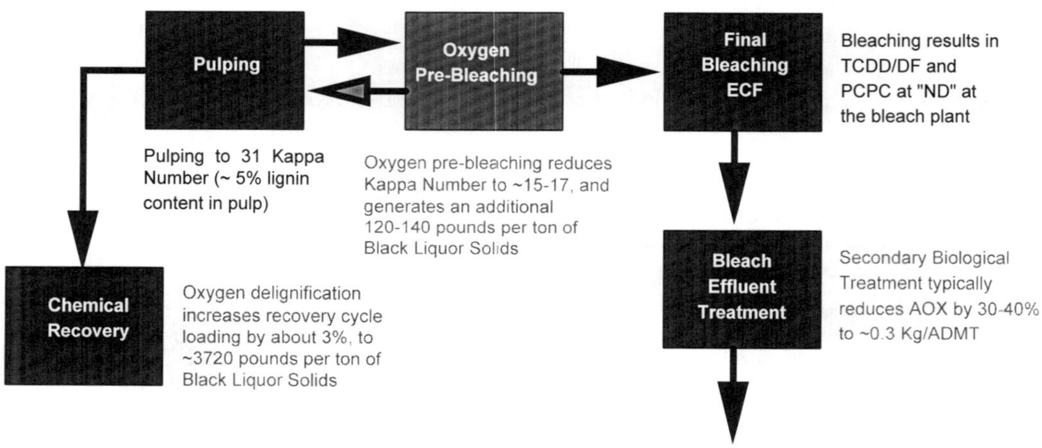

Figure 20.7. EPA BAT proposal.

It is clear that a comprehensive, thorough analysis is inherently complex and mill-specific. We chose International Paper's Moss Point, Mississippi, mill to illustrate the major issues associated with retrofitting an existing mill. The mill was built in the late 1920s and uses batch digesters, conventional drum washers, a pressurized closed screen room, and two conventional bleach lines. The pulp mill makes both hardwood and pine bleached kraft pulps used to manufacture bristol, folding carton, and fast-food bag paper grades. Bleaching operations were converted to ECF in November, 1993; and since that time neither the bleach plant sewer nor treated mill effluents contain detectable levels of TCDD, TCDF, or PCPC.

Using average cost figures for the industry, we have calculated that the Moss Point Mill will have to invest in new equipment to meet the proposed cluster rule regulations. These changes will increase the annual operating cost by $28.9 million/year (Figure 20.9), accounting for the net change in operating costs and the cost of capital. A large part of this increased cost is attributable to the construction of a new low-odor design recovery furnace which may be required to meet the air pollutant reductions as part of the cluster rule known as MACT II. The increase in annual operating cost falls to $9.9 million/year if this and other air pollution requirements are relaxed, such as:

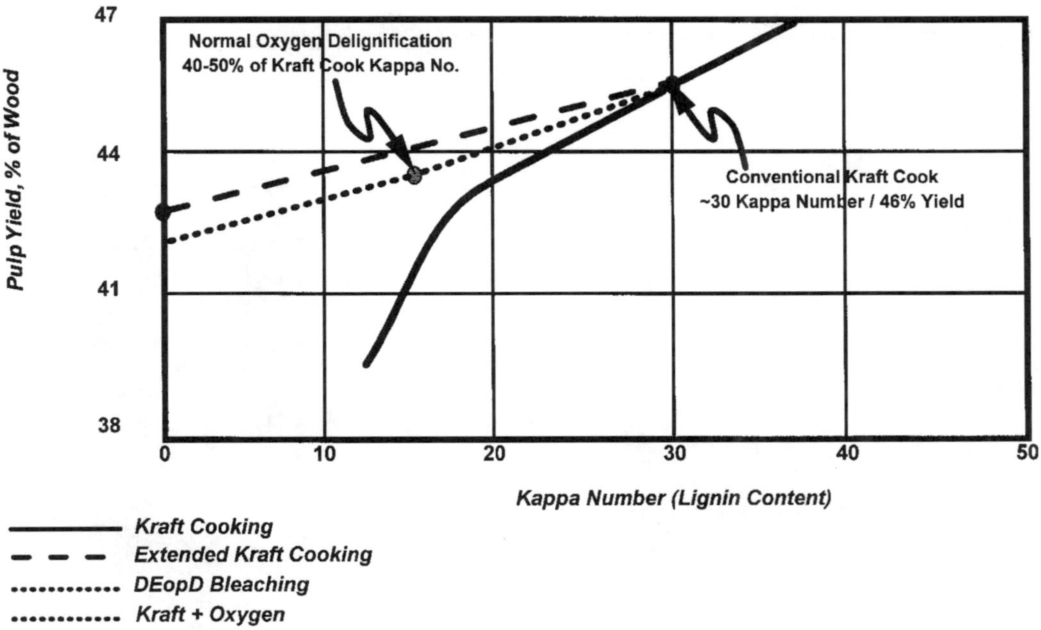

Figure 20.8. Pulping/bleaching selectivity.

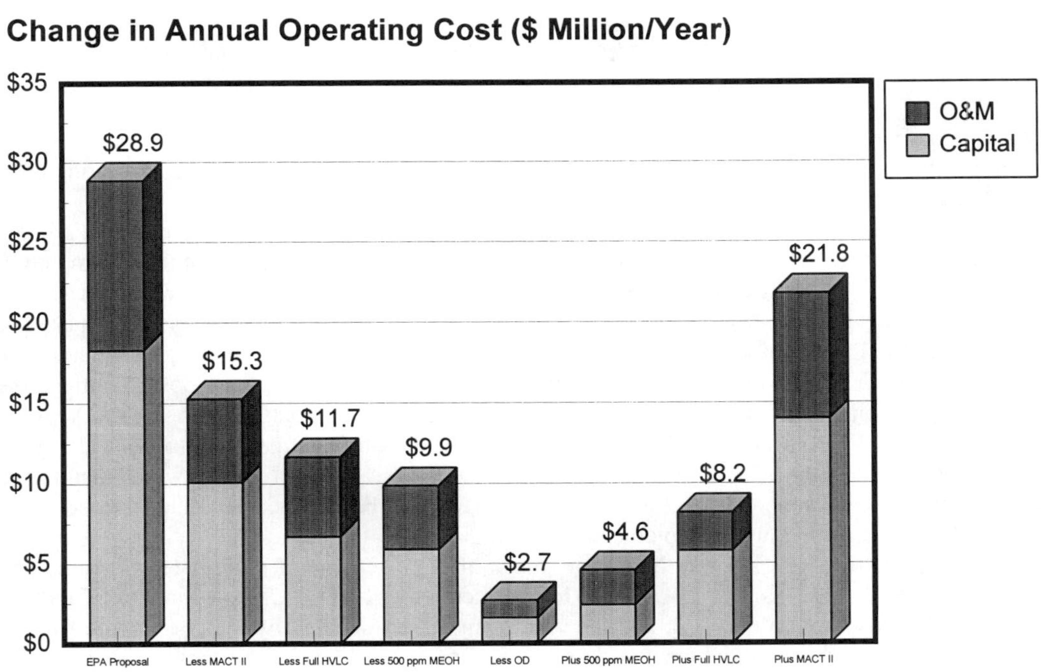

Figure 20.9. Operating cost impact at International Paper, Moss Point, Mississippi.

- less than total collection and incineration of high-volume, low-concentration (HVLC) gases.
- the increase of the floor concentration of methanol for obligatory steam stripping of contaminated condensates.

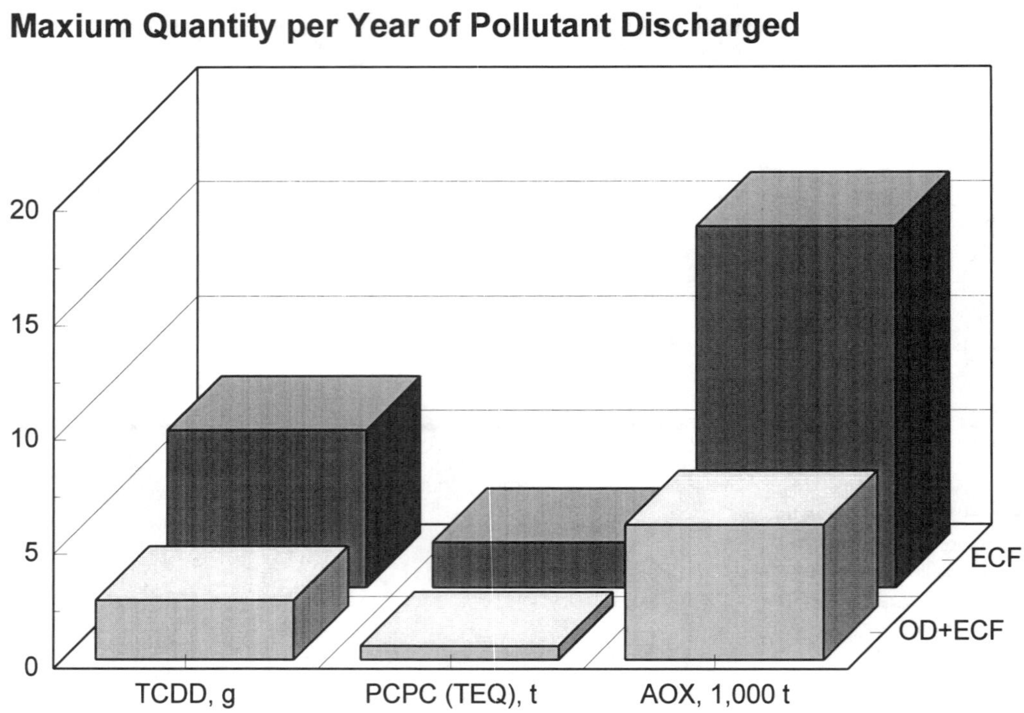

Figure 20.10. Impact of oxygen delignification on discharged pollutants.

If we drop the requirement for OD, the cost penalty decreases to \$2.7 million/year. Are the environmental benefits gained commensurate with the \$7.2 million in increased annual operating costs contributed by inclusion of oxygen delignification?

As stated earlier, the Moss Point Mill effluent, operating with ECF bleaching, does not contain detectable quantities of TCDD, TCDF, or PCPC. Let's assume for the moment that these compounds are present in ECF effluent at their detection limits; and let's further assume they are reduced after OD+ECF in direct proportion to the reduction in the use of total available chlorine between these two options. The results, expressed as the maximum quantity of each pollutant discharged annually, are shown in Figure 20.10. If, for the sake of comparison, we calculate the prorated cost of eliminating the discharge of one ton of pollutant by implementing oxygen delignification, reducing AOX would cost \$36,200/ton, PCPC a substantial \$423 million/ton, and dioxin an extraordinary \$47.2 trillion/ton (Figure 20.11). Given the evidence we have discussed here, these prohibitive costs would result in no or insignificant environmental benefits. Significant costs will be incurred with no increase in the margin of safety afforded by the implementation of ECF alone.

Expanding the analysis beyond the narrow focus of the discharge of chlorinated organic compounds, it is apparent (Figure 20.12) that the lower selectivity of OD+ECF compared to ECF alone increases the consumption of wood, and the various measures dictated by the cluster rules have a profound and often unexpected impact on fossil fuel usage and the ensuing net carbon dioxide emission. This illustrates the need for a truly holistic approach to the cost-benefit analysis of environmental protection measures and away from narrowly focused issues such as chlorinated organic compounds.

FUTURE DEVELOPMENTS IN BLEACHING TECHNOLOGY

It is often argued that oxygen delignification is a necessary step to the total elimination of chlorinated organic compounds either using totally chlorine-free (TCF) bleaching sequences, or ECF bleaching combined with any one of several proposals to achieve a totally effluent-free (TEF) bleached

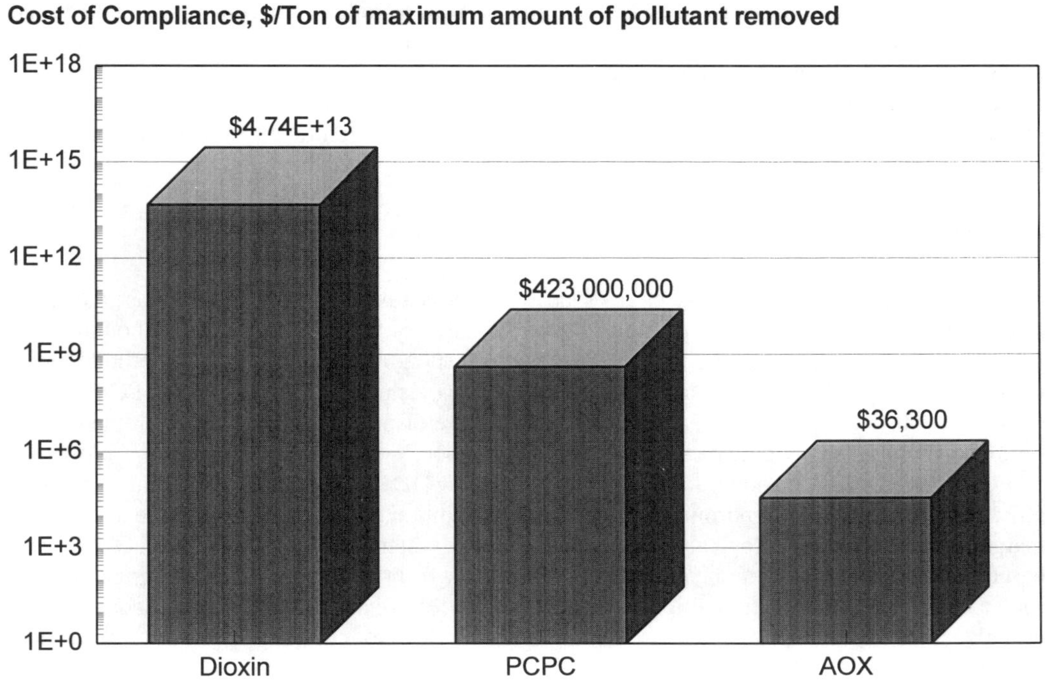

Figure 20.11. The cost of incremental reductions is extraordinary.

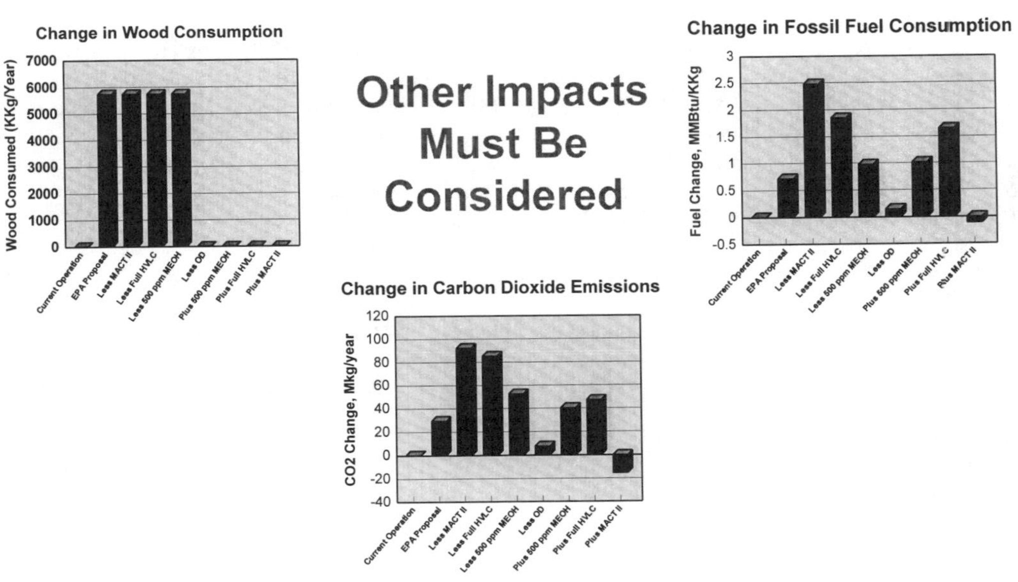

Figure 20.12. Impact analysis of cluster rule options, International Paper, Moss Point, Massachusetts.

mill. Assuming that those are desirable goals, there is mounting evidence that emerging technologies could provide a more effective way of decreasing residual lignin than oxygen delignification. The Lignozym process,[9] for example, uses an enzyme, a laccase, and a mediator (hydroxybenzotriazol) to achieve more extensive and more selective delignification under milder, more environmentally friendly conditions than OD. Ozone, dimethyldioxirane[10] and polyoxymetalates[11] offer new hope of selectively removing the lignin remaining on the fibers after cooking. To mandate the implementation of OD could be condemning the industry to invest a large amount of capital in a process that might be proved obsolete in the next few years without providing an immediate significant environmental improvement over ECF bleaching alone.

CONCLUSIONS

Oxygen delignification is an effective process to remove residual lignin and decrease the amount of chlorine-containing bleaching chemicals used in a subsequent ECF sequence. Although this decrease in chlorine dioxide usage translates into a corresponding decrease in AOX production, the toxic, bioaccummulable chlorinated organic compounds (such as TCDD, TCDF, and PCPC) are not decreased substantially. Moreover, the costs involved are disproportionate to the incremental environmental benefits that would be gained.

ECF conversion was shown to virtually eliminate TCDD, TCDF, and PCPC, and to discharge effluents with no adverse environmental impact. The addition of OD does not provide an additional margin of safety for the ecological quality of the receiving water. The decision to install OD should be based on economic factors alone and will therefore vary from mill to mill. Emerging technologies, such as enzyme delignification, might provide a more effective and environmentally compatible alternative than OD to further decrease the formation and release of chlorinated organic compounds in the future.

REFERENCES

1. Summary of Data Reflected of Pulp and Paper Industry Progress in Reducing TCDD/TCDF. NCASI, New York, Special Report 93-08, August 1993.
2. U.S. EPA/Pulp and Paper Industry Cluster, Proposed Regulations, 40 CFR Parts 63 and 430, *Federal Register,* 58 (No. 241), 66077-66216, December 17, 1993.
3. Carey, J.H., P.V. Hodson, K.R. Murkittrick, and M.R. Servos. 1993. "Recent Canadian Studies on the Physiological Effects of Pulp Mill Effluents on Fish," Environment Canada and Fisheries and Oceans.
4. Caron, R.J. and D.W. Reeve. 1992. *Pulp and Paper Canada,* 93(8):T209.
5. Hwang, H.M., R.E. Hodson, and R.F. Lee. 1986. *Environ. Sci. Technol.,* 20(10):1002.
6. NCASI. March 1995. New York, Data Presented to the U.S. EPA.
7. Tsai, T.Y., J.J. Renard, and R.B. Phillips. 1994. *TAPPI Journal,* 77(8):149.
8. Graves, J.W., H. Jameel, and T.W. Joyce. North Carolina State University, submitted for publication in *TAPPI Journal.*
9. Call, H.P. and L. Mücke. Paper presented at the 1995 Non-Chlorine Bleaching Conference, Amelia Island, FL, March 5–9, 1995.

10a. Lee, C.L., K. Hunt, and R.W. Murray. 1992. Proceedings, TAPPI Pulping Conference, TAPPI Press, Book 3, p. 1237.

10b. Ragauskas, A.J. 1993. *TAPPI Journal,* 76(7):87–90.

11. Weinstock, I.A., U.P. Agarwal, J.L. Minor, R.H. Atalla, and R.S. Rainer. Proceedings of the 1993 TAPPI Pulping Conference, November 1-3, 1993, TAPPI Press, Atlanta, GA, Book 2, 519–532.

Chapter 21

A Comparison of Chlorinated Organic Material Produced by Chlorine and Chlorine Dioxide Bleaching of Kraft Pulp

A.B. McKague, X. Shen, and D.W. Reeve

ABSTRACT

A review of the amount and nature of the chlorinated organic material produced by chlorine and chlorine dioxide bleaching of kraft pulp shows AOX and EOX levels in effluent, the EOX level in pulp, and the number of chlorinated organic compounds produced decrease dramatically when chlorine dioxide is substituted for chlorine. A study of the chlorinated organic material in pulp bleached with 100% chlorine dioxide showed trace levels of chlorinated fatty and resin acids, and chlorinated stilbenes were present.

INTRODUCTION

Chlorine dioxide is rapidly replacing chlorine as the pulp bleaching agent of choice in North America because use of chlorine dioxide results in a dramatic decrease in the amount of chlorinated organic by-products produced. Bleaching with chlorine dioxide but in the absence of chlorine is referred to as ECF (elemental chlorine–free) bleaching. AOX (adsorbable organic halide) and EOX (solvent-extractable organic halide) concentrations in effluent decrease 5- to 10-fold[1] and an even larger decrease in the amount of EOX remaining in the pulp occurs.[2–5] There is also a dramatic decrease in the number of chlorinated organic compounds formed. Less is known about the types of compounds remaining in pulp after chlorine dioxide bleaching. This chapter reviews the amount and nature of chlorinated organic compounds produced by these two types of bleaching agents and presents results of an investigation into the nature of chlorinated material present in pulp bleached with chlorine dioxide.

EXPERIMENTAL

Two samples of ECF-bleached southern pine kraft pulp were used in the investigation of chlorinated organic matter in bleached pulp. The bleaching sequence for one sample was DEDED while the other was ODEOD. The pulps were fluffed and extracted on a 500-g scale with acetone in a 12-L Soxhlet extractor. The extracts were evaporated using a rotary evaporator, dissolved in ether, and washed with water. After drying with $MgSO_4$, the ether was evaporated. The EOX of the pulps was determined by analyzing the extracts using a Mitsubishi Organic Halogen Analyzer. The chlorine content of the extracts was also determined by elemental analysis by Galbraith Laboratories, Knoxville, Tennessee. The extract was analyzed by gas chromatography (GC) after methylation with diazomethane and individual compounds identified by gas chromatography/mass spectrometry (GCMS) and comparison with standards. A Hewlett Packard 5890 Gas Chromatograph equipped with a 25-m HP-1 capillary column was used for GC with temperature programming from 100–280°C at 8°/min. GCMS was performed on the same column coupled to a VG analytical ZAB-SE high-mass high-resolution mass spectrometer.

RESULTS AND DISCUSSION

Chlorinated Organic Material in Bleaching Effluent and Pulp

A large number of studies have been carried out on the effect of chlorine dioxide substitution on the characteristics of bleach plant effluent. Because of environmental and market pressures, most of the attention has been directed toward the effects on formation of chlorinated organic material. A comparison of the levels of chlorinated organic material, measured as AOX and EOX, in effluent from conventional chlorine bleaching and 100% chlorine dioxide bleaching (ECF bleaching) is shown in Table 21.1. A 5- to 10-fold decrease in the levels of these parameters occurs when chlorine dioxide is used. Table 21.1 also shows that the number of chlorinated compounds identified to date in ECF effluent is substantially less than in effluent from conventional chlorine bleaching.

Table 21.2 shows that solvent extractable organochlorine (EOX) in ECF pulps is extremely low compared with that of pulps bleached with chlorine. Although only a limited number of studies have been done on the characterization of chlorinated organic material in bleached pulp, a number of compounds have been identified in chlorine-bleached pulp (Table 21.2). In a recent report,[9] chlorinated compounds were found to be absent from pulps bleached with chlorine dioxide, so the number and types of compounds present in these pulps is unknown, and is the subject of work presented later in this chapter.

Types of Chlorinated Organic Compounds Produced in Pulp Bleaching

The types of chlorinated organic compounds found in bleach plant effluent can be divided into different classes, as shown in Table 21.3. Chlorinated acetic acids are present at the highest levels of any single group of compounds, ranging up to 500 g/tonne in conventional chlorine bleaching effluent[10] and up to 100 g/tonne in ECF effluent.[2,11,12] Chlorophenols and chlorophenyl ethers constitute a large percentage of compounds identified in effluent from chlorine dioxide bleaching because of efforts made by the industry to demonstrate the benefits of chlorine dioxide bleaching in reducing levels of these compounds to below that proposed in new EPA regulations.

The types of chlorinated compounds identified in chlorine-bleached pulps are lipophilic in character, and thus have a tendency of preferentially staying in the pulp rather than dissolving in the effluent. Compounds identified are chlorinated fatty acids, chlorinated squalene/prenols, chlorinated sterols, chlorinated betulin (birch pulp), chlorinated phenols and PCDDs/PCDFs.[2,3,7–9] Some compounds found in ECF pulp from this study are presented later in this chapter.

Although it is clear that the levels of chlorinated organic compounds produced by bleaching are minimized by the use of chlorine dioxide, an interesting question is "should different types of compounds be produced?" Different products have been obtained from various model compounds.[13,14] Figures 21.1 and 21.2 show a simplified comparison of the fundamental reactions of chlorine and chlorine dioxide with lignin. Side chain displacement during chlorination and chlorine displacement in the E stage are other important reactions not shown. Although there are similarities in the reactions and nature of the end products, muconic acid structures should be more prominent products of chlorine dioxide bleaching. Simple muconic acids have now been identified in ECF bleaching effluent and others will likely follow.[15] ECF effluents should also contain more oxidized compounds because chlorine dioxide is a more powerful oxidant than chlorine.

Characterization of Extractives from ECF Pulp

Concentrations of EOX were low (Table 21.4) in accordance with data shown in Table 21.2. The percent chlorine in the extracts was also low compared with values of 6.2–8.4 reported for mixtures of softwoods containing predominantly jackpine bleached with low levels of chlorine dioxide substitution.[16] The higher EOX and extract chlorine content of the oxygen prebleached sample may not be significant since only single large-scale extractions were made.

A gas chromatogram of the methylated extract obtained from Pulp Sample B and the structures of the compounds identified are shown in Figure 21.3.

Table 21.1. Chlorinated Organic Material in Total Bleach Plant Effluent.[1]

Parameter	Bleaching	
	Chlorine	Chlorine Dioxide
AOX (kg/t)	3.7–6.8	0.9–1.7
EOX (g/t)	8–70	0.9–10
Number of chlorinated compounds	322	59

Table 21.2. Chlorinated Organic Material in Pulp.[2–9]

Parameter	Bleaching	
	Chlorine	Chlorine Dioxide
EOX (g/t)	40–2000	1 – <25
Number of chlorinated compounds	> 100[a]	?

[a] Estimated.

Table 21.3. Types and Number of Chlorinated Compounds Found in Bleach Plant Effluent.[1]

Type of Compound	Number Reported	
	Chlorine	Chlorine Dioxide
Acids, esters, anhydrides, furanones and pyrones	68	8
Phenols and phenyl ethers	49	17
Aldehydes and ketones	64	10
Hydrocarbons	83	18
Alcohols	19	0
Dioxins and furans	22	1
Miscellaneous	17	5
Total	322	59

Components 38a, 39, and 40 are chlorinated derivatives of oleic acid whose structures are consistent with reported information on products of reaction with chlorine dioxide.[17,18] Component 42b, chlorodehydroabietic acid, is a common product of chlorine bleaching. Components 33, 41, and 43 were identified by mass spectrometry as chlorinated stilbenes that apparently resulted from reaction of pinosylvin or its methyl ether with chlorine dioxide. A rough estimation of the concentration of the chlorinated compounds in the pulp sample ranges from 0.4–10 μg/g pulp. Similar compounds were found in the extract from Pulp Sample A.

CONCLUSIONS

A 5- to 10-fold decrease in the AOX and EOX of bleaching effluent occurs when chlorine dioxide is substituted for chlorine in the bleaching of kraft pulp. An even larger reduction occurs in the level of EOX in the pulp. Over 300 chlorinated compounds have been identified in effluent from chlorine-based bleaching, and a much smaller number in effluent from chlorine dioxide bleaching. Chlorinated fatty acids, chlorinated squalene/prenols, chlorinated betulin, chlorinated phenols and PCDDs/PCDFs have been identified in pulps bleached with chlorine. Traces of chlorinated wood extractives may also be found in pulp bleached with chlorine dioxide.

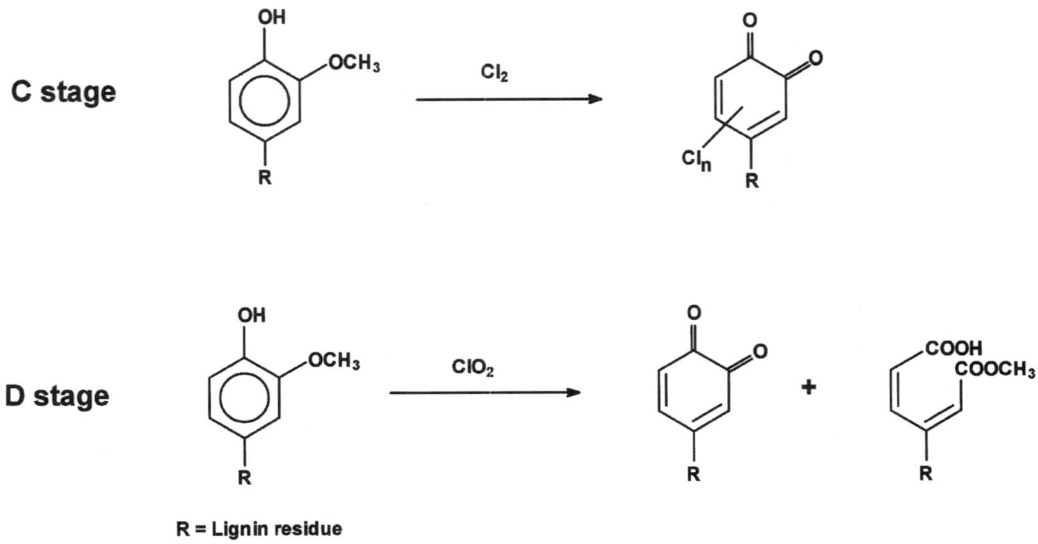

Figure 21.1. Comparison of chlorine and chlorine dioxide bleaching chemistry.

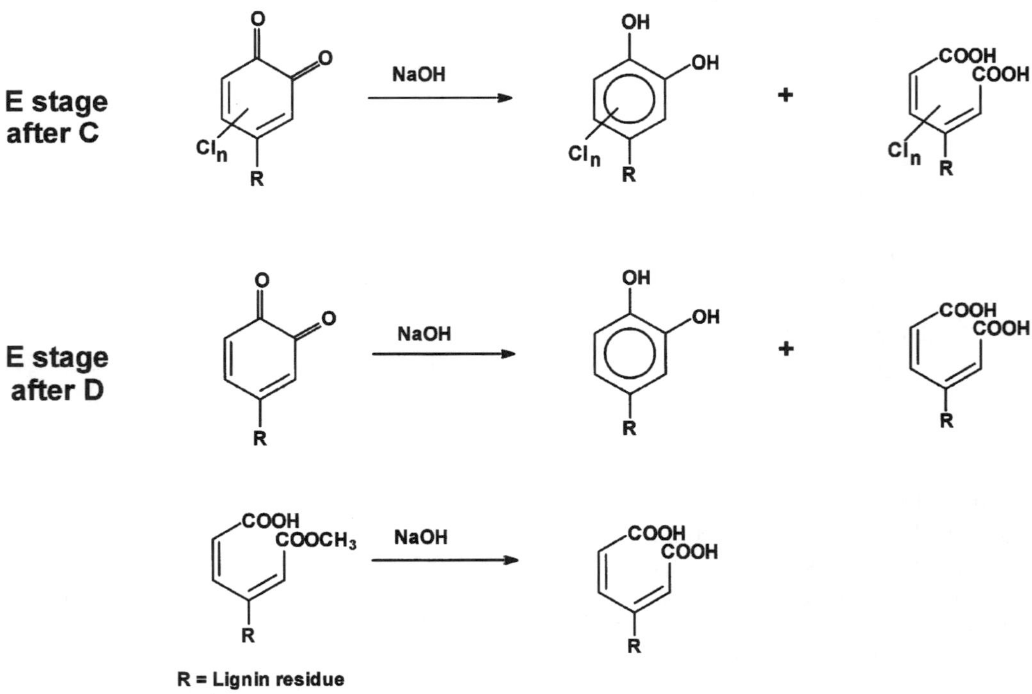

Figure 21.2. E-stage reactions during bleaching.

Table 21.4. Softwood Kraft Pulp Bleached with Chlorine Dioxide.

Pulp Sample	Wood Species	Bleaching Sequence	EOX (g/t)	Percent Chlorine In Extract
A	Southern pine	DEDED	1	0.8
B	Southern pine	ODEOD	4	3.0

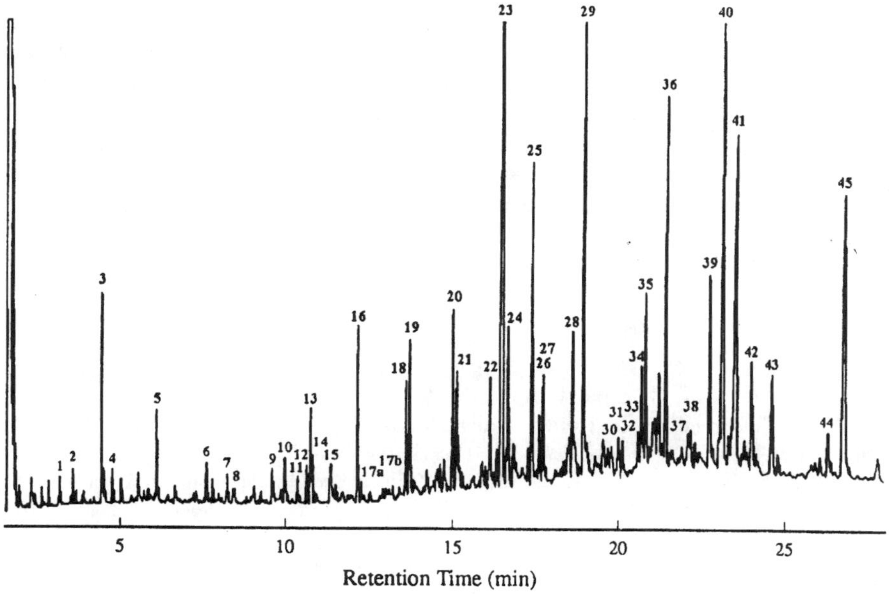

1. $CH_3COCH_2CH_2COOCH_3$
4. $CH_3(CH_2)_4COOCH_3$
5. $CH_3(CH_2)_5COOCH_3$
6. $CH_3(CH_2)_6COOCH_3$
9. $CH_3(CH_2)_7CH{=}CHCOOCH_3$
12. $CH_3(CH_2)_8COOCH_3$
13. $CH_3OOC(CH_2)_7COOCH_3$
16. $CH_3(CH_2)_{14}CH_3$
17. $CH_3(CH_2)_{11}COOCH_3$
18. $CH_3(CH_2)_{15}CH_3$
19. $CH_3(CH_2)_{12}COOCH_3$
20. $CH_3(CH_2)_{16}CH_3$
21. $CH_3(CH_2)_{13}COOCH_3$
22. $CH_3(CH_2)_{17}CH_3$
23. $CH_3(CH_2)_{14}COOCH_3$
25. $C_{16}H_{33}COOCH_3$
26. $CH_3(CH_2)_{18}CH_3$
27. $CH_3(CH_2)_{15}COOCH_3$
28. $CH_3(CH_2)_7CH{=}CH(CH_2)_7COOCH_3$
29. $CH_3(CH_2)_{16}COOCH_3$
30. $CH_3(CH_2)_7CO(CH_2)_8COOCH_3$
31. $CH_3(CH_2)_{19}CH_3$
32. $CH_3(CH_2)_{17}COOCH_3$
34. $CH_3(CH_2)_{20}CH_3$
35. $CH_3(CH_2)_{18}COOCH_3$
37. $CH_3(CH_2)_{21}CH_3$
38a. $CH_3(CH_2)_7COCHCl(CH_2)_7COOCH_3$
38b. $CH_3(CH_2)_{19}COOCH_3$
39. $CH_3(CH_2)_7CHClCHOH(CH_2)_7COOCH_3$
40. $CH_3(CH_2)_7CHClCHCl(CH_2)_7COOCH_3$
44. $CH_3(CH_2)_{24}CH_3$
45. $CH_3(CH_2)_{22}COOCH_3$

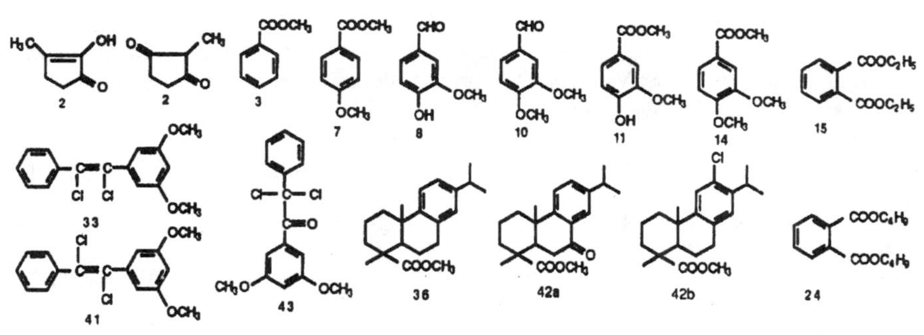

Figure 21.3. Gas chromatogram and compounds identified in extract from Pulp Sample B.

ACKNOWLEDGMENTS

This work was part of a research consortium on "Organochlorine in Pulp and Paper Products, Effluent Treatment and the Environment." Funding was from the Ontario Ministry of Universities and Colleges, University Research Incentive Fund and the following companies: Aracruz Celulose S.A.; Champion International Corporation; CRS Sirrine, Inc.; Fletcher Challenge Canada Ltd.; International Paper; ITT Rayonier Incorporated; James River Corporation; Kymmene Corporation; Nippon Paper Industries Co., Ltd.; Noranda Inc.; The Procter & Gamble Cellulose Co.; Sterling Pulp Chemicals Ltd. and Weyerhaeuser Company. The work was also part of a research consortium on "Characterization, Treatment and Fate of Bleached Plant Effluents." Funding was from Aracruz Celulose S.A.; Boise Cascade Corporation; Champion International Corporation; Georgia-Pacific Corporation; International Paper; James River Corporation; Kymmene Corporation; Nippon Paper Industries Co., Ltd.; Potlatch Corporation; Sterling Pulp Chemicals Ltd. and Weyerhaeuser Company.

NMR and mass spectral services were provided by Dr. A.A. Grey and Dr. H. Pang, respectively, at the University of Toronto Carbohydrate Research Centre which receives funding from the Medical Research Council of Canada.

REFERENCES

1. McKague, A.B. and G. Carlberg. 1996. Effluent Characteristics and Composition, in *Pulp Bleaching — Principles and Practice*, Dence, C.W. and D.W. Reeve, Eds., TAPPI Press, Atlanta.
2. Brunsvik, J.J., R. Grundlius, R. Kordes, and B. Swan. 1989. Environmental Compatible Bleaching of Chemical Pulp, in *1989 Pulping Conference Proceedings*, TAPPI Press, Atlanta, pp. 411–414.
3. Annergren, G., F. Österberg, and P. Lindblad. 1990. Chlorinated Organic Matter in Bleached Kraft Pulp — Effect of Important Process Variables, in *1990 Pulping Conference Proceedings*, TAPPI Press, Atlanta, pp. 185–194.
4. Wilson, R., J. Swaney, D.C. Pryke, C.E. Luthe, and B.I. O'Connor. 1992. Mill Experience with Chlorine Dioxide Delignification, *Pulp Pap. Can.* 93(10):T275–283.
5. Pryke, D.C., P. Winter, G.R. Bourree, and C. Mickowski. 1994. The Impact of Chlorine Dioxide Delignification on Pulp Manufacture and Effluent Characteristics at Grande Prairie, AB. *Pulp Pap. Can.* 95(6):T230–236.
6. Reeve, D.W. and K.M. Weishar. 1990. Chlorinated Organic Matter in Bleached Chemical Pulp Production, Part IV: The Occurrence of Chlorinated Organic Matter in Bleached Pulp. *J. Pulp Pap. Sci.* 16, pp. J118–125.
7. McKague, A.B. and D.W. Reeve. 1991. Identification of Chlorinated Compounds in Bleached Pulp Extracts. *Nordic Pulp Pap. Res. J.* 6, pp. 35–39.
8. Björklund Jansson, M., O. Dahlman, K. Månsson, A. Rutquist, and J. Wickholm. 1991. On the Nature of Chlorinated Organic Matter in Bleached Kraft Pulp, *International Pulp Bleaching Conference*, Stockholm, pp. 123–136.
9. Björklund Jansson, M., P. Wormald, and O. Dahlman. 1995. Reactions of Wood Extractives during ECF and TCF Bleaching of Kraft Pulp. *Pulp Pap. Can.* 96(4):T134–137.
10. McKague, A.B., M. Jarl, and K. Kringstad. *An Up-to-date List of Compounds Identified in Bleaching Effluent as of January 1989*, "SSVL-Miljö 90," project Bleaching, Stockholm, 1988. Rep. available from ÅF-IPK, Box 8309, 104-20, Stockholm.
11. Smith, T.J., P.H. Wearne, and A.F.A. Wallis. 1994. Characteristics of the Chlorinated Organic Substances in Filtrates from Bleaching of Oxygen-Delignified Eucalypt Kraft Pulp. *Water Sci. Technol.* 29, pp. 61–71.
12. O'Connor, B.I., T.G. Kovacs, R.H. Voss, P.H. Martel, and B. Van Lierop. 1994. A Laboratory Assessment of the Environmental Quality of Alternative Pulp Bleaching Effluents. *Pulp Pap. Can.* 95(3):T99–108.
13. McKague, A.B., G.J. Kang, and D.W. Reeve. 1993. Reaction of a Lignin Model Dimer with Chlorine and Chlorine Dioxide. *Holzforschung*, 47, pp. 497–500.
14. McKague, A.B., G.J. Kang, and D.W. Reeve. 1994. Reactions of Lignin Model Dimers with Chlorine Dioxide. *Nordic Pulp Pap. Res. J.* 9, pp. 84–87, 128.
15. Vilen, E., D.W. Reeve, and A.B. McKague. 1996. Identification of Muconic Acids in Bleach Plant Effluent. *Holzforschung*, 50, pp. 575–578.
16. Reeve, D.W. and A.B. McKague. 1991. Chlorinated Organic Matter in Bleached Chemical Pulp Production. Part VII: Characterization of Extractives. *J. Pulp Pap. Sci.* 17, pp. J115–118.
17. Lindgren, B.O. and C.M. Svahn. 1966. Reactions of Chlorine Dioxide with Unsaturated Compounds II. Methyl Oleate. *Acta Chem Scand.* 20, pp. 211–218.
18. Owusu-Gyima, P.K. 1993. "*Chlorine Dioxide Reactions with Unsaturated Fatty Acids,*" thesis presented to the University of Toronto, in partial fulfillment of the requirements for the degree of Master of Applied Science.

Chapter 22

Reconciling the Collision of Visions: The "Cluster Rules"

K.M. Bennett

This book's consideration of the questions surrounding the use of chlorine in the pulp and paper industry will not be complete without considering the regulatory context. The U.S. Environmental Protection Agency's (EPA's) "cluster proposals,"[1] — draft rules — are central to industry's planning concerning its use of chlorine and compounds, just as chlorine is central to EPA's deliberations on the cluster rules.

My task is to describe the chlorine issue for the pulp and paper industry as it relates to the cluster proposals. To establish the context, I will first present a brief profile of the economics of the industry, followed by a short synopsis of the industry's environmental achievements and future challenges. In light of this background, we will consider EPA's future vision of the industry as embodied in the cluster proposals.

Next, we will explore the industry's vision of the future and its agenda for achieving it, focusing in particular on the environmental challenges. Finally, we will discuss how these competing visions could be reconciled.

PROFILE OF THE INDUSTRY

Economics

To understand the controversy concerning regulation of the pulp and paper industry's pulping and bleaching process requires an understanding of the industry's economics. The pulp and paper industry is a national asset, using a highly productive workforce of 700,000 employees in 42 states to turn abundant raw materials into a wide variety of safe and efficient products in environmentally sound and energy efficient plants throughout the United States.[2] The industry contributes positively to the U.S. balance of trade, providing about $10 billion in exports annually.[3]

Pulp and paper is among the most capital intensive of U.S. industries, investing more than $120,000 in plant and equipment for every industry employee. Data from the American Forest & Paper Association show that more than $130 billion of capital have been invested since 1980. The industry predicts that more than $17 billion will have been spent between 1988 and the year 2000 to increase paper recycling capacity alone.[4] The industry invests heavily in plant and equipment which, being large and expensive, must be optimized for maximum productivity to remain a viable investment. See Figures 22.1 and 22.2.

Pulp and paper is a global industry, characterized by markets that respond to international pressures and opportunities. The U.S. industry represents 16% of the pulp mills but 35% of world pulp production; 12% of the world's paper mills but 35% of world paper and paperboard production. While the U.S. economy began to emerge in 1991 from the worldwide recession, the pulp and paper industry's emergence from its earnings slump began only in 1994. Even during the recession, regardless that revenues were not being generated to raise the funds, considerable capital spending had to continue, just to sustain operations. As a result, the industry's debt profile achieved its least favorable ratio in memory as cash flow fell short of spending in eleven of the last fourteen years.[5] See Figures 22.3 and 22.4.

In late 1994 and the first part of 1995 prices rebounded and cash flow improved. Many industry companies seem to be using this period to pay down debt and, to some extent, to return to spending programs for incremental expansion, competitiveness, and productivity increases, rather than exclusively for essential maintenance.

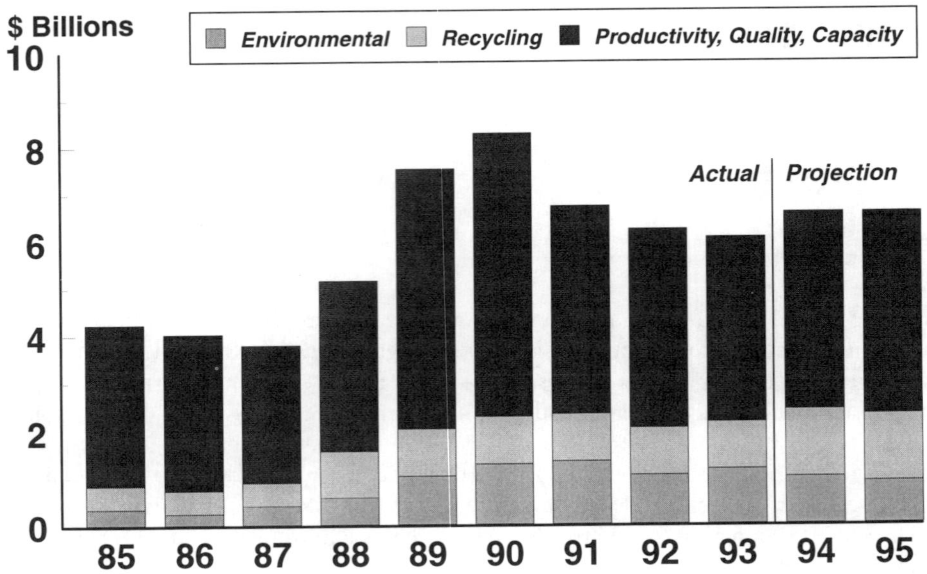

Figure 22.1. Capital expenditures by the primary pulp, paper and paperboard industry. *From:* Bureau of the Census/NCASI/AF&PA.

Figure 22.2. Paper industry capacity utilization.

Given the industry's capital intensiveness, it is essential to the health of the industry to remain capable of producing an attractive return for investors. The U.S. pulp and paper industry must compete for funds not only against the global paper industry, but also against alternative investment opportunities such as other heavy industry, electronics, financial vehicles, and many more. Yet the paper industry's capital intensiveness by nature means that the range of returns will tend to be lower than those of other less capital intensive industries. This drives the industry to optimize production

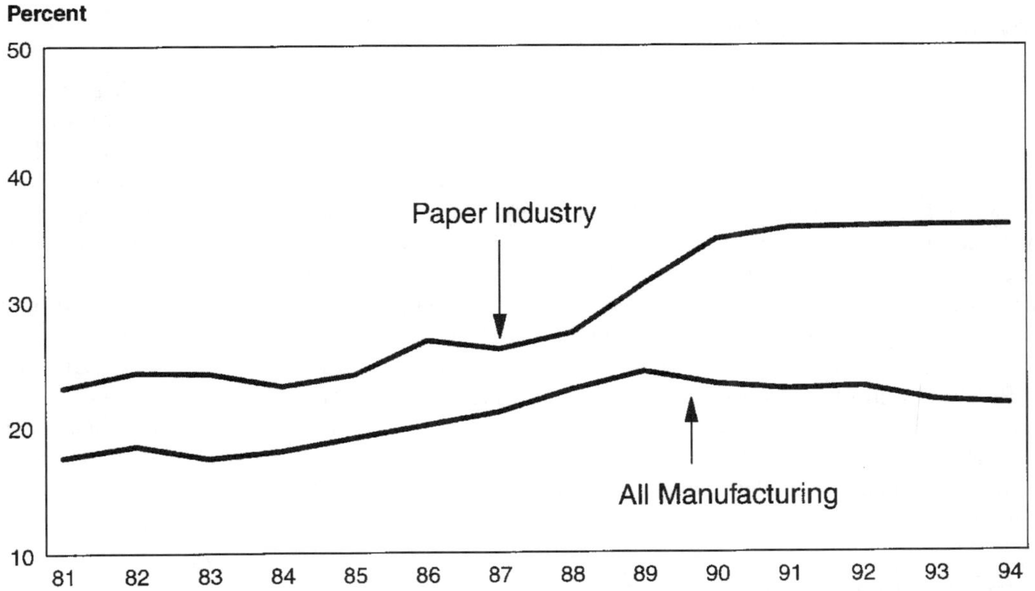

Figure 22.3. Long-term debt as a percentage of assets. Paper industry vs. all manufacturing. *From:* Bureau of the Census.

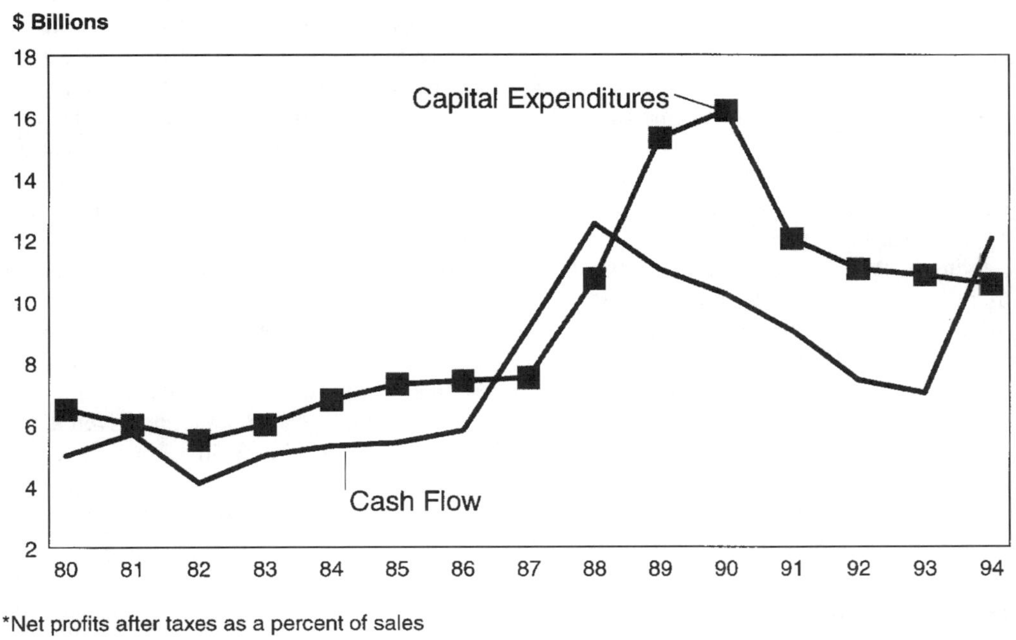

Figure 22.4. Cash flow and capital expenditures — paper and allied products. Net profits after taxes as a percent of sales. *From:* Bureau of the Census.

from its equipment in order to maintain cash flow, to recover its investment, and to generate funds for reinvestment. It means that investment decisions must be made thoughtfully, and that existing mills cannot replace capital equipment frequently — certainly not before it has been fully depreciated — without serious financial consequences. Companies in this industry must ensure that any capital they invest will be productive, cost-effective, and profitable; otherwise, they cannot endure.

Environmental Performance

Despite these economic facts of life — or perhaps because of them — the industry has achieved an exemplary record of environmental protection. For the last ten years, 14% of the industry's spending — $8 billion — has gone to environmental protection and enhancement. These expenditures have enabled the industry to reduce water consumption and effluent discharge by 70% since 1959. Discharges of biochemical oxygen demand are down by 90% in 30 years. Discharges of observable color into waterways have been reduced by almost a third. Emissions of odorous sulfur compounds have been decreased by 90% since the late 1960s. The paper industry had the highest participation and success rate of any industry in the EPA's "33/50" pollution prevention program. The proportion of wastes going to landfills has been reduced by 16% while paper recovery for recycling has exceeded the 40% rate enroute to a new goal — recovery of one half of all paper used in the U.S. by the year 2000.[6]

Relevant to this book is that chlorine use by the paper industry has been reduced by more than a third in the last ten years in spite of increased bleached pulp production. Moreover, the unwanted generation of dioxin in pulp bleaching has been reduced a whopping 92% since 1988; dioxin cannot be measured (i.e., it is not detectable at 10 parts per quadrillion) in the wastewater of well over 90% of today's mills.[7]

INTRODUCTION TO THE "CLUSTER" PROPOSALS

The "cluster proposals" refer to proposed EPA regulations that would affect certain wastewater and airborne discharges from segments of the pulp and paper industry. The EPA's upcoming revision to the pulp and paper industry's wastewater discharge guidelines will have radical implications for pulping and bleaching equipment used in the industry. The EPA wisely foresaw that its hazardous air pollutant guidelines for the industry under the Clean Air Act Amendments of 1990 would affect the same processes and equipment as the wastewater rules would. So the proposals are referred to as a "cluster" for two reasons: first, because the EPA has viewed the affected processes as a cluster, and second, because the EPA has held a series of public meetings to gather "clusters" of affected parties for discussions of the possibilities and implications of the upcoming rules.

The full text of the proposals appeared in the *Federal Register* of December 17, 1993.[8] The regulations represent the largest package of regulatory requirements ever proposed by the EPA for a single industry. While the EPA's initial estimate was that the proposals would cost the industry about $4 billion in capital costs, the industry's estimate is closer to $12 billion. The annualized costs are estimated by the EPA and the industry at $1 and $2 billion, respectively. The agency and the industry have been working for several years to understand the divergent cost estimates and narrow their differences. In light of the economics of the industry described above, these potential costs and their financial and economic impacts on the industry are a critical factor in decisionmaking. Even under the initial EPA estimate, the costs would be a significant challenge to the health of such a capital intensive industry. The EPA appeared willing to impose this high capital burden to achieve speculative benefits; the industry cannot afford $12 billion without economic return, and is reluctant to invest in the absence of a reasonable relationship between costs and benefits.

VISION OF THE FUTURE

EPA's Vision

The EPA's vision for the pulp and paper industry, of course, is what it seeks for all industries and other activities — minimal or, at least, environmentally benign discharges. At the outset of the cluster proposal development process, some EPA staff intended to require the industry to arrive at that benign result in specified ways, at governmental direction: using little or no chlorine. In fact, serious consideration was given to seeking the elimination not only of chlorine but also of chlorine compounds. This stemmed from the belief that the elimination of chlorine and its compounds was necessary to achieve the elimination of dioxin and other toxic compounds.

The EPA expressed concern not only about dioxin but about certain named chlorinated phenolic compounds. Some EPA officials believed that the use of chlorine could result in unnamed and undetected toxic compounds which could have adverse effects on the environment; faced with the inability to identify and address each such possible compound with traditional regulation, the agency decided to regulate the broad parameter "AOX" to very low levels.

Aimed at controlling the industry's use of chlorine as much as at regulating direct impacts on the environment, the proposals would regulate discharges of dioxin, chlorinated compounds, color, COD, and AOX — a measure of total adsorbable halogens — among other things.

Under a key provision guiding development of this rule, the EPA is required by Congress to establish "best available technology" for "toxic pollutants" in wastewater discharges, and then determine the discharge limitations that would be achievable using that technology. In this case, the EPA elected to base the regulations on process technology, rather than on pollution control technology. It concerns the continuum of lignin removal from wood fibers that occurs in pulping and bleaching operations. The agency's December 1993 proposals reflected a determination that the use of an oxygen stage to delignify wood pulp, coupled with elimination of the use of molecular chlorine in pulp bleaching, would represent "best available technology."[9]

Industry's Vision

The industry's vision of its future, and the path to achieve it, reflects a different approach. An expression of the industry's vision can be found in "Agenda 2020," a joint research agenda developed by the industry and the government under the leadership of the Department of Energy, in collaboration with members of the academic community. The agenda identifies key points of the joint vision, then identifies gaps which must be filled through research in order for the vision to be achieved. The hope is that this agenda can lead to a pooling of resources, not a series of prescriptive steps, to bridge the gaps.

Among the key points of the vision are that the U.S. pulp and paper industry be a global leader in providing safe and essential products based on forest resources; an example of sustainability, with renewable raw material growing in productive, well-managed forests. The industry seeks to increase its productivity and improve its capital efficiency (and to decrease its capital intensiveness). Its environmental vision calls for manufacturing plants to operate in harmony with the environment and the communities in which they are located, to make low use of water resources, and to closely control discharges to air and water so as to minimize any potential adverse impact on the environment. Even in light of the industry's environmental performance to date, the technical challenges to these goals remain considerable and the commitment strong.

RECONCILING THE COLLISION OF VISIONS: PATH FORWARD

For the industry's vision to be realized, the 1993 cluster proposals would require adjustment. In the industry's view, some of the requirements have been shown to be environmentally counterproductive. The huge capital costs would have the effect of freezing technology, further increasing capital intensiveness with no corresponding benefit to the environment. Serious impacts on the industry's global competitiveness would result from such nonproductive investments.

For illustration, consider the potential results if oxygen delignification had been required across the board in the 1970s. Many significant developments in bleaching technology and wastewater treatment, which could prove important in the larger search for closed mill technology, might have been stifled. Instead, because multiple technological paths were left available, we have increased our understanding and sophistication across a range of techniques the full potential of which may not yet be predictable. Incremental change, not mandated full-step high-capital changes, will ultimately lead to the widest number of possibilities.

The industry therefore disagrees with the 1993 proposition that oxygen delignification represents "best available technology" for all existing bleached kraft mills. The industry believes that the use of complete chlorine dioxide substitution, in lieu of molecular chlorine, represents a more cost-effective

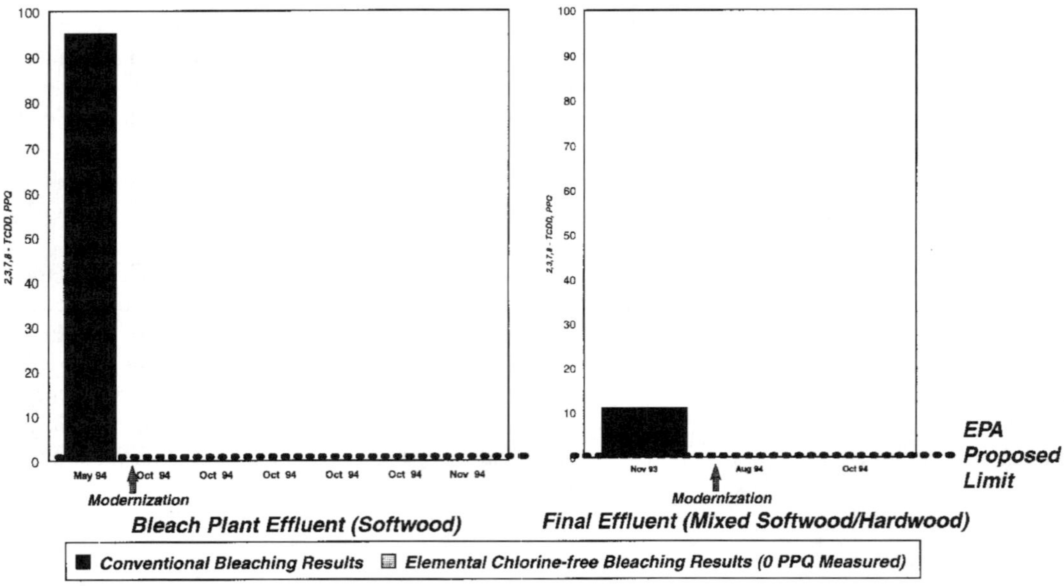

Figure 22.5. Northern softwood/hardwood kraft mill. Dioxin data (2,3,7,8-TCDD).

way to achieve a similar level of environmental protection at existing mills, without oxygen delignification. A significant body of evidence has been amassed since the 1993 proposal to demonstrate the results of chlorine dioxide bleaching, as contrasted to traditional high-chlorine processes, with or without the use of oxygen delignification.[10] The EPA has heavily weighed this evidence in developing its final regulations, which were expected in mid- to late 1996.

The industry's recommended reconciliation of these colliding visions appears to be one gaining support in many quarters, including the EPA. There are encouraging indications, if no official documentation, that the EPA will accept chlorine dioxide bleaching in the final rule. The industry recommends that the EPA adopt complete chlorine dioxide technology as "best available technology" for existing chemical pulp bleaching mills, without requiring all existing mills to install oxygen delignification. Ample data are now available to demonstrate that complete chlorine dioxide substitution eliminates the unintentional generation of dioxin (see, for example Figure 22.5), and of highly chlorinated phenolic compounds (containing four or more chlorine atoms) of potential concern to the EPA. Example data sets from James River mills are presented in a paper being given by Stinchfield and Woods at this Symposium, and others have been published as well.[11]

The industry also advocates the abandonment of AOX as a meaningful regulatory measure. AOX is not a compound but a test method, the principal advantage of which is its ease of performance. It has been shown to correlate poorly with toxic chlorinated compounds.[12] The proponents of AOX regulation believe it provides a surrogate for unnamed but problematic chlorinated compounds, and that it may be most useful as a measure of chlorine use by the regulated mills. However, the nature of the chlorine dioxide molecule and its reaction behavior make it very unlikely that toxic aromatic compounds are being formed at undetectable but threatening levels at mills practicing complete chlorine dioxide bleaching. Use of AOX as a regulatory parameter would create a worse predicament than that of the drunk who looks for his car keys under the streetlight, even though he dropped them in the alley, because the light is better: in the case of the drunk, at least the car keys are known to exist.

Regardless of the appropriateness of AOX as a regulated parameter, conversion from traditional chlorine bleaching sequences to complete chlorine dioxide substitution in fact dramatically reduces generation of AOX, as illustrated in Figure 22.6. Given this coincidence, the industry has stated that it would accept an AOX requirement established at a level achievable with complete chlorine dioxide substitution plus efficient biological wastewater treatment, without oxygen delignification.

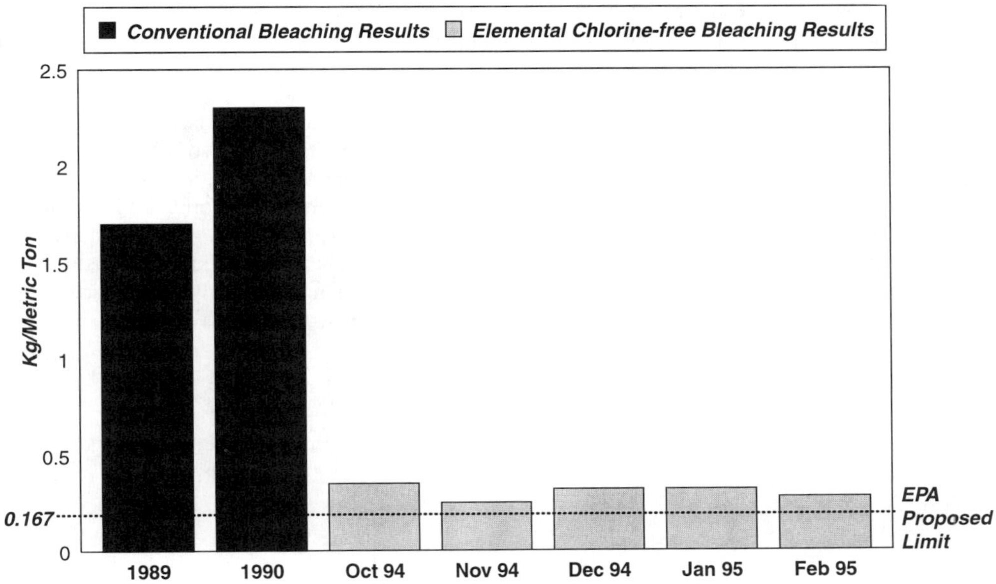

Figure 22.6. Northern softwood/hardwood kraft mill. Final effluent AOX data.

The pulp and paper industry members believe that such a compromise on chlorine compound bleaching, coupled with the other aspects discussed under the auspices of the cluster proposals, would represent a financially manageable, though challenging, approach to achieving significant environmental progress. Under the industry approach, dioxin and the chlorinated phenolic compounds of concern to the EPA would be nondetectable in industry wastewater discharges. AOX would be reduced by a third to a half. The changed chemical composition of the discharges would further alleviate concern about the potential presence of toxic chlorinated aromatic compounds.

The industry's approach would yield equivalent protection, meeting the broad objectives of the EPA's vision for the industry. It would be more consistent with the industry's global economic considerations, imposing more realistic impacts on capital intensity. It would provide for coherent technological evolution toward a vision of the "mill of the future" in which we could all take pride.

The apparent conflict in initial vision of the industry and of some at the EPA will have been reconciled by focusing on the desired environmental result, and not on the means to achieve it.

REFERENCES

1. See 58 FR 66078 *et seq.*, December 17, 1993.
2. "The Facts Behind the Headlines: Paper Products in the Business Cycle," American Forest & Paper Association, Washington, DC, Spring, 1995.
3. Special Report: "Export-Import Trends in the U.S. Paper Industry," American Forest & Paper Association, Washington, DC, March 1994.
4. "Facts," *supra.*
5. *Ibid.*
6. "A Survey of Pulp and Paper Industry Environmental Protection Expenditures — 1992," Special Report #93-10, National Council for Air & Stream Improvement, Inc., New York/Raleigh, NC, August, 1993.
7. Progress in Reducing the TCDD/TCDF Content of Effluents, Pulps and Wastewater Treatment Sludges from the Manufacturing of Bleached Chemical Pulp," National Council of the Paper Industry for Air & Stream Improvement, Inc., Special Report #94-08, New York/Raleigh, NC, August, 1994.
8. 58 *Federal Register* 66078 *et seq.*, December 17, 1993.
9. Chlorine would also be regulated under the hazardous air pollutant rules proposed as part of the "cluster." The regulations would require stack gas scrubbers to be installed on bleaching stages. This chapter does not discuss these proposals.

10. See, e.g., "Reduction of Chlorinated Organic Chemicals from Bleached Kraft Pulp Mills," Stinchfield and Woods, Symposium on Chlorine and Chlorine Compounds in the Paper Industry, American Chemical Society, Chicago (August, 1995).
11. See, for example, "Characterization of Effluent Quality at Seven Canadian Kraft Mills Operating Complete Substitution Bleach Plants," Special Report #95-09, the National Council for Air & Stream Improvement, Inc., New York/RTP, NC, July, 1995; "Characterization of Adsorbable Organic Halide (AOX) Discharge Rates at Kraft Mills Employing Complete Chlorine Dioxide Substitution," Technical Bulletin #667, the National Council for Air & Stream Improvement, Inc., New York/RTP, NC, June, 1994; also, Special Report #94-08, cited above; proceedings of the Emerging Technologies Workshop presented by North Carolina State University and the Technical Association of the Pulp and Paper Industry, Raleigh, North Carolina, May, 1995; the docket for EPA's cluster rule making; and many other sources.
12. Fleming, B.I. "The Organochlorine Spectrum: Mills, Public Must Discern Toxic, Nontoxic," *Pulp & Paper Magazine,* April, 1992, p. 59, *ff.*

Part 5

Toxicology and Mechanisms

Chapter 23

Toxicology and Regulatory Aspects of Chlorine and Chloroorganic Compounds

S.H. Henry, M. Bolger, and T. Troxell

INTRODUCTION

In recent years concerns have been growing that chlorinated organic compounds may be involved in the etiology of adverse human effects, including cancer and reproductive effects in the offspring of both humans and wildlife. This chapter will briefly summarize the information on the toxicology of the major classes of chlorinated organic compounds, and discuss relevant issues in the use of these compounds in the U.S. food and drinking water supply. In addition, regulatory aspects of the chlorinated organic compounds from the point of view of the U.S. Food and Drug Administration (FDA) will be discussed.

USES AND SOURCES OF CHLORINATED COMPOUNDS

Chlorine is extensively used as an industrial chemical. The Environmental Protection Agency's (EPA) Toxics Release Inventory estimated total U.S. chlorine emissions to be between 107 and 140 million pounds for each of the years from 1987 to 1990 (Report of the Expert Panel, 1994). In commerce today there are about 15,000 chlorinated compounds; these include pharmaceuticals, pesticides, disinfectants for drinking water and food, and many modern consumer products, such as items made of polyvinyl chloride (PVC) (Hileman, 1993). Chlorine intermediates are used in reaction pathways leading to many nonchlorinated products, such as silicon for electronic chips. In addition, nearly 85% of the pharmaceuticals manufactured today require chlorine at some stage of their production.

In addition to anthropogenic sources, there are natural sources of chlorine (Report of the Expert Panel, 1994). Volcanic eruptions have been estimated to contribute from 0.4 to 11 million tons of chlorine per year on a global basis, primarily in the form of hydrogen chloride (or hydrochloric acid). It is estimated that the photolysis of chlorides in sea salt to molecular chlorine (Cl_2) in the atmosphere results in an annual deposition of 2 to 50 kg of chlorine per hectare of land on a global basis, mainly in the form of hydrochloric acid or other chlorides, as Cl_2 reacts with various atmospheric constituents. Thus the natural sources of chlorine (and indirectly hydrochloric acid formed from chlorine) released into the environment are estimated to *exceed the total quantities released from anthropogenic sources by an order of magnitude or more.*

Inorganic chlorine ion (Cl^-) is an essential component of all biological systems; in extracellular fluid the concentration of chloride ions must be maintained at approximately 100 meq/L for normal body functions. In the body, chloride is necessary for the transmission of signals in the nervous system, regulation of the acid-base balance in body fluids, transport of carbon dioxide, and exchange of bicarbonate in the ileum and large intestine (Guyton, 1986). Chloride is necessary to maintain normal function of the immune system; neutrophils use chloride ion in the production of endogenous hypochlorous acid for the destruction of bacteria or other foreign agents entering the body (Weiss, 1989). In discussing the need for chloride in the human body, it should not be assumed that there is a deficiency of chloride in the U.S. diet. The National Academy of Sciences (1977) has estimated that the minimum dietary chloride requirements for healthy persons range from 180 to 750 mg/day (infants through adults).

TOXICOLOGY OF BY-PRODUCTS OF CHLORINATION OF DRINKING WATER

The hazards/risks of by-products of the chlorination of drinking water are of particular interest to the FDA, since drinking water is used in the preparation and processing of food. The compounds of relevance have been summarized in Table 23.1.

When unconcentrated drinking water samples collected from disinfected spring, river, and lake waters were tested, no evidence of genotoxicity or mutagenicity was observed in mammalian *in vitro* genetic tests; however, when preconcentration techniques were employed, increases in sister chromatid exchange, chromosomal aberration, and micronucleus induction in Chinese hamster ovary cells were reported. The addition of *in vitro* metabolic systems to the preconcentrated test samples resulted in a loss of mutagenic activity. No information regarding the genotoxic activity of either unconcentrated or preconcentrated chlorinated drinking waters in *in vivo* mammalian bioassays has been identified in the literature (IARC, 1991).

Several chronic/carcinogenicity studies have been conducted with laboratory animals receiving extracts of water previously disinfected with chlorine. Truhaut et al. (1979) exposed female CFLP Han mice and Sprague-Dawley rats to extracts of chlorinated river water in the diet at doses of 1.2 and 2.4 mg/kg body weight/day for 104 weeks. The doses represented 100 to 200 times the estimated daily human intake based on consumption of 3 L of water/day by a 60-kg person. Increased mortality in both sexes of rats and mice was noted, as well as an increased incidence of mammary gland ovarian carcinomas in female mice. However, results of the study must be interpreted with caution, since few tumor-type details were given and no control group tested the extraction procedure.

In another chronic bioassay, male and female Wistar rats were treated at dose rates from 0 to 97 mg extract/kg body weight/day (males) and from 0 to 165 mg extract/kg b.w./day for females by using organic extracts of disinfected surface water dissolved in purified water (Kool et al., 1985). The doses corresponded to 0 to 68 times the normal human intake of organic compounds via drinking water. A slight treatment-related increase in mortality of treated animals was noted, but no evidence of carcinogenicity was observed. Unchlorinated water extracts were not studied.

These studies did not include a control group given extracts of raw or unchlorinated water and therefore causal relationships cannot be inferred (IARC, 1991). In 1991, the International Agency for Cancer Research (IARC) concluded that there was inadequate evidence for the carcinogenicity of chlorinated drinking water in experimental animals.

A number of epidemiology studies have evaluated the potential effect of the consumption of water disinfected with chlorine/hypochlorite on the risks for certain adverse effects, including cancer and poor reproductive outcome. These studies have been reviewed by the National Research Council (NRC) (1987) and IARC (1991) and their findings are summarized below.

Several studies were done with either groundwater or surface water as the exposure variable and cancer incidence as the outcome variable, under the assumption that surface waters were more highly chlorinated than groundwaters. These studies compared the cancer rates of populations having potable water supplies derived from surface water sources with the rates of populations having water supplied from groundwater sources; most studies focused on U.S. populations, especially Louisiana and the Midwest. The correlations analyzed showed some positive associations between surface water consumption and cancers of the bladder, stomach, rectum, and lung. However, these results must be interpreted cautiously because potable water sources may not provide an adequate surrogate from which to measure chlorination and also exposure to other contaminants in the water that are unrelated to chlorination may be occurring. Factors known to affect cancer mortality, such as diet and smoking habits, were not adequately controlled in these studies.

Other epidemiology studies involving populations in the United States and the Netherlands have attempted to measure correlations between measured concentrations of trihalomethanes (THMs) in drinking water and cancer rates (NRC, 1987; IARC, 1991). Positive associations have been observed between cancers of the urinary bladder, colon, and rectum and the concentrations of THMs in the finished drinking water. However, the data presented were inadequate for inferring a causal association; concentrations of THMs were poorly defined and various confounding factors were not controlled.

Table 23.1. Potentially Carcinogenic[a] By-Products of Drinking Water Disinfectants.

By-Product	Disinfectant
Trihalomethanes	HOCl, $ClNH_2$
trichloro-, bromodichloro-, dibromochloro-	
tribromo-	O_3, HOCl, $ClNH_2$
Haloacetates	HOCl. $ClNH_2$
dichloro, trichloro	
Aldehydes	O_3, HOCl, $ClNH_2$
formaldehyde, acetaldehyde	
Miscellaneous	
bromate	O_3
hydrogen peroxide	O_3
trichloroacetaldehyde	HOCl, $ClNH_2$
2,4,6-trichlorophenol	HOCl
chloropicrin	HOCl

[a] Potentially carcinogenic means that the compound has been demonstrated to be carcinogenic in at least one laboratory animal species.

Morris et al. (1992) performed a *meta*-analysis of pooled data from 10 epidemiology studies evaluating the effects of consumption of chlorinated drinking water on cancer; the analysis indicated that exposure to chlorination by-products through drinking water was associated with a relative risk of cancer of 1.15. When only the studies that enabled statistical adjustment for confounding factors, such as age, sex, population density, smoking, occupation, etc. were used, Morris et al. (1992) found no appreciable difference between the relative risk estimate and that obtained by using the whole data set. A dose-response relationship was observed for results from studies that stratified exposure to chlorinated drinking water into "low," "intermediate," and "high" exposure groups. The authors concluded that the increased statistical power of the *meta*-analysis technique and the results it produced suggest an association between consumption of chlorinated surface waters and increased incidence of bladder and rectal cancer. However, as pointed out in the Report of the Expert Panel (1994), the possibility that consumption of chlorinated surface waters may be a surrogate for some other etiological agent has not been ruled out in the Morris et al. study. No chlorinated by-products have been identified in chlorinated drinking waters that have been associated with cancers of the rectum and bladder based on the weight of available toxicological evidence.

In addition, there are several major concerns with the *meta*-analysis technique (Report of the Expert Panel, 1994). First, such an analysis is a currently controversial technique for the combination of epidemiological studies that were conducted independently, and it does not overcome the weaknesses of the individual studies on which it is based. Secondly, the *meta*-analysis of the available case-control and cohort epidemiology studies that consider the effect of consumption of chlorinated drinking water on cancer reported relative risks greater than 1.0 for all cancer sites examined. This lack of specificity makes it difficult to establish a causal relationship between chlorinated drinking water and cancers observed. Third, the relative risks observed were small (less than 2.0); such small relative risks could be due to chance or confounding of the statistical analysis by various factors in the studies.

Several studies have examined the possible association between poor reproductive outcomes and consumption of chlorinated drinking water. Bove et al. (1992a) conducted an epidemiological study of four New Jersey counties to evaluate possible associations between THMs, other organic compounds, the source of the drinking water supply, and 13 different adverse reproductive outcomes, including stillbirths, low birth weight, very low birth weight, prematurity, and malformations. Results showed a statistically significant association between THM concentrations in the drinking water supply and low birth weight. The International Life Sciences/EPA review (ILSI, 1993) concluded that these associations should not be interpreted as causal relationships since the analysis of the study could have been confounded by exposure misclassification (no individuals were interviewed) and variables known to affect reproductive outcomes (e.g., use of tobacco and alcohol, specific occupational factors, etc.) were not controlled.

A case-control study of the same New Jersey births was performed for babies with cardiac defects, neural tube defects, oral clefts, and low and very low birth weights (Bove et al., 1992b). Interviews were used to gain information on exposure and confounding variables; however, only 52% of the subjects selected were interviewed and those not interviewed differed sharply from the subjects interviewed in terms of race, education, and adequacy of prenatal care visits. A statistically insignificant correlation was found between THM concentration in excess of 80 μg/L and neural tube defects. The authors of the study and the ILSI/EPA reviewers of the study agreed that THM and other organic chemicals could not be causally related to adverse reproductive outcomes on the basis of this study.

Hertz-Piccitto et al. (1989) examined the relationship, using survival analysis methods, between the rate of spontaneous abortions as related to the consumption of tap water in the Santa Clara, California area. Authors were able to statistically control for certain confounding factors, including smoking, alcohol consumption, previous pregnancy losses, and maternal age. The risk of spontaneous abortion was greater in women who consumed mainly tap water versus women who consumed mainly bottled water. However, these results did not implicate tap water as the causal agent since the authors found evidence of recall bias, such as differential recall on the part of the subjects following a miscarriage, which could have biased the reported associations.

When all of the epidemiological data are examined, the evidence for increased risks of cancer associated with the consumption of chlorinated drinking water (especially consumption of chlorinated surface water versus consumption of unchlorinated groundwater) is strongest for cancer of the urinary bladder (IARC, 1991). However, as IARC has pointed out, many of the epidemiological studies did not control for a number of potentially confounding variables; hence, the epidemiological data are insufficient to indicate a causal link for human cancer. In addition, when all the data on reproductive outcome are examined, there is not a clear causal link between consumption of THMs and other organic compounds and adverse reproductive outcome.

HAZARDS ASSOCIATED WITH CHLORINE DISINFECTION BY-PRODUCTS

The by-products receiving the most toxicological study are those in the haloacetic acid, haloacetonitrile, and trihalomethane classes. These compounds have several toxicological effects, but their ability to induce cancer in laboratory animals is considered to be most important. The potentially carcinogenic by-products of drinking water disinfectants are shown in Table 23.1, as adapted from Bull (1993).

A look at the classes of disinfectant by-products shown in Table 23.2 is in order. With chlorine, a large number of by-products have been identified. For the classes which occur in the highest con-

Table 23.2. Classes of Chlorinated Organic Compounds.

Chlorates and chlorites
Chlorinated alkanes
    e.g., carbon tetrachloride, chloroform, dichloroethane, methylene chloride, trichloroethane
Chlorinated alkenes
    e.g., dichloroethylene, trichloroethylene, tetrachloroethylene
Chlorinated organic acids
    e.g., dichloroacetic acid, trichloroacetic acid
Chlorinated benzenes
    e.g., dichlorobenzenes, trichlorobenzenes, hexachlorobenzene
Chlorinated phenolics
    e.g., mono-, di-, and trichlorophenols
Chlorinated nitrogen compounds
Polychlorinated biophenyls, dioxins, and furans

centrations (trihalomethanes, haloacetates, and haloacetonitriles), in general (except for bromodichloroacetate and dibromochloroacetate), substantial toxicological databases exist and some compounds have been identified as possible causes of concern. Therefore, research information which is most needed is that which would allow an estimate of risks at the low concentrations found in drinking water. This issue needs to be resolved before chlorination by-products that occur at lower concentrations can be considered a serious concern (Bull, 1993).

The by-products of chloramines are probably similar to those of chlorine, but this area needs more systematic research. Chlorine dioxide and ozone present major sources of uncertainty because few of their by-products have been completely identified.

There are fewer uncertainties with chlorine than with the alternatives. Simply put, much more information is available on the nature and concentrations of by-products, the mechanisms of toxicity and their relevance to humans, the effective dose, and the interactions between contaminants for chlorine disinfection than for other types of disinfection (Bull, 1993).

SOME BENEFITS TO HUMAN HEALTH FROM DISINFECTION OF DRINKING WATER WITH CHLORINE

In 1854 Dr. John Snow of London was able to prove that the drinking water supply was the vehicle of communication of cholera. Chlorine was first used to disinfect local drinking water supplies around the turn of the century; Jersey City, New Jersey, had a commercial water treatment plant operating in 1908 (Report of the Expert Panel, 1994).

Water treatment, largely due to chlorine, has been responsible for the dramatic reduction in typhoid fever cases in the United States. Other diseases controlled by water treatment to varying degrees include cholera, amoebic dysentery, bacterial gastroenteritis, shigellosis, salmonellosis, *Campylobacter* enteritis, and certain viral infections, such as hepatitis A (Report of the Expert Panel, 1994).

Worldwide more than 30,000 deaths per day are attributed to waterborne diseases such as typhoid, cholera, dysentery, etc. Recent outbreaks of new strains of cholera have become increasingly common in Central and South America, where there is often inadequate chlorination of the water supply. The 1991 Peruvian outbreak of cholera which resulted in nearly 300,000 cases was caused partly by poor water purification facilities and partly by lack of chlorine disinfection due to inordinate concern over possible harmful disinfection by-products as reported by international news media (Salazar-Lindo et al., 1993). Current mortality in the United States resulting from infectious diseases transmitted through contamination of potable water supplies is negligible. Most disease outbreaks associated with water in the United States since the early 1900s have been associated with viruses and protozoan cysts, such as *Giardia* and *Cryptosporidia*, which are generally more resistant to disinfection than bacteria (Report of the Expert Panel, 1994).

In addition, a new threat to drinking water safety has been recently reported. Lambert et al. (1994) discuss the microcystins, a class of extremely potent hepatotoxins produced by several species of common planktonic cyanobacteria found in surface water and associated with several incidents of human and animal illness. Conventional water treatment processes, such as coagulation, sedimentation, and chlorination, have generally proved ineffective at removing microcystin toxins from drinking water; activated carbon, ozone, and free chlorine have been effective in removing microcystins to below measurable concentrations in laboratory experiments.

REGULATORY FRAMEWORK OF CHLORINATED ORGANIC COMPOUNDS

The regulation of chlorinated organic compounds and by-products in food products under FDA jurisdiction is determined primarily by the adulteration provisions of the Federal Food, Drug, and Cosmetic (FDC) Act. There are two basic regulatory categories of potential consideration of these compounds: poisonous and deleterious substances, and food additives.

Food Additives

Section 201(s) of the act defines a food additive as "any substance the intended use of which results or may reasonably be expected to result, directly or indirectly, in its becoming a component or otherwise affecting the characteristics of any food ..., if such substance is not generally recognized, among experts qualified by scientific training and experience to evaluate its safety, as having been adequately shown through scientific procedures (or, in the case of a substance used in food prior to January 1, 1958, through either scientific procedures or experience based on common use in food) to be safe under the conditions of its intended use; ...." Since the definition of food in the act includes water, in principle, chlorine disinfectants intended for use in water for drinking or food processing would be subject to the food additive provisions.

However, the FDA does not apply these provisions so broadly. The Safe Drinking Water Act (SDWA) of 1974 gave the EPA primary authority over drinking water. In a 1979 memorandum of understanding between the FDA and EPA, the agencies determined that the passage of the SDWA implicitly repealed the FDA's authority under the FDC Act over water used for drinking purposes. The agencies agreed that chlorine and other substances added to a public drinking water system before the water enters a food processing plant would not be considered food additives. However, water used in food is still food and chlorine and its by-products in that water are still added substances and subject to the relevant non-food additive provisions of the act.

It should be noted that bottled water is also a food that is under FDA jurisdiction. Section 410 of the act provides for the FDA to issue regulations (quality standards) for bottled water in response to EPA issuance of primary drinking water regulations or to publish a notice in the *Federal Register* giving the reasons for not amending its regulations. The FDA usually adopts the EPA's maximum contaminant levels (MCLs) as part of its bottled water quality standard. These standards include levels for the chlorine by-product-derived trihalomethanes as well as a variety of other chlorinated organic compounds such as trichloroethylene and 1,1,1-trichloroethane. As the EPA completes its rule making on disinfection by-products in drinking water, the FDA will consider adopting the new MCLs as bottled water quality standards.

Food additives require premarket approval before they may be used in foods or food processing. This is done through a petitioning process and issuance of a regulation after a determination that the substance is safe for the intended use. "Safe" in the food additive legislation means "reasonable certainty of no harm." It is important to recognize that the statute does not permit a consideration of benefits or relative risk, e.g., the microbial risk of not using the disinfectant versus the chemical risk associated with using it.

The food additive provisions also include the Delaney clause, which prohibits issuance of a regulation permitting the use of a food additive that has been found to induce cancer when ingested by man or animal. Chlorine disinfectants produce a number of carcinogenic by-products which must be considered during agency safety evaluation of the additive. In 1982 the FDA initiated its "constituents" policy to address carcinogenic nonfunctional components of additives, which would include disinfection by-products. This policy distinguishes between the additive as a whole and its chemical constituents. The Delaney prohibition would apply to the additive as a whole and its functional constituents. Carcinogenic nonfunctional constituents are evaluated for safety by using quantitative risk assessment to determine if the risk from the presence of the carcinogenic entity can be shown to be insignificant.

There are several food additive regulations for chlorine and chlorine dioxide. These include use of chlorine for bleaching food starch (21 CFR § 172.892), use of chlorine dioxide in water for poultry processing (§ 173.69), use of chlorine to wash fruits and vegetables (§ 173.315), use of chlorine and chlorine dioxide in several sanitizing solutions for food-contact surfaces (§ 178.1010), and use of chlorine to bleach paper and paperboard (§ 176.170). [Chlorine and chlorine dioxide are listed in the standards of identity for various types of flour as bleaching agents (21 CFR Part 137).]

The statutory definition of a food additive cited above also provides that substances may be considered generally recognized as safe (GRAS). The agency recognizes certain unpublished uses of chlorine as GRAS. For example, the FDA considers chlorination of water for poultry and fish processing as GRAS based on history of use. The FDA also has permitted the use of chlorine dioxide at

5 ppm and 1 ppm on uncut fruits and vegetables, and on cut and peeled potatoes, respectively, provided such use is followed by a potable water rinse.

Poisonous and Deleterious Substances

The general adulteration provisions of the act cover all occurrences of chlorinated organic compounds and by-products, whether present as contaminants, in the water before it enters a food processing plant, or intentionally added in the food processing plant (food additives). The main adulteration provisions that apply to such chemicals in foods are Section 402(a)(1) and Section 406.

Section 402(a)(1) provides that a food shall be deemed to be adulterated if it bears or contains any poisonous or deleterious substance which "may render it injurious to health." The courts have interpreted "may" to connote a "reasonable possibility." In an enforcement action the burden of proof is on the agency to demonstrate a violation.

Section 406 of the act provides for establishment of tolerances for poisonous or deleterious substances added to food when they are required in the production of the food or cannot be avoided by good manufacturing practice. While the regulation specifying the tolerance must protect the public health, this provision allows the FDA to consider the avoidability of the substance as well as other exposures to the substance. This analysis also includes a consideration of the economic consequences of a tolerance.

For years the FDA has used action levels in lieu of issuing tolerances, especially in those situations in which information on control and toxicology was still evolving (21 CFR 109.4 and 109.6). Action levels are used as informal guidelines by the FDA's field staff on when to consider enforcement action. They can be issued relatively quickly. However, unlike tolerances, the FDA must be able to prove all aspects of the statutory violation in each enforcement action.

The FDA more recently has provided for the establishment of regulatory limits based on the Section 402(a)(1) provision (55 FR 20785; May 21, 1990; 21 CFR 109.4 and 109.6). These limits would be established by notice-comment rule making and would not be delayed by administrative hearings. Regulatory limits would have the force and effect of law, and the FDA would need to prove only that the limit was exceeded to take an enforcement action.

The FDA also occasionally uses advisories to provide guidance to industry on contaminant levels. An advisory would indicate the level of a chemical in a specific food(s) below which the FDA believes there would not be a significant health hazard. Thus, the FDA has a range of regulatory risk management options to ensure that chlorinated organic compounds do not present a health hazard to consumers through foods.

CONCLUSIONS — RISKS IN PERSPECTIVE

Table 23.3 summarizes the efficacy, advantages, and disadvantages of major disinfection methods for drinking water (as adapted from Report of the Expert Panel, 1994).

Since various chlorinated organic chemicals have been identified as by-products of the chlorination of drinking water and some have been shown to have carcinogenic activity in laboratory animals, concerns have been expressed about the human cancer risks associated with chlorinated drinking water. It is necessary to put into perspective, or to quantify, the risks to human health from consuming chlorinated drinking water compared with the risks from consuming raw or unchlorinated drinking water. In addition, the risks and benefits of alternative means of disinfecting water, such as ozonation, should be considered. These comparisons will be difficult. There are presently insufficient monitoring data to estimate the quantity of the various types of microroganisms likely to be present in untreated water. In addition, there is a lack of dose-response information from which to determine the concentration of microorganisms necessary to cause infection. These choices need to be made in an informed, rational manner. The public will want to be involved and it is incumbent upon scientists to do the research that will provide the rationale for these choices and to explain clearly the nature of the choices.

Table 23.3. Advantages and Disadvantages of Major Disinfection Methods for Drinking Water.

Disinfection	Comments on use and efficacy against bacteria, viruses, protozoan cysts, persistence of residual
Chlorine (as HOCl)	Widespread use in drinking water disinfection. Excellent biocidal activity vs. bacteria and viruses, moderate vs. protozoan cysts. Good residual. Produces chlorinated organic by-products. Relatively inexpensive.
Chlorine (as OCl^-)	Widespread use in drinking water disinfection. Good biocidal action vs. bacteria, moderate vs. viruses, no data *re* protozoan cysts. Good persistence of residual. Produces chlorinated organic by-products. Relatively inexpensive.
Ozone	Widespread use in Europe and Quebec for drinking water treatment. Excellent biocidal activity vs. bacteria, viruses, and protozoan cysts. No residual persistence possible. Effects of by-products poorly characterized. Expensive.
Chlorine dioxide	Widespread use in Europe as disinfectant for drinking water. Used in U.S. for taste and odor control. Excellent biocidal activity vs. bacteria, viruses, good activity vs. protozoan cysts, fair to good residual persistence. Concern over toxicity of by-products of chlorate and chlorite.
Chloramines	Used by about 20% of surface water treatment plants in U.S. Moderate biocidal activity vs. bacteria and viruses, fair biocidal action vs. protozoan cysts. Excellent residual persistence. Some concern over the toxicity of chloramine residuals in drinking water.
UV radiation	Limited to small and pilot-scale operations. Removal of organics from raw water before disinfection increases effectiveness of disinfection and reduces by-product formation. Good biocidal activity vs. bacteria and viruses, moderate vs. protozoan cysts. No residual possible. Biocidal activity highly dependent on quality of source water.

REFERENCES

Bove, F.J., M.C. Fulcomer, J.B. Klotz, J. Esmart, E.M. Dufficy, and R.T. Zagraniski. 1992a. *Population-Based Surveillance and Etiological Research of Adverse Reproductive Outcomes and Toxic Wastes. Report on Phase IV-A: Public Drinking Water Contamination and Birthweight, Fetal Deaths and Birth Defects, A Cross-Sectional Study*. New Jersey Department of Health.

Bove, F.J., M.C. Fulcomer, J.B. Klotz, J. Esmart, E.M. Dufficy, R.T. Zagraniski, and J.E. Savrin. 1992b. *Population-Based Surveillance and Etiological Research of Adverse Reproductive Outcomes and Toxic Wastes. Report on Phase IV-B: Public Drinking Water Contamination and Birthweight, and Selected Birth Defects. A Case-Control Study*. New Jersey Department of Health.

Bull, R.J. 1993. Toxicology of disinfectants and disinfectant by-products. In Craun, G. (Ed). *Safety of Water Disinfection: Balancing Chemical and Microbial Risks.* International Life Sciences Institute Press, Washington, DC.

Guyton, C. 1986. *Textbook of Medical Physiology.* 7th ed. Saunders, Philadelphia.

Hileman, B. 1993. Concerns broaden over chlorine and chlorinated hydrocarbons. *Chem. Eng. News,* April 19.

International Agency for Research on Cancer (IARC). 1991. *IARC Monographs on the Evaluation of Carcinogenic Risks to Humans. Chlorinated Drinking-water; Chlorination By-Products; Some Other Halogenated Compounds; Cobalt and Cobalt Compounds.* Vol. 52, IARC, Lyon, France.

International Life Sciences Institute (ILSI). 1993. *Report of the Panel on Reproductive Effects of Disinfection By-products in Drinking Water*. ILSI, Washington, DC and U.S. EPA, Health Effects Research Laboratory, Research Triangle Park, NC.

Kool, H.J., F. Kuper, H. Van Haeringen, and J.H. Koeman. 1985. A carcinogenicity study with mutagenic organic concentrates of drinking water in the Netherlands. *Sci. Total Environ.* 37:229–256.

Lambert, T.W., C.F.B. Holmes, and S.E. Hrudey. 1994. Microcystin class of toxins: Health effects and safety of drinking water supplies. *Environ. Rev.* 2:167–186.

Morris, R.D., A.-M. Audet, I.F. Angelillio, T.C. Chalmers, and F. Mosteller. 1992. Chlorination by-products and cancer: A *meta*-analysis. *Am. J. Public Health* 82(7):955–963.

National Academy of Sciences (NAS). 1977. *Drinking Water and Health.* Safe Drinking Water Committee, NAS, Washington, DC.

National Research Council (NRC). 1987. Drinking Water and Health: Disinfectants and Disinfectant By-Products, Vol. 7, NRC, National Academy Press, Washington, DC.

Piccitto, I., S.H. Swan, R.R. Neutra, and S.J. Samuels. 1989. Spontaneous abortions in relation to consumption of tap water: An application of methods from survival analysis to a pregnancy follow-up study. *Am. J. Epidemiol.* 130(1):79–93.

Report of the Expert Panel. 1994. Interpretive review of the potential adverse effects of chlorinated organic chemicals on human health and the environment. *Regul. Toxicol. and Pharmacol.* 20(1):S1–1057.

Salazar-Lindo, E., M. Alegre, M. Rodriguez, P. Carrion, and N. Razzeto. 1993. In Craun, G. (Ed.), *Safety of Water Disinfection: Balancing Chemical and Microbial Risks.* International Life Sciences Institute Press, Washington, DC.

Truhaut, R., J.C. Gak, and C. Graillot. 1979. Studies on the risks that may result from chemical pollution of drinking water I. Study of the long-term toxicity in rats and mice of organic micropollutants extracted from drinking water with chloroform. (Fr.) *Water Res.* 13:689–697.

Weiss, S.J. 1989. *N. Engl. J. Med.* 320:365. Cited in Utrecht, 1992.

Appendices

Classes of Chlorine and Chloroorganics Most Likely to be Found in Foods, and Their Toxic Endpoints

APPENDIX A

CHLORINE

Elemental chlorine is a strong oxidizing agent; direct exposure to sufficient concentrations causes extreme irritation to tissues of exposed organisms. Moist tissues are most susceptible since Cl_2 gas in contact with water at physiological pH will form HCl and HOCl, which are irritating and corrosive (Report of the Expert Panel, 1994).

When aqueous solutions of several labeled chlorine compounds, including chlorate ($^{36}ClO_2$), monochloramine ($NH_2{}^{36}Cl$), hypochlorite ($HO^{36}Cl$) and sodium chloride ($Na^{36}Cl$) were given to rats, all were metabolized to the Cl^- ion, except chlorate, which was also metabolized to ClO_2 and ClO_3 ions (Abdel-Rahman et al., 1984). None of these compounds were eliminated during the 120 hours of the study, and they were distributed to all major organs via the circulatory system.

Lethal effects of chlorine upon inhalation, expressed as LC_{50} values, range from 256 ppm for mice to 414 ppm for rats and 650 ppm for dogs. Lethal inhalation potency, expressed as LT_{50}, (or time required to produce 50% mortality) were 28 and 53 minutes at 1000 ppm chlorine for mice and rats, respectively; causes of death were pulmonary edema and hemorrhage.

Several nonlethal effects of chlorine have been demonstrated in various species, including irritation of eyes and respiratory tracts, growth effects, such as decreased body weight, alterations in various biochemical parameters, such as elevated hematocrit values and white blood cell counts, increased heart weights and decreased cardiovascular function, and immune effects. Teratogenic and reproductive effects have generally not been observed in laboratory rodents at levels not associated with maternal toxicity.

Chlorine, hypochlorite, and hydrochloric acid are powerful oxidants and would be expected to result in tissue irritation and oxidative effects when directly applied in sufficient quantities to cells *in vitro*. However, in various mutagenesis study systems, chlorine, sodium hypochlorite, and chlorine dioxide have shown mixed results (Report of the Expert Panel, 1994).

Concerns have been raised whether chlorine-generated mutagens arise during the treatment of food during food processing. The results discussed in a report by the Chlorine Institute (1990) indicate a low probability of this problem.

Cl_2, NaOCl, and ClO_2, but not monochloramine, have been shown to cause cellular hyperplasia, but not neoplasms, when applied to mouse skin (Robinson et al., 1986). Chlorite ion did promote mouse skin tumors initiated with dimethylbenzanthracene (Kurokawa et al., 1984); sodium hypochlorite did not promote or initiate tumors in the SENCAR mouse epidermis (Kurokawa et al., 1986).

The carcinogenic potentials of hypochlorite and chloramine have been studied by the NTP in recently completed 2-year bioassays in mice and rats (1992). F344/N rats and B_6C_3F1 mice of each sex received doses of hypochlorite in their drinking water at concentrations of 0. 70, 140, or 275 ppm for 2 years. The study gave no conclusive evidence to suggest that hypochlorite in water is carcinogenic to experimental animals, even at high levels of exposure.

In the NTP study of chloramine, F344/N rats and $B_6C_3F_1$ mice received doses of chloramine in their drinking water at levels of 50, 100, or 200 ppm. Again, the NTP Peer Review Panel concluded that there was no conclusive evidence to indicate that water containing chloramine is carcinogenic to experimental animals.

A study sponsored by the Chemical Institute of Toxicology investigated the chronic toxicity of inhaled chlorine in nonhuman primates and found no increase in tumor incidence; however, there were increased incidences of hyperplasia and metaplasia in the nasal and tracheal epithelia of the test animals (IRDC, 1984).

Chlorine is widely used to bleach flour, to disinfect meat and seafood, and in disinfecting preparation areas and utensils. No relationship was found between tumor incidence in rats or mice fed cake made from chlorinated flour; however, there were mortality problems with the mouse study, shortening the time for observation of tumors (Fisher et al., 1983; Ginnochio et al., 1983).

Much of the data on extreme exposures of humans to chlorine comes from World War I exposures or from accidental releases of chlorine in peacetime. In general, it may be concluded that a number of short-term effects, such as irritation, altered lung functionality, transitory increases in leukocytes, and minor liver changes, can result from brief exposures on the order of 10 to 60 ppm, but there was no evidence of permanent adverse effects in these studies (Report of the Expert Panel, 1994).

APPENDIX B

CHLORINATED INORGANIC CHEMICALS — CHLORATES AND CHLORITES

Chlorates and chlorites can enter foods because they are formed in drinking water as the result of treatment with chlorine dioxide and hypochlorination. Chlorine dioxide and hypochlorination are potential alternatives to using gaseous chlorine for water disinfection; the use of chlorine dioxide, for example, decreases the formation of chlorinated phenolics, which may result in unpleasant odors and tastes in drinking water supplies. Also, since chlorine dioxide is used as a replacement for chlorine gas in the bleaching of wood pulp for paper production, chlorates and chlorites could also be present in pulp and paper mill wastewater (Report of the Expert Panel, 1994).

Mammalian Toxicology

The primary site for toxicity is the hematopoietic system (Bercz et al., 1982, Abdel-Rahman et al., 1980 a,b).

- Decreased Hb level and erythrocyte counts in monkeys. (Doses of 25–400 mg/L DW for 60 days.)
- Reduced red blood cell counts in chickens and rats. (Doses of 10–100 mg/L DW for 9 months.)
- Teratogenicity effects unclear.
- Reproductive effects — High doses may affect testicular function; significance to humans unknown.
- Genotoxicity — Results in Ames, etc. tests equivocal.
- Carcinogenicity — Little if any carcinogenic potential in experimental animals of chlorite. Chlorate has not been studied.
- Epidemiology — A number of case reports exist due to the extensive use of chlorate salts in weed killer formulations. Acute lethal dose in humans is approximately 218–1453 mg/kg body weight. Symptoms include cyanosis, anuresis, jaundice, much reduced Hb. Other epidemiology studies are either negative or inconclusive due to confounding factors (e.g., tobacco/alcohol usage, socioeconomic status).

APPENDIX C

CHLORINATED ALKANES

Chloroalkanes are also by-products of the chlorination process as used in the disinfection of drinking water and wastewater, and would therefore be found in foods (Report of the Expert Panel, 1994).

Chlorinated alkanes are formed as by-products in chemical processes involving water chlorination, such as bleaching of pulp and antifouling of cooling water. Chlorinated alkanes are also used in the manufacture of a variety of products, including refrigerants, solvent degreasers, fumigants, paint removers, synthetic fibers, pharmaceuticals, and plastics. Chloroform is used in cough syrups, toothpastes and liniments. Chloroform releases to soil may occur as a result of improper landfilling of wastes containing chloroform or from the transformation of other chlorinated wastes in soil to chloroform; chloroform may then enter groundwater by leaching through soil. Carbon tetrachloride was used as a grain fumigant until prohibited in 1986; this use resulted in the release of large amounts of carbon tetrachloride into the environment. Dichloromethane is formed during the chlorination of water and has been found in about 8% of U.S. finished water supplies (NAS, 1977).

Chlorinated organic chemicals, primarily chloromethane, chloroform, and carbon tetrachloride, are produced by natural processes. For example, the global emission rate of chloromethane has been estimated as 5 million tons per year from biogenic or natural sources, compared to 26,000 tons from industrial activities (Rasmussen et al., 1980).

Chloroform (Report of the Expert Panel, 1994)

Metabolism and Bioavailability

Rapidly absorbed when administered orally, 50–77% of inhaled chloroform in humans is absorbed dermally. It is metabolized by cytochrome P450 enzymes in liver and kidneys. Rapidly excreted from all species tested with elimination half-lives in humans of 14–90 min for vessel-rich tissues and lean body mass and >24 hr for adipose tissue. Chloroform is largely excreted from the lungs after inhalation.

Mammalian Toxicology

LC_{50} values vary widely with species. Mice: 4100–12,500 ppm over 2 hr. Humans: lethal dose 211–628 mg/kg bw. Primary target organs include the central nervous system, liver, and kidneys. No observed adverse effect levels (NOAELs) for toxicity to mouse liver for acute, subchronic, chronic oral exposure: 60, 40, and 17 mg/kg bw. Lowest observed adverse effect level (LOAEL) of 12.9 mg/kg/bw/day in dogs fed $CHCl_3$ for 7.5 yr, based on observation of fatty cysts and alterations to SGPT activity. Nephrotoxicity induced by $CHCl_3$ is indicated by renal tubular cell necrosis, at levels generally associated with hepatotoxicity.

Teratogenicity

$CHCl_3$ is teratogenic in rats, mice, and rabbits at rates of exposure that are toxic to dams. Fetotoxicity is not relevant at rates of exposure to $CHCl_3$ that are not associated with maternal toxicity.

Reproductive Effects

$CHCl_3$ is fetotoxic at doses associated with maternal toxicity.

Genotoxicity

The majority of *in vivo* and *in vitro* tests show that $CHCl_3$ is not genotoxic.

Carcinogenicity

Chronic exposure to large doses of $CHCl_3$ increases the incidence of liver tumors in male and female mice and kidney tumors in the males of both mice and rats (NCI, 1976; Roe et al., 1979;

Jorgenson et al., 1985). Liver tumors in mice were reported to occur only when the route of exposure was via corn oil gavage and in the $B_6C_3F_1$ mouse (ATSDR, 1987). Increased incidence of tumors of liver and kidney of rodents after treatment with overtly toxic doses of $CHCl_3$ has been attributed to the metabolism of $CHCl_3$ to phosgene and its subsequent binding to macromolecules such as DNA. Chloroform may act as a promoter of carcinogenesis through the increased cell proliferation associated with high-dose toxicity. Carcinogenic action of $CHCl_3$ could also be associated with the depletion of cellular glutathione.

Epidemiology

There have been no extensive epidemiology studies conducted concerning the occupational risks of $CHCl_3$ exposure. The limited studies reported indicated that the various symptoms experienced by the exposed workers were related to effects of CHC13 on the central nervous system. The current occupational safety and health administration permissible exposure limit OSHA PEL is 50 ppm (ACGIH, 1993). Epidemiology studies of chlorinated water containing $CHCl_3$ are discussed later.

1,1- and 1,2-Dichloroethane (Report of the Expert Panel, 1994)

Metabolism and Bioavailability

There is no information available on the bioavailability of 1,1-dichloroethane (1,1-DCE) in mammals; based on information for other chlorinated alkanes, it would be expected to be highly absorbed from all routes of exposure. 1,2-Dichloroethane (1,2-DCE) is rapidly absorbed into the bloodstream of rodents after dermal, oral, or inhalation administration. 1,1-DCE is metabolized by mixed-function oxidase enzyme systems mediated through cytochrome P450. 1,2-DCE is metabolized by two major pathways involving cytochrome P450 and glutathione S-transferase.

Mammalian Toxicology

Acute exposure to 1,1-DCE: little information available, relatively low capacity to cause liver or kidney damage.

Acute exposure to 1,2-DCE: CNS depression, respiratory and eye irritation. Degeneration, necrosis, hemorrhagic lesions to liver, kidneys, spleen, lungs, respiratory tract, and gastrointestinal tract.

LD_{50} (oral) for 1,1-DCE: 14.1 g/kg bw in rats.

LD_{50} for 1,2-DCE: 489 and 423 mg/kg bw for male and female CD-1 mice; 680-850 mg/kg bw in rats; 2500 mg/kg bw in dogs.

Immunosuppressive effect seen in rabbits (100 mg/m^3, 3 hr/day, 6 days/week for 8 months) were possibly secondary to other toxic effects.

Teratogenicity

1,2-DCE appears not be a potent teratogen. Studies with 1,2-DCE have indicated minimal potential for producing adverse effects on reproduction and fetal development except at maternally toxic levels.

Genotoxicity

1,1-DCE has been shown to be mutagenic in a number of systems and also has been shown to bind covalently to DNA to form adducts. 1,2-DCE has also been found to be mutagenic in a variety of *in vitro* test systems, with responses generally enhanced in the presence of metabolic activation systems. Also DNA-reactive.

Carcinogenicity

A study of the carcinogenicity of 1,1-DCE in rats and mice gave equivocal results due to high mortality and signs of overt toxicity.

Available data indicate that in laboratory animals, 1,2-DCE can induce a carcinogenic response when administered overtly toxic doses via corn oil gavage. The significance of this is unclear, since inhalation exposure to lower doses of 1,2-DCE did not induce tumors in either rats or mice.

Epidemiology

Adverse effects in humans have focused largely on 1,2-DCE and the gastrointestinal and nervous systems. Short-term exposure is associated with headaches, dizziness, weakness, muscle spasms, cyanosis, vomiting, diarrhea, and epigastric pain, with occasional unconsciousness and death. Occupational exposure studies have precluded the establishment of a causal relationship between the cancers and any particular chemical, since exposure was to mixed chemicals.

Exposure Limits

For occupational exposure to 1,1-DCE, current OSHA PEL is 100 ppm and for 1,2-DCE 50 ppm has been proposed to replace the previous 1 ppm.

Carbon Tetrachloride (Report of the Expert Panel, 1994)

Metabolism and Bioavailability

Studies in laboratory animals have indicated rapid, nearly complete absorption of carbon tetrachloride (CCl_4) from the gut. Absorption from the lungs in various laboratory species has been measured at 40–60%. Metabolism of CCl_4 occurs mainly in the liver by a specific, NADPH-dependent isozyme of cytochrome P450; most of the reactive intermediates were destroyed rapidly during the metabolic process. Disposition of CCl_4 is related to tissue fat content; nervous tissue, adrenals, and bone marrow show greater concentrations of CCl_4 than blood.

Mammalian Toxicology

Oral LD_{50} for CCl_4: 2800 mg/kg bw in rats; 12,100 to 14,400 mg/kg bw in mice and 2300 mg/kg bw in dogs.

High doses in animals cause death by central nervous system depression; smaller doses result in death from liver damage, which is the main target organ (destruction of the endoplasmic reticulum and loss of activity of its associated enzymes, inhibition of protein synthesis, impaired metabolism of triglycerides with resultant fat accumulation).

Teratogenic and Reproductive Effects

Significant maternal toxicity and complete fetal resorption in some rats at 1400 mg CCl_4/kg bw. No adverse effects on reproduction at doses of 6 and 15 mg CCl_4/kg bw in male and female rats for five generations.

Genotoxicity

CCl_4 has been generally found to be nonmutagenic in a range of tests with *Salmonella typhimurium* and *Escherichia coli*, in a liver microsome system, in chromosomal damage tests in rat liver cells *in vitro,* and in unscheduled DNA synthesis in rat liver hepatocytes.

Carcinogenicity

As reviewed by IARC (1979), CCl_4 was considered to be a carcinogen in mice, rats, and hamsters (hepatocellular carcinomas); however, some studies show survival problems. There is a dearth of epidemiological information to indicate the carcinogenicity of CCl_4 in humans. CCl_4's primary mechanism of action appears to be hyperproliferation of hepatocytes in response to cellular necrosis.

Epidemiology

CCl_4 exposures in air to humans of 1000 to 2000 ppm produce central nervous system depression, dizziness, vertigo, mental confusion, abdominal pain, nausea, vomiting, and diarrhea; symptoms are exacerbated by alcohol. Several single case studies exist which suggest an association between CCl_4 exposure and liver cancer; however, these do not provide sufficient evidence for the carcinogenicity of CCl_4 in humans. Several more comprehensive retrospective studies of mortality in laundry and dry cleaning workers exposed to CCl_4 have suggested possible excess mortality due to leukemia, cancers of the lung, cervix, colon, bladder, kidney, pancreas, and liver. It is difficult to establish a relationship between these cancers and CCl_4 exposure because of the simultaneous exposures of the workers to multiple chemicals, and poor definition of the degree and duration of exposure. IARC (1979) has nevertheless concluded that, given the reports of liver cancer in animals and the suggestive single case reports, that CCl_4 should be regarded as a carcinogen in humans.

Dichloromethane or Methylene Chloride (Report of the Expert Panel, 1994)

Metabolism and Bioavailability

Dichloromethane (DCM) appears to be rapidly absorbed through the intestinal mucosa; given DCM's high volatility, inhalation is the most likely route of exposure. DCM is rapidly absorbed through the lungs and distributed throughout the body. *In vivo*, DCM is metabolized by two pathways — a microsomal cytochrome P450 pathway and a glutathione S-transferase pathway. *In vitro* studies of metabolism of DCM conducted in mouse, rat, human, and hamster liver fractions have indicated that the metabolic rate of the glutathione pathway is significantly higher in the mouse than in the rat, and that metabolism by this pathway was lower still in both the hamster and man. DCM has an affinity for body fat, perhaps 8-fold higher than other tissues.

Mammalian Toxicology

DCM is not highly acutely toxic. LC_{50} for 7 hr in mice: 16,000–56,000mg/m^3; LC_{50} in rats: 57,000 ppm for 15 min; LC_{50} in guinea pigs for 6 hr: 11,600 ppm; oral LD_{50} in mice: 6400 mg/kg bw. Central nervous system, liver, and kidney are major organ systems affected by DCM.

Teratogenic and Reproductive Effects

Teratogenic effects (delayed ossification of the sternebrae, rudimentary lumbar ribs) have been observed at maternally toxic doses in mice and rats. Exposure of rats to DCM at levels as high as 1500 ppm had no adverse effect on reproduction.

Genotoxicity

The weight of evidence indicates that DCM is not mutagenic or genotoxic in mammalian systems, but does have mutagenic activity in bacteria.

Carcinogenicity

Several inhalation and drinking water laboratory studies to assess the chronic toxicity/carcinogenicity of DCM have been conducted. Inhalation studies in the hamster (air concentrations of 0, 500, 1500, and 3500 ppm) showed no treatment-related tumorigenic response, as did an inhalation study in rats (air levels of 500 to 4000 ppm for two years). A drinking water study in mice showed no dose-related tumorigenic response. Both benign and malignant tumors were elevated in an NTP inhalation study (air levels 2000 or 4000 ppm) in mice; the relevance of these results is questionable given the sensitivity to xenobiotics of this strain of mice and the lack of comparability to other species, particularly primates and humans. DCM's association with lung tumors appears to be unique to mice; rats and hamsters do not show such a response.

The mechanism of toxicity of DCM has been postulated to be nongenotoxic through acceleration of the progression of cells already initiated by other processes to development of tumors. The development of lung tumors in mice may be related to histological damage by DCM on Clara cells in mouse lung as well as the fact that Clara cells in mouse lung differ significantly from the lungs of humans or other animals in number, distribution, and ultrastructural morphology. Several human risk assessments for DCM have been conducted attempting to incorporate metabolic and pharmacokinetic differences between laboratory animals and humans.

Epidemiology

Two major studies have been conducted of workers involved in the production of photographic film, where DCM has been used since about 1970 as a solvent; the comparison group used was upstate New York males. A comparison of death rates of the exposed workers with those in either the general population or the unexposed workers showed nonsignificant deficits in the rates of lung and liver cancer and ischemic heart disease among the exposed workers. No relationship was observed between deaths due to pancreatic cancer and DCM exposure or between DCM exposure and cancer of the lung or ischemic heart disease.

Two studies have observed the mortality of employees exposed to DCM and other chemicals in two fiber production plants located in South Carolina and Maryland. In the South Carolina study, an excess of biliary cancer mortality was reported in one follow-up period, but was judged not significant because it did not occur in another later follow-up, biliary cancer had not been observed in animal studies, and the study did not control for such personal risk factors as smoking, use of oral contraceptives, or mixed chemical exposure to the workers. In the Maryland study an excess in mortality due to prostate cancer was observed relative to total U.S. and Maryland death rates, but not relative to death rates in the county where the plant was located; this comparison was judged most likely to account for the potential confounding effects of social, economic, ethnic, and cultural factors.

Studies of workers in chemical production plants have been difficult to interpret due to mixed chemical exposures, poor documentation of exposure, lack of observed increase in incidence of cancer, and small numbers of subjects.

Exposure Limits

The current OSHA PEL is 500 ppm. The Report of the Expert Panel recommends that it is most appropriate to determine an exposure limit for DCM based on a threshold-type dose-response relationship. An RfD of 50 μg DCM/kg bw/day was obtained by the application of 100-fold safety factor to the no-observed-adverse-effect-level (NOAEL) of 5 mg DCM/kg bw/day for hepatocellular toxicity in rats.

Based on its estimates of risk, the FDA has banned the use of DCM as a propellant in hair sprays and allowed the use of DCM in decaffeinating coffee and tea.

1,1,1-Trichloroethane (Report of the Expert Panel, 1994)

Metabolism and Bioavailability

1,1,1-Trichloroethane (TCE) is readily absorbed through the lungs and the gastrointestinal tract and is poorly absorbed through the skin. Its low systemic toxicity has been suggested to be related to its poor metabolism in animals and humans. The majority of TCE administered to animals is excreted or exhaled unchanged. The small amount of TCE which is metabolized is believed to be handled primarily by hepatic microsomal and nuclear cytochrome P450 or by the peroxisome oxidase-catalase system. TCE does not accumulate in the tissues of animals.

Mammalian Toxicology

TCE is a weak hepatotoxic agent in mice and dogs. The principal toxic effect of excessive exposure to TCE, either by inhalation or orally, is depression of the central nervous system, leading to death from respiratory shock or arrest. LD_{50} values (oral) in rats, mice, rabbits, and guinea pigs are 12,300, 11,240, 5660, and 9470 mg/kg bw, respectively.

Teratogenic and Reproductive Effects

TCE has not been demonstrated to be teratogenic in laboratory animals. No adverse effects on fertility, length of gestation, or viability or teratogenicity of pups has been observed in mice exposed to TCE.

Genotoxicity

The EPA, in reviewing the available data, has concluded that most mutagenicity testing with TCE has been inadequate because the high volatility or low water solubility of the compound has not been taken into account. When adequate exposure has been assured, TCE is weakly mutagenic in *Salmonella* systems, genotoxic to mouse hepatocytes, and able to induce oncogenic transformation in F344 rat and in Syrian hamster embryo cells.

Carcinogenicity

The EPA has classified TCE as "unclassifiable as to human carcinogenicity" because the available carcinogenicity bioassays have been found to be inadequate for a conclusive assessment. TCE's low toxic potential has been attributed to its lack of metabolism; hepatotoxic effects at extreme exposure are apparently associated with its general narcotic effects.

Epidemiology

Human fatalities resulting from acute exposure to TCE have been reported. A few studies of occupationally and environmentally exposed populations suggest that there is no association between exposure to TCE and lung cancer. A case control study in California of persons exposed to TCE and other chemicals from contaminated wells showed no causal relationship between congenital cardiac anomalies in infants born in the area.

Exposure Limits

The OSHA PEL is 350 ppm. The EPA has a risk assessment under review.

APPENDIX D

CHLORINATED ALKENES

Chlorinated alkenes, such as 1,1-dichloroethylene, trichloroethylene, and tetrachloroethylene, are used as organic solvents in various industrial and manufacturing applications, or as reactive intermediates in the production of polymers and other chemicals. The major uses of chlorinated alkenes in human activities are in dry cleaning operations and in the manufacture of a variety of products, including fumigants, refrigerants, solvents, metal degreasers, and paint removers. As such, these compounds would not be expected to be important in food or drinking water, and will not be reviewed in this chapter.

APPENDIX E

CHLORINATED ORGANIC ACIDS

Halogenated organic acids are by-products of the chlorine disinfection of drinking water; under laboratory conditions, dichloro- and trichloroacetic acids are produced as major reaction products in the chlorination of humic substances extracted from soil and water.

Dichloroacetic Acid (Report of the Expert Panel, 1994)

Metabolism and Bioavailability

About half of the dose of dichloroacetic acid (DCA) is absorbed from the gastrointestinal tract following oral gavage single doses. In rats about 25% of the dose was recovered in expired air and 25% in urine; in mice, about 25% was recovered in the urine and very little in expired air. No information is available on dermal or inhalation exposure. DCA, which is a metabolite of both TCE and trichloroacetic acid (TCA) is metabolized to an intermediate via the enzymatic hydroxylation of the C-H bond, followed by spontaneous dehydrodechlorination. Based on the rapid absorption and metabolism of DCA, bioaccumulation in tissues is not expected to occur; however, no information is available on tissue disposition.

Mammalian Toxicology

Oral LD_{50} for DCA: in rats 2820 to 4480 mg/kg bw; in mice 4140–5520 mg/kg bw.

Primary target organ is the liver, as well as the kidney, testes, and brain. In rats 1100 mg/kg bw/day in drinking water for 90 days caused morphological changes in the liver, lung, testes, and brain, and decreased body weight and organ weights.

Teratogenic and Reproductive Effects

In rats doses of DCA above 14 mg/kg bw/day caused a significant treatment-related increase in numbers of fetal anomalies; maternal toxicity of DCA, including reduced body weight gain and/or increased liver kidney and spleen weights, was observed at doses above 140 mg/kg bw/day. Reproductive effects have been reported in male rats and in male dogs administered DCA at doses of 500 and 2000 mg DCA/kg bw/day and doses greater than 12 mg/kg bw/day, respectively.

Carcinogenicity

DCA in the drinking water for 61 weeks has been shown to induce liver neoplasms in mice ($B_6C_3F_1$) at doses of 1000 mg/kg bw/day; in another mouse study ($B_6C_3F_1$) 200 to 400 mg/kg bw/day DCA in the drinking water induced a significant increase in the number of liver tumors with liver toxicity. In a third study, male $B_6C_3F_1$ mice developed liver tumors at doses of 93 mg/kg bw/day. In rats (F344) at doses of 0.05–5.0 mg/L DCA for 60–104 weeks, no significant increase in hepatic neoplasia was observed, but the incidence of altered foci, hyperplastic nodules, and hyperproliferative liver lesions was significantly increased in all but the lowest-dose group.

Liver neoplasms induced by DCA are associated with liver toxicity; a threshold-type dose-response relationship for tumorigenic effects has been proposed for DCA.

Epidemiology

DCA at doses of about 50 mg/kg bw/day for the treatment of diabetes and familial hyperlipidemia significantly reduced fasting blood glucose levels, plasma lactate, alanine, plasma cholesterol, and triglyceride levels, and increased ketone bodies. One patient receiving daily doses of 50 mg DCA/kg bw/day developed symptoms of neuropathy after 16 weeks.

Exposure Limits

There are no official exposure limits promulgated by North American regulatory agencies for DCA due to the uncertainty regarding the relevance to humans of the liver tumors in mice. NRC (1987) has suggested a suggested-no-adverse-response-level (SNARL) for DCA in water of 420 μg/L, assuming a NOAEL for DCA of 60 mg/kg bw/day.

Trichloroacetic Acid (Report of the Expert Panel, 1994)

Metabolism and Bioavailability

TCA is well absorbed from the gastrointestinal tract, as has been demonstrated in mice and dogs. Metabolism has been reported to proceed primarily through reductive dehydrochlorination involving the P450 mixed-function oxygenase enzyme system. Information is not available on tissue disposition, but based on TCA's relatively high water solubility, TCA would be expected to widely distribute throughout the body.

Mammalian Toxicology

TCA is not acutely toxic; the reported LD_{50} is 3000 to 5500 mg/kg bw. The main target organ is the liver and to a secondary extent the kidney. In male Sprague-Dawley rats, the NOAEL for 90 days was 4.1 mg/kg bw/day; rats receiving 355 mg/kg bw/day showed increased relative liver and kidney weights, significantly increased hepatic *beta*-oxidation activity and glycogen accumulation in enlarged hepatocytes. In another rat study, male rats administered TCA in water via gavage at about 300 mg/kg bw/day displayed decreased body weight and alterations to renal enzyme activity; at 25 mg/kg bw/day there was no effect seen. Studies with $B_6C_3F_1$ mice receiving 1–5 g TCA/L drinking water showed increased incidences of hepatic neoplasms in conjunction with liver toxicity.

Teratogenic and Reproductive Effects

In a teratogenesis study with TCA in Long-Evans rats, maternal toxicity (e.g., reduced weight gain, altered relative liver and spleen weights) resulted at 800, 1200, and 1800 mg/kg bw/day, but not at 330 mg. Fetal resorptions occurred at the three highest dose groups; study authors concluded that it was difficult to differentiate maternal toxicity from teratogenic effects. No information is available on the potential of TCA to induce reproductive toxicity in laboratory studies.

Genotoxicity

In vitro mutagenicity studies with TCA in bacterial systems have generally been negative, including tests with metabolic activation. No significantly increased induction of DNA strand breaks was observed in mice and rats.

Carcinogenicity

TCA treatment is associated with a significant increase in liver tumors and overt liver toxicity in $B_6C_3F_1$ mice, but not in rats. Doses used were 1000 mg/kg bw/day in the mice and 5, 50, and 500 mg/kg bw/day in rats. In another mouse ($B_6C_3F_1$) study, 200 and 400 mg TCA/kg bw/day for 52 weeks induced a significant increase in the number of liver tumors, including hyperplastic nodules, adenomas, and carcinomas. Liver tumors were observed to occur with liver toxicity. The mechanism of toxicity has been hypothesized to involve the induction of peroxisome proliferation. Peroxisomes are ultramicroscopic particles that contain enzymes which generate hydrogen peroxide in cells (especially hepatocytes); chemicals known as peroxisome proliferators cause a characteristic group of effects, including hepatomegaly, peroxisome proliferation, induction of hepatic P450, induction of enzymes involved in fatty acid oxidation and hypolipidemia. A strong association has been shown between cell proliferation and liver carcinogenesis. The minimum daily dose of TCA required to induce peroxisome proliferation in mice has been shown to be 0.8 mmol/kg and 4.2 mmol/kg in rats.

APPENDIX F

CHLORINATED KETONES

Chlorinated ketones and furanones are produced during chlorination processes from precursor unchlorinated ketones and furanones naturally present in water from humic substances. Chlorinated ketones which have been found in drinking water and surface waters included 1,1-dichloropropanone, 1,1,1-trichloropropanone, and 3-chloro-4-[dichloromethyl]-5-hydroxy-2(5H)-furanone (MX).

Metabolism and Bioavailability

No information is available on bioavailability, metabolism, or bioaccumulation of haloketones.

Mammalian Toxicology

Oral LD_{50} in male Wistar rats is 1550 mg/kg bw; dermal LD_{50} in albino rabbits 2980 mg/kg bw; LC_{50} 660 and 360 ppm in albino rats after 3- and 6-hr inhalation exposures, respectively.

Reproductive and Teratogenic Effects

Maternal toxicity and fetotoxicity were observed in mice receiving a dose of 15 mg 1,1',3,3'-tetrachloroacetone via gavage on Days 6 to 15 of gestation.

Genotoxicity

In the *Salmonella* system, monochloropropanone was nonmutagenic, while 1,1-dichloropropanone and 1,3-dichloropropanone were both mutagenic without metabolic activation. In another study, 1,1-dichloropropanone, 1,1,3,3-trichloropropanone, and MX were mutagenic in *Salmonella* tests without metabolic activation. Trichloropropanone was mutagenic to some strains of *Salmonella*, as was pentachloroacetone and hexachloroacetone. MX has been shown to induce chromosomal aberrations in Chinese hamster ovary cells without metabolic activation; MX appears to be less mutagenic both to *Salmonella* and to Chinese hamster ovary cells after metabolic activation. MX has been found to account for 20 to 50% of the mutagenic potential of chlorinated drinking water.

Carcinogenicity

Limited data are available on carcinogenicity of these compounds. Short-term assays such as lung adenoma assay and skin painting initiation-promotion assays have been negative for 1,1-dichloroacetone, 1,1,1-trichloroacetone, and 1,1'3,3'-tetrachloacetone, and positive for 1,3-dichloroacetone. The available information suggests that the mechanism of action of these compounds involves the direct interaction of haloketones with cell macromolecules, possibly including DNA.

APPENDIX G

CHLORINATED BENZENES

Chlorobenzenes are used in the manufacture of a variety of products, including insecticides, herbicides, dyes, dielectric fluids, and nitroso and styrene rubbers. Chlorobenzenes may be released into the environment, and hence possibly into food and drinking water, during their production and transport, during use as solvents, air deodorizers, flame retardants, chemical intermediates, pesticides, or in industrial waste generated from their manufacture from the electrical industry. In the case of 1,4-dichlorobenzene, an estimated 70 to 90% of the total annual production would be released into the air and less than 5% into water (Report of the Expert Panel, 1994).

Dichlorobenzenes (Report of the Expert Panel, 1994)

Metabolism and Bioavailability

Dichlorobenzenes are absorbed from the gastrointestinal tract after oral administration; the majority of orally administered dichlorobenzene is accounted for by urinary metabolites. Dichlorobenzenes have been demonstrated to be absorbed from the lungs in animals following inhalation exposure. The dichlorobenzenes are primarily metabolized to dichlorophenols, which are subsequently conjugated with glucuronic and sulfuric acids prior to excretion. Metabolism occurs primarily in the liver through hydroxylation catalyzed by P450 enzymes. A significant portion of the metabolites result from enterohepatic circulation. After oral administration dichlorobenzenes are primarily distributed to the adipose tissue and liver with some accumulation in kidney tissue and lung. After inhalation exposure, the dichlorobenzenes appear to be distributed throughout the body tissues in a pattern similar to that after oral administration.

Mammalian Toxicology

Oral LD_{50} for 1,2-dichlorobenzene: 500, 2138, 1875, and 3375 mg/kg bw for rat, rabbit, mouse and guinea pig, respectively. Oral LD_{50} for 1,4 dichlorobenzene: 500 and 2812 mg/kg bw in rats and rabbits, respectively. Large doses of chlorobenzenes induce central nervous system depression and liver and kidney damage; liver and kidney are primary sites for toxicity. Male rats dosed at 1200 and 1500 mg/kg bw day 1,4-dichlorobenzene showed decreased survival, necrosis and damage to hepatocytes, hypoplasia of the bone marrow, and depletion of lymphoid cells of the spleen and thymus. Mice dosed orally for 13 weeks at 600 mg 1,4-dichlorobenzene/kg bw/day showed liver toxicity (hepatocellular degeneration and cytomegaly). F344/N rats exposed to 1,4-dichlorobenzene in corn oil gavage at doses of 0, 150, 300, and 600 mg/kg bw/day for 2 years, showed a dose-dependent increase in renal nephropathy. Toxicity of 1,2-dichlorobenzene has been shown to be similar to 1,4-dichlorobenzene.

Teratogenic and Reproductive Effects

At doses of 1,4-dichlorobenzene of 500 mg/kg bw/day and greater to rats, there was a dose-related increase in the incidence of maternal toxicity and an increase in the frequency of an extra rib (possibly related to maternal toxicity). With 1,2-dichlorobenzene, maternal toxicity was observed at all inhalation dose levels (100, 300, and 400 ppm) in rats and rabbits. 1,4-Dichlorobenzene via inhalation at doses of 400 and 800 ppm in rats and rabbits showed slight maternal toxicity.

Genotoxicity

Dichlorobenzenes appear not to be mutagenic in bacterial systems and in all systems tested appear to have relatively limited genotoxic potential. Positive results following intraperitoneal injection (*in vivo* mouse micronucleus assay) may indicate that the data derived from *in vitro* tests may not be adequate to reflect the genetic activity of the dichlorobenzenes in multicellular organisms.

Carcinogenicity

The carcinogenicity of 1,2-dichlorobenzene has been studied by the NTP; F344/N rats and $B_6C_3F_1$ mice received 42.5 and 85 mg/kg bw/day in corn oil gavage. In the treated animals, there was no significantly increased incidence over controls of any malignancy; NTP concluded that the compound was not a carcinogen in laboratory animals. In the NTP 1,4-dichlorobenzene study, F344/N rats and $B_6C_3F_1$ mice were dosed by gavage with 0, 150, or 300 mg/kg bw/day (rats) and 0, 300, and 600 mg/kg bw/day (mice). High-dose male rats showed an increased incidence of renal tubular cell adenoma and adenocarcinoma. In the mice the combined incidences of hepatocellular adenoma and carcinoma were significantly increased in the upper dose groups of both sexes. In the livers, both sexes of mice showed significant hepatocellular degeneration; high dose mice showed significantly increased incidences of certain rare tumors (hepatoblastoma and phaeochromocytoma). NTP concluded that 1,4-dichlorobenzene was a renal carcinogen in male rats and a liver carcinogen in male and female mice.

The mechanism of toxicity appears to be mediated through several routes, including one species-specific mechanism. In male rats, kidney carcinogenicity seems to be related to *alpha-2-mu*-globulin accumulation in renal cells, which is believed to induce cell toxicity followed by cell proliferation. In other sex/species, liver and kidney toxicity is likely associated with oxidative metabolism. A correlation between the extent of formation of reactive metabolites and hepatotoxic potential has been established for the dichlorobenzenes.

Epidemiology

Accidental poisoning and long-term occupational exposure to dichlorobenzenes cause toxic effects mainly in the organs of the blood-forming system. Toxicity to the liver, central nervous system, respiratory tract, and reticuloendothelial system have also been reported. In one lymphocyte karyotype analysis study of workers exposed to 1,2-dichlorobenzene, the number of altered cells having chromosomal aberrations was significantly greater in the exposed workers; the significance of this finding is unclear.

Trichlorobenzenes (Report of the Expert Panel, 1994)

Metabolism and Bioavailability

1,2,4-Trichlorobenzene appears to be well absorbed following oral exposure. A high percentage of the dose is excreted in the urine, indicating high systemic bioavailability from the gastrointestinal tract. 1,2,4-Trichlorobenzene appears not to accumulate in fetuses of pregnant rats. In mammals,

1,2,4-trichlorobenzene is metabolized through oxidation catalyzed by P450 monooxygenase enzymes in the liver. Species differences have been demonstrated between rats and monkeys. Based on the available data, 1,2,4-trichlorobenzene appears to be widely distributed throughout the body, with a preference for lipid-rich tissues.

Mammalian Toxicity

When studied in several oral and inhalation subchronic toxicity studies, 1,2,4-trichlorobenzene has been shown to alter, usually by increasing, the activity of enzymes involved in the metabolism of xenobiotic compounds, induce lesions in the liver and kidneys, and alter metabolism of thyroid hormones. Oral doses demonstrated to have these effects include 10 to 40 mg/kg bw/day in rats.

Teratogenic and Reproductive Effects

The available evidence suggests that 1,2,4-trichlorobenzene has very little teratogenic potential even at doses that can induce maternal toxicity; in addition, the compound has limited, if any, potential to affect reproduction absent maternal toxicity.

Genotoxicity

In *Salmonella* systems, 1,2,4-trichlorobenzene has been reported to show entirely negative results, both with and without metabolic activation. Negative results have also been reported in tests with Chinese hamster ovary cells and micronuclei in the bone marrow of mice.

Carcinogenicity

No evidence of carcinogenicity has been observed in several subchronic to chronic studies with 1,2,4-trichlorobenzene, but a definitive study has not been performed. Liver and kidney toxicity observed in toxicity studies is likely associated with products of oxidative metabolism, including formation of an epoxide which would arylate cellular macromolecules and cause cytotoxicity.

Hexachlorobenzene (Report of the Expert Panel, 1994)

Metabolism and Bioavailability

The bioavailability of hexachlorobenzene (HCB) seems to depend on the vehicle of administration; when dissolved in an oil solution, approximately 80% of the dose was absorbed in rats, whereas only about 20% was absorbed when the vehicle was water. The use of high-fat, low-carbohydrate diets has been reported to accelerate the absorption of hexachlorobenzene from the gastrointestinal tract compared to that of a high-carbohydrate, low-fat diet. No information is available on the absorption of HCB by inhalation or dermally.

HCB is metabolized more slowly than the less chlorinated benzenes, chlorinated phenols, etc.; metabolites largely result from the work of P450 mixed-function oxidases. HCB tends to partition to the lipid-rich tissues, such as adipose tissue, bone marrow, skin, and adrenal cortex.

Mammalian Toxicology

Many toxicology studies have been conducted on HCB. Key points will be presented here. The LD_{50} for HCB in mice, rats, rabbits, and cats has been reported as 4000, 3500 to 10,000, 260, and 1700 mg/kg bw. Porphyria, a marker for HCB exposure, has been observed in rats exposed for 15 weeks to 2.0 mg/kg bw/day with 0.5 mg/kg bw/day being the NOAEL. HCB causes overt hepatotoxicity, with the NOAEL being 1.6 ppm in the diet for rats in a 130-week study. Kidney toxicity has also been reported following chronic exposure to HCB at doses equivalent to 10 mg/kg bw/day in rats. HCB affects the immune system in neonatal rats and mice at levels ranging from 50 to 150 ppm in rats and to 167 ppm in mice.

Teratogenic and Reproductive Effects

HCB appears to be teratogenic in mice at 100 mg/kg bw/day on gestation day 7–16, resulting in an increased incidence of fetal malformations. HCB is apparently not teratogenic in rats and mink; fetotoxicity resulted only at maternally toxic levels.

Genotoxicity

HCB has limited mutagenic and/or genotoxic potential. Negative results have been reported in *Salmonella* strains and in dominant lethal assays in mice; positive results have been reported in *Saccharomyces*.

Carcinogenicity

HCB, when administered orally to mice, hamsters, and rats, has been demonstrated to produce cancer in the liver, thyroid, kidney, parathyroid, and adrenal gland. The mechanism of toxicity for HCB is not well understood. A recent hypothesis is that HCB induces liver toxicity and carcinogenesis similarly to dioxin though the aryl hydrocarbon hydroxylase (Ah) receptor.

Epidemiology

A Turkish population accidentally poisoned when HCB used as a fungicide was ingested 1954–1959 has been studied. Exposures ranged from 50 to 200 mg/day. Adverse effects included cutaneous lesions, porphyria, and death; infants exposed through mother's milk died in 95% of the cases. In a follow-up 20 to 25 years later, researchers found that most individuals still showed porphyria and thyromegaly, but cancer incidence appears not to be elevated in the exposed population. Other epidemiological studies have looked at occupational exposure in the U.S. and increased risk of lung and renal cancer; a causal relationship has not been demonstrated.

Exposure Limits

The EPA has determined a HCB exposure limit of 0.8 μg/kg bw/day based on the NOAEL of 0.08 mg/kg bw/day.

APPENDIX H

CHLORINATED PHENOLICS

Chlorinated phenols in the environment, and possibly in food and drinking water, arise from wood processing plants, wood treatment and wood preservation plants, kraft pulp mills, sewage treatment plants, and pesticide manufacturing plants. Mono-, di-, and trichlorophenols may be formed from the interaction of aqueous chlorine with certain organic molecules, including natural phenols, during the chlorination of drinking water and wastewater. Chlorophenols are also produced by natural sources, such as microbes, burning of wood, or forest fires; it is not possible with the current extent of information to quantitate these amounts.

Metabolism and Bioavailability

Pentachlorophenol (PCP) is readily absorbed by the gastrointestinal tracts of rats, mice, monkeys, and humans, as indicated by urinary excretion data. About half of the dose of PCP is excreted as the parent compound and about 6% as conjugated PCP. Some other chlorophenols are conjugated with glucuronic acid and/or sulfuric acid to facilitate their excretion. Generally, chlorophenols do not have a high bioaccumulation potential in humans.

Mammalian Toxicology

PCP is moderately acutely toxic in mammals; the oral LD_{50} in the rat ranges from 80 to 170 mg/kg bw. The minimum lethal dose in humans has been estimated to be 29 mg/kg bw. Oral LD_{50} values for 2,4,6-tetrachlorohydroquinone ranges from 276 to 820 mg/kg bw; LD_{50} for 2,4-dichlorophenol ranges from 580 to 4000 mg/kg bw. Chlorinated guaiacols appear to be less acutely toxic than their chlorinated phenol counterparts.

Teratogenicity and Reproductive Effects

Teratogenic studies of the chlorophenols have been complicated by the presence of other chemicals in the test preparations. Fetotoxic and teratogenic effects have been observed at greater than 30 mg/kg bw/day in rats following oral administration of PCP. The NOAEL for reproduction, neonatal growth, survival, or development has been estimated to be 3 mg PCP/kg bw/day in the rat.

Genotoxicity

Chlorophenols have been shown to be largely negative in mutagenesis studies. PCP and certain isomers of the trichlorophenols, trichlorocatechols, and trichloroguaiacols were genotoxic in various test systems.

Carcinogenesis

2-Chlorophenol and 2,4-dichlorophenol did not induce an increased incidence of tumors in rats. 2,4,6-Trichlorophenol administered to F344 rats resulted in a dose-related increase in incidence of lymphomas or leukemias. Increases in the incidence of hepatocellular carcinomas and adenomas were observed following dietary exposures of $B_6C_3F_1$ mice to 2,4,6-trichlorophenol. PCP was not carcinogenic following chronic oral exposures in laboratory studies with two strains of mice, but two technical-grade PCP products were carcinogenic in male and female $B_6C_3F_1$ mice (adrenal and hepatocellular neoplasms and hemangiosarcomas). Toxic effects of the chlorophenols have been explained by their actions as uncouplers of oxidative phosphorylation. The potency of chlorophenols as oxidative phosphorylation uncouplers decreases with decreasing degree of chlorination. Lesser chlorinated phenols are less potent and tend to produce more general cellular narcotic or anesthetic type activity.

Epidemiology

Several studies have suggested possible associations between chlorophenol and cancers of the soft tissues non-Hodgkin's lymphoma. However, these studies have been confounded by lack of exposure specifically to chlorophenols.

Exposure Limits

EPA oral reference dose (RfD) for 2,4-dichlorophenol: 3 μg/kg bw/day. Health and Welfare Canada acceptable daily intake (ADI): 0.1 mg/kg bw/day. ADI for PCP: 6 μg/kg bw/day.

APPENDIX I

CHLORINATED NITROGEN CHEMICALS

Chloroacetonitriles have been used as fumigants and chemical intermediates and may enter the environment during production, use, or disposal. The only measurements of these chemicals in the environment have been as by-products during the chlorine disinfection of drinking water; hence, they could find their way into food.

Dichloroacetonitrile (Report of the Expert Panel, 1994)

Metabolism and Bioavailability

Dichloroacetonitrile (DCAN) was almost completely absorbed when given orally to mice or rats. Most halonitriles are metabolized to the corresponding cyanohydrins, to cyanide, and then to thiocyanate. About 20% of an orally administered dose of DCAN is retained by rats with the following tissue distribution in decreasing order: blood, muscle, skin, and liver.

Mammalian Toxicology

Oral gavage LD_{50} of DCAN in rats: 339 and 330 mg/kg bw male and female, respectively; 270 and 279 mg/kg bw for male and female mice. No signs of toxicity were observed when male and female CD rats were treated with 12 and 90 mg DCAN/kg bw/day over 14 days.

Teratogenic and Reproductive Effects

Long-Evans rats exposed to DCAN by gavage at 55 mg/kg bw/day on gestation days 7 to 21 showed reduced body weight gain, increased fetal resorptions, and decreased litter size. The no-observed-effect-level (NOEL) for reproductive effects appeared to be 15 mg/kg bw/day in the rat.

Genotoxicity

DCAN induced some mutations in some strains of *Salmonella* both with and without exogenous metabolic activation. DCAN has also been reported to cause sister chromatid exchange in Chinese hamster ovary cells and DNA strand breaks in a human cell line. *In vivo* mouse micronuclei and sperm morphology assays have generally been negative.

Carcinogenicity

No long-term studies designed to evaluate the complete carcinogenic potential of DCAN were found. Short-term carcinogenic screening assays, such as lung adenoma assays, have been negative. Haloacetonitriles are biologically reactive either directly or after metabolic conversion, and may be direct-acting alkylating agents. However, short-term cancer screening assays do not support this finding; possibly DCAN is so reactive that is does not survive long enough in biological systems to cause the type of damage necessary to induce tumors.

Epidemiology

No data available.

Exposure Limits

A suggested no-adverse-response-level of 0.056 mg/L for DCAN in drinking water has been developed, based on a no-observed-adverse-effect-level of 8 mg/kg bw/day in rats.

Nitrochloroform (Report of the Expert Panel, 1994)

Metabolism and Bioavailability

No studies on the metabolism of nitrochloroform (NCF) by mammals were found. Bacteria are known to metabolize NCF by cytochrome P450 enzymes. No data are available on tissue disposition.

Mammalian Toxicology

LD_{50} in rats for NCF is 250 mg/kg bw. Chronic exposure to male and female $B_6C_3F_1$ mice to 33.3 and 66.7 mg/kg bw/day for 78 weeks had significantly reduced survival at the high dose and showed loss of weight.

Genotoxicity

Mixed results have been obtained on the mutagenic potential of NCF in various *in vitro* short-term tests.

Carcinogenicity

A 78-week oral gavage dosing study in rats exposed male rats to 24.5 and 25.7 mg/kg bw/day and female rats to 20.3 and 21.5 mg/kg bw/day. Results were equivocal; in some cases, treated animals had fewer tumors than controls. NCF is, at most, a weak carcinogen, likely operating due to local irritant effects at the site of contact.

Epidemiology

NCF was used as a toxic antipersonnel agent during World War I; concentrations of 0.3 to 3.7 ppm of NCF were found to induce closing of the eyelids and 4 ppm rendered men unfit for combat.

APPENDIX J

CHLORINATED ALDEHYDES

Chlorinated aldehydes are formed as by-products from the chlorination of humic substances extracted from soil and water, and hence could be found in food.

Metabolism and Bioavailability

Chloral hydrate is absorbed at the level of approximately 40 to 60% from the gastrointestinal tract. Metabolism of these compounds proceeds through an NADPH-dependent P450 enzyme pathway. Considering their relatively high water solubility, chloral, chloral hydrate, and their associated metabolites are likely to be widely distributed throughout the body.

Mammalian Toxicology

Oral LD_{50} in rats and mice for chloral hydrate: 285 and 1100 mg/kg bw, respectively. Oral LD_{50} in mice 1440 and 1260 mg/kg bw, respectively. Main site of chloroaldehyes is the liver. The immune system is also affected.

Teratogenic and Reproductive Effects

Chloral hydrate does not appear to be teratogenic in mice. No relevant information is available on reproductive effects.

Genotoxicity

Most of the chlorinated aldehydes act as direct mutagens in bacterial test systems without metabolic activation. Chloral hydrate was negative in the liver tissue of mice and rats in a test designed to detect DNA strand breaks.

Carcinogenicity

Chloral hydrate is generally considered to be a hepatocarcinogen in mice based on the similarity between chloral hydrate and trichloroacetic acid in terms of chemical structure and metabolism.

Epidemiology

No epidemiology studies are available. Liver and kidney damage and suppression of immune function have been reported in humans exposed to chloral hydrate.

Exposure Limits

The EPA has reported a reference dose for chloral to be μg/kg bw/day, based on a lowest-observed-adverse-effects-level (LOAEL) of 15.7 mg/kg bw/day for liver toxicity in mice.

APPENDIX K

POLYCHLORINATED DIBENZO-*PARA*-DIOXINS AND POLYCHLORINATED DIBENZOFURANS

These compounds find their way into the food chain through various routes of environmental contamination. The toxicology of these compounds is too complex and lengthy to be covered here and has also been discussed earlier.

TOXICOLOGY OF BY-PRODUCTS OF CHLORINATION OF DRINKING WATER

The hazards and risks of by-products of the chlorination of drinking water are of particular interest to the FDA since drinking water is used for the preparation and processing of food. The compounds of relevance and their classes of disinfectant by-products have been summarized in Tables 23.1and Table 23.2.

When unconcentrated drinking water samples collected from disinfected spring, river, and lake waters were tested, no evidence of genotoxicity or mutagenicity has been observed in mammalian *in vitro* genetic tests; however, when preconcentration techniques were employed, increases in sister chromatid exchange, chromosomal aberration and micronucleus induction in Chinese hamster ovary cells were reported. The addition of *in vitro* metabolic systems to the preconcentrated samples resulted in a loss of mutagenic activity. No information regarding the genotoxic activity of either whole or preconcentrated chlorinated drinking waters in *in vivo* mammalian bioassays has been identified in the literature (IARC, 1991).

There have been several chronic/carcinogenicity studies conducted with laboratory animals receiving extracts of water previously disinfected with chlorine. Truhaut et al. (1979) exposed female CFLP Han mice and Sprague-Dawley rats to extracts of chlorinated river water in the diet at doses of 1.2 and 2.4 mg/kg body weight/day for 104 weeks. The doses represented 100 to 200 times the estimated daily human intake based on a 60-kg person consuming 3 liters of water/day. Increased mortality in both sexes of the rats and mice was noted as well as an increased incidence of mammary gland ovarian carcinomas in female mice. However, results of the study must be interpreted with caution since few tumor type details were given and there was no control group to test the extraction procedure.

In another chronic bioassay, male and female Wistar rats were treated at dose rates from 0 to 97 mg extract/kg body weight/day (males) and 0 to 165 mg extract/kg bw/day for females using organic extracts of disinfected surface water dissolved in purified water (Kool et al., 1985). The doses corresponded to 0 to 68 times the normal human intake of organics via drinking water. A slight treatment-related increase in mortality of treated animals was noted, but no evidence of carcinogenicity was observed. Unchlorinated water extracts were not studied.

These studies did not include a control group given extracts of raw or unchlorinated water and therefore causal relationships cannot be inferred (IARC, 1991). In 1991, the International Agency for Cancer Research (IARC) also concluded that there was inadequate evidence for the carcinogenicity of chlorinated drinking water in experimental animals. Attempts to demonstrate an effect of chlorinated water consumption on lipid metabolism and on thyroid function in experimental animals have also proved negative or inconclusive.

A number of epidemiology studies have appeared in the scientific literature which have evaluated the potential effect of the consumption of water disinfected with chlorine/hypochlorite on the risks for certain adverse effects, including cancer and poor reproductive outcome. These studies have been reviewed by NRC (1987) and IARC (1991) and findings are summarized below.

Several studies were done with either groundwater or surface water as the exposure variable and cancer incidence as the outcome variable, assuming that surface waters were more highly chlorinated than groundwaters. These studies compared the cancer rates of populations having potable water supplies derived from surface water sources with the rates of populations having water supplied from groundwater sources, with most studies focusing on populations in the U.S. especially Louisiana and the midwest. The correlations analyzed showed some positive associations between surface water consumption and cancers of the bladder, stomach, rectum, and lung. However, these results must be interpreted cautiously since potable water sources may not provide an adequate surrogate from which to measure chlorination and also, exposure to other contaminants in the water unrelated to chlorination may be occurring. Factors known to affect cancer mortality, such as diet and smoking habits, were not adequately controlled in these studies.

Other epidemiology studies involving populations in the U.S. and the Netherlands have attempted to measure correlations between measured concentrations of trihalomethanes (THMs) in drinking water and cancer rates (NRC, 1987; IARC, 1991). Positive associations have been observed between cancers of the urinary bladder, colon, and rectum and the concentrations of THMs in the finished drinking water. However, the data presented were inadequate for inferring a causal association; concentrations of THMs were poorly defined and various confounding factors were not controlled.

Morris et al. (1992) performed a *meta*-analysis of pooled data from 10 epidemiology studies evaluating the effects of consumption of chlorinated drinking water on cancer; the analysis indicated that exposure to chlorination by-products through the drinking water was associated with a relative risk of cancer of 1.15. When using only the studies that enabled statistical adjustment for confounding factors, such as age, sex, population density, smoking, occupation, etc., Morris et al. (1992) found no appreciable difference in the relative risk estimate from that obtained using the whole data set. A dose-response relationship was observed based on results from studies that stratified exposure to chlorinated drinking water into "low," "intermediate," and "high" exposure groups. The authors concluded that the increased statistical power of the *meta*-analysis technique and the results it produced suggest an association between the consumption of chlorinated surface waters and increased incidence of bladder and rectal cancer. However, as pointed out in the Report of the Expert Panel (1994), the possibility that consumption of chlorinated surface waters may be a surrogate for some other etiological agent has not been ruled out in the Morris et al. study. No chlorinated by-products have been identified in chlorinated drinking waters which have been associated with cancers of the rectum and bladder based on the weight of available toxicological evidence.

In addition, there are several major concerns with the *meta*-analysis technique (Report of the Expert Panel, 1994). First, such an analysis is a currently controversial technique for the combination of epidemiological studies that were conducted independently and does not overcome the weaknesses of the individual studies on which it is based. Secondly, the *meta*-analysis of the available case-control and cohort epidemiology studies considering the effect of consumption of chlorinated drinking water on cancer reported relative risks greater than 1.0 for all cancer sites examined. This lack of specificity makes it difficult to establish a causal relationship between chlorinated drinking water and cancers observed. Third, the relative risks observed were small (less than 2.0); such small relative risks could be due to chance or confounding of the statistical analysis by various factors in the studies.

Several studies have examined the possible association between poor reproductive outcomes and consumption of chlorinated drinking water. Bove et al. (1992a) conducted an epidemiological study of four New Jersey counties to evaluate possible associations between trihalomethanes (THMs), other organics, the source of drinking water supply, and 13 different adverse reproductive outcomes including stillbirths, low birth weight, very low birth weight, prematurity, and malformations. Results showed the presence of a statistically significant association between THM concentrations in the drinking water supply and low birth weight. The International Life Sciences/Environmental Protection Agency review (ILSI, 1993) concluded that these associations should not be interpreted as causal relationships since the analysis of the study could have been confounded by exposure misclassification (no individuals were interviewed) and variables known to affect reproductive outcomes (e.g., use of tobacco and alcohol, specific occupational factors, etc.) were not controlled.

A case-control study of the same New Jersey births was performed of babies with cardiac defects, neural tube defects, oral clefts, and low and very low birth weights (Bove et al., 1992b). Interviews were used to gain exposure and confounding variables information; however only 52% of the subjects selected were interviewed and the group not interviewed differed sharply from the subjects interviewed in terms of race, education, and adequacy of prenatal care visits. A statistically imprecise correlation was found between THM concentration in excess of 80 μg/L and neural tube defects. The authors of the study and the ILSI/EPA reviewers of the study agreed that THM and other organic chemicals cannot be causally related to adverse reproductive outcomes on the basis of this study.

Hertz-Picciotto et al. (1989) examined, using survival analysis methods, the rate of spontaneous abortions as related to the consumption of tap water in the Santa Clara, California area. Authors were able to statistically control for certain confounding factors, including smoking, alcohol consumption, previous pregnancy losses, and maternal age. The risk of spontaneous abortion was greater in women who consumed mainly tap water versus women who consumed mainly bottled water. However, these results did not implicate tap water as the causal agent since the authors found evidence of recall bias, such as differential recall on the part of the subjects following a miscarriage, which could have biased the reported associations.

When all of the epidemiological data are examined, the evidence for increased risks of cancer associated with the consumption of chlorinated drinking water (especially consumption of chlorinated surface water versus consumption of unchlorinated groundwater) is strongest for cancer of the urinary bladder (IARC, 1991). However, as IARC has pointed out, many of the epidemiological studies did not control for a number of potentially confounding variables; hence, the epidemiological data are insufficient to indicate a causal link for human cancer. In addition, when all of the reproductive end-point data are examined, there does not appear to be a clear causal relationship between THMs and other organic chemicals and adverse reproductive outcomes.

HAZARDS ASSOCIATED WITH CHLORINE DISINFECTION BY-PRODUCTS

The by-products receiving the most toxicological study are those in the haloacetic acid, haloacetonitrile and trihalomethane classes. These compounds have several toxicological effects, but their ability to induce cancer in laboratory animals is considered to be most important. Some of the potentially carcinogenic by-products of drinking water disinfectants are shown in Table 23.2 earlier in this chapter (Bull, 1993).

A look at the classes of disinfectant by-products shown in Table 23.1 is in order. With chlorine, a large number of disinfectant by-products have been identified. For the classes which occur in the highest concentration, (trihalomethanes, haloacetates, and haloacetonitriles), in general (except for bromodichloracetate and dibromochloroacetate), substantial toxicological databases exist and some compounds have been identified as possible causes of concern. Therefore, the research information which is most needed is that which would allow an estimate of risks at the low concentrations found in drinking water. Practically speaking, this issue needs to be resolved before we become too seriously concerned about chlorination by-products that occur at lower concentrations (Bull, 1993).

The by-products of chloramines are probably similar to those of chlorine, but this area needs more systematic research. Chlorine dioxide and ozone present major sources of uncertainty because their by-products have not been completely identified.

There are fewer uncertainties with chlorine than the alternatives. More information is available on the nature and concentrations of by-products, the mechanisms of toxicity and their relevance to humans, the effective dose, and interactions between contaminants for chlorine disinfection than for other types of disinfection (Bull, 1993).

REFERENCES

Abdel-Rahman, .S., D. Couri, and J.D. John. 1980a. Chlorine dioxide metabolism in rats. *J. Environ. Pathol. Toxicol.* 3:421–430.

Abdel-Rahman, S., D. Couri, and R.J. Bull. 1980b. Kinetics of ClO_2, ClO_2^- and ClO_3^- in drinking water on blood glutathione and hemolysis is rat and chicken. *J. Environ. Pathol. Toxicol.* 3:431–449.

Abdel-Rahman, M.S., D. Couri, and R.J. Bull. 1984. Toxicity of chlorine dioxide in drinking water. *J. Am. Coll. Toxicol.* 3:277–284.

Agency for Toxic Substances and Disease Registry (ATSDR) 1987. *Toxicological Profile for Methylene Chloride.* ATSDR, Oak Ridge National Laboratory, DOE Interagency Agreement 1425–1425-A1.

American Conference of Governmental Industrial Hygienists (ACGIH). 1993. *Threshold Limit Value for Chemical Substances and Physical Agents and Biological Exposure Indices. 1993.* ACGIH. Cincinnati, OH.

Bercz, J.P., L. Jones, L. Garner, D. Murray, A. Ludwig, and J. Boston. 1982. Subchronic toxicity of chlorine dioxide and related compounds in drinking water in the non-human primate. *Environ. Health Perspect.* 46:47–55.

Bove, F.J., M.C. Fulcomer, J.B. Klotz, J. Esmart, E.M. Dufficy, and R.T. Zagraniski. 1992a. *Population-Based Surveillance and Etiological Research of Adverse Reproductive Outcomes and Toxic Wastes. Report on Phase IV-A: Public Drinking Water Contamination and Birthweight, Fetal Deaths and Birth Defects, A Cross-Sectional Study.* New Jersey Department of Health.

Bove, F.J., M.C. Fulcomer, J.B. Klotz, J. Esmart, E.M. Dufficy, R.T. Zagraniski, and J.E. Savrin. 1992b. *Population-Based Surveillance and Etiological Research of Adverse Reproductive Outcomes and Toxic Wastes. Report on Phase IV-B: Public Drinking Water Contamination and Birthweight, and Selected Birth Defects. A Case-Control Study.* New Jersey Department of Health.

Bull, R.J. 1993. Toxicology of disinfectants and disinfectant by-products. In Craun, G. (Ed). *Safety of Water Disinfection: Balancing Chemical and Microbial Risks.* International Life Sciences Institute Press, Washington, DC.

Chlorine Institute (1990). *Toxicity Summary for Chlorine and Hypochlorite, and Chlorine in Drinking Water.* Prepared for the Chlorine Institute, Inc., Washington, DC.

Fisher, N., R. Berry, and J. Hardy. 1983. Short-term study in rats of chlorinated cake flour. *Food Chem. Toxicol.* 21:423–426.

Ginnochio, A.V., N. Fisher, J.B. Hutchinson, R. Berry, and J. Hardy. 1983. Long-term toxicity and carcinogenicity studies of cake made from chlorinated flour. 2. Studies in mice. *Food Chem. Toxicol.* 4:435–439.

Guyton, C. 1986. *Textbook of Medical Physiology.* 7th ed. Saunders, Philadelphia.

Hertz-Piccitto, I., S.H. Swan, R.R. Neutra, and S.J. Samuels. 1989. Spontaneous abortions in relation to consumption of tap water: An application of methods from survival analysis to a pregnancy follow-up study. *Am. J. Epidemiol.* 130(1):79–93.

Hileman, B. April 19, 1993. Concerns broaden over chlorine and chlorinated hydrocarbons. *Chem. Eng. News.*

International Agency for Research on Cancer (IARC). 1979. *IARC Monographs on the Evaluation of Carcinogenic Risks to Humans. Some Halogenated Hydrocarbons.* Vol. 20. IARC, Lyon, France.

International Agency for Research on Cancer (IARC). 1991. *IARC Monographs on the Evaluation of Carcinogenic Risks to Humans. Chlorinated Drinking-water; Chlorination By-Products; Some Other Halogenated Compounds; Cobalt and Cobalt Compounds.* Vol. 52, IARC, Lyon, France.

International Life Sciences Institute (ILSI). 1993. *Report of the Panel on Reproductive Effects of Disinfection Byproducts in Drinking Water.* ILSI, Washington, DC. and U.S. EPA, Health Effects Research Laboratory, Research Triangle Park, NC.

IRDC, 1984. Chronic Inhalation Toxicity Study on Chlorine in Non-Human Primates. International Research and Development Corporation, Chemical Industry Institute of Toxicology, Research Triangle Park, NC. Report 480-001.

Jorgenson, T.A., E.F. Meierhenry, C.J. Rushbrook, R.J. Bull, and M. Robinson. 1985. Carcinogenicity of chloroform in drinking water to male Osborne-Mendel rats and female $B_6C_3F_1$ mice (SRI International study). *Fund. App. Toxicol.* 5:760–769.

Kool, H.J., F. Kuper, H. Van Haeringen, and J.H. Koeman. 1985. A carcinogenicity study with mutagenic organic concentrates of drinking water in the Netherlands. *Sci. Total Environ.* 37:229–256.

Kurokawa, Y., N. Takamura, Y. Matsushima, T. Imazawa, and Y. Hayashi. 1984. Studies on the promoting and complete carcinogenic activities of some oxidizing chemicals in skin carcinogenesis. *Cancer Lett.* 24:299–304.

Kurokawa, Y., N. Takamura, Y. Matshushima, Y. Konishi, Y. Hiasa, S. Ssahina, M. Takahashi, A. Maekawa, and Y. Hayashi. 1986. Long-term *in vivo* carcinogenesis tests of potassium bromate, sodium hypochlorite, and sodium chlorite conducted in Japan. *Environ. Health Perspect.* 69:221.

Lambert, T.W., C.F.B. Holmes, and S.E. Hrudey. 1994. Microcystin class of toxins: health effects and safety of drinking water supplies. *Environ. Rev.* 2:167–186.

Morris, R.D., A.-M. Audet, I.F. Angelillio, T.C. Chalmers, and F. Mosteller. 1992. Chlorination by-products and cancer: a *meta*-analysis. *Am. J. Public Health* 82(7):955–963.

National Academy of Sciences (NAS). 1977. *Drinking Water and Health.* Safe Drinking Water Committee, NAS, Washington, DC.

National Cancer Institute (NCI). 1976. *Report on Carcinogenesis Bioassay of Chloroform.* NCI, Carcinogenesis Program, Division of Cancer Cause and Prevention, Bethesda, MD.

National Research Council (NRC). 1987. *Drinking Water and Health: Disinfectants and Disinfectant By-Products,* Vol. 7, NRC, National Academy Press, Washington, DC.

NTP. 1992. Toxicology and Carcinogenesis Studies of Chlorinated and Chlorinated Deionized and Charcoal Filtered Water (Case Nos. 7782-50-5, 7681-52-9, and 10599-90-3) in F344/N Rats and $B_6C_3F_1$ Mice. Drinking Water Studies. National Toxicology Program; Research Triangle Park, NC. NTP-TR-392. NIH 92-2847 (draft report).

Rasmussen, R.A., L.E. Rasmussen, M.A.K. Khalil, and R.W. Dalluge. 1980. *J. Geophysics Res.* 85:7350.

Report of the Expert Panel, 1994. Interpretive review of the potential adverse effects of chlorinated organic chemicals on human health and the environment. *Reg. Tox. and Pharm.* 20(1):S1–1057.

Robinson, M., R.J. Bull, M. Schamer, and R.E. Long. 1986. Epidermal hyperplasia in mouse skin following treatment with alternative drinking water disinfectants. *Environ. Health Perspect.* 69:293–300.

Roe, F.J.C., A.A.K. Palmer, A.N. Worden, and N.J. Van Abbe. 1979. Safety evaluation of toothpaste containing chloroform. I. Long-term studies in mice. *J. Environ. Pathol. Toxicol.* 2:799–819.

Salazar-Lindo, E., M. Alegre, M. Rodriguez, P. Carrion, and N. Razzeto. 1993. In Craun, G. (Ed). *Safety of Water Disinfection: Balancing Chemical and Microbial Risks.* International Life Sciences Institute Press, Washington, DC.

Truhaut, R., J.C. Gak, and C. Graillot. 1979. Studies on the risks that may result from chemical pollution of drinking water I. Study of the long-term toxicity in rats and mice of organic micropollutants extracted from drinking water with chloroform. (Fr.) *Water Res.* 13:689–697.

Utrecht, J.P. 1992. The role of leukocyte-generated reactive metabolites in the pathogenesis of idiosyncratic drug reaction. *Drug Metab. Rev.* 24(3):299–306.

Weiss, S.J. 1989. *N. Engl. J. Med.* 320:365. Cited in Utrecht, 1992.

Chapter 24

Removal of Persistent Toxic Bioaccumulative Chemicals from Pulp and Paper Mill Effluent Streams: Implications for Fish Advisories

S.E. Holm

ABSTRACT

Several organizations have called for the reduction or elimination of Persistent Toxic Bioaccumulatives (PTBs) from industrial waste streams. Since the late 1980s when the EPA discovered that the prevailing pulp bleaching process, based on elemental chlorine, formed inadvertent trace amounts of 2,3,7,8-tetrachlorodibenzodioxin (TCDD), the pulp and paper industry has been developing and implementing technology to eliminate dioxin. Data are presented for effluent, sediment, and fish to show that dioxin has been virtually eliminated using chlorine dioxide technology. As a result of this elimination, fish body burdens have declined significantly. Many states have reevaluated potential health risks resulting from ingesting dioxin-contaminated fish. Recent EPA data presented herein show substantial reductions in fish advisories downstream from pulp and paper mills.

INTRODUCTION

Several organizations have called for the reduction or elimination of Persistent Toxic Bioaccumulatives (PTBs) from industrial waste streams because of the potential environmental consequences associated with this class of chemicals when they are released into the environment. These compounds are persistent because of their resistance to biological degradation, bioaccumulate in the fatty tissue of organisms, and are toxic to aquatic species at relatively low levels. PTBs may be produced commercially, such as for use as a pesticide or herbicide, or as inadvertent by-products from automobile emissions, incinerators, during pulp bleaching using chlorine (Rappe, 1992), and from outboard boat motor emissions (Deardorff, 1994).

Since the late 1980s, when the EPA discovered that the prevailing pulp bleaching process, based on elemental chlorine, inadvertently formed trace amounts of TCDD, facilities incorporating this process have been identified as a major source of TCDD in surface waters and in freshwater fish (U.S. EPA, 1990). Because human body burdens are highly associated with ingestion of TCDD-contaminated foodstuffs and because levels of TCDD in fish have been of sufficient concentration, health departments in many states have issued fish consumption advisories for TCDD. The advent of chlorine dioxide as a replacement for elemental chlorine has virtually eliminated TCDD in effluents and, as a consequence, fish body burdens have declined dramatically.

In a recent evaluation of chlorine dioxide bleaching, an expert panel (Solomon et al., 1993) reached the conclusion that chlorine dioxide and chlorine react with lignin by different chemical processes and produce different organochlorine products. The laboratory and mill studies show consistently that substituting chlorine dioxide for chlorine results in an approximate five- to tenfold reduction in organochlorine discharge and a decrease in the degree of chlorine substitution. These changes generally reduce persistence and toxicity and the potential for bioaccumulation. In particular, the formation of highly hydrophobic chemicals such as dioxin is substantially reduced or virtually eliminated by increasing chlorine dioxide substitution to 100%.

METHODS

Georgia-Pacific's Leaf River Mill is a state-of-the-art kraft pulp mill located in southern Mississippi that began operating in 1984. Elemental chlorine was used in its bleaching process from 1984 to 1989. Since 1989, the mill has operated using complete chlorine dioxide substitution.

Effluent sampling has occurred in accordance with the facility's National Pollutant Discharge Elimination System (NPDES) permit. Sediment samples were collected in July and August of 1994. Sampling protocols followed those reported by the U.S. EPA (1990), Rappe et al. (1995), and Fiedler et al. (1995). Fish sampling and analysis has occurred yearly since 1989 in cooperation with the Mississippi Department of Environmental Quality.

Detailed protocols were written for all field and laboratory operations. These protocols combined U.S. EPA methods and recommendations agreed upon at a European Community workshop on QA/QC.

Sediment

Extraction and cleanup procedures followed a method by Kjeller et al. (1990) and reported by Rappe et al. (1995). The final extracts were analyzed by HRGC/HRMS using a polar 60 m Trx-2330 for all 2,3,7,8-substituted polychlorinated dibenzodioxins (PCDDs) and polychlorinated dibenzofurans (PCDFs) and a nonpolar DB-5 column for all homologues and the hepta- and octa-CDDs and CDFs. The GC columns were interfaced to a VG 70-250S double focusing mass spectrometer operating at a resolution of 8000–10000. The instrument was operated in a selected ion monitoring (SIM) mode. A blank sample was analyzed for every batch of 4–6 samples.

Data reported are by a loss on ignition (LOI) determination. This was accomplished by heating the sample at 500°C for 2 hours. Most sediment data in the literature have been reported on a dry weight basis. Where, however, the compounds of interest, like PCDDs and PCDFs, are highly lipophilic and probably formed on or bound to organic particles, the results can be reported based on the content of organic compounds as determined by the LOI.

Fish

Fish were generally collected at a total of three sites: one upstream, one near the mill outfall, and one downstream. Five fish which exceeded minimum length and weight requirements were collected from each site. TCDD concentrations were determined from muscle tissue (skinless fillets).

RESULTS

Effluent Data

Effluent samples from the Leaf River mill have been below quantifiable levels for TCDD since the mill underwent full substitution with chlorine dioxide in 1989. Data from February, 1991 until November, 1994 are presented in Table 24.1.

Table 24.1. Leaf River Effluent TCDD Analysis.

Date	Concentration (ppt)
Sept. 1991	ND[a]
Sept. 1992	ND
Nov. 1993	ND
June 1994	ND
Nov. 1994	ND

[a] ND = Nondetect @ 0.01 ppt.

Sediment Data

Sediments are known sinks for PTBs such as TCDD. The Leaf River mill is located on the Leaf River between sampling points 505 and 506. Sediment data for this location (Figures 24.1 and 24.2,

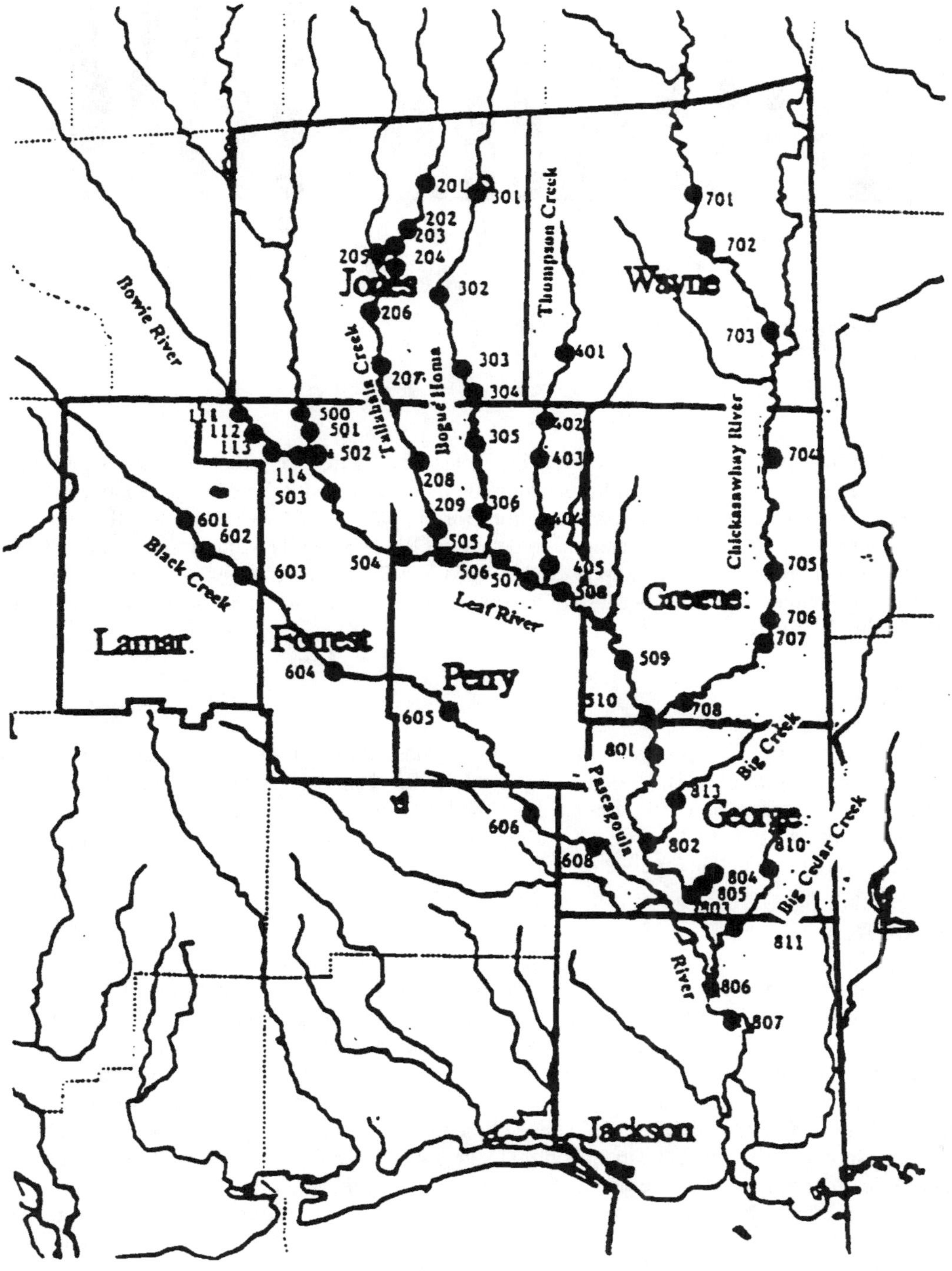

Figure 24.1. Sediment sampling sites in the Leaf, Pascagoula and Chickasawhay river drainage areas.

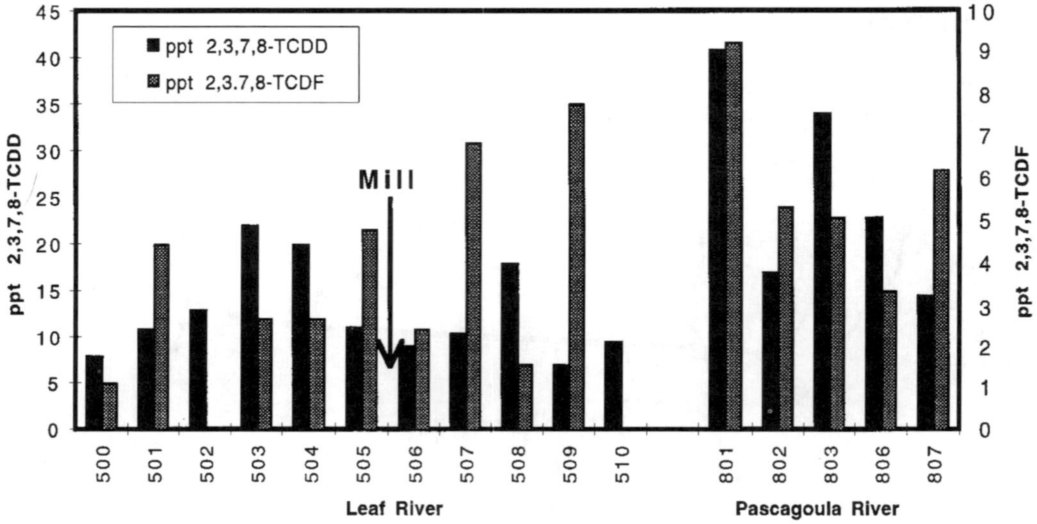

Figure 24.2. Concentration of 2,3,7,8-TCDD and 2,3,7,8-TCDF in Leaf River and Pascagoula River sediment.

modified from Rappe, 1995) is reported on a LOI basis. The figure shows that the highest concentrations were found in the Pascagoula River below the confluence with the Chickasawhay River. No differences could be ascertained in concentrations for samples taken above and below the pulp mill. Consequently, the mill is not a point source into the Leaf River. Several potential sources of TCDD to river sediments are discussed in Rappe et al. (1995) and Fiedler et al. (1995). Future research to discern these sources is planned because of the unusually high concentrations of chlorinated PCDDs found in this relatively pristine area.

Fish Data

Monitoring of three fish species — flathead catfish, channel catfish, and blue catfish — has shown declines in concentrations of TCDD since chlorine dioxide substitution technology was incorporated. In all species monitored, concentrations have declined from approximately three times above the EPA Fish Consumption Advisory level of 7 ppt to approximately one order of magnitude below this level. Data for fish are presented in Figure 24.3.

DISCUSSION

The Leaf River mill was considered a source of TCDD prior to the incorporation of full chlorine dioxide substitution (U.S. EPA, 1990). It is noteworthy that these sediment data collected in 1994 show no observable difference between upstream and downstream locations. This is important for two reasons: (1) chlorine dioxide substitution has resulted in no detectable TCDD differences in sediment, a sensitive environmental marker, and (2) TCDD concentrations in fish resulting from exposure to historical TCDD deposited in this sediment reservoir can no longer be ascertained. Moreover, corresponding median fish data in bottom feeders such as catfish show concentrations approximately one order of magnitude below the EPA advisory level of 7 ppt.

The data presented herein are not unique to the Leaf River mill. For example, a very recent study of the Grande Prairie mill in Alberta, Canada, has shown similar responses when implementing full substitution technology (Pryke et al., 1995). In addition to significant declines in TCDD, in another study (Pryke et al., 1994) it was reported that utilization of chlorine dioxide decreases the formation of another class of PTBs, chlorinated phenolic compounds. Testing of influent to secondary treatment showed

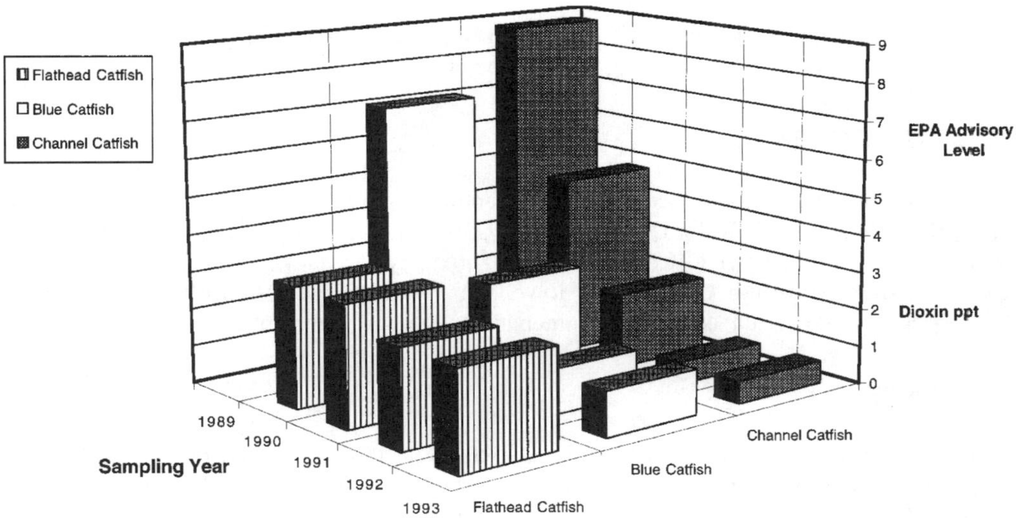

Figure 24.3. Leaf River fish study. Median dioxin concentrations (ppt).

the tri-, tetra-, and pentachlorinated phenols to be below detectable limits (0.1 ppb). This is 25–50 times lower than the U.S. EPA's proposed minimum levels for bleach plant effluent (EPA, 1993).

Fish Advisory Status

Food is estimated to result in approximately 98% of the body burden of TCDD in humans (U.S. EPA, 1994). Consequently, concentrations of dioxin in fish resulting from exposure to mill effluent are related to a predicted human health risk which was based on extrapolation from chronic rodent bioassays. Since the late 1980s, the increased use of chlorine dioxide as a bleaching agent has markedly reduced dioxin discharges. In the Southeast, according to data gathered from the EPA's Fish Consumption Database, Bans and Advisories File, as well as from individual state health and environmental agencies, a total of 17 dioxin advisories were issued for water bodies downstream of U.S. pulp mills since 1983. However, since 1991, 10 states have lifted 10 of the 17 advisories.

In Mississippi, the state Department of Environmental Quality's (DEQ) Office of Pollution Control first issued a dioxin advisory for the Leaf River in November 1989. In February of 1994, the DEQ reduced the extent of the fish consumption advisory to just a 16.5 mile stretch of the river. After additional monitoring, the DEQ lifted the advisory on April 17, 1995. State authorities generally lift advisories once the contamination levels drop and remain below state standards. The Mississippi DEQ sets a standard advisory level of 5 ppt.

This recovery of the ecosystem is not limited to the southeastern U.S. Nationwide, since 1991, 13 states have lifted a total of 17 dioxin advisories on water bodies downstream of U.S. pulp mills. In fact, dioxin advisories for U.S. water bodies have dropped by more than 50% over the last four years. As the industry continues its conversion to chlorine dioxide, the number of water bodies under a dioxin advisory will continue to decline. An EPA study of current dioxin advisories downstream of U.S. pulp mills shows that all of the advisories could be lifted following implementation of the agency-proposed Best Available Technology (BAT) guidelines, which are based on the complete substitution of chlorine dioxide for chlorine gas in pulp bleaching.

CONCLUSION

Virtual elimination of TCDD and other PTBs has been demonstrated at the Leaf River mill following complete chlorine dioxide substitution. If further reductions of TCDD in fish are considered of value, unknown point sources will need to be discovered and controlled. The fish advisory status may be an important mechanism in targeting future research.

ACKNOWLEDGMENT

The author gratefully acknowledges the contributions supplied by the following individuals: C. Rappe, R. Andersson, S.E. Kulp, Umea University, Sweden; M. Bonner, Bonner Analytical Testing Co., Hattiesburg, Mississippi; K. Cooper, Rutgers University, Piscataway, New Jersey; H. Fiedler, C. Lau, University of Bayreuth, Germany; F. Howell, University of Southern Mississippi, Hattiesburg, Mississippi; D. Pryke, Alliance for Environmental Technology, Erin, Ontario.

REFERENCES

Deardorff, T. 1994. Generation of Polychlorinated Dioxins (PCDD) and Furans (PCDF) by Boat Motor Operation, and Impact on Aquatic Sediments. *2nd International Conference on Environmental Fate and Effects of Bleached Pulp Mill Effluents*. Vancouver, British Columbia, Canada. p. 102.

Fiedler, H., C. Lau, K. Cooper, R. Andersson, S.-E. Kulp, C. Rappe, F. Howell, and M. Bonner. 1995. PCDD/PCDF Levels in Sediments from a River System in Southern Mississippi. *Dioxin, '95*. pp. 349–352.

Kjeller, L.O., C. Rappe, K.C. Jones, and A.E. Johnston. 1990. Evidence for Increases in the Environmental Burden of Polychlorinated Dibenzo-*p*-dioxins and Furans (PCDD/PCDFs) Over the Last Century. *Dioxin '90*. pp. 433–436.

Pryke, D.C., G.R. Bourree, P. Winter, and C. Mickowski. 1994. The Impact of Chlorine Dioxide Delignification on Pulp Manufacturing and Effluent Characteristics at Grande Prairie, Alberta. *Pulp & Paper Canada*. 95(6):T230–T236.

Pryke, D.C., G.R. Bourree, S.M. Swanson, J.W. Owens, and P. Kloepper-Sams. 1995. Environmental Improvements at Grande Prairie and Ecosytem Response. *Pulp & Paper Canada*. 96(11):T377–385.

Rappe, C. 1992. Sources of PCDDs and PCDFs. Introduction. Reactions, Levels, Patterns, Profiles and Trends. *Chemosphere*. 25:41–44.

Rappe, C., R. Andersson, M. Bonner, K. Cooper, H. Fiedler, F. Howell, S.E. Kulp, and C. Lau. 1995. PCDDs and PCDFs in Sediments in a River System in Southern Mississippi, USA. *Dioxin '95*. pp. 273–280.

Solomon, K., H. Bergman, R. Huggett, D. Mackay, and B. McKague. 1993. A Review and Assessment of the Ecological Risks Associated with the Use of Chlorine Dioxide for the Bleaching of Pulp. Prepared for the Alliance for Environmental Technology.

U.S. EPA. 1990. USEPA/Paper Industry Cooperative Dioxin Study "The 104 Mill Study" Statistical Findings and Analyses. Office of Water Regulations and Standards, Washington, DC, EPA 440/2-88-025.

U.S. EPA. 1993. 40 CFR Parts 63 and 430: Effluent Limits Guidelines, Pretreatment Standards, and New Source Performance Standards: Pulp, Paper and Paperboard Category; National Emission Standards for Hazardous Air Pollutants for Source Category: Pulp and Paper Production; Proposed Rule. Fed. Reg. Vol. 58, No. 241, pp. 66078–66216. December 17.

U.S. EPA. 1994. Estimating Exposure to Dioxin-Like Compounds. Office of Research and Development. Washington, DC. EPA/600/6-88/005Ca.

Chapter 25

The Effect of Chlorine and Chlorine Dioxide on Microbial Populations in Drinking Water

M.A. Rouf

INTRODUCTION

Use of disinfectant to treat drinking water supplies is the most important step to prevent the spread of waterborne diseases. Improperly disinfected water, if the source water is contaminated, could give rise to epidemic diseases. The importance of water as a vehicle for infection is documented by the drinking water disease outbreak report of 1993 by CDC.[17] Most of the outbreaks are gastroenteritis of unknown etiology (Figure 25.1 and Table 25.1). However, it is suspected that these outbreaks are due to enteric viruses and some protozoa. One of the main problems are assessing the effectiveness of a disinfectant, based on laboratory data, is in its application in practical field situations involving many more parameters.

In the United States[82] and in other Western countries, government edicts regulate water treatment. Chlorine has been used as a means of controlling infections since the 1800s. However, the large-scale use of chlorine for water treatment was not practiced until the early 1900s when the city of Chicago used chloride of lime in 1908. Disinfection of drinking water, swimming pools, and treated sewage effluents has been recently reviewed by Hurst, 1991.[37] Chlorine or its derivatives have undergone tremendous evolution in their uses in a variety of disinfection processes and in medicine, water treatment, food, textile, paper, and bleaching purposes in industries, among others. The primary use of chlorine compounds in drinking water is to kill microbes and viruses.

The antimicrobial activity of chlorine, chlorine dioxide, hypochlorites, and the inorganic and organic chloramines are all due to the liberation of free chlorine. Dychdala, 1991[27] and Sobsey, 1989[75] have reviewed chlorine and chlorine compounds and have summarized the biocidal effect of free available chlorine and chlorine dioxide. Generally, all the chlorine compounds are considered to be oxidants, but exactly how they exert their microbe-killing effect is not clearly understood. The health risk and safety concerns associated with different disinfectants vary widely and may result from pH, residual disinfectant, by-products of the disinfectant, the presence of humic materials, etc.[12]

The purpose of this chapter is to briefly review the use of chlorine and chlorine dioxide as drinking water disinfectants, provide a general understanding of the nature of chlorine and chlorine dioxide and their biocidal activity, and give an overview of the health risk associated with their uses.

NATURE OF CHLORINE AND CHLORINE DIOXIDE

Chlorine

In nature it is found as sodium chloride in seawater, salt lakes, salt mines and igneous rocks. Chlorine gas (Cl_2) is prepared industrially by electrolysis of brine and is a greenish-yellow diatomic gas with a suffocating odor. It is highly reactive, and combines readily with all elements except nitrogen and rare gases.

Chlorine is marketed in the form of compressed gas in steel cylinders. It is soluble in water and forms Cl_2, HOCl, and Cl^-. There are many commercially produced chlorine compounds that are available on the market.[27]

Theoretically, chlorine is capable of forming a number of oxycompounds, but only four of these are considered to be stable chlorine oxide molecules: Cl_2O, ClO_2, ClO_3, Cl_2O_7.[52]

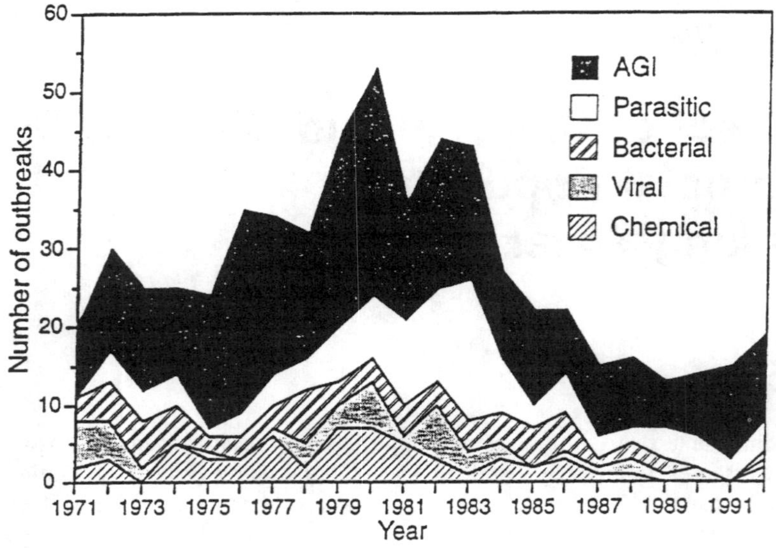

Figure 25.1. Waterborne outbreaks, by year and etiologic agent — United States, 1971–1992 (N=609). AGI=Acute gastrointestinal illness of unknown etiology. From CDC, 1993, with permission.

Table 25.1. Outbreaks Associated with Water Intended for Drinking, by Etiologic Agent and Type of Water System—United States, 1991–1992.

	Type of Water System							
	Community		Noncommunity		Individual		Total	
Agent[a]	Outbreaks	Cases	Outbreaks	Cases	Outbreaks	Cases	Outbreaks	Cases
AGI	3	10,077	19	3,252	1	38	23	13,367
Giardia	2	95	2	28	0	0	4	123
Cryptosporidium	2	3,000	1	551	0	0	3	3,551
Hepatitis A	0	0	0	0	1	10	1	10
Shigella sonnei	0	0	1	150	0	0	1	150
Nitrate	0	0	0	0	1	1	1	1
Fluoride	1	262	0	0	0	0	1	262
Total	**8**	**13,434**	**23**	**3,981**	**3**	**49**	**34**	**17,464**
(Percent[b])	(24)	(77)	(68)	(23)	(9)	(< 1)	(100)	(100)

[a] AGI = acute gastrointestinal illness of unknown etiology.
[b] The percentage of 34 outbreaks, or of 17,464 cases.
From of CDC, 1993. With permission.

Chlorine Dioxide

Chlorine dioxide is a strongly oxidizing, yellow to reddish yellow gas at room temperature. It has an unpleasant odor similar to that of chlorine and to some extent like nitric acid. It is unstable in light and reacts violently with organic materials.

It is made on a fairly large scale when, where, and as required. Chlorine dioxide is soluble in water and solutions of 8 g/L are stable in the dark, but in the light it slowly decomposes to HCl and $HClO_3$.

Commercially, ClO_2 is made by exothermic reaction of sodium chlorate in 4 to 4.5 M sulfuric acid containing 0.05 to 0.25 M chloride ion with sulfur dioxide:

$$2NaClO_3 + SO_2 + H_2SO_4 \rightarrow 2\ ClO_2 + 2NaHSO_4$$

MECHANISMS OF BIOCIDAL ACTIVITY

Although it is not clearly understood how chlorine and chlorine compounds kill microbes, there are plausible concepts and evidence to explain the biocidal activity.

First and foremost is the fact that all chlorine compounds are powerful oxidants which can cause a variety of damages to cells.

When chlorine gas is added to water, the following reactions happen:

$$H_2O$$

$$Cl_2(g) \rightleftarrows Cl_2(aq)$$

$$Cl_2(aq) + H_2O \rightleftarrows HOCl + H^+ + Cl^-$$

$$HOCl \rightleftarrows H^+ + OCl^-$$

If hypochlorite is added to water, then hypochlorous acid is formed in this way:

$$Ca(OCl)_2 + 2H_2O \leftarrow Ca(OH)_2 + 2HOCl$$

The dissociation of hypochlorous acid depends on pH, which also determines the disinfection efficiency. The oxidation capacity of chlorine dioxide is represented by this equation and thus it is 2.5 times that of chlorine:

$$ClO_2 + 5e = Cl^- + 2O^{2-}$$

Also, the redox potential of this reaction is unknown. However, $ClO_2 + e = ClO_2^-$ has a redox potential (25°C) of 0.95 V. The oxidation capacity and redox potentials are dependent on pH and temperature. The reactions of chlorine dioxide in water and its impurities can be numerous and are too complex to be given in a few simple steps. But it is known that chlorine dioxide in water gives rise to ClO_2^-, Cl^-, HOCl, $ClHO_3$, $ClHO_4$, and nascent oxygen. The stability of chlorine dioxide in water is dependent on its purity. Further, chlorine dioxide, like chlorine, has a variety of reactions with many inorganic and organic compounds.

There seems to be a lack of recent studies about the specific mechanism of cell death. This is perhaps due to the fact that a variety of reaction products are formed when chlorine compounds react with organisms in water. Controlled parameter experiments are difficult with live cells and chlorine compounds in water due to numerous chemicals and chemical complexes present and formed in water and in living cells.

However, some general conclusions can be made from the available information about the mechanism of cell death:

1. Strong oxidation capacity and power of chlorine compounds causing destruction or damage to all types of cellular and molecular constituents; precipitation or coagulation of cell organic matter.
2. A host of reactions with organic compounds — in part, direct combination of chlorine with enzymes, proteins, and cell membranes. Also, formation of oxidation products, N-chloro-compounds, halogenation of certain amino acids and inhibition of key enzymatic reactions such as sulfahydral enzymes and destruction of permeability of cell membrane, and inactivation of hydrogen transport enzymes, etc., all play roles in killing cells. The pH of the water plays a very important role in keeping compounds such as HOCl in undissociated form, which can readily enter and subsequently damage cells.

Table 25.2. Water-Transmitted Human Pathogens.

Group	Pathogen	Disease Contracted
Bacteria	*Salmonella* (>2000 types)	Typhoid, paratyphoid, salmonellosis
	Shigella (4 spp.)	Bacillary dysentery
	Enteropathogenic *Escherichia coli*	Gastroenteritis
	Yersinia enterocolytica	Gastroenteritis
	Campylobacter jejuni	Gastroenteritis
	Vibrio cholerae	Cholera
	Leptospira	Weil's disease
	Legionella pneumophila	Acute respiratory
Protozoa	*Entamoeba histolytica*	Amoebic dysentery, liver abscess, colonic ulceration
	Giardia lamblia	Diarrhea, malabsorption
	Cryptosporidium	Diarrhea
	Naegleria flowleri	Meningoencephalitis
Helminths	*Acaris lubricoides* (roundworm)	Ascariasis
	Ancyclostoma duodenale (hookworm)	Anemia
	Necator americanus (hookworm)	Anemia
	Taenia saginata (tapeworm)	Taeniasis
	Trichuris (whipworm)	Abdominal pain, diarrhea
	Toxicara (roundworm)	Fever, abdominal pain
	Strongyloides (threadworm)	Abdominal pain, nausea, diarrhea
Enteroviruses	Poliovirus	Meningitis, paralysis, fever
	Echovirus	Meningitis, diarrhea, rash, fever, respiratory disease
	Coxsackievirus A	Meningitis, herpangina, fever, respiratory disease
	Coxsackievirus B	Myocarditis, congenital heart anomalies, pleurodynia, respiratory disease, fever, rash, meningitis
	New enteroviruses (types 68-71)	Meningitis, encephalitis, acute hemorrhagic conjunctivitis, fever, respiratory disease
	Hepatitis type A, Enterovirus 72	Infectious hepatitis
	Norwalk virus	Diarrhea, vomiting, fever
	Calcivirus	Gastroenteritis
	Astrovirus	Gastroenteritis
	Reovirus	Not clearly established
	Rotovirus	Diarrhea, vomiting
	Adenovirus	Respiratory disease, eye infections, gastroenteritis
	Snow-Mountain Agent	Gastroenteritis
	Epidemic, non-A, non-B, hepatitis	Hepatitis

From Bull et al., *Critical Reviews in Environmental Control.* CRC Press, Boca Raton FL, 1990. With Permission.

WATER TREATMENT

The importance of water treatment for disinfection becomes very obvious when one looks at the various human pathogens (Table 25.2) that are transmitted by water,[12] and the number of outbreaks associated with drinking water by year and etiological agents in the U.S. (Figure 25.1, Table 25.1).

Chlorine and its derivatives are the disinfectants of choice for water treatment in the United States. A recent survey conducted by the American Water Works Association Research Foundation[1] found that over 82% of the large utilities used chlorine as a disinfectant. The survey found that of the 263 large utilities, 216 used chlorine, 67 chloramines, 16 chlorine dioxide, and 1 ozone. The survey also indicated that disinfectant other than chlorine continues to be on the rise. A prior survey[55] also estimated that over 95% of the water treatment facilities in the United States used chlorine as the disinfectant.

All chlorine compounds are strong oxidizing agents. Their effectiveness is highly influenced by organic or humic materials, pH, temperature, hardness, etc.

A liquid form of compressed chlorine as a gas is added to water for disinfection. In water it forms hypochlorous acid (HOCl). Even in a minute amount chlorine has fast germicidal activity. Hypochlorites are widely used in the food and dairy industries and are available in liquid or powder form as salt of calcium, sodium, or lithium. They are also employed in households, hospitals, and institutions as a bleach.

Since the activity of chlorine is considerably affected by the presence of organic matter, in the disinfection of water it is necessary to determine the chlorine demand. It is a normal practice to add sufficient chlorine to the water supply to meet the chlorine demand of the water and simultaneously maintain enough residual chlorine for proper disinfection. When the chlorine survives the chlorine demand of the water, the residual chlorine — as Cl_2, HOCl, and OCl^- — is termed "free chlorine." Also, in some plants "free chlorination" is used as a first step, followed by combined residual chlorination (by ammoniation) to maintain a stable residual in the entire distribution system for efficient germicidal action.

The concern about the formation of chlorinated organic chemicals as a by-product of chlorine disinfection has led to the increased interest and uses of chlorine dioxide as a water and wastewater disinfectant. The use of chlorine dioxide in drinking water disinfection offers many advantages. It is a very effective oxidant and a residual can be maintained to control contamination of the distribution system. It does not react readily with natural aquatic humus and other organic compounds and therefore forms less trihalomethanes as compared to chlorine.[37] However, chlorine dioxide exerts an oxygen demand on water similar to chlorine, indicating formation of trihalomethanes from natural aquatic humus.[78]

Other advantages include the breakdown of phenolic compounds, phenolic odor, lack of reaction with ammonia or chlorophenols and removal of cyanides, sulfides, aldehydes, and mercaptans. It also has greater bacteriocidal and sporicidal activity, possibly due to its oxidation capacity involving five electrons. The oxidation capacity of chlorine dioxide is about 2.5 times that of chlorine.[38,52] There are, however, some concerns that relate to the formation of chlorate and chlorite ions from chlorine dioxide treatment of drinking water.[30] The microbicidal effect of chlorine dioxide is unaffected at pH 6 to 10.[68]

In the past, chlorine dioxide naturally contained some chlorine, raising health concerns similar to that of chlorine. Recently, some producers have developed high purity chlorine dioxide at a reasonable cost.

SENSITIVITY AND RESISTANCE OF VARIOUS MICROBES TO FREE CHLORINE

Thomas Alcock in London, England, in 1827, published his book on the use of sodium and calcium hypochlorites and advocated their use in disinfecting and deodorizing both animate and inanimate objects. The Hungarian physician, Semmelweis (between 1847 and 1849) usually receives the credit for using chlorine as a disinfectant in reducing death from puerperal fever. Thus chlorine has a long history of usage as an antimicrobial agent. Even in a minute quantity it is an effective germicide.

Reviews by Dychdala[27] and Sobsey[75] have compiled the sensitivity of various microbes to free available chlorine (Tables 25.3, 25.4, and 25.5), and these tables show that the most common problem organisms belonging to non-acid fast bacteria, viruses, and protozoa can be inactivated with 0.5 to 5 mg/L of free chlorine. There are, however, some organisms which show fairly high resistance to free chlorine.

Bacteria

The order of resistance of various microbes to disinfectant can be generalized as: non-spore forming vegetative bacteria< viruses< unicellular algae< spore forming bacterial cells< bacterial spores< acid fast bacteria< protozoan cysts.

Table 25.3. Inactivation of Common Waterborne Organisms by Free Chlorine.

Organism	pH	Temp. (°C)	Time	$AvCl_2$ (mg/L)	Biocidal Results	References
Bacteria						
Achromobacter metalcaligenes	6.0	21	15 sec.	5.0	100%	Hays et al., 1963
Bacillus anthracis	7.2	22	120 min.	2.3–2.4	100%	Brazis et al., 1958
B. globigii	7.2	22	120 min.	2.5–2.6	99.99%	Brazis et al., 1958
Clostridium botulinum toxin type A	7.0	25	30 sec.	0.5	100%	Brazis et al., 1959
Escherichia coli	7.0	20–25	1 min.	0.055	100%	Butterfield et al., 1943
E. typhosa	8.5	20–25	1 min.	0.1–0.29	100%	Butterfield et al., 1943
Microcystic aeruginosa	8.2	22	—	2.0	Growth controlled	Palmer & Maloney, 1955
Mycobacterium tuberculosis	8.4	50–60	30 sec.	50	100%	Costigan, 1936
Listeria monocytogens	9.5	20	30 sec.	100	99.999%	Lopes, 1986; El-Kest & Marth, 1988
Pseudomonas fluorescens IM	6.0	21	15 sec.	5.0	100%	Hays et al., 1963
Shigella dysenteriae	7.0	20–25	3 min.	0.046–0.055	100%	Butterfield et al., 1943
Staphylococcus aureus	7.2	25	30 sec.	0.8	100%	Dychdala, 1960; Bolton, et al., 1988
Streptococcus faecalis	7.5	20–25	2 min.	0.5	100%	Stuart & Ortengio, 1964
All vegetative bacteria	9.0	25	30 sec.	0.2	100%	Snow, 1956
Yersinia enterocolytica	9.0	20	5 min.	100	99.99%	Orth & Mrozek, 1989
Protozoa						
Entamoeba histolytica Cysts	7.0	25	150 min.	0.08–0.12	99–100%	Clarke et al., 1956
Viruses						
Purified adenovirus 3	8.8–9.0	25	40–50 sec.	0.2	99.8%	Clarke et al., 1956
Purified Coxsackie A_2	6.9–7.1	27–29	3 min.	0.92–1.0	99.6%	Clarke & Chang, 1959
Purified Coxsackie B_1	7.0	25	2 min.	0.31–0.40	99.9%	Kelly & Sanderson, 1958
Purified Coxsackie B_5	7.0	25–28	1 min.	0.21–0.30	99.9%	Clarke & Chang, 1959
Infectious hepatitis	6.7–6.8	room	30 min.	3.25	protected 12 volun.	Clarke & Chang, 1959
Purified poliovirus (Mahoney)	7.0	25–28	3 min.	0.21–0.30	99.9%	Clarke & Chang, 1959
Purified poliovirus (Lensen)	7.4–7.9	19–25	10 min.	1.0–0.5	protected 164 inoc. mice	Clarke & Chang, 1959
Purified poliovirus III (Sankett)	7.0	25–28	2 min.	0.11–0.2	99.9%	Clarke & Chang, 1959
Purified Theiler's	6.5–7.0	25–27	5 min.	4–6	99%	Clarke & Chang, 1959
Simian rotavirus	6.0	5	15 sec.	0.5	99.99%	Berman & Hoff, 1984

From Dychdala, *Disinfection, Sterilization and Preservation,* edited by Seymour S. Block, Lea and Febiger, Philadelphia, 1991. With permission.

Table 25.4. Inactivation of Health-Related Bacteria, Viruses, and Protozoan Cysts in Water by Free Chlorine.

Organism	Water	Av. Cl_2 (mg/L)	Temp. (°C)	pH	Time (min)	Reduction	Reference
Bacteria							
E. coli	BDF[a]	0.2	25	7.0	15	99.998-	Ram and Malley, 1984
HPC	treated[b]	0.25–1.3	24	8.0	15	99.82	Wolfe et al., 1984
C. jejuni	BDF	0.1	4	8.0	5	99.98-	Blaser et al., 1986
L. pneumophila	tap	0.25	20	7.7–7.8	58	99	Kuchta et al., 1985
L. pneumophila	tap	0.25	20	7.7–7.8	4	99	Kuchta et al., 1985
M. chelonei	BDF	0.3	25	7.0	60	40	Carson et al., 1978
M. fortuitum	BDF	0.15	?	7.0	60	0	Pelletier and Du Moulin, 1987
Viruses							
Parvo-H-1	PBS[c]	0.2	20	7.0	6	99.9	Chum et al., 1984
Rota, WA	treated	0.75	22	8.3–8.6	60	94.3	Raphael et al., 1987
Rota, SA11	BDF	0.1	4	8.0	0.5	99.9	Vaughn et al., 1986
Rota, Wa	BDF	0.1	4	8./0	0.65	99.9	Vaughn et al., 1986
Hepatitis A	BDF	0.5	5	6.0	6.5	99.99	Sobsey et al., 1988
Coliphage MS2	BDF	0.5	5	6.0	1.2	99.99	Sobsey et al., 1988
Protozoan Cysts							
G. lamblia	BDF	1–4	5	7.0	3–32	90	Jarroll et al., 1981
G. lamblia	BDF	0.2–3.0	5.0	6.0	—	99	Hibler et al., 1987
G. muris	BDF	23.8–78.5	5	7.0	5.7–42.6	99	Leahy et al., 1987
N. gruberi	BDF	2.64	25	*5.0*	2.8	99	Rubin et al., 1983
Cryptosporidium	gut homog in PBS	30,000	4	7.0	18 hr	<95	Campbell et al., 1982

[a] BDF—buffered demand free.
[b] Treated—conventionally treated surface water.
[c] PBS—phosphate buffered saline.
From Sobsey, *Water Scicnce Technology,* 1989. With permission.

Table 25.5. Inactivation of Health-Related Microbes in Water by Chlorine Dioxide.

Organism	Water	Av. Cl_2 (mg/L)	Temp. (°C)	pH	Time (min)	Reduction	Reference
Bacteria							
E. coli	BDF[a]	0.3–0.8	5	7.0	0.6–1.8	99	Cronier, 1977
E. coli	PBS[b]	0.4–0.8	?	7.0	5	>99.999	Fujioka et al., 1986
Fec. colif.	effl.	1.9	?	?	10	99.94	Fujioka et al., 1986
Fecal strep.	effl.	1.9	?	?	10	99.5	Fujioka et al., 1986
C. perfringens	effl.	1.9	?	?	10	0	Fujioka et al., 1986
L. pneumophila	BDF	0.5–0.35	23	?	15	99.9–99.99	Berg et al., 1982; 1985
L. pneumonia	BDF	ca. 0.12	23	?	15	99.3–99.7	Harakeh et al., 1985
Viruses							
Coliphage f2	BDF	1.5	5	752	2	99.994	Olivieri et al., 1985
Polio 1	BDF	2.0	5	7.2	5–10	99–99.96	Olivieri et al., 1985
Polio 1	PBS	0.4–0.8	?	7.0	5	99.9–99.97	Fujioka et al., 1986
Rota SA 11:							
diversed	BDF	0.5–1.0	5	6.0	0.2–0.6	99	Berman and Hoff, 1984
cell-assoc.	BDF	0.45–1.0	5	6.0	1.2–4.8	99	Berman and Hoff, 1984
Hepatitis A	BDF	0.14–0.23	5	6.0	8.4	99	Sobsey et al., 1990
Coliphage MS2	BDF	0.15	5	6.0	34	99	Sobsey at al., 1990
Coliphage MS2	BDF	0.15	5	—	<0.33	>99.95	Sobsey et al., 1990
Protozoan Cysts							
N. gruberi	BDF	0.8–1.95	5	7.0	7.8–19.9	99	Chen et al., 1985
G. muris	BDF	0.1–5.55	5	7.0	1.3–168	99	Leahy et al., 1985; 1987
G. muris	BDF	0.21–1.12	25	7.0	3.3–28.9		Rubin, 1988

[a] BDF = buffered demand free.
[b] PBS = phosphate buffered saline.
From Sobsey, *Water Science Technology*, 1989. With permission.

There are certain pathogens, such as *Yersinia enterocolytica*[59] and *Listeria monocytogens,*[28] that are considerably more resistant to chlorine, requiring 100 mg/L of free chlorine to disinfect. Fungi are also generally more resistant to chlorine. For example, *Aspergillus niger,* and *Rhodotorula flava* require 100 mg/L of free chlorine and 30–60 min to 5 min, respectively, of exposure time to kill these organisms.[27]

Not only do the different types of bacteria, fungi, viruses, and protozoa show varying degrees of resistance, but the antecedent growth conditions also play an important role in the resistance of microbes. However, this selective resistance of microorganisms may be compensated for by adjusting pH, temperature, chlorine concentrations, etc.

Many investigators[6,32,47,65] reported 90–99.99% reduction of *E. coli* at pH 7–10 at temperature 4 to 25°C within 3.5 to 60 min at a concentration of 0.2 to 1.5 mg/L of free chlorine under different experimental conditions. *Listeria monocytogens,*[28,51] and *Legionella pneumophila*[43,56] can be inactivated at 100 mg/L of free chlorine at 30 sec and at 0.25 to 6 mg/L in 4–52 min, respectively. The ordinary intestinal bacteria, including *E. coli,* can be disinfected using 0.1 to 0.6 mg/L of free chlorine under normal water disinfecting conditions. The generally accepted specifications for disinfecting drinking water supplies require a free chlorine residual of 1 to 2 mg/L for 1 to 2 h at a pH of less than 8 and turbidity of less than 1 unit for bacteria and viruses, etc.[39]

Coliforms, and particularly *E. coli,* are generally found to be more resistant to chlorine than most other vegetative common waterborne pathogenic bacteria; therefore, it has been used as the test organism for determining the effectiveness of chlorine treatment of water. However, reliable prediction of microbial disinfection is difficult due to inherent differences in susceptibility and the resistance of bacteria, viruses, and cysts to disinfection. Also, physiological state, such as growth condition, prior disinfection, and antecedent growth conditions of a microbe may greatly influence its response to disinfectant.[75]

Berg and co-workers in 1982 reported that *Escherichia coli* grown at lower temperature with reduced growth rate was more sensitive to disinfection by chlorine dioxide. Similar results with antecedent growth conditions were also reported in *Yersinia enterocolytica, Klebsiella pneumonia* and *Legionella pneumophila* with chlorine and chlorine dioxide.[4,33,43] Others[2,50] reported higher degrees of resistance for organisms that were previously exposed to chlorine compounds. For example, *E. coli* that received previous exposure required 55-fold greater chlorine concentration to achieve the same degree of inactivation as unexposed cells.[50] Also, differences in sensitivity to disinfection by different isolates of the same species may vary greatly. For example, a sixfold difference in resistance was observed among the six isolates of *E. coli.*[84] A review of inactivation of health related microorganisms in water disinfection by Sobsey[75] identified the varieties of factors that interfere with disinfection, influencing disinfection efficiency.

Acid fast bacilli offer different levels of resistance to chlorine compounds. *Mycobacterium tuberculosis* required 30 sec at pH 8.4 to kill at a concentration of 50 mg/L of free chlorine;[22] *M. chelonei* showed a 40% reduction in number in 60 min at pH 7.00 in 0.3 mg/L of free chlorine. On the other hand, *M. fortuitum,* when subjected to similar conditions at half the chlorine amount, had zero reduction in number. *M. fortuitum* was also found to be the most resistant at pH 6, 8, and 10 at the various levels of free chlorine tested by Grabow et al.[31]

The review by Sobsey[75] lists inactivation of vegetative bacteria such as *E. coli* and other coliforms, *Salmonella* spp., *Shigella* spp., *Campylobacter jejuni* and *Legionella* spp. and other spp., viruses, and protozoa by free chlorine and chlorine dioxide (Tables 25.4 and 25.5).

Filamentous bacteria, such as *Actinomycetes* bacteria attached to organic matter, and clumping bacteria, show greater resistance to chlorine than nonassociated organisms.[57] The gram positive cocci, such as *Staphylococcus aureus,* are readily killed at pH 7.2 in 30 sec at a free chlorine concentration of 0.8 mg/L.[9] Others reported 99.99% death of *S. aureus* at pH 8.5 to 8.9 in less than five sec in the presence of 15 mg/L, whereas for the similar death rate at pH 11.2, the same concentration of chlorine required a contact time of 1.85 min.[86] *Streptococcus fecalis* die in 2 min at pH 7.5 at a free chlorine concentration of 0.5 mg/L.[77] Grabow[31] noted that *S. fecalis* counts decreased rapidly during the first minute, after which a few organisms persist in slowly declining numbers. Their observation was based at pH 6.8 and 10 at a free chlorine concentration ranging from 0.03 to 44 mg/L.

Bacterial spores are more resistant than vegetative cells to chlorine compounds. Several investigators studied the sporicidal activity of hypochlorites[23,71,85] at various concentrations of available chlorine and at different pH. It was observed that at a concentration of 500 mg/L of free chlorine, time required to kill at pH 10 was 31 min,[71] whereas at 25 mg/L it was 121 min. Also, at lower pH values such as 6–9, time required to kill spores was reduced to 2.5 min at a concentration of 25 mg/L vs. 121 min.[85] Another reviewer found that to achieve a 90–99% kill of bacterial spores in a reasonable time of 5 min at a pH value of 6–9, it would require about 500 mg/L of available chlorine.[81]

Viruses

Viruses, in general, are more resistant than common vegetative bacterial cells. Chlorine resistance of common water-transmitted human viral pathogens has been studied by many investigators. (Tables 25.3, 25.4, and 25.5).

There are many water-transmitted human viruses that are responsible for various diseases (Table 25.1). Of these, hepatitis A virus is used as the index organism for checking drinking water quality.[39] Hepatitis A virus can be safely inactivated by 0.14 to 0.5 mg/L of free chlorine residual at pH 6–10, at a temperature 5 to 25°C in less than 9 min.[31,76] Grabow et al.[31] also found that hepatitis A virus, under certain conditions, was more resistant than coliphage MS2, reovirus type 3, *Escherichia coli* and *Streptococcus fecalis*; and at pH 6, 8, and 10 more sensitive than coliphage V1, poliovirus type 2 and *Mycobacterium fortuitum*. At lower pH, viruses are more resistant than coliforms to inactivation by chlorine dioxide.[75] Increased resistance to inactivation by chlorine compounds has also been observed for bacteria and viruses that are associated with certain particles.[46] However, viruses associated with inorganic particles such as kaolinite, aluminum oxide, or aluminum phosphate were inactivated at a similar rate whether dispersed or adsorbed.[8] Dissolved organic compounds also offer a certain amount of protection to viruses and microorganisms against disinfectants in general.

The inactivation rates of poliovirus in a cell maintenance medium and in phosphate buffer with chlorine dioxide and chlorate ion at 0.4, 1.0, and 2.0 mg/L of available chlorine, pH 7, and 20°C were recently studied.[79] The residual chlorine dioxide (ClO_2) and chlorate ion (ClO_2^-) formed in the maintenance medium and in phosphate buffer were determined. In lower concentration of chlorine dioxide (0.4 mg/L) there was a slower inactivation of viruses in maintenance medium. At 2.0 mg/L of ClO_2, 99% of the virus was inactivated in 2 min in both virus suspensions. Chlorite ion (2.0 mg/L) has little effect on the inactivation of the virus in 15 min contact. Further, the inactivation rate with 2.0 mg/L of ClO_2 was almost similar to 0.4 mg/L of HOCl due to equal oxidizing capacities. Generally, viruses can be inactivated by the routinely accepted specifications of free chlorine residuals used for disinfection of drinking water supplies.

Protozoa

Among protozoans, *Giardia* is most commonly associated with waterborne disease outbreak. However, *Cryptosporidium* is also becoming a major problem in drinking water supplies. *Giardia* and *Cryptosporidium* are widely dispersed in the aquatic environment. Among 66 surface water treatment plants in 14 states and one Canadian province, *Giardia* spp. were detected in 81% and *Cryptosporidium* spp. were found in 87% of the raw water samples.[48] The difficulty in recognizing protozoan diseases is the detection and identification of the organisms. *Giardia* and all other protozoan cysts are more resistant to inactivation by chlorination than enteric bacterial pathogens. *Giardia lamblia* cysts can be inactivated by 0.2 to 3 mg/L of free chlorine at 5°C, pH 7.00, in 3 to 32 min.[36,40] *Giardia muris* cysts require considerably more free chlorine and chlorine dioxide as well as more time than *G. lamblia* to cause 99% reduction in survival.[44,45,70] Amoebic dysentery–causing *Entamoeba histolytica* cysts can be inactivated in 150 min at 0.08–0.12 mg/L of free chlorine.[20]

Cryptosporidium parvum produces highly resistant oocysts as the infective stage in its life cycle and has been recently identified as a contaminant in many public water supplies and swimming pools, causing diarrhea in humans.[17] There were 3551 cases of *Cryptosporidium* outbreaks in 1991–1992. The importance of waterborne cryptosporidiosis was recently underscored by an outbreak in Milwaukee, Wisconsin in March–April 1993 when an estimated 403,000 had watery diarrhea.[17] Water analysis of 66 U.S. and Canadian filtered water treatment plants has revealed a low level of *Cryptosporidium* oocysts in up to 27% of the drinking water supplies. However, the method used does not assess the viability or infectivity of the oocysts.[49] Recently a matched case-control study of a community-wide outbreak of Cryptosporidiosis revealed that a wave pool was responsible for the outbreak.[53] The sedimentation and filtration processes remove most oocysts, but a small proportion may still remain, and the routine chlorine concentration used in drinking water treatment does not kill the oocysts.

Pretreatment with chlorine is thought to enhance excystation in *Cryptosporidium.* However, a recent study[64] showed that pretreatment with chlorine at Ct values of 75 and 150 and untreated *Cryptosporidium* oocysts do not show any difference in excystation. Also, the viability of oocysts stored in drinking water after chlorine treatment was not significantly reduced.

TOXICITY AND HEALTH RISK OF DISINFECTION WITH CHLORINE COMPOUNDS

There are many potential health risks associated with chlorine and chlorine dioxide treatment of drinking water. However, when one considers the risk versus the benefit of chlorine treatment of water, the benefits far outweigh the potential risks to humans. Bull et al.[12] have recently reviewed the overall risk associated with disinfection of drinking water with chlorine compounds, ozone, and iodine.

Considerable literature exists on the by-products of chlorination. They are hypochlorite, trihalomethanes, chlorinated acids and aldehydes, haloacetonitrites, chlorophenols and chlorinated ketones, etc. The principal by-products of chlorine disinfection are organochlorines, particularly trihalomethanes (THMs). Chlorine dioxide also may produce THMs.[78] Not much is known about the specific by-products of ClO_2 with humic materials. Additional health concerns from ClO_2 relate to the formation of chlorate and chlorite ions.[12] Sodium hypochlorite does not appear to be a carcinogen, but a number of by-products of chlorination, such as THMs, warrant much attention.

Studies of health risk associated with chlorine and chlorine dioxide and their by-products have been done with pigeons, rats, and mice.[12,41,67] Presently available information provides little justification to conclude that these results apply to humans.[12] Although chlorine dioxide does not readily react with organic compounds, forming THMs, its use is complicated by the toxicological effects of chlorite and chlorate, both of which can produce methemoglobinemia. Chlorite also produces hemolytic anemia at a low dose.[35]

Reaction products of chlorine dioxide and hypochlorous acid were mutagenic to *Salmonella typhimurium.*[72] Chlorine dioxide also rapidly reacts with cysteine, tryptophan, tyrosine, histidine, hydroxyproline, proline, and with certain peptides L-glycyl-L-tryptophan, L-trytophyl glycine. Mutagenicity studies with *Salmonella* microsomes assay with the six amino acids and the peptides showed hydroxyproline and tyrosine exerted mutagenic activity.[80] Nonvolatile reaction products of chlorine and chlorine dioxide with L-tryptophan were also shown to be direct-acting mutagens to *S. typhimurium.* Several of the fluorescent bands obtained after thin-layer chromatographic separation were shown to be more mutagenic than the reaction mixtures using Ames *Salmonella* microsome assay.[60]

At a pilot-scale study of drinking water treatment in Evansville, Indiana in 1993, it was observed that the levels of mutagenic activity in samples treated with liquid or gaseous ClO_2 followed by Cl_2 were essentially the same. However, the addition of NH_4Cl following gaseous ClO_2 reduced the level of mutagenic activity by 50% compared to the samples treated with either liquid ClO_2 and Cl_2 or Cl_2 alone.[62] The authors concluded that the substitution of NH_4Cl for Cl_2 as secondary disinfectant may be beneficial in drinking water treatment.

Although some epidemiological evidence suggests a small increase in incidence of cancer in bladder, colon, and rectum,[24] attempts to associate the development of cancer with trihalomethanes has not been successful. In a case-control study to test the hypothesis that chronic ingestion of THMs formed as by-products of chlorination in drinking water carries a risk of colon cancer, it was found that THM in Wisconsin drinking water did not pose a significant colon cancer risk.[88] Another major study[15] on bladder cancer rates in whites concluded that chlorinated drinking water is responsible for bladder cancer but has a latent period of 40 years and is apparently related to non-THMs.

CONCLUSIONS

Chlorine disinfection for water purification is still the most widely used means of treating drinking water. It is cost-effective and if properly applied can be germicidal for a vast array of organisms. Considerable complications, however, arise in assessing the effectiveness of disinfectants since germicidal activity depends on many factors, such as pH, temperature, inorganic, organic, and humic matters, water source, etc. Due to health risks from by-products of chlorine and chlorine dioxide, and government regulations, the popularity of alternative disinfectants such as chloramines, ozone, and use of reducing agents has come into the picture. A survey in 1987[54] showed that among 263 large utilities, 216 used chlorine, 67 chloramines, 16 chlorine dioxide and 1 ozone. Obviously, chlorine is the disinfectant of choice, although alternative disinfectants are becoming more common.

The largest number of waterborne diseases outbreaks in the United States in 1991–92 and 1971–92 are associated with acute gastrointestinal illness of unknown etiology.[17] Very likely the causative agents for these outbreaks are viruses and protozoa. The interest and expertise of the laboratories, physicians, and investigators influences the identification of etiological agents. Among protozoans, *Cryptosporidium* continues to gain importance as a waterborne pathogen due to its tremendous resistance to chlorine and the size of the oocysts that can escape a routine filtration process.

Chlorine dioxide is a stronger oxidizing agent than chlorine and offers certain advantages over chlorine. It does not react readily with natural humus and other organic materials and therefore forms less trihalomethanes. It also breaks down phenolic compounds and odor, and can remove cyanides, sulfides, aldehydes, and mercaptans. However, the formation of chlorite and chlorate has toxicological implications. There is limited information on the occurrence and concentration of the toxic by-products of chlorine dioxide. Possibility of regulation of total residual concentration of chlorine dioxide and its inorganic species should provide impetus for investigation of various disinfection alternatives.

REFERENCES

1. AWWARF. 1987. National Trihalomethane Survey Report, prepared by Decision Research under the supervision of the Metropolitan Water District of Southern California.
2. Bates, R.C., P.T.B. Shaffer, and S.M. Sutherland. 1977. Development of poliovirus having increased resistance to chlorine inactivation. *Appl. Environ. Microbiol.,* 33:849–853.
3. Berg, J.D., A. Matin, and P.V. Roberts. 1982. Effect of antecedent growth conditions on sensitivity of *Escherichia coli* to chlorine dioxide. *Appl. Environ. Microbiol.,* 44:814–819.
4. Berg, J.D., J.C. Hoff, P.V. Roberts, and A. Matin. 1985. Disinfection resistance of *Legionella pneumophila* and *Escherichia coli* grown in continuous and batch culture. In: *Water Chlorination: Chemistry, Environmental Impact and Health Effects,* Vol. 5, R.L. Jolley (Ed.). Lewis Publishers, Boca Raton, FL, pp. 603–613.
5. Berman, D. and J.C. Hoff. 1984. Inactivation of simian rotavirus SA11 by chlorine, chlorine dioxide and monochloramine. *Appl. Environ. Microbiol.,* 48:317–323.
6. Berman, D., E.W. Rice, and J.C. Hoff. 1988. Inactivation of particle-associated coliforms by chlorine and monochloramine. *Appl. Environ. Mcrobiol.,* 54:507–512.
7. Blaser, M.J., P.F. Smith, W.-L.W. Wang, and J.C. Hoff. 1986. Inactivation of *Campylobacter jejuni* by chlorine and monochloramine. *Appl. Environ. Microbiol.,* 51:307–311.
8. Boardman, G.D. and O.J. Sproul. 1977. Protection of viruses during disinfection by adsorption of particulate matter. *J. Water Poll. Control Fed.,* 49:1857–1861.
9. Bolton, K.J., C.E.R. Dadd, G.C. Mead, and W.M. Waites. 1988. Chlorine resistance of strains of *Staphylococcus aureus* isolated from poultry processing plants. *Appl. Microbiol.,* 6:31–34.
10. Brazis, A.R., J.E. Leslie, P.W. Kabler, and R.L. Woodward. 1958. The inactivation of *Bacillus globigii* and *Bacillus anthracis* by free available chlorine. *Appl. Microbiol.,* 6:338–342.
11. Brazis, A.R., et al. 1959. Effectiveness of halogens or halogen compounds in detoxifying *Clostridium botulinum toxins. J. Am. Water Works Assoc.,* 51:902–912.
12. Bull, R.J., C. Gerba, and Trussell. 1990. Evaluation of the health risk associated with disinfection. *Critical Reviews in Environmental Control.,* 20:77–113.
13. Butterfield, C.T., E. Wattie, S. Megregian, and C.W. Chambers. 1943. Influence of pH and temperature on the survival of coli-forms and enteric pathogens when exposed to free chlorine. *Public Health Rep.,* 58:1837–1866.

14. Campbell, I., S. Tzipori, G. Hutchison, and K.W. Angus. 1982. Effect of disinfectants on survival of Cryptosporidium oocysts. *Vet. Rec.,* 111:414–415.
15. Cantor, K.P., R. Hoover, P. Harge, T.J. Mason, D.T. Silverman, R. Altman, D.F. Austin, M.A. Child, C.R. Key, L.D. Marrett, M.N. Myers, A.S. Narayana, L.I. Levin, J.W. Sullivan, G.M. Swanson, D.B. Thomas, and D.W. West. 1987. Bladder cancer, drinking water source, and tap water consumption: a case-control study, *J. Natl. Cancer Inst.,* 79:1269.
16. Carson, L.A., N.J. Petersen, M.S. Favero, and S.M. Aguero. 1978. Growth characteristics of atypical mycobacteria in water and their comparative resistance to disinfectants. *Appl. Environ. Microbiol.,* 36:839–846.
17. CDC. 1993. Surveillance for Waterborne Diseases Outbreaks — United States 1991–92. *MMWR 1993.* 42(SS-5):1–22.
18. Chen, Y.S.R., O.J. Sproul, and A.J. Rubin. 1985. Inactivation of *Naegleria gruberi* cysts by chlorine dioxide. *Water Res.,* 19:783–789.
19. Chum, C.C., C.D. Boardman, and R.C. Bates. 1984. The inactivation kinetics of H-1 parvovirus by chlorine. *Water Res.,* 18:195–203.
20. Clarke, N.A., R.E. Stevenson, and P.W. Kabler. 1956. The inactivity of purified type 3 Adenovirus in water by chlorine. *Am. J. Hyg.,* 64:314–319.
21. Clarke, N.A. and S.L. Chang. 1959. Enteric viruses in water. *J. Am. Water Works Assoc.,* 51:1299–1317.
22. Costigan, S.M. 1936. Effectiveness of hot hypochlorites of low alkalinity in destroying *Mycobacterium tuberculosis. J. Bacteriol.,* 32:57–63.
23. Cousins, C.M. and C.D. Allan. 1967. Sporicidal properties of some halogens. *J. Applied Bact.,* 168–174.
24. Craun, C.F. 1985. Epidemiologic considerations for evaluating associations between disinfection of waster and cancer in humans, in *Water Chlorination: Chemistry, Environmental Impact and Health Effects.* Vol. 5, Jolley, R.L., Ed., Lewis Publishers, Boca Raton, FL, p. 1 133.
25. Cronier, S.D. 1977. *Destruction by Chlorine Dioxide of Viruses and Bacteria in Water.* M. Sc. Thesis, Department of Civil and Environmental Engineering, Univ. of Cincinnati, OH.
26. Dychdala, G.R. 1960. Studies on bactericidal effectiveness of calcium hypochlorite stabilizers for swimming pool use. Pennwalt Corp, unpublished.
27. Dychdala, G.R. 1991. Chlorine and chlorine compounds. In *Disinfection, Sterilization and Preservation.* Edited by S.S. Block, Lea and Febiger, Philadelphia. pp. 131–151.
28. El-Kest, S.E. and E.H. Marth. 1988. Inactivation of *Listeria monocytogenes* by chlorine. *J. Food Protect.,* 51(7):520–524, 530.
29. Fujioka, R.S., M.A. Dow, and B.S. Yoneyama. 1986. Comparative disinfection of indicator bacteria and poliovirus by chlorine dioxide, *Water Sci. Tech.,* 8(10):125–132.
30. Gordon, G., et al. 1990. Minimizing chlorite ion and chlorate ion in water treated with chlorine dioxide. *J. Am. Water Works Assoc.,* 82:160–165.
31. Grabow, W.O.K., et al. 1983. Inactivation of hepatitis A virus and indicator organisms in water by free chlorine residuals. *Appl. Environ. Microbiol.,* 46:619–624.
32. Haas, C.N., M.G. Keralius, D.M. Brncich, and M.A. Zapkin. 1986. Alteration of chemical and disinfectant properties of hypochlorite by sodium, potassium, and lithium. *Environ. Sci. Technol.,* 20:822–826.
33. Harakeh, M.S., J.D. Berg, J.C. Hoff, and A. Matin. 1985. Susceptibility of chemostat-grown *Yersinia enterocolytica* and *Klebsiella pneumonia* to chlorine dioxide. *Appl. Environ. Microbiol.,* 49:69–72.
34. Hays, H., P.R. Eliker, and W.E. Sandine. 1963. Effect of acidification on stability and bactericidal activity of added chlorine in water supplies. *J. Milk Food Technol.,* 26:147–149.
35. Hefferman, W.P., C. Guion, and R.J. Bull. 1979. Oxidative damage to the erythrocyte induced by sodium chlorite *in vivo. J. Environ. Pathol. Toxicol.,* 2:1487.
36. Hibler, C.P., C.M. Hancock, L.M. Perger, J.G. Wegrzn, and K.D. Swabby. 1987. Inactivation of *Giardia* cysts with chlorine at 0.5 to 5.0°C. Amer. Water Works Assoc. Res. Foundation, Denver, CO.
37. Hurst, C.J. 1991. Disinfection of drinking water, swimming pool water, and treated sewage effluents. In *Disinfection, Sterilization and Preservation.* Edited by S.S. Block, Lea and Febiger, Philadelphia, pp. 713–729.
38. Ingols, R.S., H.A. Wyckoff, T.W. Kethley, H.W. Hodgden, E.L. Fincher, J.C. Hildebrand, and J.E. Mandel. 1953. Bacteriocidal studies of chlorine. *Ind. Eng. Chem.* 45:996–1000.
39. International Association on Water Pollution Research and Control Study Group on Water Virology. 1982. The health significance of viruses in water. *Water Res.* 17:121–132.
40. Jarroll, E.L., A.K. Bingham, and E.A. Meyer. 1981. Effect of chlorine on *Giardia lamblia* cyst viability. *Appl. Environ. Microbiol.,* 41:483–487.
41. Jorgenson, T.A., E.F. Meierhenry, C.J. Rushbrook, R.J. Bull, and M. Robinson. 1985. Carcinogenicity of chloroform in drinking water to male Osborne-Mendl rats and female B6C3F1 mice. *Fundam. Appl. Toxicol.,* 5:760–769.

42. Kelly, S. and W. Sanderson. 1958. The effect of chlorine in water on enteric viruses. *Am. J. Public Health,* 48:1323–1334.
43. Kuchta, J.M., S.J. States, J.E. McGlauchlin, J.H. Overmeyer, R.M. Wadowsky, A.M. McNamara, R.S. Wolford, and R.B. Yee. 1985. Enhanced chlorine resistance of tap water-adapted *Legionella pneumophila* as compared with agar medium-passaged strains. *Appl. Environ. Microbiol.* 50:21–26.
44. Leahy, J.G. 1985. Inactivation of *Giardia muris* cysts by chlorine and chlorine dioxide. M.Sc. Thesis, Ohio State University Department of Civil Engineering, Columbus, p. 115.
45. Leahy, J.G., A.J. Rubin, and O.J. Sproul. 1987. Inactivation of *Giardia muris* cysts by free chlorine. *Appl. Environ. Microbiol.,* 53:1448–1453.
46. LeChevallier, M.W. and G.A. McFeters. 1984a. Recent advances in coliform methodology for water analysis. *J Env. Health,* 47:5–9.
47. LeChevallier, M.W., T.S. Hassenauer, A.K. Camper, and G.A. McFeters. 1984b. Disinfection of bacteria attached to granular activated carbon. *Appl. Environ. Microbiol.,* 48:918–923.
48. LeChevallier, M.W., W.D. Norton, and R.G. Lee. 1991a. Occurrence of *Giardia* and *Cryptosporidium* spp. in Surface Water Supplies. *Appl. Environ. Microbiol.* 57:2610–2616.
49. LeChevallier, M.W., W.D. Norton, and R.G. Lee. 1991b. *Giardia* and *Cryptosporidium* spp. in filtered drinking water supplies. *Appl. Environ. Microbiol.* 57:2617–2621.
50. Leyval, C., C. Arz, J.C. Block, and M. Rizet. 1984. *Escherichia coli* resistant to chlorine after successive chlorinations. *Environ. Technol. Lett.,* 5:159–364.
51. Lopes, J.A. 1986. Evaluation of dairy and food plant sanitizers against *Salmonella typhimurium* and *Listeria monocytogenes. J. Dairy Sci.,* 69:2791–2796.
52. Masschelein, W.J. 1979. *Chlorine Dioxide: Chemistry and Environmental Impact of Oxychlorine Compounds.* Ann Arbor Science Publishers, Inc., Ann Arbor, MI.
53. McAnulty, J.M., D.W. Fleming, and A.H. Gonzalez. 1994. A community-wide outbreak of cryptosporidiosis associated with swimming at a wave pool. *J. Am. Med. Soc.* 272(20):1597–1600.
54. McGuire, M. and R. Meadows. 1987. American Water Works Association Research Foundation Trihalomethane Survey — A program report, presented at the Conference on Current Research in Drinking Water Treatment, Cincinnati, OH, March 24–26, 1987.
55. Morris, J.C. 1971. Chlorination and Disinfection — State of the Art. *J. Am. Water Works Assoc.,* 63:769–774.
56. Muraca, P., E.J. Stout, and V.L. Yu. 1987. Comparative assessment of chlorine, heat, ozone, and UV light for killing *Legionella pneumophila* within a model plumbing system.
57. Nagy, L.A. and B.H. Olson. 1982. The occurrence of filamentous fungi in drinking water distribution systems. *Can. J. Microbiol.,* 28:667–671.
58. Olivieri, V.P., F.H. Hauchman, C.I. Noss, and R. Vasl. 1985. Mode of action of chlorine dioxide on selected viruses. In: *Water Chlorination: Chemistry, Environmental Impact and Health Effects,* J.L. Jolley (Ed.), Lewis Publishers, Boca Raton, FL, pp. 651–666.
59. Orth, R. and H. Mrozek. 1989. Is the control of *Listeria, Campylobacter* and *Yersinia* a disinfection problem? *Fleischwirtshaft.* 69(10):1575–1576.
60. Owusu-Yaw, J., W.B. Wheeler, and C.I. Wei. 1991. Genotoxicity studies of the reaction of chlorine dioxide with L-tryptophan. *Toxicology Letter* (Amsterdam) 56(1-2):213–228.
61. Palmer, C.M. and T.E. Maloney. 1955. Preliminary screening for potential algicides. *Ohio J. Sci.,* 55:1–8.
62. Patterson, K.S., S.I. Garner, R. Millers, and B. Lykins, Jr. 1993. Mutagenicity of drinking water samples predisinfected with chlorine dioxide. Environmental and Molecular Mutagenesis, Supplement 22. 21:62.
63. Pelletier, P.A. and G.C. Du Moulin. 1987. Mycobacteria in public water supplies: comparative resistance to chlorine. In: *Emerging Issues in Effluent Disinfection Workshop,* Water Pollution Control Federation, Washington, DC.
64. Quinn, C.M. and W.B. Betts. 1993. Longer term viability status of chlorine-treated *Cryptosporidium* oocysts in tap water. *Biomed. Lett.,* 48(192):315–318.
65. Ram, N.M. and J.P. Malley, Jr. 1984. Chlorine residual monitoring in the presence of N-organic compounds. *J. Am. Water Works Assoc.,* 76:74–81.
66. Raphael, R.A., S.A. Sattar, and V.S. Springthorpe. 1987. Lack of human rotavirus inactivation by residual chlorine in municipal drinking water systems. *Revue International des Sciences de l'Eau,* 3:67–69.
67. Revis, N.W., P. McCauley, R. Bull, and G. Holdsworth. 1986. Relationship of drinking water disinfectants to plasma cholesterol and thyroid hormone levels in experimental studies. *Proc. Natl. Acad. Sci. USA.,* 83:1485.
68. Ridenour, G.M. and R.S. Ingols. 1947. Bacteriocidal properties of chlorine dioxide. *J. Am. Water Works Assoc.,* 39:561–567.

69. Rubin, A.J., J.P. Engel, and O.J. Sproul. 1983. Disinfection of amoebic cysts in water with free chlorine. *J. Water Poll. Control Fed.,* 55:1174–1182.
70. Rubin, A.J. 1988. Factors affecting the inactivation of *Giardia* cysts by monochloramine and comparison with other disinfectants. In: *Proceedings: Conference on Current Research in Drinking Water Treatment,* U.S. EPA, EPA/600/9-88/004, Cincinnati, OH, pp. 224–229.
71. Rudolph, A.S. and J. Levine. 1941. Factors affecting the germicidal efficiency of hypochlorite solutions. Bull. 150, *Eng. Exp. Sta.,* Iowa State College.
72. Sen, A.C., J. Owusu-Yaw, W.B. Wheeler, and C.I. Wei. 1989. Reactions of aqueous chlorine and chlorine dioxide with tryptophan, N-methyltryptophan and 3 IAA kinetic and mutagenic studies. *J. Food Sci.* 54(4):1057–1060.
73. Snow, W.B. 1956. Recommended chlorine residuals for military water supplies. *J. Am. Water Works Assoc.,* 48:1510–1514.
74. Sobsey, M.D., T. Fuji, and P.A. Shields. 1988. Inactivation of hepatitis A virus and model viruses in water by free chlorine and monochloramine. In: *Proceedings of the International Conference for Water and Wastewater Microbiology,* IAWPRC, Pergamon Press, NY.
75. Sobsey, M.D. 1989. Inactivation of health related microorganisms in water by disinfection processes. *Water Sci. Tech.* 21(3):179–195.
76. Sobsey, M.D., T. Fugi, D. Batt, and R.M. Hall. 1990. Inactivation of cell associated and dispersed hepatic A virus by free and combined chlorine and chlorine dioxide. *Proceedings of Water Quality Technology Conference,* Denver, CO. pp. 167–179.
77. Stuart, L.S. and L.F. Ortenzio. 1964. Swimming pool chlorine stabilizers. *Soap Chem. Spec.,* 40:79–82; 112–113.
78. Symons, J., A. Stevens, R. Clark, E. Geldrich, O.T. Love, and J. DeMarco. 1981. Treatment techniques for controlling trihalomethanes in drinking water, EPA 600/2-81-156, U.S. EPA.
79. Tachikawa, M., K. Saita, M. Tezuka, and R. Sawmura. 1993. Inactivation of Poliovirus with chlorine dioxide. *Japanese J. Toxicol. Env. Health* 39(6):572–576.
80. Tan, H.K., W.B. Wheeler, and C.I. Wei. 1987. Reaction of chlorine dioxide with amino acids and peptides kinetics and mutagenicity studies. *Mutat. Res.* 188(4):259–266.
81. Trueman, J.R. 1971. The halogens. In: *Inhibition and Destruction of the Microbial Cell,* edited by Hugo, W.B., Academic Press, London, pp. 137–183.
82. U.S. Environmental Protection Agency. 1989. Final rule for national primary drinking water regulations. *Federal Register,* 54(124):27486–27541.
83. Vaughn, J.M., Y.-S. Chen, and M.Z. Thomas. 1986. Inactivation of human and simian rotaviruses by chlorine. *Appl. Environ. Microbiol.,* 48.508–514.
84. Ward, N.R., R.L. Wolfe, and B.H. Olson. 1984. Effect of pH, application technique, and chlorine-to-nitrogen ratio on disinfectant activity of inorganic chloramines with pure culture bacteria. *Appl. Environ. Microbiol.,* 48:508–514.
85. Weber, G.P. and M. Levine. 1944. Factors affecting germicidal efficiency of chlorine and chloramine. *Am. J. Publ. Health* 34:719–728.
86. Weber, G.R. 1950. Effect of concentration and reaction (pH) on germicidal activity of chloramine - T. *Public Health Rep.,* Wash. 65, 503–512.
87. Wolfe, R.L., N.R. Ward, and B.H. Olson. 1984. Inactivation of heterotrophic bacterial populations in finished drinking water by chlorine and chloramines. *Water Res.,* 19:1393–1403.
88. Young, T.B., D.A. Wolf, and M.S. Kandrek. 1987. Case-control study of colon cancer and drinking water trihalomethanes in Wisconsin. *Int. J. Epidemiol.,* 16(2):190–197.

Chapter 26

Immunotoxicity of Dioxin: Is the TEF Concept Premature?

P.T. Thomas, L.L. Aylward, and N.J. Karch

INTRODUCTION

Of all the mechanisms for xenobiotic transformation, that involving the Ah receptor is one of the best understood (Parkinson, 1995). Various classes of chemicals are known to contain congeners that bind appreciably to the Ah receptor (AhR), among which the halogenated hydrocarbons have received the most attention. Once bound and activated following phosphorylation, this intracellular receptor binds to the Ah receptor nuclear translocator protein (Arnt). Enhanced transcription of Cytochrome P-4501A1 (CYP1A1) and other genes (and subsequently gene products) by the AhR-Arnt complex results in the many downstream biological effects attributed to the AhR, including effects on the immune system. Although many halogenated hydrocarbon classes have congeners that bind appreciably to the AhR, the polychlorinated dibenzo-*p*-dioxins (PCDDs), polychlorinated dibenzofurans (PCDFs), and polychlorinated and polybrominated biphenyls (PCBs and PBBs) have received the most regulatory attention. Among all of these chemicals, 2,3,7,8-tetrachlorodibenzo-*p*-dioxin (TCDD) binds with the greatest affinity and has been studied most extensively. Numerous reports in animals establish the toxic effects of TCDD and other congeners on the immune system and have been reviewed extensively (Kerkvliet and Burleson, 1995; Thomas and Faith, 1985; Thomas et al., 1990; Holsapple et al., 1991; Holsapple and Smialowicz, 1995).

The EPA and various international organizations established the Toxic Equivalency Factor approach as a practical policy measure for performing risk assessments on complicated mixtures of polychlorinated dioxins and furans (PCDDs and PCDFs) (U.S. EPA 1989). This approach assumes that all PCDDs and PCDFs share a mechanism of action of binding to the Ah receptor and cause the same toxic effects as TCDD, varying only in potency. In this system, TCDD is assumed to be the most toxic congener, and is assigned a toxic potency value of 1. Other PCDDs and PCDFs are assigned relative potencies, or Toxic Equivalency Factors (TEFs) such as 0.5, 0.1, etc. The endpoints used to establish these relative potencies vary considerably. The toxicity of a mixture of various PCDDs and PCDFs can be expressed in terms of TCDD toxic equivalents (TEQs) by multiplying the quantity of each congener by its TEF, and summing the results. TEFs remain a convenient and practical approach to risk assessments of mixtures of PCDDs and PCDFs, although recent investigations have shown them to overestimate risks (Harper et al., 1995).

The EPA's reassessment of the risks of TCDD and structurally related compounds extended the TEF concept, previously applied only to PCDDs and PCDFs, to PCBs. Also, noncancer effects of TCDD received much greater emphasis, and immunotoxicity appears to be an endpoint for which the EPA appears likely to apply the TEF concept. This chapter addresses the scientific basis for extending TEFs, and the underlying science in the role of the AhR, to immunotoxicity.

Evidence for AhR-Independent Mechanisms

The most widely observed effect of TCDD on the animal and cellular systems in which it has been studied is induction of Cytochrome P450 enzymes, most notably CYP1A1. This effect is mediated by the AhR and is one of the few effects for which the biochemical steps stemming from binding to the AhR are known. However, enzyme induction is not thought to be correlated with the toxicity observed in laboratory rodents. In mice, functional changes in immunity segregate with Ah receptor "responsiveness" and, under certain conditions, immune changes occur at exposure levels approaching those which result in enzyme induction. However, evidence is lacking that CYP1A1 induction or any other events in gene expression directly mediates immunotoxicity of TCDD. In fact, the following evidence supports AhR-independent mechanisms of immunotoxicity.

TCDD Alters Early Activational Events Independent of Nuclear Translocation–Dependent Pathways

Although to date no studies have investigated the effect of TCDD on gene expression in an immunocompetent cell model, in cultured hepatoma cells, TCDD induces a rise in mRNAs of immediate early genes, such as c-fos and c-jun, and a large increase in transcription factor AP-1 (Puga et al., 1992). Compared to induction by phorbol esters, induction by TCDD of these two parameters in genetic expression is delayed but eventually reaches comparable levels. Saurosporine, a protein kinase C inhibitor, inhibits both processes. Such impacts on gene expression are also seen in cultured hepatoma cells which are defective in the AhR nuclear translocator protein. This suggests that protooncogene expression and other early events induced by TCDD at the molecular level are independent of nuclear translocation of the AhR (Bombick et al., 1985, 1988; Matsumura et al., 1984). Recent studies (e.g., Enan and Matsumura, 1995) using microsomal fractions prepared from guinea pig adipose tissue suggest that TCDD is capable of interacting with the AhR without nuclear translocation and activating certain protein kinases important in growth factor signal transduction.

AhR-Mediated Immune Effects Seen In Vitro Often Are Not Reproduced In Vivo

Of all immune parameters studied in animals after exposure to TCDD, the antibody response to sheep red blood cells (SRBCs) is consistently suppressed. Davis and Safe (1991) compared the *in vitro* structure activity relationships for PCDDs and PCDFs, which show a wide difference *in vivo* immunotoxic potency. All congeners appeared to be potent *in vitro* and produced concentration-dependent suppression of the *in vitro* anti-SRBC response using cells from either Ah responsive or nonresponsive strains. The anti-SRBC response is also modulated by other factors. For example, Morris et al. (1991) reported that the *in vitro* effects of TCDD on the anti-SRBC plaque-forming cell (PFC) response were critically dependent on the type and concentration of serum used in culture. Furthermore, and more significantly, in serum free cultures, TCDD caused a fifteen-fold enhancement of the response. This lack of concordance between *in vitro* observations and those seen *in vivo* warrants further investigation.

In vitro studies have also suggested that immunosuppression is possible in the absence of strong affinity for Ah receptor binding. Studies by Tucker et al. (1986), using the *in vitro* Michell Dutton culture system, demonstrate comparable suppression of the anti-SRBC PFC response with TCDD and the relatively weak ligand-binding isomer 2,3,7-triCDD. In addition to *in vitro/in vivo* differences, recent studies by Harper et al. (1995), using the Ah-responsive $B_6C_3F_1$ mouse strain, suggest that the immunosuppressive activities of certain PCB congeners are nonadditive. Thus, for some Ah receptor-mediated responses, the TEF approach for PCBs overestimates toxicity. Other studies support this. Silkworth et al. (1993) demonstrated that TCDD-induced immunotoxicity can be modified by co-exposure to other organic compounds in complex mixtures via Ah receptor–independent processes.

Immune Response Profiles of Animals Exposed to TCDD, Including Ah Low-Responsive DBA/2 Mice, May Be Dependent Upon the Nature and Duration of Exposure

Morris et al. (1992) reported that the sensitivity of DBA/2 mice to TCDD-induced immunosuppression was increased approximately 10 times following subchronic as opposed to acute dosages of equal amounts of the compound. The immunosuppression seen was not accompanied by hepatomegaly, a striking effect seen in the Ah-responsive $B_6C_3F_1$ strain. Furthermore, the degree of thymic involution was enhanced in the DBA/2 strain following subchronic exposure. These observations, however, appear to conflict with earlier studies (Vecchi et al., 1983) where, relative to C57BL/6 mice,

DBA/2 animals remained insensitive to TCDD-induced immunosuppression even after 8 weekly oral doses of 0.5 μg TCDD/kg, an observation consistent with low AhR affinity and responsiveness of this strain. Other studies have reported that subchronic exposure of $B_6C_3F_1$ mice to the low-affinity ligand 2,7-diCDD induced suppression of the anti-SRBC antibody-forming cell response *in vivo* in a manner comparable to that seen with TCDD in the absence of liver enzyme aryl hydrocarbon hydroxylase induction (Holsapple et al., 1986). Again, this suggests an aspect of the mechanism that is independent of aryl hydrocarbon hydroxylase toxicity. White et al. (1985) also reported that the potent AhR agonist 7,12-dimethylbenz(a)anthracene induced comparable immunotoxicity in $B_6C_3F_1$ and DBA/2 mice, suggesting further that immunosuppressive effects may be independent of the AhR. Neubert et al. (1990, 1992) reported a pronounced decrease in the percentage and absolute number of T-helper/inducer cells following acute exposures of 10 ng/kg TCDD to marmosets. By contrast, a clear increase in this cell type was noted in animals exposed to multiple, lower doses of 0.3 ng/kg daily for up to 24 weeks. This exposure regimen resulted in animals achieving approximately 85% of the apparent steady state level. All of these results suggest that different mechanism(s) for different effects may operate depending upon the exposure regimen.

In summary, this evidence suggests that the mechanisms of TCDD immunotoxicity may have AhR-independent components. Since the concept of the TEF rests largely on the assumption that binding to the AhR is a critical step in the toxicity (and immunotoxicity) of 2,3,7,8-substituted PCDD congeners, applying the TEF concept to predicting immunotoxicity may be premature.

Early Studies of TCDD-Induced Immunosuppression at Low Doses Have Been Difficult to Reproduce

Early work by Clark et al. (1981) demonstrated that 2,3,7,8-TCDD significantly suppressed the generation of cytotoxic T-lymphocytes (CTL) in Ah-responsive C57BL/6 mice at doses as low as 0.004 μg/kg. They concluded that a population of suppressor cells prevented the formation of CTLs *in vitro*. The dose that inhibited the response by 50% (ID_{50}) was significantly lower than that shown by many investigators for the anti-SRBC response (< 1.0 μg/kg). Recent studies by Hanson and Smialowicz (1994), however, failed to demonstrate suppression in either *in vivo* or *in vitro* CTL responses at doses as high as 3 μg/kg at weekly intervals for 4 weeks. In these studies, CTL sensitization began 1 week following the last TCDD dose. More recently, De Krey and Kerkvliet (1995) reported dose-related suppression of specific CTL response following a single oral dose of TCDD given one day before CTL sensitization. The authors attribute the differences in CTL effects to timing of tumor cell sensitization relative to exposure, as well as to the sex of the experimental animal. It is apparent from these studies that confirming the effects of TCDD on an important correlate of cell-mediated immunity has been difficult to accomplish and may be highly dependent on the experimental design used.

Species-Specific Effects Are Not Consistent and May Differ in Mechanism

From an immunologic perspective the antibody PFC response is generally considered to be of potential relevance to humans. That is, mechanisms responsible for the generation of antibody-forming cells in response to challenge in rodents are similar to those in humans. Remarkably similar results have been reported for this test in TCDD-exposed Ah-responsive mice. Although not yet confirmed, Narasimhan et al. (1994) suggested that the PFC response may be suppressed in mice at doses lower than those capable of inducing significant EROD activity.

The effects on this response that appear to segregate with the Ah locus in mice, however, are not consistently seen in other species, suggesting different mechanisms. For example, Smialowicz et al. (1994) demonstrated that Fisher 344 rats exposed to up to 30 μg/kg TCDD (single dose) had significantly enhanced PFC responses. This effect occurred at TCDD doses that led to induction of hepatic CYP1A1 and CYP1A2 enzymes to comparable levels of activity in both rats and mice. The authors interpret this immunotoxic effect as one altering splenic helper and suppressor lymphocyte subpopulations, a mechanism not postulated for mice. Species differences in TCDD-induced susceptibility to

Trichinella spiralis infection have also been recently reported by Luebke et al. (1995). In nonhuman primate studies, Neubert et al. (1990) demonstrated a significant reduction in numbers of CD4+ helper/inducer T-cells in marmosets following a single injection of 10 ng TCDD/kg; the NOEL was 3 μg/kg. Due to the long lag phase between exposure and measurement (up to 15 weeks in the 10 ng/kg group), the authors suggested that the effect was a primary one on the thymus. The animals did not, however, exhibit increases in infections, which would suggest immunosuppression on a functional level. Recent investigations by the same researchers on a population of chemical plant workers with known body burdens of PCDD/PCDFs do not corroborate the effects observed in marmosets (Neubert et al., 1993, 1995). Hong et al. (1989) evaluated immune function in adult and young rhesus monkeys up to four years before evaluation. The dosages administered were considered to be equivalent to "average industrial exposure levels" ranging from 5–25 ppt in the diet. The authors observed no increase in the rates of infection and concluded that there was no evidence of immune deficit in offspring of adult animals with high TCDD body burdens (300–1,400 ppt). In summary, although effects on the immune system are reproducible in mice and segregate with the Ah locus, the data suggesting similar effects in other species are not consistently seen.

Mechanisms for Changes in Host Resistance Following TCDD Exposure Are Not Understood

The effects of TCDD exposure in experimental animals of induced infection or tumor challenge have drawn attention. Although great differences exist in experimental design, the challenge agent, and animal model, TCDD exposure appears to impair host resistance in mice. However, the mechanism(s) responsible for this impairment is not clear. Clark et al. (1983) reported that mice were susceptible to *Herpes simplex* type II virus following four weekly injections of TCDD at dose levels as low as 40 ng/kg. Significantly increased mortality was seen in mice treated with a single injection of as little as 100 ng TCDD/kg and challenged with influenza A2/Taiwan virus (House et al., 1990). These two studies are significant because the reported ID_{50} for TCDD is below the TEF value for the murine PFC response (0.7 μg/kg). Recently published studies by Burleson et al. (1996) have reproduced the influenza morbidity findings of House et al. (1990). In spite of this replication, indicators of other aspects of susceptibility to influenza in mice, which would normally accompany the morbidity findings (e.g., increases in lung consolidation and weight due to inflammation), were not seen (Burleson et al, 1996). This suggests that factors other than immunosuppression may be responsible for reducing host defense. In rats, however, immunologic mechanisms may play a role in resistance to influenza. TCDD treatment has been demonstrated to increase significantly pulmonary influenza virus titers, lung weights, and reduce viral specific NK cell activity (Yang et al., 1994).

Under certain conditions, TCDD exposure in mice increases susceptibility to *Listeria monocytogens* infection. Hinsdill et al. (1980) reported that adult Swiss Webster mice fed up to 50 μg/kg TCDD in the diet exhibited increased bacteremia and mortality. Furthermore, Luster et al. (1980) reported increased susceptibility to *Listeria* infection after perinatal TCDD exposure of $B_6C_3F_1$ mice to 5 μg/kg on day of gestation –14 and 7 and 14 postpartum. However, Vos et al., (1987) reported that oral administration of 50 μg TCDD/kg once a week for four weeks to Swiss mice did not alter the number in the spleen of viable *Listeria* recovered. This suggests that TCDD did not affect nonspecific phagocytosis and killing. Recently, House et al. (1990) reported that a single oral dose of up to 10 μg/kg of TCDD did not enhance mortality of adult $B_6C_3F_1$ mice to *Listeria* challenge administered seven days later. Of interest is that 10 μg/kg of TCDD significantly suppresses the anti-SRBC response in this study. Kerkvliet and Oughton (1993) reported that a single oral dose of 5 μg/kg TCDD induced an influx of neutrophil and macrophage inflammatory cells in the peritoneum in response to SRBC injection. Thus, TCDD exposure could *augment* host defenses. In spite of these cellular changes, this study and that of Dooley and Holsapple (1988) suggest that TCDD does not alter macrophage activation or antigen presentation — two potential mechanisms that would explain why the SRBC response is suppressed.

In summary, since TCDD exposure appears simultaneously immunosuppressive to T-cell immune function and yet may augment the inflammatory process, the overall impact of TCDD exposure on host defense may vary depending upon a number of factors, including mechanisms of pathogenesis and clearance. Furthermore, differences in study design, TCDD dose, dose regimen, animal strain, sex, and age may contribute to differences in observed effects on host defense.

TCDD-Induced Immunosuppression in Animals is Not Observed in Humans

As outlined above, under the right conditions, immunotoxicity and impaired host defenses can be demonstrated in both adult and perinatally exposed laboratory animals. However, similarly substantial effects in humans have not been observed. For example, no abnormalities in serum immunoglobulin levels and mitogen responses were noted in children from the Seveso, Italy exposures (Reggiani, 1978). A follow-up study of TCDD-exposed children several years later revealed a statistically, temporary and not clinically significant increase in complement protein levels, peripheral blood lymphocytes and *in vitro* lymphoproliferative responses (Tognoni and Bonaccorsi, 1982) with no change in disease prevalence.

Hoffman et al. (1986) reported decreased T-cell numbers and suppressed cell-mediated immunity (including anergy as measured by skin testing and "relative anergy") in residents of a mobile home park in Missouri contaminated with TCDD. However, the data were not reproducible, due to improper interpretation of the skin test results (Evans et al., 1988) or heat-compromised antigen preparations. In studies by Webb et al. (1989), individuals in Missouri with documented TCDD tissue levels exhibited subtle modulation of lymphocyte subpopulations, which was correlated with TCDD body burden. However, no effect of exposure was seen on cellular function or associated clinical disease. A study by Mocarelli et al. (1991) of heavily exposed individuals from the Seveso area revealed that with the exception of chloracne, only minimal, if any, health effects remained. Recent studies by Neubert et al. (1993, 1995) of chemical plant workers with moderate body burdens of PCDDs and PCDFs revealed no biologically significant changes in immunity.

The EPA appears to consider effects in humans and animals on the immune system to be similar. One complicating factor may be that the use of body burden as the dosimetric for comparison is not the correct one, and that some measure of area under the curve or concentration above a threshold is more accurate (Finley and Paustenbach, 1995). Regardless of which dosimetric is used, the clinically significant effects of PCDD exposure in animals are not in concordance with those observed in humans.

SUMMARY AND CONCLUSIONS

Under defined laboratory conditions, TCDD at low levels modulates both the immune response and host resistance to induced infection in laboratory animals. The one immunologic parameter that is reproducibly altered at dose levels in the range of 1 μg/kg and appears to segregate with the Ah locus in mice is the splenic antibody PFC response. This observation has led to a proposal to derive TEFs for PCDDs based upon this assay. Given the uncertainties of the role that the Ah receptor plays in TCDD immunotoxicity, the lack of consistent and specific effects across species, and the lack of definitive evidence that exposure to TCDD causes human immunosuppression, the reliance on TEFs to predict immunotoxic risk to humans is too uncertain and premature to be of practical use at this time.

ACKNOWLEDGMENTS

The authors wish to thank Drs. Nancy Kerkvliet and Norbert Kaminski for their comments in the preparation of this manuscript.

REFERENCES

Bombick, D., J. Jankun, K. Tullis, and F. Matsumura. 1988. 2,3,7,8-Tetrachlorodibenzo-*p*-dioxin causes increases in expression of c-*erb*-A and Levels of protein-tyrosine kinases in selected tissues of responsive mouse strains. *Proc. Natl. Acad. Sci. USA.* 85:4128–4132.

Bombick, D.W., B.V. Madhukar, D.W. Brewster, and F. Matsumura. 1985. TCDD (2,3,7,8-tetrachlorodibenzo-*p*-dioxin) causes increases in protein kinases particularly protein kinase C in the hepatic plasma membrane of the rat and the guinea pig. *Biochem. Biophys. Res. Commun.* 127:296–302.

Burleson, G.R., H. Lebrec, Y.G. Yang, J.D. Ibanes, K.N. Pennington, and L.S. Birnbaum. 1996. Effect of 2,3,7,8-tetrachlorodibenzo-*p*-dioxin (TCDD) on influenza virus host resistance in mice. *Fund. Appl. Toxicol.* 29:40–47.

Clark, D., J. Gauldie, M. Szewczuk, and G. Sweeny. 1981. Enhanced suppressor cell activity as a mechanism of immunosuppression by 2,3,7,8-TCDD. *Proc. Soc. Exp. Biol. Med.* 186:290–299.

Clark, D.A., G. Sweeney, S. Safe, E. Hancock, D.G. Kilburn, and J. Gauldie. 1983. Cellular and genetic basis for suppression of cytotoxic T-cell generation by haloaromatic hydrocarbons. *Immunopharmacol.* 6:143–153.

Davis, D. and S. Safe. 1991. Halogenated aryl hydrocarbon induced suppression of the *in vitro* plaque-forming cell response to sheep red blood cells is not dependent upon the Ah receptor. *Immunopharmacol.* 21:183–190.

De Krey, G. and N. Kerkvliet. 1995. Suppression of cytotoxic T-lymphocyte activity by 2,3,7,8-tetrachlorodibenzo-*p*-dioxin occurs *in vivo*, but not *in vitro*, and is independent of corticosterone elevation. *Toxicology*. 97:105–112.

Dooley, R. and M. Holsapple. 1988. Elucidation of cellular targets responsible for tetrachlorodibenzo-*p*-dioxin (TCDD)-induced suppression of antibody responses: The role of the B-lymphocyte. *Immunopharmacol.* 16:167–180.

Enan, E. and F. Matsumura. 1995. Evidence for a second pathway in the action mechanism of 2,3,7,8-tetrachlorodibenzo-*p*-dioxin (TCDD). *Biochem. Pharmacol.* 49:249–261.

Evans, R., K. Webb, A. Knutson, S. Rodman, D. Roberts, J. Bagby, W. Garrett, and J. Andrews. 1988. A medical follow-up of the health effects of long-term exposure to 2,3,7,8-tetrachlorodibenzo-*p*-dioxin (TCDD). *Arch. Environ. Health,* 43(4):273–278.

Finley, B.L. and D.J. Paustenbach. 1995. Human vs. Animal Sensitivity to the Immunological Effects of Dioxin: A Preliminary Analysis, presented at the 15th International Symposium on Chlorinated Dioxins and Related Compounds in Edmonton, Alberta, Canada, August 21–25.

Hanson, C. and R. Smialowicz. 1994. Evaluation of the effect of low-level 2,3,7,8-TCDD exposure on cell mediated immunity. *Toxicology*, 88:213–224.

Harper, N., K. Connor, M. Steinberg, and S. Safe. 1995. Immunosuppressive activity of polychlorinated biphenyl mixtures and congeners: Nonadditive (antagonistic) interactions. *Fund. Appl. Toxicol.* 27:131–139.

Hinsdill, R., D. Couch, and R. Speirs. 1980. Immunosuppression in mice induced by dioxin (TCDD) in feed. *J. Environ. Pathol. Toxicol.* 4:425.

Hoffman, R.E., P.A. Stehr-Green, K.B. Webb, and R.G. Evans. 1986. Health effects of long-term exposure to 2,3,7,8-tetrachlorodibenzo-*p*-dioxin. *J. Amer. Med. Assn.* 255:2031–2038.

Holsapple, M., J.A. McCay, and D.W. Barnes. 1986. Immunosuppression without liver induction by subchronic exposure to 2,7-dichlorodibenzo-*p*-dioxin in adult female B6C3F1 mice. *Toxicol. Appl. Pharmacol.* 83:445.

Holsapple, M., N. Snyder, S. Wood, and D. Morris. 1991. A review of 2,3,7,8-tetrachlorodibenzo-*p*-dioxin induced changes in immunocompetence: 1991 update. *Toxicology.* 69:219–251.

Holsapple, M. and R. Smialowicz. 1995. Immunotoxicity of Halogenated Aromatic Hydrocarbons, in *Experimental Immunotoxicology*. CRC Press, Boca Raton, pp. 257–297.

Hong, R., K. Taylor, and R. Abonour. 1989. Immune abnormalities associated with chronic TCDD exposure in rhesus. *Chemosphere.* 18(1–6):313–320.

House, R.V., D.L. Lauer, M.J. Murray, P.T. Thomas, J.P. Ehrlich, G.R. Burleson, and J.H. Dean. 1990. Examination of immune parameters and host resistance mechanisms in B6C3F1 mice following adult exposure to 2,3,7,8-tetrachlorodibenzo-*p*-dioxin. *Toxicol. Environ. Health* 31:203–215.

Kerkvliet, N. and G. Burleson. 1995. Immunotoxicity of TCDD and Related Halogenated Aromatic Hydrocarbons. In *Immunotoxicology and Immunopharmacology,* 2nd Edition. J. Dean, M. Luster, A. Munson, and I. Kimber, Eds., Raven Press, NY, pp. 97–122.

Kerkvliet, N. and J. Oughton. 1993. Acute inflammatory response to sheep red blood cell challenge in mice treated with 2,3,7,8-tetrachlorodibenzo-*p*-dioxin (TCDD): phenotypic and functional analysis of peritoneal exudate cells. *Toxicol. Appl. Pharmacol.* 119(2):248–257.

Kerkvliet, N. and J. Brauner. 1987. Mechanisms of 1,2,3,4,6,7,8-heptachlorodibenzo-*p*-dioxin (HpCDD)-induced humoral immune suppression: Evidence of primary defect in T-cell regulation. *Toxicol. Appl. Pharmacol.* 87:18–31.

Luebke, R., C. Copeland, and D. Andrews. 1995. Host resistance to *Trichinella spiralis* infection in rats exposed to 2,3,7,8-tetrachlorodibenzo-*p*-dioxin (TCDD). *Fund. Appl. Toxicol.* 24:285–289.

Luster, M., G. Boorman, J. Dean, M. Harris, R. Luebke, M. Padarathsingh, and J. Moore. 1980. Examination of bone marrow, immunologic parameters and host susceptibility following pre- and postnatal exposure to 2,3,7,8-tetrachlorodibenzo-*p*-dioxin (TCDD). *Int. J. Immunopharmacol.* 2:301–310.

Matsumura, F., D.W. Brewster, B.V. Madhukar, and D.W. Bombick. 1984. Alteration of rat hepatic plasma membrane functions by 2,3,7,8-tetrachlorodibenzo-*p*-dioxin (TCDD). *Arch. Environ. Contamin. Toxicol.* 13:509–515.

Mocarelli, P., L. Needham, A. Marocchi, D. Patterson, P. Brambilla, P.M. Gerthoux, L. Meazza, and V. Carreri. 1991. Serum concentrations of 2,3,7,8-tetrachlorodibenzo-*p*-dioxin (TCDD) and test results from selected residents of Seveso, Italy. *Toxicol. Appl. Pharmacol.* 72:169.

Morris, D., S. Jordan, and M.P. Holsapple. 1991. Effects of 2,3,7,8-TCDD on humoral immunity: I. Similarities to *Staphylococcus aureus* Cowan strain I (SAC) in the *in vitro* T-dependent antibody response. *Immunopharmacol.* 21:159–170.

Morris, D., N.K. Snyder, V. Gokani, R.E. Blair, and M.P. Holsapple. 1992. Enhanced suppression of humoral immunity in DBA/2 mice following subchronic exposure to 2,3,7,8-tetrachlorodibenzo-*p*-dioxin (TCDD). *Toxicol. Appl. Pharmacol.* 112:128–132.

Nagarkatti, P.S., G.D. Sweeney, J. Gauldie, and D.A. Clark. 1984. Sensitivity to suppression of cytotoxic T cell generation by 2,3,7,8-tetrachlorodibenzo-*p*-dioxin (TCDD) is dependent on the Ah genotype of the murine host. *Toxicol. Appl. Pharmacol.* 72:169.

Narasimhan, T., A. Craig, L. Arellano, N. Harper, L. Howie, M. Menache, L. Birnbaum, and S. Safe. 1994. Relative sensitivities of 2,3,7,8-TCDD-induced Cyp1a-1 and Cyp1a-2 gene expression and immunotoxicity in female B6C3F1 mice. *Fund. Appl. Toxicol.* 23:598–607.

Neubert, R., U. Jacob-Muller, R. Stahlmann, H. Helge, and D. Neubert. 1990. Polyhalogenated dibenzo-*p*-dioxins and dibenzofurans and the immune system. 1. Effects on peripheral lymphocyte subpopulations of a non-human primate *(Callithrix jacchus)* after treatment with 2,3,7,8-tetrachlorodibenzo-*p*-dioxin (TCDD). *Arch. Toxicol.* 64:345–359.

Neubert R., G. Golor, R. Stahlmann, H. Helge, and D. Neubert. 1992. Polyhalogenated dibenzo-*p*-dioxins and dibenzofurans and the immune system. 4. Effects of multiple-dose treatment with TCDD on peripheral lymphocyte subpopulations of a non-human primate *(Callithrix jacchus). Arch. Toxicol.* 66:250–259.

Neubert, R., L. Maskow, J. Webb, U. Jacob-Muller, A. Nogueira, I. Delagdo, H. Helge, and D. Heubert. 1993. Chlorinated dibenzo-*p*-dioxins and dibenzofurans and the human immune system. 1. Blood cell receptors in volunteers with moderately increased body burdens. *Life Sci.* 53:1995–2006.

Neubert, R., L. Maskow, I. Delgado, H. Helge, and D. Neubert. 1995. Chlorinated dibenzo-*p*-dioxins and dibenzofurans and the human immune system. 2. *In vitro* proliferation of lymphocytes from workers with quantified moderately-increased body burdens. *Life Sci.* 56(6):421–436.

Parkinson, A. 1995. Biotransformation of xenobiotics, in *Cassarett & Doull's Toxicology*. C. Klaassen, Ed. McGraw-Hill, NY, pp. 113–186.

Puga, A., D. Neubert, and F. Carrier. 1992. Dioxin induces expression of c-fos and c-jun proto-oncogenes and a large increase in transcription factor AP-1. *DNA Cell Biol.* 11:269–281.

Reggiani, G. 1978. Medical problems raised by the TCDD contamination in Seveso, Italy. *Arch. Toxicol.* 40:161.

Silkworth, J., D. Cutler, P. O'Keefe, and T. Lipinskas. 1993. Potentiation and antagonism of 2,3,7,8-tetrachlorodibenzo-*p*-dioxin effects in a complex environmental mixture. *Toxicol. Appl. Pharmacol.* 119:236–247.

Smialowicz, R., M. Riddle, W. Williams, and J. Dilberto. 1994. Effects of 2,3,7,8-TCDD on humoral immunity and lymphocyte subpopulations: differences between mice and rats. *Toxicol. Appl. Pharmacol.* 124:248–256.

Thomas, P.T., W. Busse, N. Kerkvliet, M. Luster, A. Munson, M. Murray, D. Roberts, M. Robinson, J. Silkworth, R. Sjoblad, and R. Smialowicz. 1990. Immunologic Effects of Pesticides, in *The Effects of Pesticides on Human Health. Advances in Modern Environmental Toxicology, Vol. XVIII.*

Thomas, P. and R. Faith. 1985. Adult and perinatal immunotoxicity induced by halogenated aromatic hydrocarbons, in *Immunotoxicology and Immunopharmacology.* 1st Edition, Raven Press, New York, pp. 305–313.

Tognoni, G. and A. Bonaccorsi. 1982. Epidemiological problems with TCDD (a critical view). *Drug Metab. Rev.* 13:447–469.

Tucker, A.N., S.J. Vore, and M.I. Luster. 1986. Suppression of B-cell differentiation by 2,3,7,8-tetrachlorodibenzo-*p*-dioxin (TCDD). *Mol. Pharmacol.* 29(4)372–377.

U.S. Environmental Protection Agency (EPA). March 1989. Interim procedures for estimating risks associated with exposures to mixtures of chlorinated dibenzo-*p*-dioxins and dibenzofurans (CDDs and CDFs) and 1989 Update. *Risk Assessment Forum.* EPA/625/3-89/016.

Vecchi, A., M. Sironi, M. Canegrati, M. Recchia, and S. Garattini. 1983. Immunosuppressive effects of 2,3,7,8-tetrachlorodibenzo-*p*-dioxin in strains of mice with different susceptibility to induction of aryl hydrocarbon hydroxylase. *Toxicol. Appl. Pharmacol.* 68:434–441.

Vos, J., J. Kreeftenberg, H. Engel, A. Minderhoud, and L. Van Noorle Jansen. 1987. Studies on 2,3,7,8-tetrachlorodibenzo-*p*-dioxin-induced immune suppression and decreased resistance to infection: endotoxin hypersensitivity, serum zinc concentrations and effect of thymosin treatment. *Toxicol.* 9:75–86.

Webb, K.B., R.G. Evans, A.P. Knutsen, S.T. Roodman, D.W. Roberts, W.F. Schramm, B.B. Gibson, J.S. Andrews, L.L. Needham, and D.G. Patterson. 1989. Medical evaluation of subjects with known body levels of 2,3,7,8-tetrachlorodibenzo-*p*-dioxin. *J. Toxicol. Environ. Health.* 28:183–193.

White, K.L., Jr., H.H. Lysy, and M.P. Holsapple. 1985. Immunosuppression by polycyclic aromatic hydrocarbons: a structure-activity relationship in B6C3F1 and DBA/2 mice. *Immunopharmacol.* 3:155–64.

Yang, Y., H. Lebrec, and G. Burleson. 1994. Effect of 2,3,7,8-tetrachlorodibenzo-*p*-dioxin (TCDD) on pulmonary influenza virus titer and natural killer (NK) activity in rats. *Fund. Appl. Toxicol.* 23(1):125–131.

Chapter 27

A Mechanistic Approach to Assessing Cancer Risks of Chloroform

B.E. Butterworth, M.V. Templin, D.C. Wolf, S.J. Borghoff, G.L. Kedderis, and R.B. Conolly

Many carcinogens induce cancer by reacting with the DNA to produce mutations in growth control genes. Risk assessments for such genotoxic agents are very conservative and assume a linear response even at vanishingly small doses. In contrast, many agents, such as chloroform, produce cancer by a nongenotoxic-cytotoxic mode of action. Tumors are induced secondarily to events associated with toxicity and regenerative cell proliferation (Butterworth et al., 1992, 1995). No increased risk of cancer would be expected below doses that do not produce toxicity.

Chloroform represents one of the best-studied examples of a nongenotoxic-cytotoxic carcinogen. Chloroform is a common environmental pollutant produced by some industrial processes and in trace amounts from the chlorination of drinking water. Inhalation and ingestion are the most important routes of exposure. Chloroform induces liver cancer in male and female $B_6C_3F_1$ mice when administered by gavage, kidney cancer in male Osborne-Mendel rats when given by gavage or in the drinking water, and kidney cancer in male BDF_1 mice and liver cancer in female BDF_1 mice when administered by inhalation (National Cancer Institute, 1976; Jorgenson et al., 1985; Matsushima, 1994; Yamamoto et al., 1994). No cancer was induced in either male or female F344 rats exposed to chloroform by inhalation (Matsushima, 1994; Yamamoto et al., 1994). The doses employed in the chloroform cancer studies have been criticized as being excessive and not relevant to common environmental exposures.

Chloroform has been subjected to a substantial number of genotoxicity assays. In the most complete and recent review of the literature it was concluded that neither chloroform nor its metabolites appear to directly interact with DNA or possess genotoxic activity (International Programme on Chemical Safety, 1994). Perhaps the large number of cancer studies themselves are the best indication that chloroform is not directly genotoxic. For example, when chloroform is given to $B_6C_3F_1$ mice by gavage, liver regenerative cell proliferation and eventually liver cancer are induced (National Cancer Institute, 1976, Larson et al., 1994a). When similar daily doses are given in the drinking water, neither liver cell proliferation nor cancer are observed (Jorgenson et al., 1985, Larson et al., 1994a). Complete loss of carcinogenic activity in a given target organ by simply changing the vehicle and dose rate would not be expected of a mutagenic carcinogen. It must be remembered that the mice in the drinking water study were exposed chronically to massive amounts of chloroform. Were chloroform directly mutagenic, irreversible mutations would accumulate even if they were formed at a slower rate, as is the case for mutagens and radiation. The signature of a mutagenic carcinogen is that it produces cancer in multiple tissues and species (Ashby and Tennant, 1988). Chloroform was administered in very large amounts by inhalation to male and female F-344 rats for two years with no observed tumor induction (Matsushima, 1994; Yamamoto et al., 1994). This is not the mark of a genotoxic carcinogen.

In every case of chloroform-induced cancer examined thus far, there has been a preceding cytotoxic response. For example, the $B_6C_3F_1$ mouse liver tumor response disappears if the same daily gavage dose of chloroform is given instead in the drinking water (National Cancer Institute, 1976; Jorgenson et al., 1985). Cytotoxicity and regenerative cell proliferation are seen in the $B_6C_3F_1$ mouse liver when chloroform is given by gavage but not when given in the drinking water, contributing to the body of evidence that tumor formation is secondary to these effects (Larson et al., 1994a). Dosimetry studies suggest that the reason that toxicity is not observed when chloroform is given in the drinking water is because the compound is not delivered or converted to toxic metabolites at a rate sufficient to kill hepatocytes.

Chloroform given by gavage at the doses that produced kidney cancer in male Osborne-Mendel rats also induces renal toxicity and regenerative cell proliferation in the kidney in this strain of rat (National Cancer Institute, 1976; Templin et al., 1996b). A weak tumor response in the Osborne-

Mendel male rat kidney was also seen when chloroform was given in the drinking water, but only at the highest concentration of 1800 ppm, which also resulted in a 60% decrease in water consumption and a 25% loss in body weight (Jorgenson et al., 1985). Since the highest concentration exceeded the maximum tolerated dose (MTD) and since the nutritional status of this high-dose group was uncertain, extrapolating the tumor response to lower doses is questionable. Effects secondary to water deprivation, altered nutrition, chloroform nephrotoxicity, or a combination of these factors may have also been involved in tumor formation, which complicates interpretation of these results and the ability to conduct meaningful mechanistic studies.

In a long-term study in which male and female BDF_1 mice were exposed to chloroform atmospheres of 5, 30, or 90 ppm for 5 days/wk, the only significant induction of tumors was observed at the highest concentration in the male mouse kidney (Matsushima, 1994; Yamamoto et al., 1994), which is particularly sensitive to the toxic effects of chloroform. In fact, the male BDF_1 mice in the cancer study had to have exposure concentrations increased stepwise starting at 5 ppm over a period of weeks to adapt to the 30 and 90 ppm concentrations. Exposure of male BDF_1 mice directly to atmospheres of 30 or 90 ppm of chloroform resulted in significant deaths from acute kidney failure (Templin et al., 1996a). Thus, both the 30 and 90 ppm exposures to male BDF_1 mice exceeded any reasonable criteria for a maximum tolerated dose and the data from such a study are questionable. In any case, exposure under bioassay conditions produces nephrotoxicity and regenerative cell proliferation at this target site in male but not female BDF_1 mice, consistent with the pattern of tumor formation (Templin et al., 1996a).

Genetic background is a critical consideration in interpreting rodent cancer studies. $B_6C_3F_1$ mice are a cross between C57BL6 and C3H mice. C3H mice and $B6C3F_1$ mice carry the hepatocarcinogen sensitivity (*Hcs*) genes (Drinkwater, 1994). Thus, hepatotoxic doses of chloroform produce liver cancer in the $B_6C_3F_1$ strain of mouse (National Cancer Institute, 1976; Larson et al., 1994a, 1994b). BDF_1 mice are a cross between C57BL6 and DBA mice. These parental strain mice do not carry the *Hcs* genes. Instead, male DBA and BDF_1 mice are particularly sensitive to the nephrotoxic effects of chloroform (Hill et al., 1975). While the nephrotoxic response occurs exclusively in male mice, high enough doses of chloroform will induce hepatotoxicity in both sexes. Thus, hepatotoxic doses of chloroform produce a lower incidence of liver cancer in BDF_1 mice than in $B_6C_3F_1$ mice, but produce a strong nephrotoxic and modest kidney cancer response in the male but not female BDF_1 strain. Nephrotoxicity in male mice has been attributed to higher levels of cytochrome P450 enzymes in the kidneys of male as compared to female mice. Cytochrome P450 enzyme levels in liver appear to be similar between the sexes (Smith et al., 1984). Thus, the pattern of chloroform-induced tumor formation in $B_6C_3F_1$ and BDF_1 mice is consistent with what would be expected from the genetic background of these two strains.

Chloroform administered to $B6C3F_1$ mice and F-344 rats by inhalation for up to 90 days produced concentration-dependent responses for toxicity and regenerative cell proliferation that are consistent with and predictive of the patterns of induced tumors in a recently completed two-year chloroform inhalation study (Matsushima, 1994; Yamamoto et al., 1994; Larson et al., 1996, Templin et al., 1996c). Chloroform also produces toxicity and cell proliferation in the nasal turbinates of rats, but these lesions do not progress to cancer (Méry et al., 1994; Matsushima, 1994; Yamamoto et al., 1994).

Integration of the above data with knowledge of dosimetry yields a cohesive explanation of the patterns of chloroform-induced cancer that is appropriate to each route of administration and consistent with a nongenotoxic-cytotoxic mode of action. Chloroform serves as an example for formulating appropriate risk assessments for carcinogens acting through a nongenotoxic/cytotoxic mode of action. Quantitative linking of dosimetry, cytolethality, regenerative cell proliferation, and tumor formation provide a framework for testing potential cause-and-effect relationships and for deriving appropriate risk assessments. The linearized multistage model, when applied to the mouse liver tumor data from the chloroform gavage study, estimates an increased lifetime risk of cancer of one in a million at an airborne exposure concentration of 0.000008 ppm (U.S. EPA, 1985, 1994). Assuming that chloroform-induced female mouse liver cancer is secondary to events associated with necrosis and regenerative cell proliferation, then no increases in liver cancer in female mice would be predicted at the NOEL of 10 ppm or below, based on the inhalation toxicity and cell proliferation study (Larson et al., 1996). Applying an uncertainty factor of 1,000 yields an estimate of a virtually safe airborne concentration of 0.01 ppm. This estimate relies on inhalation data and is more consistent with the mode of action of chloroform.

Thus, predictions of cancer risk from chloroform based on these mechanistic data are lower by orders of magnitude than predictions based on default risk assessment procedures. No increased cancer risk would be expected with the trace amounts of chloroform commonly found in the air or drinking water. Risk assessments are the basis of regulatory standards and should be based on the best science available.

In every case of chloroform-induced cancer examined thus far, there has been a corresponding association with a cytotoxic response. The only unanswered questions are minor and involve cancer studies of questionable validity. Critical examination of both the genotoxicity data and the patterns of toxicity leads to the conclusion that chloroform produces cancer through a nongenotoxic-cytotoxic mode of action (Butterworth et al., 1995). The experimental findings with chloroform represent an unusually rich database. If these data are insufficient to support an altered risk assessment, then what incentive is there to conduct any mechanistic research in the hope of bringing better science to the process?

REFERENCES

Ashby, J. and R.W. Tennant. 1988. Chemical structure, *Salmonella* mutagenicity and extent of carcinogenicity as indicators of genotoxic carcinogenesis among 222 chemicals tested in rodents by the U.S. NTP/NCI. *Mutat. Res.* 204:17–115.

Butterworth, B.E., J.A. Popp, R.B. Conolly, and T.L. Goldsworthy. 1992. Chemically-induced cell proliferation in carcinogenesis. In "Mechanisms of Carcinogenesis in Risk Identification," H. Vainio, P.N. Magee, D.B. McGregor, and A.J. McMichael, Eds., IARC Scientific Publication Number 116, pp. 279–305.

Butterworth, B.E., R.B. Conolly, and K.T. Morgan. 1995. A strategy for establishing mode of action of chemical carcinogens as a guide for approaches to risk assessments. *Cancer Lett.* 93:129–146.

Drinkwater, N.R. 1994. Genetic control of hepatocarcinogenesis in C3H mice. *Drug Metabol. Rev.* 26:201–208.

Hill, R.N., T.L. Clemens, D.K. Lui, and E.S. Vesell. 1975. Genetic control of chloroform toxicity in mice. *Science,* 190:159–161.

International Programme on Chemical Safety. 1994. Chloroform, Environmental Health Criteria 163, World Health Organization, Geneva.

Jorgenson, T.A., E.F. Meierhenry, C.J. Rushbrook, R.J. Bull, and M. Robinson. 1985. Carcinogenicity of chloroform in drinking water to male Osborne-Mendel rats and female B6C3F1 mice. *Fund. Appl. Toxicol.* 5:760–769.

Larson, J.L., D.C. Wolf, and B.E. Butterworth. 1994a. Induced cytotoxicity and cell proliferation in the hepatocarcinogenicity of chloroform in female $B6C3F_1$ mice: Comparison of administration by gavage in corn oil vs. ad libitum in drinking water. *Fund. Appl. Toxicol.* 22:90–102.

Larson, J.L., D.C. Wolf, and B.E. Butterworth. 1994b. Induced cytolethality and regenerative cell proliferation in the livers and kidneys of male $B6C3F_1$ mice given chloroform by gavage. *Fund. Appl. Toxicol.* 23:537–543.

Larson, J.L., M.V. Templin, D.C. Wolf, K.C. Jamison, J.R. Leininger, S. Méry, K.T. Morgan, B.A. Wong, R.B. Conolly, and B.E. Butterworth. 1996. A 90-day chloroform inhalation study in female and male $B6C3F_1$ mice: Implications for cancer risk assessment. *Fund. Appl. Toxicol.* 30:118–137.

Matsushima, T. 1994. An inhalation carcinogenesis study of chloroform. Initial Report. Japan Bioassay Laboratory, Japan Industrial Safety and Health Association, 2445 Ohshibahara Hirasawa Hando Kanagawa, 257 Japan.

Méry, S., J.L. Larson, B.E. Butterworth, D.C. Wolf, R. Harden, and K.T. Morgan. 1994. Nasal toxicity of chloroform in male F-344 rats and female $B6C3F_1$ mice following a 1-week inhalation exposure. *Toxicol. Appl. Pharmacol.* 125:214–227.

National Cancer Institute (NCI). 1976. Carcinogenesis Bioassay of Chloroform, Bethesda, MD, National Tech. Inform. Service No. PB264018/AS.

Smith, J.H., K. Maita, S. Sleight, and J.B. Hook. 1984. Effect of sex hormone status on chloroform nephrotoxicity and renal mixed function oxidases in mice. *Toxicology* 30:305–316.

Templin, M.V., K.C. Jamison, C.S. Sprankle, D.C. Wolf, B.A. Wong, and B.E. Butterworth. 1996a. Chloroform-induced cytotoxicity and regenerative cell proliferation in the kidneys and liver of BDF_1 mice. *Cancer Lett.* 108:225–231.

Templin, M.V., K.C. Jamison, D.C. Wolf, K.T. Morgan, and B.E. Butterworth. 1996b. Comparison of chloroform-induced toxicity in the kidneys, liver, and nasal passages of male Osborne-Mendel and F-344 rats. *Cancer Lett.* 104:71–78.

Templin, M.V., J.L. Larson, B.E. Butterworth, Jamison, B.A. Wong, and D.C. Wolf. 1996c. A 90-day chloroform inhalation study in F-344 rats: Profile of toxicity and relevance to cancer studies. *Fund. Appl. Toxicol.* 32:109–125

U.S. EPA. 1985. Health Assessment Document for Chloroform, EPA/600/8-84/004F.

U.S. EPA. 1994. Integrated Risk Information System (IRIS) Database, May 3.

Yamamoto, S., S. Aiso, N. Ikawa, and T. Matsushima. 1994. Japan Bioassay Laboratory, Japan Industrial Safety and Health Association, 2445 Ohshibahara Hirasawa Hando Kanagawa, 257 Japan; Carcinogenesis studies of chloroform in F344 rats and BDF_1 mice (abstract), Proceedings of the Fifty-third Annual Meeting of the Japanese Cancer Association.

Chapter 28

Environmental Endocrine Modulators and the DES Paradigm

R. Golden

INTRODUCTION

There is currently a great deal of attention directed at the hypothesis that exposures particularly *in utero*, to environmental endocrine modulators might be associated with a number of adverse human health consequences. This hypothesis, implicating many organochlorine and other synthetic compounds, has appeared numerous times over the past several years in the media and at several scientific conferences. The recent Dioxin Reassessment also suggests that dioxin (e.g., TCDD) and related compounds have adverse effects on the endocrine system (USEPA, 1994).

While there is no question that hormones are potent modulators of biochemical and physiological function, the implication that *in utero* exposures to environmental endocrine modulators have the capability to produce clinically detectable effects is much less certain. The subject of potential adverse effects following exposure to exogenous, hormonally active substances (whatever their source) in humans is extraordinarily complex. While endocrine modulators may involve any hormonal system, the principal focus of the current hypothesis is on estrogen-mediated effects. As used in this chapter, the term environmental estrogen refers to chemical substances that exhibit some degree of estrogen-like activity (i.e., estrogenic). Numerous terms have been used in this regard, including environmental endocrine disruptors, environmental endocrine modulators, environmental hormones, environmental estrogen, and xenoestrogens. However, in order to be accurate, the terms must apply to both naturally occurring (e.g., phytoestrogens) as well as to synthetic estrogenic compounds.

In general, whether considering naturally occurring exogenous estrogenic substances, such as those produced by certain plants or some synthetic organochlorine chemicals, binding to the estrogen receptor is an essential characteristic defining the ability of a compound to act as an environmental estrogen. Many chemicals share this common trait — the ability to mimic the action of estrogen through binding with the estrogen receptor. The estrogen receptor is sometimes described as being "permissive" because of the wide variety of chemical structures which can bind to it and induce an estrogen response.

HEALTH ENDPOINTS CLAIMED TO BE ASSOCIATED WITH ENVIRONMENTAL ENDOCRINE MODULATORS

A variety of specific adverse health consequences to humans have been hypothesized as possibly associated with exposure to environmental endocrine modulators (primarily environmental estrogens) including:

- Breast cancer
- Endometriosis
- Decreased sperm count
- Male and female infertility
- Undescended testicles
- Decreased penis size
- Alterations of sexual behavior
- Learning disability or delay
- Testicular cancer

- Prostate cancer
- Immune system effects
- Goiter

With respect to the idea that environmental estrogens could adversely affect humans, there is biological plausibility in this hypothesis for most of the above endpoints. Estrogens, whether exposure occurs *in utero* (i.e., prenatally) or in adulthood can have profound biochemical, physiological, and specific target organ effects on development, reproduction, behavior, and metabolism. Consequently, it is reasonable to hypothesize that exposure to estrogen-like compounds, whatever their source, could adversely affect human health. However, biological plausibility alone is not a sufficient basis for concluding that environmental estrogens have adversely affected humans.

Exogenous vs. Endogenous Hormones

A pivotal issue regarding the potential effects of exogenous estrogenic substances on human physiology concerns the natural background hormonal milieu. Whether in males or females, there are normal background levels of circulating estrogen. Estrogen (i.e.,17ß-estradiol, the main estrogen) levels in adult reproductive age women range from 50–300 pg/mL depending on the menstrual phase, while the range in adult males is from 10–60 pg/mL. While under tight physiological control, estradiol levels still fluctuate and background levels vary during puberty, pregnancy, lactation, menopause, and even during exercise. That exogenous hormones can affect the natural balance to cause effects is demonstrated by the use of supraphysiologic doses of estrogen in birth control pills to cause infertility and postmenopausal use of estrogen replacement therapy to maintain premenopausal levels. Against this natural background exposure to estrogen, exogenous estrogen must occur in amounts sufficient to overcome normal physiological regulatory and control mechanisms in order to exert an effect. A concern for certain exogenous compounds identified as having weak estrogenic activity is their relatively long half-lives and ability to bioaccumulate. Whether such compounds can bioaccumulate in humans to levels capable of causing clinically detectable effects is a critical question which must be answered.

Assessment of Biologic Plausibility — The DES Paradigm

In assessing the likelihood that exposure to environmental estrogens might be etiologically linked to any particular adverse effect, consideration of biological plausibility is paramount; i.e., does it make sense with what is known about how endogenous or exogenous estrogen cause effects? Assessment of biological plausibility can be based, in part, upon the extensive data in both experimental animals and humans on the effects of *in utero* exposure to diethylstilbestrol (DES) on adult offspring. In conjunction with these data, factors such as dose-response, potency, and the overall weight of the evidence must also be considered for individual effects.

The DES data can be used to infer potential adverse effects of *in utero* exposure to environmental estrogens; however, in the context of this discussion, DES is not considered an environmental estrogen. The effects observed in both animals and humans following *in utero* exposure to DES are consistent both with basic principles of dose-response as well as with the presence of no-effect levels. Critical reliance upon the abundant DES database provides a useful tool for assessing many of the endpoints of concern. The DES paradigm can be used to assess the following scenarios:

- If an effect hypothesized as possibly resulting from *in utero* exposure to environmental estrogen has not been reported following *in utero* exposure to DES, it may be biologically implausible that such an effect (even if real) is mediated via an estrogen mechanism of action.
- If an effect hypothesized as possibly resulting from *in utero* exposure to environmental estrogen is similar to an effect reported following *in utero* exposure to DES, the likelihood that such an effect is mechanistically linked to environmental estrogen must be evaluated based upon comparative potency and dose-response considerations.

DES as a Model Compound to Judge Potential Effects of Exogenous Estrogen

DES is a synthetic estrogenic compound which is several times more potent than 17ß-estradiol (Verdeal and Ryan, 1979; McLachlan, 1981) and hundreds to thousands of times more potent than other exogenous estrogens (not including oral contraceptives or estrogen replacement therapy), whether synthetic organochlorine compounds (e.g., certain metabolites of DDT, selected PCB congeners, aldrin, dieldrin, endrin, etc.) or naturally occurring phytoestrogens from plants. DES exerts its effects not only because it mimics estrogen by binding to the estrogen receptor to generate an estrogen gene response, but also because, as a nonsteroid, it crosses the placenta and is not metabolized by the enzyme systems designed to inactivate steroids on the fetal side of the placenta. Likewise, most compounds identified as environmental estrogens are also nonsteroids and would likely behave similarly.

The large difference in potency between DES and environmental estrogens must be considered when inferring potential effects from *in utero* exposure to environmental estrogens. Based on substantial differences in estrogenic potency, if a particular effect has not been reported in conjunction with *in utero* exposure to doses of DES which are not sufficient to produce that effect, the likelihood that exposure to environmental estrogens, from whatever source, would be sufficient to produce such an effect must be assessed.

Effects of DES in Animals

DES has been studied extensively in a variety of animal species and the effects of *in utero* exposure have been well characterized. Numerous studies in mice demonstrate that *in utero* exposure to sufficient doses of DES during the critical period of reproductive tract organogenesis produces a broad spectrum of adverse effects in male and female offspring, including various reproductive tract abnormalities and cancers of the uterus, vagina, and prostate (Newbold and Bullock, 1990; Newbold et al., 1985a,b; McLachlan et al., 1975; Newbold 1993, 1995). Table 28.1 summarizes these effects.

The experimental mouse model with DES has served to predict a number of adverse outcomes in humans (Newbold, 1993, 1995). Many, but not all, of the effects of DES in animals have been reported in humans. In the mouse model (as well as for human exposures), besides the total administered dose, the most important issue for DES-induced effects is the timing of the dose. In studies with mice, DES is typically administered on days 9 through 16 of gestation, the period of major reproductive tract development. The concept of a critical window of exposure is an important factor in both animal studies as well as in human exposure situations. If, for example, DES is administered after the reproductive tract has formed, the likelihood that adverse developmental effects would occur is reduced.

Effects of DES in Humans

DES was widely prescribed to between 2 and 4 million pregnant women from the late 1940s until 1971 in the belief (now known to be mistaken) that it was useful in preventing miscarriage (Stillman, 1982). The use of DES was discontinued in the United States in 1972 with the discovery that a small number of women exposed *in utero* later developed vaginal clear cell adenocarcinoma (Herbst et al., 1972). It has subsequently been determined that *in utero* exposure to DES is associated with numerous other adverse effects on the reproductive tracts of both males and females. In males, these effects include epididymal cysts, microphallus, cryptorchidism, testicular abnormalities, decreased sperm count, and increased incidence of abnormal sperm (Stillman, 1982; Bibbo and Gill, 1977; Whitehead and Leiter, 1981; Henderson et al., 1976, Gill et al., 1977, 1979). In females, adverse effects include adenosis, clear cell adenocarcinoma, structural defects of the cervix, vagina, uterus, and fallopian tubes, infertility, and irregular menstrual cycles (Stillman, 1982; Herbst et al., 1972, 1975, 1981; Bibbo and Gill, 1977).

Because of the widespread clinical use of DES, there is substantial dose-response data on the effects in both male and female offspring exposed *in utero*. During the peak years of DES use, in

Table 28.1. Effects in Male and Female Mice Following *In Utero* Exposure to DES.

Effects in Male Offspring	Effects in Female Offspring
Sperm abnormalities	Ovarian cysts
Decreased sperm counts	Ovarian tumors
Epididymal cysts	Structural abnormalities of uterus
Cryptorchid testes	Malformations of cervical canal
Microphallus	Vaginal adenocarcinoma
Hypospadias	Vaginal adenosis
Prostatic inflammation	Salpingitis of the oviduct
Prostatic tumors	Uterine tumors
Infertility	Infertility
Immune dysfunction	Immune dysfunction

different medical centers, different DES dosing regimens were used. This resulted in large groups of women receiving substantially different total DES doses during pregnancy. Such data permits rough approximations of the total maternal dose of DES required to produce some adverse effects on offspring exposed *in utero*. In both sexes, *in utero* exposure to sufficient maternal doses of DES has resulted in a number of adverse effects, including anatomic abnormalities of the reproductive system.

However, there are also clinical studies which identify no effect [total] maternal dose levels of DES. The study by Leary et al. (1984) demonstrates the existence of an apparent dose threshold for certain effects attributable to *in utero* DES exposure. This study was designed to determine if males exposed *in utero* to DES had a higher frequency of urogenital abnormalities than an unexposed control group. Comparison of 265 DES-exposed males with 274 controls revealed that *in utero* exposure to DES did not increase the risk of any urogenital abnormalities, including penile length or diameter, testis length or width, epididymal cysts, or varicocele. There were also no adverse effects on sperm volume, density, motility, morphology, or fertility. In this study, the median total administered maternal dose of DES was 720 mg, with a median duration of exposure of 101 days; the timing of DES exposure was not provided, although it was usually administered during the first trimester of pregnancy. While it is possible that more subtle effects might have been present, only the effects noted above were studied.

Estrogenic Activity of Selected Compounds

A number of organochlorine compounds have been tested for estrogenic (e.g., uterotropic) activity. The data are extensive and only a brief overview is presented. For example, Gellert (1978) tested a series of PCB Aroclors (1221, 1242, 1254, and 1260) and reported that only Aroclor 1221 had *in vivo* uterotrophic activity, roughly one millionth the estrogenic activity of 17ß-estradiol. In contrast, Jansen et al. (1993) reported that Aroclor 1242 exhibited estrogenic activity as measured by increased uterotrophic activity at doses between 10,000 and 100,000 times higher than 17ß-estradiol. Other studies with commercial PCB mixtures show no estrogenic activity in breast cancer cells (Krishnan and Safe, 1993).

For DDT and related metabolites, *in vitro* test results demonstrate that both *o,p′*-DDT and technical grade DDT have estrogenic activity approximately one million times less than 17ß-estradiol (Soto et al., 1994). *In vivo* studies with DDT and some of its structural homologues also demonstrate dose thresholds for estrogenic effects. Gellert et al. (1974) tested the ability of graded doses of *o,p'*-DDT (10, 50, 100, and 500 g/day) to produce estrogenic effects in young rats after administration for 27 days. Only a dose of 500 g/day produced significant effects. Numerous other studies attest to the *in vivo* estrogenic potential of DDT and related homologues (Heinrichs et al., 1971; Gellert and Heinrichs, 1975; Duby et al., 1971). However, when graded doses are tested, there are no-effect levels for estrogenic effects. While it is difficult to translate short-term *in vivo* dosing results to potential effects from chronic exposure, the issue of dose-response and no-effect levels is important.

A similar situation exists for other weakly estrogenic organochlorine compounds, although many have not been adequately tested *in vivo*. It is important to note that *in vitro* results may not necessarily be indicative of *in vivo* activity. For example, in an *in vitro* assay, dieldrin was reported to have approximately one-millionth the activity of 17ß-estradiol (Soto et al., 1994). However, in an *in vivo* assay dieldrin (in addition to aldrin and mirex) was devoid of any activity (Gellert et al., 1978). A dose of dieldrin one million times larger than 17ß-estradiol failed to produce a uterotropic response. While the differences between *in vivo* and *in vitro* activity may appear slight, such empirical data demonstrate that uncritical reliance upon convenient *in vitro* assays may not always provide the most accurate assessment of potential estrogenic activity. Accurate *in vivo* and *in vitro* testing of compounds (both singly and as mixtures) with suspected estrogenic activity to characterize potency and dose-response is a critical research priority.

Comparative Potency: DES and Environmental Estrogens

Based on the study by Leary et al. (1984), it appears that a total median maternal dose of 720 mg DES represents an approximate no-effect exposure level for *in utero* DES-induced effects on the male reproductive tract. From this approximate no-observed-effect-level (NOEL) a rough potency comparison can be made with compounds identified as environmental estrogens. For example, in an *in vitro* test with human breast cancer cells, Soto et al. (1992) reported that technical grade DDT has an estrogenic potency 0.0001 of 17ß-estradiol, while DES has 10 times the potency of 17ß-estradiol. For the purpose of this potency comparison, DES is assumed to be only 1000 times more potent than DDT. Therefore, if a total dose of 720 mg of DES did not produce an effect, a similar no-effect total estrogenic dose of DDT would be:

$$720 \text{ mg} \times 1 \times 10^3 = 7.2 \times 10^5 \text{ mg or } 7.2 \times 10^2 \text{ g}$$

$$7.2 \times 10^2 \text{ g} \times 454 \text{ g/lb} = 1.6 \text{ lb}$$

This rough calculation is not meant to imply that a DDT "dose" of 1.6 lb would be without any toxicity. Rather, it is only intended to illustrate that, based on comparative estrogenic potency, a total maternal intake of 1.6 lb of DDT (or any other estrogenic compound or mixture of compounds with similar potency) over a period of 101 days would not be sufficient (based on an estrogenic mechanism of action) to produce urogenital abnormalities in adult male offspring exposed *in utero*.

The above scenario is based upon the total DES dosage administered during pregnancy. However, as previously noted, adverse effects resulting from *in utero* exposure to DES (or other estrogenic substances) may be a consequence of exposure during a critical window of sensitivity during development. If this is correct, a comparison based upon a single daily exposure may be more appropriate in judging the likelihood that an *in utero* exposure might result in an adverse effect in the offspring. Since Leary et al. (1984) report a median duration of exposure of 101 days, the hypothetical daily equivalent "dose" of DDT would be 1.6 lb/101 or 0.016 lb. This rough calculation suggests that, based on comparative potency, a single maternal dose of 0.016 lb of DDT (or any other estrogenic compound or mixture of compounds with similar potency) during a window of sensitivity would not be sufficient to produce urogenital abnormalities (based on an estrogenic mechanism of action) in adult male offspring exposed *in utero*.

The above approximations demonstrate the apparent dosages of exogenous estrogen that would be required to produce effects following *in utero* exposure based on a potency comparison with DES. In light of this comparison, the following additional data are of interest. Soto et al. (1992, 1994) reported that DDT, *o,p′*-DDT, *p,p′*-DDT, *o,p′*-DDD, endosulfan, kepone, dieldrin, and toxaphene all have estrogenic potencies approximately 1/10,000,000 that of DES in an *in vitro* test system. In order to demonstrate that dietary intakes or body burdens of these compounds have a collective biological activity capable of causing endocrine-related effects following *in utero* exposure, it will be necessary to determine how a mixture of weak estrogens would somehow produce effects while significant maternal DES doses are devoid of effects. It will also be important to consider the ability of some of these compounds to bioaccumulate in adipose tissue and the resulting serum levels that are available to cause effects. Additional research will also be required to determine the extent to which such compounds are able to cross the placenta and be available in a bioactive state to the fetus.

No-Effect Levels for DES-Induced Effects in Animals

Numerous animal studies with DES have demonstrated a variety of adverse effects in both male and female offspring following *in utero* exposure. Prenatal exposure of male rats or mice to DES has produced testicular abnormalities, cryptorchidism, microphallus, and azoospermia, while similar exposure of female rats or mice has produced vaginal adenosis, adenocarcinoma, squamous cell carcinoma, hyperkeratosis, endometrial hyperplasia, and infertility (Miller, 1983). The various studies in which these effects were noted also typically demonstrate the existence of no-effect levels for *in utero* DES-induced effects. For example, in studies designed to investigate the effects of *in utero* exposure to DES on the structure and function of the male genital tract, McLachlan (1981) administered subcutaneous DES doses of 0.01, 1.0, 2.5, 5, 10, or 100 g/kg to mice on days 9 through 16 of gestation and concluded that, in the mouse model, only the highest dose was associated with noticeable changes in the reproductive tract. Infertility was only observed at the two highest DES doses.

In a study by Green et al. (1940) the dose-response nature of prenatal exposure to estrogen was explored in considerable detail. Graded doses of different estrogens, including DES, were administered during gestation. The results show clear no-effect levels for effects on both the external and internal genitalia of male rats.

Coupled with the potency of DES, the no-effect doses demonstrated in both animal and human studies appear to be extraordinary in comparison with our current understanding of typical exposures (i.e., doses) and potencies of environmental estrogenic substances either singly or in mixtures. Dose-response data suggest that no-effect levels must be considered when assessing the scientific validity of claimed effects from exposure to environmental estrogens. Other potentially important factors (e.g., bioaccumulation, free vs. bound estrogen, anti-estrogens, additivity and synergism, net estrogenic activity, etc.) will undoubtedly be involved in determining the ultimate probability of adverse effects. However, the mere presence of weakly estrogenic substances in environmental media is not a sufficient basis to conclude they can produce adverse effects.

SOME EFFECTS DUE TO EXOGENOUS ESTROGEN THAT MAY RESULT FROM PRENATAL EXPOSURE

Certain adverse effects in offspring are likely only if exposure to exogenous estrogen (e.g., DES) occurs *in utero*. These include undescended testicles (cryptorchidism), decreased penis size, and other structural abnormalities of the reproductive tract in either males or females. Other adverse effects may be influenced by *in utero* exposure to exogenous estrogen, including breast cancer, testicular cancer, immune effects, thyroid effects, decreased sperm quality, and infertility. Some examples are reviewed below to demonstrate how the DES paradigm can be used to help judge whether some of the effects attributed to environmental estrogens are likely to result from such exposures.

Effects of DES on Undescended Testicles (Cryptorchidism)

Animal studies demonstrate that prenatal exposure to DES and other estrogens can produce cryptorchidism (McLachlan et al., 1975). However, the situation in humans is much less certain. Overall, the data are not generally supportive of an association between maternal estrogen exposure and increased incidence of cryptorchidism in the offspring.

Gill et al. (1979) reported an increased incidence of cryptorchidism in 308 men following maternal use of DES. In this study, women received the first dose of DES starting at the seventh week of gestation, well before the formation of the reproductive tract organs. In reviewing this study, Beard et al. (1984), observed that the incidence rate of cryptorchidism in the comparison control group in the study by Gill et al. (1979) was probably too low, likely resulting in a spurious increased incidence of cryptorchidism in the exposed group. In contrast, Beard et al. (1984) studied the incidence rate of cryptorchidism in over 14,000 male births. While acknowledging that animal studies demonstrate a cause and effect relationship between prenatal exposure to DES and cryptorchidism, Beard et al. (1984) note that:

> ...evidence that estrogen and/or progestin exposure causes cryptorchidism in humans is less impressive.... Among the males exposed *in utero* to diethylstilbestrol who were recently studied at the Mayo Clinic, there was no difference in the number of cryptorchid individuals among those exposed and those not exposed to diethylstilbestrol.

Beard et al. (1984) concluded:

> Multivariate analysis also provided no evidence to suggest that *in utero* estrogen exposure is associated with cryptorchidism in male offspring.

The study by Beard et al. (1984) is also significant for another reason. This study (similar to that of Leary et al., reviewed below) also involved cases born to mothers who were given DES according to the Mayo Clinic protocol. The total median dose of DES administered to mothers at the Mayo Clinic was less than at other medical facilities. As reported by Leary et al. (1984), the total administered median DES dose at the Mayo Clinic was 720 mg. While there are other clinical studies in which adverse effects in both male and female offspring following *in utero* exposure to DES have been reported, total maternal doses of DES in these studies were generally much larger than 720 mg. For example, Bibbo and Gill (1977), and Gill et al. (1977) at the University of Chicago studied 168 male and 229 female DES-exposed offspring, all of whose mothers received a total mean DES dose of approximately 11,600 mg. In both of these studies, some adverse effects (including cryptorchidism) were reported in both male and female offspring while Leary et al. (1984) reported no adverse effects in males exposed *in utero* to DES.

DES and Decreased Sperm Quality

It is biologically plausible that *in utero* exposure to estrogen can affect sperm quality. In studies with mice, *in utero* exposure to DES is associated with decreased sperm production and abnormal morphology in adult offspring (McLachlan et al., 1975; McLachlan, 1981) There are a number of studies in adult men prenatally exposed to DES. However, with respect to decreased sperm quality, the results of these studies are mixed, and appear to be a function of the maternal dose of DES. Gill et al. (1979) report decreased sperm count (115 million/mL in controls vs. 91 million/mL in DES-exposed) and abnormal morphology in 18% of DES-exposed men compared to 8% in controls. On the other hand, Leary et al. (1984) examined 110 DES-exposed men and reported no differences between exposed and controls in sperm count, motility, or abnormal morphology. In discussing the differences between their results and previous studies, Leary et al. (1984) note:

> In contrast to other reported studies, ours failed to detect increased risks of anomalies among diethylstilbestrol-exposed men.... Our results, however, typically were not even suggestive of an increased risk among the exposed.... Indeed, when statistical significance was found, it usually was in the direction opposite to adverse diethylstilbestrol effects.

With respect to the increased risks for reproductive parameters reported by others, Leary et al. (1984) observed:

> ...the dose levels used at the Mayo Clinic [where this study took place] were lower than those used by the University of Chicago cohorts as reported by Gill et al.

Thus, while clinical data suggest that at some maternal dose level, *in utero* exposure to DES can result in decreased sperm quality, the existence of an apparent no-effect level for such effects must be considered. The failure of the maternal DES dosages in the Leary et al. (1984) study to cause any effects on sperm quality raises a question of how environmental estrogens (net exposure or body burdens), which are hundreds to thousands of times less potent than DES, would be capable of causing a decline in sperm quality if exposure occurs *in utero*.

DES and Male Fertility

In a small study of men exposed *in utero* to DES, the results of an *in vitro* sperm penetration assay suggested an apparent decrease in fertilization potential (Stenchever et al., 1981). However, in a larger study using the same *in vitro* assay, the same authors were unable to confirm their previous findings (Shy et al., 1984). The largest fertility study of adult men exposed *in utero* to DES has recently been published (Wilcox et al., 1995). In this study, the 253 men with *in utero* exposure to DES had three times more self-reported malformations of the genitalia than placebo-exposed controls. However, despite the increased incidence of genital malformations, the fertility of the DES-exposed men was identical to the fertility of the control group. Furthermore, there were no differences between DES-exposed men and controls with respect to sexual function as measured by frequency of intercourse or reported episodes of decreased libido. Wilcox et al. (1995) concluded:

> High doses of diethylstilbestrol did not lead to impairment of fertility or sexual function in adult men who had been exposed to the drug *in utero*.

The study by Wilcox et al. (1995) is also noteworthy for another reason. The cohort of men in this study was from the University of Chicago. The cohort studied by Wilcox et al. (1995) was, in part, the same cohort studied by Gill et al. (1979) who reported decreased sperm count and abnormal sperm morphology in DES-exposed men. Even assuming that the sperm counts in the men studied by Wilcox et al. (1995) were reduced by *in utero* exposure to DES, the reduction was not sufficient to adversely affect fertility. Wilcox et al. (1995) reported that 95% of the women in the treatment group received between 11,500 mg and 12,600 mg of DES during pregnancy. Thus, *in utero* exposure of male fetuses to a total dose of more than 10,000 mg DES, which was associated with an increased incidence of reproductive tract abnormalities (reported by Gill et al. and confirmed by Wilcox et al.) had no effect on subsequent male fertility. The failure of these maternal DES doses to cause any effects on male fertility raises a question of how environmental estrogens (net exposure or body burdens), which are hundreds to thousands of times less potent than DES, would be capable of causing adverse effects on fertility if exposure occurs *in utero*.

DES and Female Fertility

There are a number of studies on women exposed *in utero* to DES, and the results from these studies, like those on sperm quality, appear to be related to the total maternal DES dose. For example, in response to a report by Bibbo and Gill (1977) which reported statistically significant menstrual irregularities in women exposed to DES *in utero*, Barnes (1979) investigated the menstrual histories of 218 patients exposed to DES *in utero* compared to 158 control subjects. This study found no significant differences in the complaint of menstrual irregularities at either the initial or follow-up examination. Age at menarche was also the same in both groups. In comparing the results of her study with the study by Bibbo and Gill (1977) Barnes (1979) notes:

> The women in the study by Bibbo et al. received the full DES schedule. The mothers in our study...did not always start as early as 6 weeks or complete the schedule through the 36th week....patients in the Boston clinical center were exposed to a higher dose than patients in the other centers. Exposure to DES before the 18th week of pregnancy and the administration of more than 2300 mg is related to a higher frequency of vaginal epithelial changes.

The possible relationship between reproductive tract abnormalities subsequent to *in utero* exposure to DES and fertility problems was studied by Berger and Alper (1986). In this study, 44% of controls achieved pregnancy, while only 4% of the DES-exposed women became pregnant. Of particular relevance to the hypothesis that infertility in DES-exposed women is related to various structural abnormalities of the reproductive tract is the fact that 96% of the infertile DES-exposed women in this study had structural uterine abnormalities. This finding is also supported by Berger and Goldstein (1980), who reported that all 25 DES-exposed women in their study with a history of infertility or recurrent abortion had uterine abnormalities.

Senekjian et al. (1988) studied the fertility of women exposed *in utero* to DES and also identified risk factors influencing fertility. The total mean DES doses for exposed women with either primary or secondary infertility were 12,241 mg and 12,189 mg, respectively. There were significantly more reproductive tract abnormalities (e.g., cervicovaginal ridges and tubal defects) in DES-exposed women compared to controls. For primary infertility, a significantly greater number of DES-exposed women (46%) failed to conceive when compared to controls (14%). In women with primary infertility, 46% of DES-exposed had abnormal hysterosalpingograms, while there were none in the unexposed group; tubal abnormalities were found in 42% of DES-exposed women and in none of the unexposed women. The authors noted that:

> ...structural abnormalities of the cervix and vagina present at the initial examination may be associated with the subsequent occurrence of primary infertility.

These studies are consistent with a dose-response hypothesis, suggesting that unless *in utero* exposure to total maternal DES doses was sufficient to produce reproductive tract abnormalities in the female offspring, there is a decreased likelihood of fertility problems. The implications of these data to possible effects on female fertility resulting from *in utero* exposure to environmental estrogens are twofold. First, unless *in utero* exposure to estrogenic compounds is sufficient to produce reproductive tract abnormalities, there appears to be little likelihood that such exposures would result in subsequent fertility problems. Second, because of the recognized potency differences between DES and estrogenic compounds in the environment, the total dose (or the dose during a critical window of sensitivity) required of a mixture of exogenous environmental estrogen is unlikely to be sufficient to produce effects similar to those documented for *in utero* exposure to DES.

DES and Testicular Cancer

The biological plausibility of the hypothesis that exposure to exogenous estrogen is involved in the etiology of testicular cancer is not supported by the DES data. If *in utero* exposure to estrogen is a significant risk factor for testicular cancer, studies of males prenatally exposed to DES should demonstrate such an association. However, with respect to this issue, the conclusions in the most recent SEER report (NCI, 1993) are germane:

> *In utero* DES exposure, however, has not been linked to testicular cancer.

This conclusion is based on a review of several extensive studies in which the incidence of testicular cancer in sons exposed *in utero* as a result of maternal dosing with DES was investigated (Greshman and Stolley, 1988; Moss et al., 1986; and Depue et al., 1983). While the results of the available studies are not in complete agreement, the SEER report concluded that the weight of the evidence was that, to date, prenatal exposure to exogenous estrogen is not a risk factor for testicular cancer in offspring.

Testicular cancer is somewhat different than many other cancers in that the peak incidence of occurrence is in the fourth decade of life (NCI, 1993). This means that the collective studies of men exposed to DES *in utero* are likely to be a sufficient basis for demonstrating an effect, if such an effect were present. Despite the fact that the incidence of testicular cancer has been increasing, the lack of an association with *in utero* exposure to DES raises a question of how *in utero* exposure to environmental estrogens (net exposure or body burdens), which are much less potent than DES, would be etiologically associated with testicular cancer. However, to the extent that postnatal exposure to exogenous estrogen might be etiologically associated with testicular cancer, the DES data are not relevant for assessing potential effects resulting from exposures occurring postnatally.

CONCLUSIONS

The DES paradigm offers a useful clinical tool to assist in judging the biological plausibility that certain effects hypothesized as possibly associated with *in utero* exposure to environmental estrogens might actually result from such exposures. This paradigm, which is based on sound principles of dose-response and abundant animal and human data, permits scientific judgement to be used to determine if certain effects, hypothesized as associated with *in utero* exposure to environmental estrogen, are mechanistically linked to such exposures.

DES is several times more potent than 17ß-estradiol (i.e., estrogen) and hundreds to thousands of times more potent than any environmental estrogens tested to date. There are no-effect dose levels for DES-induced *in utero* effects in animals and humans. The existence of no-effect levels for a potent estrogen like DES administered during a critical period of gestation raises legitimate questions concerning the biological plausibility that typical exposures to trace levels of environmental estrogens would be capable of causing adverse effects on offspring exposed *in utero*. Potency differences and the threshold issue must be taken into account in any scientific assessment of potential effects from *in utero* exposure to environmental estrogens. Based on the DES data, for certain effects, it appears unlikely that *in utero* exposure to environmental estrogens, from whatever source, would be sufficient to produce similar effects. Additional research is required to explore both the quantitative and qualitative relationships between DES and compounds identified as environmental estrogens.

REFERENCES

Barnes, A.B. 1979. Menstrual history of young women exposed *in utero* to diethylstilbestrol. *Fertil. Steril.* 32:148–53.

Beard, M., I.J. Melton, W.M. O'Fallon, K.L. Noller, and R.C. Benson. 1984. Cryptorchism and maternal estrogen exposure. *Amer. J. Epidemiol.* 120(5):707–16.

Berger, M.J. and M.M. Alper. 1986. Intractable primary infertility in women exposed to diethylstilbestrol *in utero*. *Reprod. Med.* 31(4):231–5.

Berger, M.J. and D.P. Goldstein. 1980. Impaired reproductive performance in DES-exposed women. *Obstet. Gynecol.* 55:25.

Bibbo, M. and W.B. Gill. 1977. Follow-up study of male and female offspring of DES-exposed mothers. *Obstet. Gynecol.* 49:1–8.

Depue, R.H., M.C. Pike, and B.E. Henderson. 1983. Estrogen exposure during gestation and risk of testicular cancer. *J. Nat. Cancer Inst.* 71:1151–55.

Duby, R.T., H.F. Travis, and C.E. Terrill. 1971. Uterotrophic Activity of DDT in Rats and Mink and its Influence on Reproduction in the Rat. *Toxicol. Appl. Pharmacol.* 18:348–355.

Gellert, R.J. 1978. Uterotrophic activity of polychlorinated biphenyls and induction of precocious reproductive aging in neonatally treated female rats. *Environ. Res.* 16:123–130.

Gellert, R.J. and W.L. Heinrichs. 1975. Effects of DDT homologs administered to female rats during the perinatal period. *Biol. Neonate.* 26:283–290.

Gellert, R.J., W.L. Hendricks, and R. Swerdloff. 1974. Effects of neonatally-administered DDT homologs on reproductive function in male and female rats. *Neuroendocrinol.* 16:84–89.

Gellert, R.J., W.L. Heinrichs, and R. Swerdloff. 1978. Kepone, Mirex, Dieldrin and Aldrin: Estrogenic activity and the induction of persistent vaginal estrus and ovulation in rats following neonatal treatment. *Environ. Res.* 16:131–138.

Gill, W.B., G.F.B. Schumacher, and M. Bibbo. 1977. Pathological semen and anatomical abnormalities of the genital tract in human male subjects exposed to diethylstilbestrol *in utero*. *J. Urol.* 477–480.

Gill, W.B., G.F.B. Schumacher, and M. Bibbo. 1979. Association of diethylstilbestrol exposure *in utero* with cryptorchidism, testicular hypoplasia and semen abnormalities. *J. Urol.* 122:36–39.

Green, R.R., M.W. Burrill, and A.C. Ivy. 1940. Experimental intersexuality. The effects of estrogens on the antenatal sexual development of the rat. *Amer. J. Anat.* 67:305–45.

Greshman, S.T. and P.D. Stolley. 1988. A case-control study of testicular cancer using Connecticut tumour registry data. *Int. J. Epid.* 17:738–42.

Heinrichs, W.L. and R.J. Gellert. 1971. DDT administered to neonatal rats induces persistent estrus syndrome. *Science.* 173:642–643.

Henderson, B.E., B. Benton, and M. Cosgrove. 1976. Urogenital tract abnormalities in sons of women treated with diethylstilbestrol. *Pediatrics.* 58:505–7.

Herbst, A.L. and M.M. Hubby. 1981. Reproductive and gynecologic surgical experience in diethylstilbestrol-exposed daughters. *Am. J. Obstet. Gynecol.* 141(8):1019–28.

Herbst, A.L., D.C. Poskanzer, S.J. Robboy, I. Friedlander, and R.E. Scully. 1975. Prenatal exposure to stilbestrol: A prospective comparison of exposed offspring with unexposed controls. *N. Engl. J. Med.* 292:334–339.

Herbst, A.L., R.J. Kurman, and R.E. Scully. 1972. Clear cell adenocarcinoma of the genital tract in young females: Registry report. *N. Eng. J. Med.* 287:1259.

Jansen, H.T., P.S. Cooke, J. Porcelli, T.C. Liu, and L.G. Hansen. 1993. Estrogenic and antiestrogenic actions of PCBs in the female rat: In vitro and in vivo studies. *Reproductive Toxicol.* 7:237–248.

Krishnan, V. and S. Safe. 1993. Polychlorinated biphenyls (PCBs), dibenzo-*p*-dioxins (PCDDs), and dibenzofurans (PCDFs) as antiestrogens in MCF-7 human breast cancer cells: quantitative structure-activity relationships. *Toxicol. Appl. Pharmacol.* 120(1):55–61.

Leary, F.J., L.J. Resseguie, L.T. Kurland, P.C. O'Brien, R.F. Emslander, and K.L. Noller. 1984. Males exposed in utero to diethylstilbestrol. *J. Amer. Med. Assoc.* 252(21):2984–9.

McLachlan, J.A. 1981. Rodent models for perinatal exposure to diethylstilbestrol and their relation to human disease in the male. In: *Developmental Effects of Diethylstilbestrol in Pregnancy.* A.L. Herbst and H.A. Bern, Eds. New York, Thieme-Stratton, Inc., pp. 148–157.

McLachlan, J., R. Newbold, and B. Bullock. 1975. Reproductive tract lesions in male mice exposed prenatally to diethylstilbestrol. *Science.* 190:991–992.

Miller, R.K. 1983. Perinatal toxicology: Its recognition and fundamentals. *Am. J. Ind. Med.* 4(1-2):205–44.

Moss, A.R., D. Osmond, P. Bacchetti, F.M. Torti, and V. Gurgin. 1986. Hormonal risk factors in testicular cancer. *Amer. J. Epid.* 124:39–52.

National Cancer Institute (NCI). 1993. SEER. *Cancer Statistics Review 1973-1990.* Miller et al., Eds., U.S. Dept. Health and Human Services, Public Health Service, NIH/NCI, Pub. No. 93-2789.

Newbold, R.R. 1995. Cellular and molecular effects of developmental exposure to diethylstilbestrol: Implications for other environmental estrogens. *Environ. Health Perspect.* 103(Supp. 107):83–87.

Newbold, R.R. 1993. Gender-related behavior in women exposed prenatally to diethylstilbestrol. *Environ. Health Perspect.* 101(3):208–13.

Newbold, R.R., B.C. Bullock, and J.A. McLachlan. 1985a. Lesions of the rete testis in mice exposed prenatally to diethylstilbestrol. *Cancer Res.* 45(10):5145–50.

Newbold, R.R., B.C. Bullock, and J.A. McLachlan. 1985b. Progressive proliferative changes in the oviduct of mice following developmental exposure to diethylstilbestrol. *Teratogenesis Carcin. Mutagen.* 5(6):473–80.

Newbold, R.R. and B.C. Bullock. 1990. Uterine adenocarcinoma in mice following developmental treatment with estrogens: a model for hormonal carcinogenesis. *Cancer Res.* 50(23):7677–81.

Senekjian, E.K., R.K. Potkul, K. Frey, and L. Herbst. 1988. Infertility among daughters either exposed or not exposed to diethylstilbestrol. *Am. J. Obstet. Gynecol.* 158:493–98.

Shy, K.K., M.A. Stenchever, L.E. Karp, R.E. Berger, R.A. Williamson, and J. Leonard. 1984. Genital tract examination and zona-free hamster egg penetration tests from men exposed in utero to diethylstilbestrol. *Fertil. Steril.* 42:772–78.

Soto, A.M., T. Lin, H. Justicia, R.M. Silvia, and C. Sonnenschein. 1992. An "in culture" bioassay to assess the estrogenicity of xenobiotics (E-screen). In: *Chemically-Induced Alterations in Sexual and Functional Development: The Human/Wildlife Connection.* Princeton Scientific Publishing Co., Inc.

Soto, A.M., K.L. Chung, and C. Sonnenshein. 1994. The pesticides endosulfan, toxaphene, and dieldrin have estrogenic effects on human estrogen-sensitive cells. *Environ. Health Perspect.* 102(4):337–412.

Stenchever, M.A., R.A. Williamson, and J. Leonard. 1981. Possible relationship between in utero diethylstilbestrol exposure and male fertility. *Am. J. Obstet. Gynecol.* 140(2):186–93.

Stillman, R.J. 1982. *In utero* exposure to diethylstilbestrol: Adverse effects on the reproductive tract and reproductive performance in male and female offspring. *Am. J. Obstet. Gynecol.* 142:905–921.

U.S. Environmental Protection Agency (EPA). 1994. *Health Assessment Document for 2,3,7,8-Tetrachlorodibenzo-p-dioxin (TCDD) and Related Compounds, Volume III.* Office of Research and Development. EPA/600/BP-92/001c.

Verdeal, K. and D.S. Ryan. 1979. Naturally-occurring estrogens in plant foodstuffs — a review. *J. Food Protection* 42:577–83.

Whitehead, E.D and E. Leiter. 1981. Genital abnormalities and abnormal semen analysis in male patients exposed to diethylstilbestrol *in utero. J. Urol.* 125(1):47–50.

Wilcox, A.J., D.D. Baird, C.R. Weinberg, P.P. Hornsby, and A.L. Herbst. 1995. Fertility in men exposed prenatally to diethylstilbestrol. *N. Engl. J. Med.* 332:1411–16.

Chapter 29

EPA's Dioxin Reassessment: A Bad Example for Chlorine Compounds

T.B. Starr and A. Poland

On 13 September 1994, the U.S. EPA officially released draft documents describing its reassessment of potential human health risks arising from exposure to dioxin and related compounds that bid to the aryl hydrocarbon (Ah) receptor. The most notable of the agency's conclusions is that impacts on human health, including enzyme induction and developmental, reproductive, endocrine, and carcinogenic effects, may be occurring at or within an order of magnitude of current background body burdens of these compounds. While much has been learned in recent years about the toxicity of dioxin-like compounds, there are still major gaps in understanding that can be filled only by additional careful scientific study. The EPA's new risk characterization is thus heavily dependent upon many unproved assumptions and untested hypotheses, including proportionality of responses to fractional receptor occupancy and simple additivity of congener-specific effects in complex mixtures. As is discussed in more detail below, these and other questionable assumptions upon which the reassessment's conclusions are based must be clearly distinguished from hard scientific facts. Otherwise, the press and public will surely misinterpret the reassessment's hypothetical risks as real.

Mechanism of action and implications for low-dose risk exposure to dioxin-like compounds clearly cause a variety of toxicologic effects in laboratory animals, and scientists generally agree that most of these effects are mediated through the Ah receptor. However, the exposures estimated to give rise to human background body burdens are far smaller than those known to cause toxicity. The pivotal issue is thus whether adverse human health effects can reasonably be expected to occur at or near current background body burdens. Receptor theory predicts that binding of dioxins to the Ah receptor may be linear at very low exposures, so some binding may occur even at vanishingly small doses. However, this theory does not predict that responses will result whenever binding occurs. Thus, responses may exhibit thresholds.

HIGH-DOSE EFFECTS

Cancer

Although dioxin has been shown to cause cancer in laboratory animals, the epidemiologic link between exposure and increased human cancer risk is inconclusive at present. Where positive associations have been observed, they are weak, even among persons with presumed heavy exposure arising from chemical manufacturing processes or industrial accidents. Attributing cancer excesses observed among workers to their presumed dioxin exposures is questionable given the workers' known exposures to carcinogens such as asbestos and 4-aminobiphenyl. Furthermore, control for the influences of cigarette smoking and other cancer risk factors has been inadequate. Although the EPA emphasizes the positive associations that have been reported, conflicting evidence cannot be dismissed and alternative explanations cannot be ruled out.

Developmental/Reproductive/Endocrine Effects

When considered carefully, the reports of adverse developmental effects in rodents and humans show fundamental inconsistencies. Because high dose *in utero* exposures result in the feminization of male rat mating behavior, the EPA argues that similar effects can be expected in humans. However, sexual behavior patterning in rodents is fundamentally different from that in humans; this endpoint is therefore of questionable relevance for potential human risks. Other developmental effects have been noted in humans and other primates, but only at doses so high as to cause maternal toxicity.

Inclusion of endometriosis as a potential adverse effect is premature. The association of endometriosis with dioxin exposure has been reported in one study of monkeys, but only *a posteriori*, and alternative causes (i.e., previous exposures to other substances and surgical treatments) cannot be ruled out. Furthermore, a contradictory report showing no association between dioxin body burden and the incidence of endometriosis in monkeys is not considered in the current draft. Such selective emphasis does not accurately reflect the full weight of the scientific evidence.

Endocrine disruption has been cited as another adverse outcome of dioxin exposure, but the evidence is for high doses and even then it is very weak. For example, one study reported a significant increase in the prevalence of low total serum testosterone in the third exposure quartile, but this finding was not confirmed in the highest quartile, and the test for trend was not significant. Another report citing decreased glucose tolerance has been neither peer-reviewed nor confirmed.

These limited observations may be suggestive, but they are not sufficient to conclude that people are at risk of reproductive/endocrine or developmental effects at or near current body burdens.

Immunological Effects

There is no evidence that dioxin exposure compromises immune function in humans. In animals, effects have been noted only at high doses or with protocols that do not evaluate normal immune function (e.g., injection challenges with sheep red blood or tumor cells). Effects on immunological surrogate markers (e.g., distribution of T-cell subsets and surface marker expression) have been reported in some animal experiments, but these markers have no known significance for the immune competence of the animals. In exposed humans, such effects have not been found.

Biochemical Changes

If, as the EPA suggests, the "average" human body burden is currently within an order of magnitude of the level required to elicit adverse effects, then sensitive biochemical markers should already be elevated in the general population. A sensitive marker that has been detected in most tissues, including human placenta, is the induction of cytochrome P4501A1. Induction of this enzyme was markedly evident in placentas from highly exposed (Yusho/Yu Cheng) or smoking mothers, but it has not been detected in the general nonsmoking population. It seems implausible that adverse health effects would occur at background body burdens if these subtle biochemical changes have not been observed.

WHAT IS THE CORRECT DOSE METRIC?

The EPA has equated acute and chronic exposure patterns on the basis of projected body burdens even though the corresponding daily doses can be orders of magnitude apart. Because adipose tissue, the major storage organ for dioxin, equilibrates slowly, acute dioxin administration will produce higher peak concentrations in serum and target organs than will administration of the same total dose in small daily increments. Use of body burden as a measure of exposures is thus misleading.

TOXICOLOGIC EQUIVALENCY FACTORS (TEFs)

Compounds that bind to the Ah receptor include dioxin-like PCDFs and PCBs that are present in vastly greater concentrations than TCDD itself. In fact, most of the background body burden in humans is due to these other chemicals. The validity of using TEFs hinges on the presumption that all of these compounds act in a linear and additive manner. However, this additivity hypothesis is contradicted by evidence of antagonism among dioxin-like PCBs, and between dioxin and indole-3-carbinol, a substance which also binds to the Ah receptor and is found in most cruciferous vegetables (e.g., broccoli). (Substances are said to be antagonistic when the net effect of mixtures is less than a simple summation of TEFs.) Moreover, even if TEFs are additive for some particular endpoint, there is insufficient data to support extrapolation of equivalent TEFs and additivity to other endpoints. We urge caution regarding the use of TEFs, and recommend more clarity regarding the mixed nature of exposures, particularly in human populations.

CONCLUSION

We do not believe that there is sufficient evidence to conclude that adverse health effects should be expected at or near current human body burdens. We urge the scientific community to examine these issues carefully. A credible and scientifically defensible estimate of the potential human health risks from dioxin exposure will emerge only after a thorough and critical peer review.

Appendix

Risk Characterization Panel

B. Frank Vincent, Moderator

The Risk Characterization Panel was the concluding meeting of the Symposium. The Panel was comprised of Dr. Robert Scheuplein, former FDA toxicologist; Dr. Thomas Starr, Environ Corporation; Dr. Byron Butterworth, Chemical Industry Institute of Toxicology; Dr. Alan Poland, McArdle Laboratory at the University of Wisconsin; Dr. Robert Golden, Environmental Risk Sciences; and Mr. Stewart Holm, Georgia Pacific Corporation. The moderator was Dr. Frank Vincent, James River Corporation. Dr. Vincent also edited the Risk Characterization Panel transcript for this publication.

Dr. Vincent:

As a wrap-up to this symposium on chlorine and chlorine compounds in the paper industry which we have been discussing over the last 3-1/2 days, this panel will address some of the situations and problems that have arisen with these compounds and these types of compounds. We have assembled a distinguished group of people to talk a little bit about what the data mean, at least to them, and see if we can come up with some idea of whether there really is a significant risk from these compounds. The proceedings are being recorded because we want to provide a transcript from this session. If you have a question, I would very much appreciate it if you would use the microphone and state your name and affiliation. I also need to note that we made an extraordinary effort to get Dr. Bill Farland of the EPA to join the panel, but he was not able to participate, which disappointed us quite a bit.

Would one of the panel like to begin with a word or two about chlorine compounds in the paper industry?

Dr. Starr:

The EPA risk characterization's activity has, in effect, said that the current average body burden of dioxin is enough to cause cancer, reproductive and developmental effects, immunotoxic effects, metabolic disorders, and on and on. That is simply not true, and it's time for the toxicology community to make that statement loudly and clearly to everybody who will listen.

Dr. Butterworth:

Chloroform is another by-product of chlorination, and we've shown that the mode of action of its carcinogenicity is to cause cytolethality, thus inducing cell turnover. So although it is a carcinogen, it is a carcinogen only because you poison an organ at such massive doses that you really cause it to die and then regenerate day after day after day. The doses required to do that are many, many orders of magnitude higher than you would encounter in the environment (i.e., at the fence line in the air at the pulp and paper mill, for example). So our studies clearly show that at the doses at the environmental levels that you would encounter, chloroform would not be expected to be a human carcinogen.

Dr. Poland:

I think everything's been said. I guess as an academic, what incensed me and a lot of my ivory tower colleagues was the EPA in its wisdom disagreeing with the other government agencies (USDA, FDA) and with other countries, four or five of which have ruled on dioxin, as well as most of the academic research in this area. They rearranged the data that had been accumulated over the last 20 years to come up with their own conclusion. I think what upset us, as academics, and made some of us activists or involved in a process that I thought I'd never be involved in, is this distortion or misuse of science. Any risk assessment is a mixture of science and policy, but the two should be clearly demarcated. Science is the dispassionate pursuit of truth and the attempt to disprove your hypothesis, and I think that the agency has given us a bad name in this regard. I would hope in the future that they would depend heavily on the science. If the science is limiting then they would separate that from their subsequent actions.

Dr. Vincent:

To move things along, let me propose a subject to the panel, and while they are discussing that maybe you in the audience will have some questions of your own.

One of the things that we have heard, certainly from a number of areas, is that one of the big problems with chlorine, especially in the disinfection of drinking water, is the formation of so-called trihalomethanes and other short carbon chain chlorinated materials that can be shown to be carcinogenic under certain test conditions. The pulp and paper industry of course discharges some chlorinated compounds as well. Maybe it would be interesting for the panel to address the question, having looked at chloroform, having understood some of the implications and what almost certainly is the mechanism of action to form the tumors that have been seen, is it likely that some of these other so-called trihalomethanes operate by the same mechanism and may therefore be subject to the same sort of conclusions, namely that cytolethality is the prelude to carcinogenicity? Maybe Byron would be a good one to start off on that, and we could get some other comments on that as well.

Dr. Butterworth:

If there's bromine, sometimes you get brominated compounds. This really is a little different story because many of these are genotoxic. That is, they can react directly with the DNA to induce mutations, whereas chloroform is non-genotoxic. I think for each one of these, you're going to have to do the researching and show what your dealing with. However, these are made in very, very small amounts and usually for these kinds of carcinogens, once again, you have to reach doses where you begin to get cytolethality and cell turnover, and then the combination between mutagenic activity and cells that are dividing is very devastating. You do get a lot of mutations in cancer, so I would suspect that you would find that even for these compounds the dose-response curve is going to follow very closely the cytolethality curve. But it's a little trickier because you are, in fact, dealing with mutagenic chemicals in some cases, so you just have to do the research and see what you come up with.

Mr. Holm:

I'll put this in terms of a little bit less of a science and more of a public policy answer to it and jumping off from where Byron was. What we are trying with some of these cytotoxic non-mutagenic chemicals, as Byron stated earlier today, is get the agency away from the linearized multistage (LMS) model into a threshold type situation, and I think there's a good chance of doing that based on information from the Chemical Industry Institute of Toxicology (CIIT). I think it's a very nice data set, and I think it tells a very interesting story. On the other hand, I really don't think that the agency would, any time soon, consider a threshold type model for genotoxic chemicals and, as Byron pointed out, some of these products of ozone disinfection are, in fact, genotoxic. Therefore, if the agency moves away from the current LMS thinking for some of the cytolethal chemicals that Byron stated and also pushes for ozonation in drinking water, they could be going from a less risky to a more risky situation in terms of your number at the end of that risk assessment calculation. That's going to be a different and very difficult public policy situation for them and may underlie some of the problems we are facing now.

Question from the audience: I am Charles Berger, former chemical engineer, not in the chlorine business, but I have a general interest in wastes and wish more people shared it. I don't believe that any subject such as this can possibly be settled by a town meeting or by Congress. I think it has to be the judgment of accredited scientific people. At the same time, I am concerned that the public perception of this is that you can always find an expert, all you have to do is pay them, and that disturbs me a great deal. I don't have any answer for it. I understand you gentlemen have to be on somebody's payroll; that simply has to be. Also reading in the *Chemical Engineering News* that the chemical industry, through some kind of an accounting procedure which the industry itself monitors and supplies to the government, produces something like 1700 lb per year of what is called "toxic wastes" that are disseminated, I think, to the water and to the atmosphere. Now I don't know what they define as toxic wastes. I hope it isn't brine, for example, because that surely wouldn't, in my opinion, qualify as that, but that number is rather surprising, and it's surprisingly large. When you consider that our so-called municipal waste, if you look at it in those terms, is still increasing in spite of our recycle efforts. It runs, I think, about 4 lb per person per day and of course a lot of that stuff gets leached out and goes into our environment. I realize that I'm not concentrating on chlorine, because I don't want to condemn chlorine. It's obvious to me that it has to be used judiciously, but I'd like to know what can we do to recognize the fact that our evolution didn't grow up with most things like this? As far as we know, many of these chemicals that we're producing today and which may or may not be broken down by our biological systems were not in our evolution. That is the thing that concerns me more than anything else, but I have to leave it to scientific judgment and not to what I consider people who are rabble rousers, if you want to call it that, but somebody has to get this ball moving. Was I vague enough or did I eat any ideas?

Dr. Vincent:

Thank you, we appreciate the question. Maybe as a representative of an institution not on anybody's particular payroll, Dr. Poland, would you like to say a word to start off?

Dr. Poland:

I think that you covered a lot of areas. First of all, to deal with this simplistic idea of eliminating halogens in the environment: organic halogens are naturally occurring too. It's just like estrogens, we're immediately worrying about synthetic ones when, in fact, there are a lot of phyto-estrogens in plants and in soy and in all kinds of things. So there are a number of marine chloro and bromo compounds, and there are a number of enzymes that deal with making those and if, in fact, we couldn't activate halogens and make those, we wouldn't have a thyroid. So they started out by brominating benzene rings and moved to iodine and so forth. On the other hand, it's true that chloro-aromatics are very difficult for most mammalian degradation enzymes to get away from, and I think people in industry in general and the problems with halo-aromatics are not only a problem of the paper industry. That's a perceptual problem; they were the problem of the chemical manufacturing industry in the late sixties and seventies. We have pretty well stopped distributing all of those compounds. We have abandoned those herbicides we don't use, those PCBs are no longer used even in closed processes, etc., and most of those patents have been abandoned. Interestingly, the production of halogenated aromatics was a by-product of those chemical companies that were first founded turning brine into potash and they had a lot of chlorine left over so they started R&D at the turn of the century and found uses for this chlorine.

I agree with Dr. Butterworth that everything has to be considered as an individual case and assessed scientifically. We should not make these rash judgments to eliminate elements from the atomic chart. Such actions are naive. You really have to go case by case.

I think chloro-aromatics are really very different than chloro-alkanes. Chloro-alkanes have the problem of the halo-alkanes you've heard of and the atmospheric problems. Chloro-aromatics are not like that. The problem with dioxin is that we have made zillions of chloro-aromatics and most of them are inert. The reason they were made is their inertness to acid and alkali and their resistance biologically. They are lipophilic and so they accumulate. What we did was to use a biological system, like we use bacteria for a selection of toxicity. 99.999% of these molecules never affect the body, they just sit there in your fat and someone can measure them. It turned out that the ones that were isosterically appropriate were a selection system and people who experienced huge exposures in factories got sick and then we got interested in them. I have a couple of books on my bookshelf of just about every compound that industry has made as a chloro-aromatic. Most them are really uninteresting. They are also not made now and it's only this extraordinary sensitivity that has clued us in to these other compounds, the rest of them just sort of sit there. So I'm in favor of what Dr. Butterworth said about there being no sweeping generalities. Let science substitute for astrology. It's the only thing we have to reasonably assess things, and your point about everybody's an expert, especially if you come from more than 50 miles away with slides. That's my definition.

Dr. Golden:

Just to pick up on this point about who makes these judgments, I think that the scientific community is really the group that ought to be playing the major role here. This has been stressed over and over again, and I think over the years science has evolved. It's a system of rules, if you will, on how hypotheses are tested, verified, or rejected. There's publication, there's peer review, there is a process here that can be followed and it's followed very successfully. For instance, in science, to put a new drug on the market, it's tested, it's tested in animals, it's then tested in people, it goes through clinical trials, it's all done very objectively. There is sort of a set of rules and things aren't made up to fit a hypothesis that is not working. It ought to be the same way when chemicals are regulated: the science ought to just dictate what happens. When we get the scientific information the way Byron has gathered it on chloroform, a nice story, it all makes sense, it all hangs together, it's been peer reviewed, people accept this. This is just the way it ought to be. To me, it's the same way with the dioxin. When hypotheses are put forth, as Al pointed out so eloquently, that basically violate 150 years of research — it shouldn't be this way, there is a system here, we don't have to come up with a new system. There is one in place that works, in my view, perfectly well. That's the scientific method and we ought to go wherever that leads us.

Mr. Berger (previous questioner):

I wouldn't criticize either of those comments, but I'm putting forth the general view, not only restricted to chlorine — which I understand is where you're mainly directing it — but toward the broader aspect of what the chemical industry does.

Dr. Scheuplein:

Responding to some of those issues, I think the thing that is interesting about your question is that those concerns are relatively new in the last 25 years and something has been done about them. Industries now, when they want to make a new product, call in their scientific experts and look at what the cost is going to be in terms of getting rid of the product after it's been used. This is built into the cost of the product. A good effective product may not be the best that works. It may be the one you can get rid of the easiest, one that creates the smallest environmental problem. Something that's going on in all the countries now is life cycle analysis. You look at the material you're making, how cheap it is to produce, and the cost of getting rid of the environmental debris. All that's built in, and that's the kind of product you go with. There's a lot of public support for that, obviously. So I guess my point is that your point is well taken. We ought to be considering that, but most of the problems we're talking about now are problems of residues that were left over from 25–35 years ago. I cannot predict the future, but I don't think that's going to change. I don't think that companies are suddenly going to turn around and neglect all of this information that has altered their policies and say "Well, EPA is going to quiet down one day and FDA is going quiet down, so we'll go back to what we were doing." If you can put yourself back 25–35 years ago, there were problems with fires on the rivers here and you couldn't breathe. I think, again, we've done an enormous amount to clean up the environment and I don't see it changing.

Questioner: John Miller, University of Chicago

My question hits on the consumers in the market these days. That's the increasing price of both pulp and finished paper products. Pulp on the market, both virgin fiber and secondary fiber, has hit high levels and is up 50% in the past year. Meanwhile, the paper industry continues to go through a consolidation, pulp mills being closed, older mills being closed. How much of this is a result of environmental regulation and how much of this would be considered financial pressures on the industry, Wall Streets types just trying to maximize return on investment?

Mr. Holm:

You're giving a question to a bunch of scientists that probably have Econ 101 in their backgrounds, but that's probably been it. One of the things about the pulp and paper industry, it is a cyclical industry. If you recall about five years ago there were hardly any companies making any profit at that time. Kathleen Bennett, Vice President of James River Corporation, had some graphs on the U.S. economy coming out of the recession. The graphs compared the Gross Domestic Product and those type of things and the pulp and paper industry, which generally lagged the other trends, that is, it came up much later on. I think that what has been done over time is the pulp and paper industry has really learned a lot from that downturn, where there was a recession with really low profits for quite a stretch of time. Now that everything's geared back up and where full production and demand is still very high, it behooves us to streamline our organization as much as possible. Does this really mean anything in terms of environment? I think it may on a case-by-case basis or a mill-specific basis, but I think supply and demand is probably more a part of that than environmental regulation.

Dr. Vincent:

I think too that the operations of the companies — how they have structured themselves to respond to the economy and to the demand — probably has a lot to do with it. The companies that were stretched out a little bit prior to the business downturn certainly were the ones that would be most vulnerable to any sort of financial problem and about the only way you can respond to a financial problem if you need cash and can't borrow it is to sell something or significantly reduce your operating costs. I think that's pretty much what we're seeing here as well as in some other parts of the economy.

Dr. Butterworth:

I have a couple of comments in response to the gentlemen who spoke. The perception is that if you're from industry that you're biased and you can't be trusted, and that industry finds experts to push their cause. But there is one point of fact in this arena: everybody is biased. For example, if you're writing a research grant, you have to show the chemical you're dealing with is toxic or they won't give you any money. They don't fund research on nontoxic chemicals, so the first sentence of your research grant has to be, "This is a very dangerous chemical." Everybody has an ax to grind, everybody has carried some bias with them. That's just the way it is. It's a very difficult arena and everybody has their own points of view. Another serious problem I think we have is overpopulation. We have so many people in the United States consuming so much. You talked about the amount of waste. Well, that's only going to get worse and worse because of the numbers of people that we have. There's a population group whose motto is "Any cause is a lost cause without a reduction in the population." There's some truth to that. For example, there's concern that the environmental estrogens are causing failure to reproduce in a lot of species, but that's not the only thing out there in the environment. One of the most serious problems, if you talk to the people who work in ecotoxicology and the wildlife, is loss of habitat. More and more people are moving into the hills. They have their terrain-altering vehicles, they're everywhere now, so where you used to have wilderness and quiet you have guys buzzing through with their 4-wheel-drive vehicles, which makes it very difficult for the wildlife population to reproduce. So population dispersal is another serious problem that we have to deal with, and obvious issues with waste will only get worse even if we're doing as good a job as we can. And finally, to be really gloomy, the United States is not the only country out there. We have decided that DDT is very bad because it bioaccumulates, it's persistent in the environment, and it has very severe toxicological effects, but last year in India they built three new DDT plants. We're not the only ones. It's a tough world, and we're supposed to be taking charge of the earth and taking control of it, but sometimes I worry whether we have the knowledge, wisdom, or restraint to do that.

Victor Turoski (Symposium Chair):

Bob Scheuplein and I were talking just before the convening of the Risk Assessment Panel. I believe dioxin in the paper industry is a dead issue. I think that the industry has done a good job in eliminating the small amount that they produced. There are the natural forms of dioxins, dioxin produced from forest fires and such, which we can't do much about, but the current amount in the environment, in my view, is not going to hurt anybody. Yet we still seem to have this big concern about dioxin. Bob, I wonder if you, for the benefit of the audience, would go over some of the areas that you and I talked about. For example, why this thing is perpetuating. Perhaps you could also give your views on what we might do to put it to bed and get on with more important issues.

Dr. Scheuplein:

We were talking about some of the inertia or momentum that's built into the system now and we mentioned that there are groups of people with vested interests in keeping dioxin going. These are ordinary folks that like what they're doing and they don't look at the bigger picture necessarily, but there are lots of researchers whose grants depend on dioxin. Dioxin has its own journal now and it has its own meeting and there are not many chemicals that have their own meeting. There are calcium and magnesium, and I can understand those, but this has its own meeting every year. Thousands of people go to these meetings. One of the interesting things about dioxin is it's got all those congeners, and there are interesting things to measure with analytical chemistry so there are lots of people and it's very expensive to do that and there's a lot of profit to be made in analytical chemistry of dioxin congeners. They all need good solvents, so there's the solvent purification industry for that. Also, another thing that's sort of irrelevant to the dioxin history, but dioxin is a damn interesting molecule: it does things in cells that few molecules do, so there's just an enormous amount of intellectual interest in this too. I think when you start adding up everyone who's got a vested interest in this you get a large number.

Dr. Vincent:

Thanks, Bob.

One of the things that we heard this morning was that the EPA has developed a concept called Toxicity Equivalence Factors (TEFs) which they use to normalize the toxic potency of chemicals, and this gives them an opportunity, by applying the right factor, to take it over like an accountant to the right-most column and come up with a number for each of these materials. Then they just add up the values in this column and say this is the toxic equivalent of this mixture. There has been a great deal of comment, largely that this treatment doesn't really work right, but it's a nice way to do things, it's easy. However, the end result may or may not reflect the truth, and I think it would be instructive if the panel would address that issue. I think they have all had a good bit of experience and probably some convictions in that area. Who would like to talk about that?

Dr. Poland:

Well, it's an interesting idea. For genotoxic drugs, the only data I know about where this has been applied is radon or uranium and cigarette smoking. So we've talked about two hits of environmental concern, because in uranium miners the studies showed that the lung cancer was largely found among smokers and was greatly potentiated by smoking. So getting nuclides in your lung made it much, much worse. But in receptor-mediated compounds like estrogens, and peroxisome proliferators are going to be next there, the phthalates in plastic along with drugs such as clofibrate and so forth and the dioxins and other halogenated aromatics that bind to the Ah receptor (and no doubt we'll have more) raises other issues. This concept of multiple agonists acting on the same button is another one because it buys several orders of magnitude if you use this as default due to lack of knowledge. Now what they should probably do is say "What happens if you walk along the south side of the street and you eat peanut butter with aflatoxin and worry?" These would be the organic chemists out there, the carbon guys. It's an interesting concept that they've introduced, and this is something, by the way, that drug companies always worry about. The EPA has ignored the orthodoxy that most drugs are agonists and antagonists of a receptor. So you have the endogenous compound, whatever that hormone may be, either synergized by more epinephrine or an epinephrine blocker like propanalol or something like that. Science has developed very formal ways to deal with this and assess the relative amount of the two and the amount of additivity or response, but you can skip over the pebbles of science if you have a grand design. What we're upset with is that since there is an existing orthodoxy that a lot of scientists have spent a long time setting up here, it's annoying that it hasn't been used at all. This dioxin study was supposed to be peer reviewed, involving academics as advertised. There's not a person in that document who has written it, reviewed it the first time or reviewed it on the SAB who knows anything, who ever has published a paper that I can find on receptor theory, receptor modeling, and there are two references to minor reviews in the whole document. So I think that the agency is really seeking to bend over backwards like scientists do to disprove the hypothesis that the document was peer reviewed. As Byron said, there really are ways to do this. It's opening up of the agency. It's opening up just like, for example, it would be very interesting if you heard the National Cancer Institute (NCI) talk about phyto-estrogens. They recently had a conference on estrogens in the environment. You wouldn't believe the difference between that and the EPA. The NCI is interested in the science thereof rather than making a policy. What's happening? Are we eating too much soy? Is it protecting us or not where the phyto-estrogens are involved? I think the real problem here is that science is capable of doing things through knowledge and that there are environmental problems. I think the real problem here, though, is with the agency.

Dr. Scheuplein:

I may take it from the agency's point of view and what the EPA might say. We got TEFs because it was a practical solution to an impossible problem. You find a fish that's got dioxin but it has got 12 other different kinds of congeners in it. So the EPA has got to regulate the levels in that fish and what do they do? Do they just say "Well, it's only got dioxin in it, let's ignore the rest."? No, they don't do that. Do they say "Well, let's treat everything the same, assume the others are equally as potent as dioxin, add them all, and assume that is the way we find the right level."? No, we can't, it's extreme on the other side. So they said "Well, is there some way of relating the anticipated toxicity of these congeners to dioxin?" Then they would sort of talk in terms of how equivalent they were; 1/10 as equivalent, 1/100 as equivalent, and add that up. That's what they did, and that's the way to draw a regulatory line in the sand, the way to get something done without taking 100 years. You really would have to go through toxicity tests and long-term tests on all these studies and try to do this. So that's the way it started. I think what has happened, though, as Alan was saying, this is okay if you have to make a quick decision, but it's not okay to say I'm actually calculating the risk this way. It's okay to say "Well, the agency is going to draw this line for this particular incident because that's the data it has right now and it wants to make the most reasonable decision it can at the time." But this was an interim decision in 1987; that's when these things were put together even though they started before in 1984 when they really thought about this. So the question is why it hasn't been looked at very much since 1987, even though there are ways to examine it, as you've heard, very critically. Maybe you can't add them up, maybe they interfere with each other. What about the food, you don't eat dioxin for dinner or lunch, you eat food for dinner or lunch. There are things that also bind the receptor that are hundreds of thousands of times more in abundance than dioxin. What about the impact of that? None of that has been done. What they did is they took the '87 analysis and applied it to the '94 analysis, and that's where they come up with this extra factor of 10 that says there's now reason to be concerned. Well, as scientists point out, that probably is not true. We don't really know where the risk is, but you can't use the TEF procedure and expect to get the correct answer.

Dr. Starr:

I could say just a few more things. Maybe another way to look at that whole issue is to ask what things would have to be true for the TEF approach to actually work and what assumptions would have to hold. Among other things, all of the congeners would have to have the same kinetics; they couldn't differ in terms of their half-lives. Otherwise, mixtures would behave differently in terms of the relative abundance of the congeners as a function of duration of exposure in the body. So you've got to have equivalent pharmacokinetics. And beyond that, depending on what system you're measuring toxicity in, all of the dose-response curves would have to be at least parallel. So if it's a cell culture, an enzyme induction measure in the liver, if it's a tumor in the liver, if it's a tumor in the lung, or if it's a reproductive or developmental abnormality like cleft palate, those response curves would have to be parallel across the congeners and across the endpoints. Those things are not true. They would have to be parallel across species, and that's not true. So, for this test approach to work, there are an enormous number of assumptions that would have to be true that we know aren't true. I think it's incumbent on any agency putting forth such a procedure as a sort of stopgap measure to acknowledge that's the best it can be — a stopgap measure that really doesn't have a whole lot of scientific support at this time.

Dr. Poland:

The problems that the agency faces with receptors and xenobiotics (foreign chemicals) are exactly the central concern of pharmacology; it's exactly what the FDA has faced for a long time in registering drugs from pharmaceutical companies. There are bookshelves—whole libraries—full of books on all of the aspects of this, and it's a bit like developing a new theory of economics, voodoo economics, etc. They never bothered to even cite the original work, that they might now deviate from it and say we're extrapolating the line down. So it almost is barbarism in that you're unaware of the historical precedence. I guess that's what is so irritating to an academic. One almost gets the feeling that they're trying to squelch creativity and solutions and find new kinds of problems and say there's a rigid hierarchy. It's as if they never heard of whole disciplines. It's hard to explain our exasperation.

Dr. Golden:

I think another corollary to what Tom just said is that there has been a lack of testing, for the reason that Bob said. With TCDD it's real interesting. It causes a lot of effects and it's fascinating scientifically, but there's been a preoccupation with testing TCDD and not with testing these other congeners. The TEF system is sort of a stopgap measure, which I think is sort of clever, but for the reasons that Tom just mentioned, there are a lot of theoretical problems with it, but when someone tells me that you can add them all together and this will be a predictor for reproductive toxicity, my response is that's fine, how about testing a few? Let's test an octa, a hepta, and see if they really do cause these effects at the same fraction which the agency claims; that is, do they really cause these effects like TCDD does? That hasn't been done. That isn't a two-year feeding study, these sorts of things could be done faster, but the right kind of research, in my view, has never been conducted on these compounds. Rather, it is presumed that you can just add them all together and that bottom line, whatever that number is, will somehow be a predictor of toxicity.

Dr. Starr:

It just occurred to me that it would be useful to make you aware that one of the recommendations that seemed to be very strong, made by the Science Advisory Board in May to EPA, was to restructure the document in such a way that the risks that were attributable specifically to TCDD would be separated out from the risks attributable to other dioxins and furans. Those would be still separated again from the risks attributed to "dioxin-like" PCBs. So at least one could tell what the strength of the data was in support of those three components. Then if the agency felt compelled to add them up at the end, they would show the consequences of all the warts and the individual pieces being there and maybe the argument wouldn't be quite so alarming that effects would be occurring at body burdens that comprised only 10% TCDD and 98% these other things which are still questionable in terms of testing. So that was a strong recommendation. Whether or not EPA will actually do that, which will require a great deal of re-engineering of the documents, I don't know, but that was certainly something that was made explicit as a request.

Dr. Poland:

Let's look at the levels a different way: 7 parts per trillion translates to about 20 picamolar in the fat and about 1 picamolar in the liver. That turns out to be 3 to 6 molecules per cell that, as Tom said, the agency believes are now causing hazard in the U.S. population or within an order of magnitude, and that's a maximal estimate, that's not free compound, that's total and you have to talk about free. We're giving them all the conservative factors. I guess that's the kind of stuff that, when you really do the stoichiometry, we feel we're talking about how many angels are on the head of a pin. Of course, we're supposed to be scientists and we can measure. There's enough analytical data now for dioxin. Where genes are cloned, we have antibodies that cause downstream effects to occur. We know that cellular messages can be detected. When we have 0.1 picamolar in a rat we can detect a response which is below, we believe, a present human exposure level and we have people exposed to more than that. So, for most of the studies, we're not suggesting that they're wrong, we're suggesting that no amount of bookkeeping of these isomers or of conjecture replaces science. They started with academic science; all they had to do was the applied science going from whatever university eggheads had done on animals across species to whatever they wanted to know and use. They haven't produced any useful data.

Dr. Vincent:

Building on those comments, one of the possibilities would be to see if we could encourage some of the government programs to come up with the answers to some of the questions posed by the panel. The National Toxicology Program (NTP) is funded to look for carcinogens and, among other things, they're set up to do exposure studies. It would be interesting for the panel to consider the possibility of having the NTP carry out some of the studies we've been talking about, looking at some of the other compounds to determine whether the toxicity equivalence factor is, in fact, correct. If not, what should it be? Do these congeners even accomplish the same end effects that 2,3,7,8-TCDD accomplishes? Is that something that is doable? Anyone want to respond to that? It certainly seems that the money that we're spending in that area is something from which we may or may not be getting our money's worth. Mechanistic studies are one thing that I know Ken Olden has suggested, and this would be an interesting study for a government lab to operate.

Dr. Poland:

There's a confusion in the report in Chapter 8, and I'm sure this confusion, will continue and it's part of the NTP's program. Dr. Butterworth showed this enormous chart of all the oncogenes and all the signal transduction pathways that various things take, if you remember the metabolic chart of the cell he showed. There's a presumption that understanding mechanism at a very fine level will somehow help risk assessment, but there's no example for that in pharmacology. What I mean is the following: If you take a dose-response curve of how much insulin kills a rat or how much androgen causes weight increase, you get a fine dose-response curve of either a biochemical intermediate or toxicity or morphology. Then over the next 73 years you understand more and more about the mechanism of insulin and all the intermediates in the pathway, but that hasn't changed the dose-response curve. It's exactly the same. You may understand, you think, a lot more about how it happens. I posit that it doesn't change risk assessment, that you don't have to back off and wait for it because pharmacologists have dealt with this for a long time. We got insulin on the market, while epinephrine and acetylcholine date from the 1920s. We didn't know how they worked, but we knew they had a receptor and we knew the structure activity relationship and we knew the dose-response curves. So it flies in the face of the whole history of things, and I think Bob Scheuplein was talking the same way; it's a way of procrastinating. So I do disagree. I think mechanism is very important when it shows lack of genotoxicity as Byron pointed out, and we know dioxin acts through a receptor. For example, going through each of the *mu* globulins in the *alpha* 2 globulins in the urine of male rats, or even things which are major, such as thyroid response and thyroid cancer, won't change the dose-response relationship. But to think about knowing the entire signal transduction pathway before you make any risk assessment, I don't know of an example in pharmacology where this changed our view. I think that view has to be tempered. It's a way of getting funded until retirement or in perpetuity for the agencies, but I think you've got to act, and there's no example of this.

Dr. Golden:

Let me say something else along those lines. Al mentioned dose-response and I submit that we don't even have that for a lot of the effects of dioxin. A favorite story of mine Stewart will know because we looked at a poster together at the last toxicology meeting. Some of the effects of dioxin that people are concerned about are potential reproductive effects that follow from exposure, and I mentioned it in my talk this morning. Exposure *in utero* to the single dose that is most often used in these studies is to administer once a mg/kg dose to these animals and that causes an effect. You have to realize that a mg/kg dose is approximately a million times higher than the typical daily intake that we get in our diet. So it's a million times higher and an effect can be seen. These kinds of experiments are repeated over and over again. They do it in rats, then guinea pigs, then rats again, and it keeps doing this. Stewart and I talked to someone in front of his poster where he had used that single dose procedure and we asked "Did you try any lower doses to see where you saw the effect and when it went away?", and he said "No." To our question "Why not?" he replied "If I did that, I wouldn't be standing here showing you these results. This is government research." I said "Of course you'd be showing results, they would be whatever you're showing now. But you would also be able to show something called dose-response." That part of it, I don't understand why that's not done. I truly don't understand why the appropriate kind of studies can't be done to find out where this effect goes away. A million-fold safety factor, I can sleep real soundly with that. Is it merely a hundred thousand times, would that be the safety factor, is it a thousand? This is all doable but it's not done.

Dorothy Cannon with Technical Resources International, Rockville, Maryland:

I'd like to suggest what the moderator posed in terms of your panel making some recommendations to the NTP on which chemicals to look at would be very valuable. My work is for the National Cancer Institute looking for potential candidates for testing by the NTP, and we've been looking at a group of about 300 chemicals that have TLVs that were recommended to us by the ACGIH Committee. Among those are five polychlorinated naphthalenes. The problem in looking at the polychlorinated naphthalenes as potential testing candidates is that they're among this group of persistent lipophilic, no-longer-produced chemicals, and in order to pose chemicals for testing by the NTP, we have to develop a production exposure scenario or they'll lose interest in them. So what I'm suggesting is that input from a group like yours would be very valuable in sorting through large numbers of chemicals and attempting to find an array of chemicals that would be worth testing.

Dr. Vincent:

These are produced — not voluntarily, but they certainly are produced — so that wouldn't be a problem. Thank you for the suggestion. Any further comments on the TEFs or any of the ideas that were brought out on those? The panel, I gathered, feels that toxicity equivalence factors have perhaps served a purpose but need to be investigated, updated, more clearly determined. Is that where we come out on toxicity equivalence factors, any additions or subtractions to that statement? Is there another question from the audience?

Marquita Hill, University of Maine:

I'm not sure if it's a question but a frustration. Here again, you're all scientists, and I've enjoyed this week tremendously and learned a lot. At the same time, I know when I go back to Maine I am going to go back to the same issues of public relations. In fact, I was just talking to my husband last night about the newest upset on dioxin in Maine. It's strictly an irrational thing and at some point we're all going to have to deal with public relations issues. If any of you live in states like Maine you're going to have to do that. I don't like the idea of industries touting themselves, so to speak, but on the other hand it seems like the pulp and paper industry hasn't done this at all, and in Maine they just prefer to ignore the issue. They just don't talk about it, and meanwhile we're in the position of having to make policy (i.e., the Maine environmental priorities project this fall will be developing the issues of main concern), and it's very likely that some of those decisions will be made on the basis upon which media has the most dioxin just because the reports that have been written have emphasized dioxin so much. I guess I would like some sort of reaction, although I know you don't expect any great wisdom on the issue necessarily.

Dr. Scheuplein:

I had a similar problem. I took an oath when I joined the Food and Drug Administration about 18 years ago to try to protect the public health, and I found myself not doing that. I was working on color additives at risks of 10^{-9} and dioxins at risks that are intensely lower than that and what can I do to get the public thinking about this a little differently? The answer — it's not unique with me and it's not necessarily a good answer, was to try to compare these risks to something else that you do with dioxin, like eating. I looked at the work by Doll and Peto, and they did an analysis of the cancer risk all over the world. They came up and said there are basically three places you get cancer, from cigarettes, food, and from everything else put together. The everything else put together comprises a finite number of sources that you've heard about, but basically the two main sources are food and cigarettes. I said Oh, let's see how big the risk is from food. We don't know what the contaminant is or what the carcinogen is or what the promoter is in food at the moment, but let's see how big the risk is. Well, the risk is easy to calculate. Turns out they said about 1/3 of the cancers were due to food, and we know that the cancer rate is 1/4 or 1/5. 1/4 times 1/3 is about 1/12 or 1/15. So that's the risk according to non-extrapolation studies, just counting dead bodies and making an attribution as to where the risk came from. The risk is about 1/15 from eating food, and we're worried about 10^{-18} risks. I've gone to talks and tried to explain to people, let's try to put it into perspective. Even if EPA is right when they say the risk is 10^{-4} from dioxin at normal levels, how does that compare with 1/15? What can you do to reduce that 1/15 level? This is the real problem that I have with policy in the country because if you're really interested in making people's lives better and lengthening them, the thing you could do would be to just make a minor change to the pattern of food consumption. Try to find something that you could tell people to do that they're not doing. Whether it's eat vegetables with anticarcinogens or eat less fat or eat in some pattern. There's almost no work at all going into that. The problem is that the funds and the effort are going to things like dioxin and other things that do not really help people.

Mr. Holm:

I guess we had a similar situation in southern Mississippi, we had some people sue our company and say that they were having harm caused to them by dioxin from our facility. These people didn't eat fish and were at a location that wasn't on the river that was some 100 miles downstream from the facility. They won in the first case, and that is on appeal now, if that gives you any idea about some of what goes into these types of things. I think in the cases of Maine and Mississippi and probably around the country that scientific data is going to have to be generated or at least used in combating some of the feelings that come out of some of these groups that want to push an agenda, and I'm not going to be able to convince somebody in Maine against something that EDF or Greenpeace simply states at a public meeting without data. If I have information and I can present a cogent argument and come to some logical conclusion and if I do it over and over again, I might reach my goal. It's not going to happen tomorrow, it's not going to happen next year, but I'll get there eventually.

Dr. Vincent:

If you speak with data, usually you come out ahead, that's very true. Any other comments for Marquita? Let me ask one more question which has I think plagued a number of us. To borrow a phrase from a panel member, the imperial decision to use the linearized multistage model, based on Dick Kociba's data, has led to a "safe" intake level for 2,3,7,8-TCDD which is the lowest in the world by quite a bit. The next closest one is a couple of orders of magnitude greater than what EPA says is safe. I would like the panel to look at the linearized multistage model as the way of treating the data in light of our present knowledge of the mechanism of action of these substances.

Dr. Starr:

Well, it's probably worth pointing out that the use of the linearized multistage model has been going on for about 20 years, and the current cancer risk assessment guidelines that EPA operates with are just about 10 years old; they were issued in 1986. A revision to them is in the works right now and may actually see the light of day before the year is out as a draft for public comment. I participated as one of many people in a workshop that was held to review various aspects of the guidelines last fall. One of the interesting features about the new guidelines is that the linearized multistage model is not in them; it has been dropped as a legitimate way to analyze data. In place of it is a simple ruler extrapolation from the lowest dose at which an effect is noted. So instead of the fancy and confusing sophistication of an elaborate mathematical, statistical procedure, there is a simple, easily comprehended and obviously artificial extrapolation with a straight-line ruler down to the origin. It's not from an upper bound — it's from the point estimate of risk at some dose. So, LMS is on its way out, but still that makes the question of how best to do this even if the model is no good, what do you replace it with? The new guidelines also have an important change in philosophy. Basically what they do is make the point that you should make whatever scientific inferences you can make only down to the low end of the observable response range. So if you have tumor increases, you go down to the point where you don't see them anymore, and at that point, you ask how does this exposure level compare with the exposure level of concern to people? What is the margin of exposure that we have to work with? Is it a million times smaller, a thousand times smaller, ten times smaller? If it's a big number, you can be reasonably comfortable that you're well below exposure levels that are known to cause effects. If you want to calculate what the risk might be, you could use this ruler, but you don't try to make the extrapolation look like it has scientific credibility. The next step down the line is to be able to replace actual body counts or tumors with surrogate markers like foci in the liver, or that are known to be pre-neoplastic, or suspected of being pre-neoplastic, or specific genetic markers of mutations. If that allows you with a credible or rational argument to connect a marker to a carcinogenic endpoint and extend the observable response range below where it is for tumors, then you can go down there and ask again the question, What is the margin of exposure for humans? So there is always a place where the science will take you, but it takes you to the low end of the observable response range. As science evolves and systems become more sensitive and the connection to the carcinogenic process becomes better understood, you can push the low end of the observable response range down closer to the actual exposure conditions of interest. But never ever do you go beyond that without saying this is a straight-line extrapolation to zero without scientific merit. So it's a clear distinction between where a science stops and where the policy assumptions of risk management have to come into play. We tried to get the agency to stop going that far, to stop even going below about the ED_{10} level, the dose that would produce about a 10% response level in experimental animals. They're reluctant to do that, but they are willing to consider, on a case by case basis, just how far down they might go. So I think that's a step in the right direction. We should recognize that until we know the whole picture, we can't go down to human exposure levels except with a ruler.

Dr. Vincent:

Using that type of analysis today, about where would 2,3,7,8-TCDD fall? Do you have an idea on that?

Dr. Starr:

To be honest, the linearized multistage model is nothing much beyond a fancy ruler, so the risks that you would get making the assumptions that had been made would not be very different. The basic question is still one of what data is available to support the kind of mixed body burden that's being used to characterize human exposure levels in relationship to experimental animals. Most of the data, as Bob Golden has already pointed out, has been acquired for TCDD alone, and as almost all of us have pointed out, the body burden is not TCDD alone, it's only about 10% of that. I think if the agency were to go forward with some of the SAB recommendations and split out TCDD and other dioxins and furans and the PCBs that are like dioxin in their action, that would be a long step forward. The risks that were clearly identified that could be attributed specifically to TCDD would be shown to occur at much, much higher levels than the current body burden of TCDD itself.

Dr. Vincent:

You commented that the new procedure is not going to include the 95% upper bound confidence interval, so the ruler is going to be applied at a level different from the level used now. Is that largely true?

Dr. Starr:

The guidelines aren't completely 100% prescriptive. What they do recommend is that the analysis consider fitting a curve to the data in the observable response range and getting to some kind of standard exposure level, a reference dose, for example, that caused a 10% increase or a doubling of the background, or something like that. And then from there to use the ruler if you need to go further.

Dr. Vincent:

Any other responses or comments? Any questions or comments from the audience? If not, we will conclude the Risk Assessment Panel. I think that our major point of consensus is that TEFs need a lot of help, and I think we concluded that real science is really the key to most of these issues, and that good science is the way to go. We should not have advocates posing as scientists, and if the agency or, for that matter, any group applies science and looks for the truth that we will be far better off when we start to make public policy. Is there any disagreement with that position? (Pause)

I thank each of you for coming. I especially thank each of the panel members for their presence, for providing insights from their expertise, for their opinions, and for helping us understand a little bit better some of the issues we've discussed. Thank you very much.

Victor Turoski (Symposium Chair):

When we assembled this panel, we tried to get people from industry and government, academicians, and consultants, and we included some of the top toxicologists in this country. I'd like to say to each and every one of you that you are being heard and that you are making a difference in this country for the better. I personally and professionally thank each one of you for coming to this symposium at your own expense to share your views. I appreciate it very much.

Dr. Vincent:

Thank you. Our symposium is concluded.

Index

B

C

E

F

G

H

P

Q

R

U

V

W

X

Y

Z